Circular Steel Production

Circular Steel Production

Pathways to Net-Zero Carbon Emissions

Sandra Kiessling, Hamidreza Gohari Darabkhani, and Abdel-Hamid Soliman

WILEY-VCH

Authors

Dr. Sandra Kiessling
University of Staffordshire
College Road
Mellor Building
ST4 2DE Staffordshire
UK

Prof. Hamidreza Gohari Darabkhani
University of Staffordshire
College Road
Mellor Building
ST4 2DE Staffordshire
UK

Prof. Abdel-Hamid Soliman
University of Staffordshire
College Road
Mellor Building
ST4 2DE Staffordshire
UK

Cover Design: Wiley
Cover Image: © urfin/Shutterstock

Typesetting: Straive, Chennai, India
Printing and Binding: CPI Group (UK) Ltd, Croydon, CR0 4YY

Library of Congress Card No.: applied for

British Library Cataloguing-in-Publication Data A catalogue record for this book is available from the British Library.

Bibliographic information published by the Deutsche Nationalbibliothek
The Deutsche Nationalbibliothek lists this publication in the Deutsche Nationalbibliografie; detailed bibliographic data are available on the Internet at <http://dnb.d-nb.de>.

Print ISBN: 978-3-527-35315-6
ePDF ISBN: 978-3-527-85272-7
ePub ISBN: 978-3-527-85273-4
oBook ISBN: 978-3-527-85274-1

C9783527353156_291125

We have worked on this publication for the sole purpose of providing a comprehensive guide for heavy industry, such as the steel industry and ancillary sectors, to activate serious efforts to decarbonise steel manufacturing and related sectors.

Contents

Authors' Biographies

Dr. Sandra Kiessling

After more than three decades of working in industry and academia, Dr. Sandra Kiessling is currently focusing her efforts on research and commercial collaborations on processes and technologies which have a direct impact on reducing CO_2 emissions and promoting a zero-carbon emissions society. She is an expert with a proven track record in carbon avoidance, utilisation, capture and storage (CCS and CCUS) technologies, renewable energy systems (biomass, wind, hydro, solar and geothermal) and energy-independent buildings. She is also involved in establishing the carbon footprint of any industrial, commercial and residential setting and supports commercial entities in achieving reductions of their CO_2 emissions.

Prof. Hamidreza Gohari Darabkhani

With over 25 years of experience spanning both industry and academia, Professor Gohari Darabkhani has focused his research on advancing technologies aimed at achieving net-zero emission and energy targets by 2050. He is a recognised scientist with a proven track record in Carbon Capture and Storage (CCS) technologies, renewable energy systems and Micro Combined Heat and Power (Micro-CHP) units in smart cities. As the Director of Staffordshire Centre for Renewable and Sustainable Engineering (SCRSE), he manages three key research themes within the centre:

1. Clean Energy Technologies – generation, storage, and management;
2. Sustainable Engineering – innovative design, advanced materials, and smart manufacturing;
3. Information & Communication Technology – signal and image processing, telecommunications, and AI.

His current work emphasises the development and utilisation of low carbon fuels (e.g. biofuel, syngas, hydrogen and e-fuels) and their crucial role in driving decarbonisation, particularly in hard-to-abate sectors such as the iron and steel industries.

Prof. Abdel-Hamid Soliman

Professor Soliman has multi-disciplinary academic and research experience in digital signal processing, telecommunications, data acquisition systems, wireless sensor networks (WSN) and image and video processing. He is focused on harnessing and integrating diverse technologies to develop smart and intelligent systems to contribute to smart cities and practical applications across energy, healthcare, industry, and agriculture. In addition to his research activities, he is involved in several enterprise projects and consultancy activities for national and international companies. He is actively seeking funds and support for the research activities in his areas of research interest by determining the appropriate funds and applying to these funds through various funding organisations. Now, he is leading and involved in several Externally-funded projects.

Preface

This book contains a total of 12 chapters. Chapter 1 offers a glimpse into the history and background of the steel industry and how different technologies and methods evolved. Chapter 2 paves the way with insight into the employed methodology, and Chapter 3 takes the reader on a journey through the literature review, starting with a description of the carbon life cycle. It follows with detailing the different resources for and stages in steel production and exploring the usage of biomass as an alternative fuel in steelmaking. The total CO_2 emissions of steel production are investigated, delving into a detailed analysis of all CO_2 emissions produced along every step of resources mining, converting and utilising, which are required to produce steel. Furthermore, energy and exergy analysis are offered with greater detail about energy utilisation within the steel industry. The literature review concludes with selected decarbonisation case studies, giving a comprehensive overview of where the technical development of a decarbonised steel industry could be headed. It provides insight into how the iron and steel industries could be achieving the urgently needed CO_2 emission reductions. Chapter 4 investigates in detail the aspects of the innovative model and strategy of the Bio Steel Cycle (BiSC) and the seven steps to net-zero carbon emissions in steel production and provides details of the engineering aspects of process simulation software solutions. The findings of a thorough investigation in Chapter 5 highlight the opportunities of the BiSC implementation and components within it, such as heat loss recovery, retrofitting a variety of renewable energy technology and a thorough MCDA analysis. The MCDA adds a dimension of supporting decision-making aspects such as technical readiness levels, implementation status and cost. Chapter 6 offers flowcharts and process simulations in software solutions such as Aspen, Simul8 and Inosim. The analysis of the results in Chapter 7 applies standard mathematical procedures, data-modelling features and statistics and showcases an overview of the total CO_2 emissions along most procedures in steelmaking process. In Chapter 8, the knowledge contributions are summarised, and a PESTEL and SWOT analysis of the situation for green steel in the United Kingdom offers a lookout into future market opportunities. This is followed by a summary of CO_2 emissions for the entire steelmaking process, and finally, it pinpoints the advantages of the innovative BiSC model, concept and strategy as well as the significant opportunities of achieving a higher degree of energy independence. Chapter 9 sets out the skill sets required for the implementation of BiSC model and strategy and the opportunities for a workforce engaged in green technology. Chapter 10 highlights the future of green steel from a global perspective. Chapter 11 features an idealised timeline of BiSC strategy implementation and thus decarbonising the steel industry. Chapter 12 provides some reflective thoughts on the strategy and implementation of BiSC and green steel implementation effects and opportunities.

Acknowledgement

In researching and writing this book, we relied to a great extent on the generosity of other people, who have given us their time, advice and expertise, made space for us and put up with the different quirks of our personalities during the intense and unrelenting work the project required.

Thank you especially to the team of Wiley Publishers, whose insight, vision and great attention to detail have immeasurably improved the quality of this book.

Great thanks are extended to the expert reviewers, who so generously gave up their time to read the manuscript and extended constructive criticism by making comments and suggestions which improved the work greatly.

Thanks are also extended to the University of Staffordshire for this research opportunity and for providing the infrastructure for research and development, without which this book would not have come to be.

Nomenclature

BAT	Best available technology
BCA	Border carbon adjustment
BF	Blast furnace
BOF	Basic oxygen furnace
BS/EN/ISO	British Standard/European Standard/International Organization for Standardization [British national version of ISO Standards (International Organization for Standardization)]
CapBF	Total capacity (kg) of blast furnace
CapBOF	Total capacity (kg) of basic oxygen furnace
CapEAF	Total capacity (kg) of electric arc furnace
CAPEX	Capital expenses
CCUS	Carbon capture and utilisation or storage
CGE	Computable general equilibrium
CH4	Methane
CCS	Carbon capture and storage
CO_2	Carbon dioxide
DRI	Direct reduced iron
EAF	Electric arc furnace
Eimp	Total imported energy (kg/steel)
EmSp.El.	CO_2 emission savings/avoidance potency factor
Fe_2O_3	Hematite l hydrogen feed ratio
FeO	Wuestite
GEI	Grid emission intensity
GHG	Greenhouse gas emissions
H_2O	Water, chemical formula
HBI	Hot briquetted iron
H-DR	Hydrogen direct reduction
HHV	Higher heating value
HSLA	high-strength low-alloy steel
I4.0	Industry 4.0
IEA	International Energy Agency
IPCC	Intergovernmental Panel on Climate Change
IRENA	International Renewable Energy Agency

LHV	Lower heating value
LKAB	Luossavaara-Kiirunavaara Aktiebolag (Swedish Mining Corporation)
LC-AC	Linz–Donawitz ARBED Steel Company
LST	Tonne liquid steel (metric)
MAC	Marginal abatement cost
$MCO_{2,proc}$	On-site CO_2 emission (kg/steel)
Mind	Production rate of steel (kg) capacity
MO.Ind	Usage of oxygen on site
NG	Natural gas, fossil-derived methane
O&M	Operation and maintenance
O_2	Oxygen, chemical formula
OLP	Oxygen–lime powder process
OPC	Ordinary Portland cement
OPEX	Operating expenses
PC	Pulverised coal
PCC	Electricity import for CO_2 capture/savings process (MJ)
PEM	Proton exchange membrane
Pind	Electricity import for the industrial process (MJ)
PV	Solar photovoltaic cells
Q	Net heat transferred into the system, Q is the sum of all heat transfer into and out of the system
SEC	Specific energy consumption
SOE	Solid oxide electrolysis
SSAB	Svenskt Stål AB (Swedish Steel Corporation)
TGRBF	Top gas recycling blast furnace
W	Net work done by the system, W is the sum of all work done on/by the system
WTO	World Trade Organisation
ΔU	Change in internal energy U of the system
Ø	Sign for average

1

History of steelmaking

1.1 How it all began

Early smelting of metals began with the Arabic science from about 3500 BCE (before the Common Era) in Egypt and the Near East (Sumer), with smelting of copper, gold, mercury and alloys like bronze [1]. Initially, pure iron from meteorites was used for decoration and making spearheads. Learning to smelt iron (Fe) from its ore oxide (Fe_2O_3) became commonplace in Egypt only around 2000 BCE, and it began to replace bronze in Europe c. 1000 BCE. Iron's high melting point at 1535 °C for pure iron, in contrast to copper at 1083 °C, required smelting furnace construction to develop rapidly to reach such high temperatures. Iron ore, being the second most abundant metal in the earth's crust [2] after aluminium, heralded the commencement of the so-called Iron Age in the British Isles from about 800 BCE until the Roman invasion of 43 CE (Common Era).

Iron ore can be found in four mineral sources:

- Magnetite at 65% iron; not available in Britain [3];
- Haematite at 50% iron concentration, present as separate, blue-tinged so-called kidney iron [3];
- Blackband at ~30% concentration found in coal seams [4]; and
- Jurassic iron at only ~20% concentration discovered across England from the North Yorkshire Moors to the Cotswold Hills in the south–west [4].

Early ironmaking processes involved bloomeries, where partially enclosed fires built on exposed positions provided a sufficient blast of natural wind for the required draught [1]. The term 'bloomeries' was coined due to the putty-like mass of hot iron that was extracted from the furnace and then forged with hammer and anvil into a bloom. The output was limited to only a few kilograms of iron blooms per day, which were used to make tools and small bladed weapons. The molten iron at 4% carbon would run off into shapes moulded in the sand that resembled a sow with her sucklings, which gave rise to the term 'pig-iron', now called cast iron. The first use of cast iron in building and construction dates back to 1380 CE in Europe. This cast iron was only made in very small quantities, which were reheated – to obtain reduced surface carbon. This produced the so-called wrought or bar iron, which was malleable

Circular Steel Production: Pathways to Net-Zero Carbon Emissions, First Edition.
Sandra Kiessling, Hamidreza Gohari Darabkhani, and Abdel-Hamid Soliman.

enough to be worked after [1]. In the United Kingdom, in the 1700s, indigenous iron production accounted for only 12,000 tonnes/p.a. Eventually, to overcome the challenges of difficult supply routes for timber and coal needed in the then-still inefficient furnaces, multiple small ironmasters organised themselves and banded together in small iron-producing areas, such as South Wales. Additionally, the industry required a high labour force, which resulted in very high costs. As a direct result of these costs, most iron was therefore being imported into the United Kingdom, particularly in quantities of cheaper imports from Sweden. Back then, ca. 50% of the iron used in Britain came from Sweden [1].

1.2 First attempts at steelmaking

Steel is defined to be as a commercial *iron* (hard metal; oxidises easily when exposed to air) [5] that contains *carbon* (chemical element; key ingredient for most life on earth; forms pliable *graphite* [crystalline carbon] and also diamonds – one of the hardest materials in the world; the basis for graphene) in amounts of up to about 1.7% as an essential *alloying* constituent (a substance composed of two or more metals or of a metal/non-metal combination intricately united commonly by being fused together and dissolving in each other when molten) [6], which is malleable when exposed to suitable conditions and is distinguished from *cast iron* (molten iron filled into moulds) by its malleability and lower carbon content. The first evidence of steelmaking is from the thirteenth century and has been discovered in medieval Anatolia, Kubadabad (Türkiye) [7]. A timeline of the steelmaking process evolution is shown in Figure 1.1.

1.3 Steelmaking process development

Although evidence of early efforts in steelmaking was discovered in the thirteenth century CE [7] in the Arab world, it is argued that actually the earliest known use of the crucible steel production technique occurred in India and central Asia in the early first millennium CE [8–13].

Documented early attempts to refine the process do not seem to have had a great impact on the craft, until the nineteenth century CE. In c. *1865* [8–10], Robert F. Mushet discovered that by adding *manganese* (manganese oxide crystals consist of MnO 6 octahedral units shared by corners and/or edges) to Bessemer steel [14], casting (filling the void of a mould with molten steel) of ingots (an oblong piece of metal) free of blowholes was possible, and adding manganese also reduced the metal brittleness forced by the presence of *sulphur* (yellow crystal, often with pungent smell) [15]. Mushet [8–10, 13] produced the first commercial alloy steel in 1868 when he found that adding *tungsten* (also: Wolfram) to steel [16], particularly *after* air cooling, significantly increased its hardness, which lead to this material further on forming the basis of the subsequent development of tool steels for the machining of metals [17].

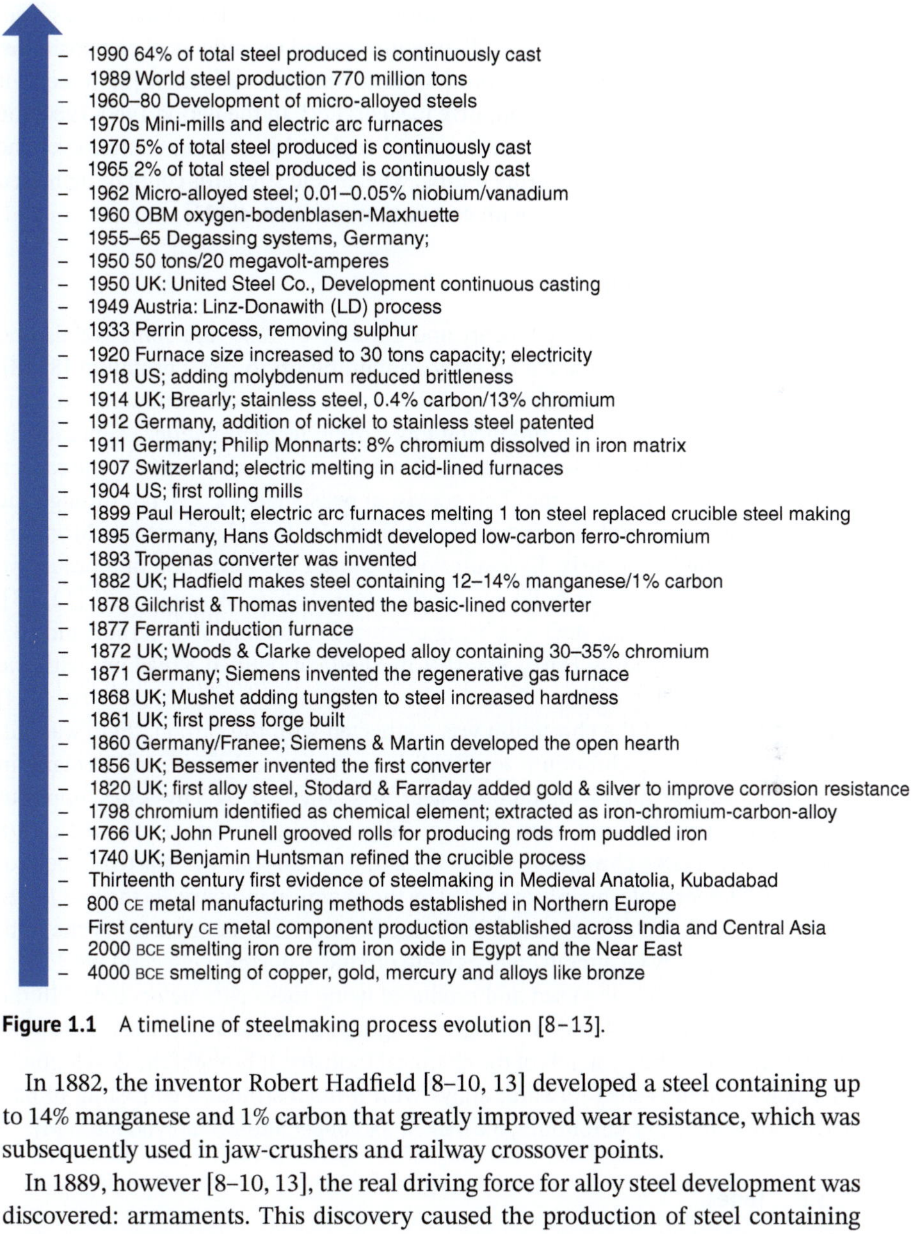

Figure 1.1 A timeline of steelmaking process evolution [8–13].

In 1882, the inventor Robert Hadfield [8–10, 13] developed a steel containing up to 14% manganese and 1% carbon that greatly improved wear resistance, which was subsequently used in jaw-crushers and railway crossover points.

In 1889, however [8–10, 13], the real driving force for alloy steel development was discovered: armaments. This discovery caused the production of steel containing only 0.3% carbon and 4% nickel, which was subsequently improved further by adding *chromium* [18]. In 1918, steel was made less brittle by adding *molybdenum* [19].

1.3.1 Micro-alloyed steels

An important development that occurred immediately after the Second World War was the improvement in steel compositions for sections that could readily be welded by increasing the manganese content to 1.5% and maintaining the carbon content below 0.25% [17]. *By 1962*, high-strength low-alloy (HSLA) steel, and subsequently the term 'micro-alloyed steel', was introduced for mild-steel compositions, which contained 0.01–0.05% *niobium* [20] and similarly *vanadium* [21] (used for making ferrovanadium or as an additive to steel) [22]. In the period 1960–1980, the development of micro-alloyed steels evolved rapidly by exerting control over the environmental temperature during rolling, where yield strengths could be increased by almost 200%, in comparison with conventional mild steel [23].

1.3.2 Stainless steels

Chromium was first used by Stodart and Faraday in *1820* and John Woods and John Clark in *1872* to produce an alloy containing up to 35% chromium [8, 13]. *In 1895*, commercial success became possible when Hans Goldschmidt in Germany [12] invented the method of producing low-carbon *ferrochromium*. Eventually, in *1911*, Philip Monnartz in Germany established the relationship between the carbon content of chromium steels and their corrosion resistance [12, 13], but through the ensuing interwar period, it became standard practice to add at least 8% chromium dissolved in the iron matrix. Exposure to air caused the formation of a protective layer of chromic oxide to build up on the steel surface. Around *1914* [12, 13], more corrosion-resistant steel was being produced by adding 0.4% carbon and 13% chromium, developed by Harry Brearley in Sheffield, United Kingdom, with the purpose of producing cutlery in mind. Although Brearley's steel trials in *1914* [24] showed that 3.5% of the chromium was connected with the carbon, there was still sufficient remaining chromium to establish corrosion resistance. In Germany in *1912* [25], adding nickel to stainless steel was patented by Thyssen-Krupp, but the compositions had not been commercially exploited before *1929*, when 18/8 austenitic steel (18% chromium, 8% nickel and 0.2% carbon) was started to be produced on a larger scale, mainly exploited by the chemical industry. By the late *1930s*, austenitic stainless steels and modified compositions were mainly used by the military, when the early jet aircraft engines were produced during the second World War [12], which are still being used and produced using these parameters today. In the *1950s*, the production of steel with high strength and corrosion/wear resistance was mainly driven by the demands of the chemical industry. It brought the development of Duplex stainless-steel forward: alloys with a microstructure consisting of 65% ferrite/austenite, 25% chromium, 5% nickel, 3% *copper* and 2% molybdenum [4].

1.3.3 Tubes

Since *1824* [8–13], when pressure butt welding of heated, curved metal strips was invented in Britain, similar processes are still commonly used today for producing seamed tubing. However, in c. *1921*, this process was immensely improved in the United States, when the seam was joined by electric resistance welding instead of pressure butt welding [9, 10]. To this day, seamed tubes are still being manufactured using this method, whereby particularly large-diameter tubes are being formed by spirally coiling a continuous strip of steel around, which then arc-welds the spiral seam.

1.3.4 Seamless tubing

Seamless tubing is a hollow cylinder of metal (often steel, stainless steel, titanium, or other alloys) made by extrusion or piercing a solid billet, without any joints or welds. This results in a smooth, continuous structure throughout the tube length. The requirement for seamless tubing meant the requirement for piercing of a round billet, which was developed in Britain in 1841 [12]. Since *1886*, when the Mannesmann company in Germany under Reinhard and Max Mannesmann [26] developed the process of rolling the billet longitudinally and at the same time forcing it onto a piercing bar called a mandrel [12], ferrous/non-ferrous metals are formed.

1.3.5 Plates and sheets

Plates are produced by hot rolling [8, 21, 25], and in order to produce steel sheets from crude plates, the steel needs to be cold-rolled in mills. These mills consist of different stands in sequence, fitted with an initial big and then decreasing-sized double rolls, where crude steel plates pass through the rolling stands sections. The first mill of this type was installed in *1904* in the United States [13].

1.3.6 Timeline of steel production processes

With the earliest known use of the *crucible* steel production technique occurring in India and Central Asia in the early first millennium CE [8–13], archaeological investigations have provided us with the evidence that these techniques applied were quite crude but efficient: wrought iron was heated in closed oven-like furnaces, together with charcoal and materials rich in carbon, and the end-product of this process was known as wootz or Damascus steel. It is thought that trading activity with Middle Eastern countries resulted in obtaining the knowledge about these techniques to reach the Northern Hemisphere, as records indicate that the crucible process was applied in 800 CE in Northern Europe – amongst other evidence, these were utilised to produce the Ulfberht Viking swords [27].

Staying for a moment with the Northern Hemisphere, it has been established that the person deserving the title of founder father of British steelmaking was the inventor and engineer Henry Bessemer [8, 13, 28], whose work was later referred to as the driving force to initiate a process by which a superior issue of cast iron for the manufacture of guns was invented in 1856 [12, 29]. Bessemer conducted a series of experiments and decided to carry out a trial using a converter: bringing air into contact with a specific amount of molten crude iron with the aim of converting it into malleable iron [8, 12, 13, 28, 29]. The first converter was built at Baxter House, St. Pancras, London, which featured a brick-lined cylindrical body, and the base was furnished with holes through which air could be blown.

At the first trial, after 7 cwt (cwt = hundredweight, which equals a hundred metric pounds) [30] of molten pig iron was poured in, ambient air at 10–15 lb. pressure was blown through the bottom holes, and after approximately 10 minutes, a violent reaction ensued, by which the oxygen in the air and the carbon in the pig iron were started sending up an exponentially increasing eruption of sparks and formed a voluminous white flame. A series of minor explosions threw molten slags and splashes of metal high up in the air – rendering the converter a veritable volcano in a state of active eruption [8, 12, 13, 28, 29]. After another 10 minutes, the violent reaction and the flame had died down, indicating the process was complete. The contents of the converter were poured into a shallow pan or ladle, forming the metal into an ingot, and upon testing, it was found to be wholly decarbonised, malleable iron – heralding the era of modern steel production [13].

As exciting as this breakthrough was, there was one apparent difficulty with Bessemer's process [13]: it could only convert pig iron low in phosphorus and sulphur into the desired malleable steel. Whilst these would have been quite easy to remove by adding a basic flux, such as lime [31], the resulting basic slag would have degraded the acidic refractory lining (bricks). However, there were good supplies of iron ores low in phosphorus in Britain and the United States [2, 3, 6], but they were far more expensive.

In *1878*, the engineers and inventors Sidney Gilchrist Thomas and Percy Gilchrist built a basic-lined converter, using calcined dolomite as the refractory material [8, 12]. This approach made it possible to use a lime-rich slag, which held phosphorus and sulphur in solution. At odds with its potential, this 'basic Bessemer' process [12] was surprisingly little used in Britain or the United States, which gave the industries of Alsace and Lorraine a competitive advantage, as they were able to use the prevalent phosphoric ores [13, 32]. This catapulted the development of the Belgian, French and German steel industries to the top position in world production of steel, which rose to about 50 million tonnes by 1900 [2, 3, 6, 13, 32].

In 1995, Webb reviewed [29] the work of German author Wengenroth, who investigated the coexisting British and German steel manufacturing industries between 1865 and 1995. At the very beginning of the steel production era, British steel was the dominating force worldwide [2, 3, 6, 13, 32]. However, German ingenuity won the better over the British steel industry eventually and developed mechanisms and processes that would lay the technological foundations for German expertise and dominance over the British steel industry over the next century [29].

Carr and Taplin [8], as well as Webb [29], also researched the processes involved in steel production and identified the following key elements:

1) Puddling: wrought iron production, converting pig iron (cast iron) into wrought by applying heat, stirring frequently in a furnace in the presence of oxidising substances [33]. Wrought iron is produced by heating iron and repeatedly hammering and shaping it using specialised tools.
2) Crucible steel processes [34] for producing fine or tool steel. Crucible steel is a high-quality steel made by melting iron and carbon in a crucible (a ceramic or metal) container.

In *1740*, Benjamin Huntsman decided to refine the process by heating small pieces of carbon steel in a closed fireclay case, which was covered in *coke* (pyrolysis process: heating coking coal to c. 1000–1100 °C in the absence of oxygen to drive off volatile compounds [17]) and set alight. When the temperature rose to 1600 °C/2900 °F, the steel started melting for the first time and produced a homogeneous metal of uniform composition, which was then used for manufacturing watch and clock springs [17].

In order to improve specific properties such as strength, wear and corrosion resistance, alloying elements were added to steels.

In *1820,* the first experimental study of alloy additions to steel was made by the Britons James Stodart and Michael Faraday, who added gold and silver to steel in an attempt to improve its corrosion resistance. The mixtures were not commercially feasible, but they initiated the idea of adding chromium to steel (see Stainless Steel) [17].

In the *1860*s, William and Friedrich Siemens in Britain and Pierre and Émile Martin in France developed the 'open hearth' steelmaking process. Combustion gases preheated air and fuel gas to 800 °C (1450 °F), whereby a flame temperature of about 2000 °C (3600 °F) could be obtained enough to melt the inserted charge. Refining was achieved by causing a reaction between the slag (to which iron ore was added) and the liquid metal in the hearth of the furnace, inducing the removal of carbon, manganese and silicon [35] from the metal, and furnace capacity gradually increased to 100 tonnes and eventually to 300 tonnes. Acid-lined furnaces were used at first, but later a basic process was developed, which facilitated the removal of phosphorus and sulphur from the charge. Sufficient heat could be produced in 12–18 hours, providing an adequate time frame for analysis of the material and adjusting its composition before it was tapped from the furnace [17]. Although Britain had better access to ore and pig iron, from sites across Britain, production was first acidic and then basic, and finally it used open-hearth Siemens–Martin furnaces that made steel cheap enough for mass production during the last third of the century [36].

After *1870,* the coke-fire furnace was replaced by the Siemens regenerative gas furnace [17], as it produced even higher temperatures than the open-hearth furnace. The Siemens furnace was furnished with a series of combustion holes, with each hole holding several *crucibles* (a vessel that does not melt easily and is capable of withstanding high temperatures; Definitions.net, 2020), and was capable of heating as many as 100 crucibles at a time. In the twentieth century, the electric furnace replaced the crucible process, which was used to produce high-quality tool steel and high-speed steel in countries where electric power was cheap [17].

Parallel to the already mentioned developments and with the increased levels of efficacy of the electric power industry towards the *end of the nineteenth century,* the use of electricity as an energy source in steelmaking became increasingly important. *Around the 1900s,* small electric-arc furnaces with a melting capacity of c. 1 tonne of steel, replacing crucible steelmaking, were in use to primarily make tool steels [17], and by 1920, furnace size had increased to a production capacity of 30 tonnes. This was achieved by furnishing electricity supply with a three-phase 7.5 megavolt-amperes, where three graphite electrodes were being fed through the roof of the furnace and the arcs forming between the electrodes and the charge in the hearth.

In *1933,* the expensive Perrin process for removing sulphur was developed, where steel was poured into a ladle containing a liquid-reducing slag; violent mixing occurred by which sulphur was transferred from the metal mass to the slag.

In the post-war period, desulphurising powders based on calcium, silicon and magnesium were injected into the liquid steel in the ladle through a lance using an inert carrier gas, which was the method first developed in Japan to initially produce steels for gas and oil pipelines [17].

In *1949*, the Linz–Donawitz (LD) process was developed in Austria [29, 37]. Using a new approach, oxygen was blown through a lance into the top of a pear-shaped vessel, whose outer shape very much resembled the Bessemer converter [17]. As there was no inert *nitrogen* gas (inert/diatomic gas: colourless, odourless and tasteless) present in the air, which would have provided a cooling effect, any heat not lost to the off-gas was being used to melt scrap added to the pig iron charge. A basic slag was then produced by adding lime to the charge, thus removing phosphorus and sulphur, establishing the known basic oxygen process (BOP). With BOP, it became a reality to produce 200 tonnes of steel from a charge consisting of up to 35% scrap in a tap-to-tap time of 60 minutes. The charges in BOP-applying furnaces were increased to 400 tonnes, and, provided low-silicon charges were being used, blowing times were reduced to less than 20 minutes [17]. In quick succession of the LD process, an improved method was developed where burnt lime along with oxygen was blown through the lance [17], later known as the LD-AC (after the ARBED steel company of Luxembourg & Centre National of Belgium) or the oxygen–lime powder (OLP) process, which led to a more effective refining of pig iron smelted from high-phosphorus European ores [17].

After the Second World War, the increasing need for improved properties of steel was a driver for the development of continuing the refining in the ladle *after* steel had been tapped from the furnace, as opposed to stirring the liquid in the ladle through means of blowing a jet of argon through the mass. Reduced variations in the temperature and composition of the metal caused solid oxide inclusions to rise to the surface and become absorbed by the slag, as well as dissolved gases such as hydrogen, oxygen and nitrogen. However, when casting large ingots, gas stirring alone could not remove hydrogen to an acceptable level. *After 1950,* and as large vacuum pumps were commercially available, ladles were placed in large, evacuated chambers, where hydrogen to less than two parts per million was removed by blowing argon as a matter of stirring the mass. In Germany between *1955 and 1965,* a variety of improved degassing systems of this type were developed, which later proved to be of great benefit to the steelmaking industry [17].

Around 1950, furnace capacity had increased to 50 tonnes and electric power to 20 megavolt-amperes [17]. As little to no refining occurred, small arc furnaces, which were lined with acidic refractories, were considered little more than melting units, whereas larger furnaces were basic-lined, and a lime-rich slag was produced by which silicon, sulphur and phosphorus were being removed from the melt. These were so efficient that they could be operated with a charge that was entirely scrap or a mixture of scrap and pig iron, and steel of excellent quality with sulphur and phosphorus contents of as low as 0.01% could be produced. *By 1950*, this electric arc process, being ideally suited for producing low-alloy steels, had almost completely replaced the basic open-hearth process in this capacity. During this time, electric arc furnaces produced c. 10% of all the steel of c. 200 million tonnes manufactured worldwide and basic electric arc furnaces accounted for almost 30% of steel production by 1989, when the world steel production accumulated to 770 million tonnes of steel [17].

In the *mid-1960s,* two concentric *tuyeres* (a nozzle/lancet through which air is pressed into a furnace, forge or smelter) [38] with a hydrocarbon gas in the outer annulus and oxygen in the centre were used in Germany and Canada, effectively constituting a return to the original bottom-blown Bessemer concept: the OBM process (for ***o****xygen* ***b****odenblasen* ***m****axhuette*, 'oxygen bottom-blowing Maxhuette'), which was in North America called Q-BOP (quick basic oxygen process; a Q-BOP furnace has no overhead lance, and the oxygen is blown in through tuyeres (pipes) at the bottom of the furnace). Eventually, all oxygen steelmaking processes replaced the open-hearth and Bessemer processes on both sides of the Atlantic [17].

1.4 Casting

Steel-forming operations were on a relatively small scale until the introduction of the Bessemer process, in which large volumes of liquid steel were produced for the first time. The liquid metal was poured from ladles into large cast-iron ingot moulds with an average size of 700 mm in square section and 1.5–2 m in length. Such an ingot would weigh about 7 tonnes. After solidifying, the ingot was stripped from the mould, reheated and then reduced in size by hot-rolling in a primary (blooming) mill to give billets about 100 mm in section. The billets were sheared into 3 to 4 m in length, and they formed the starting material for rolling into bars, beams, rods and strips [17].

This type of billet production persisted until the *1960*s, when a profound change occurred with the development of continuous-casting machines. With liquid steel going directly from the furnace into the casting machine, there was no need to pour large ingots or to reheat them with heavy energy requirements. Nor were the very expensive blooming mills required for reducing the ingots to forms that were produced directly by casting. Continuous casting was first used for non-ferrous metals in the 1930s, and in the early *1950*s, experiments were undertaken with it at steel plants in Britain, the United States and the USSR. The first production plant using continuous casting was operated at Barrow, England, by the United Steels Company. In *1965*, 2% of total steel production was continuously cast; by *1970*, this had risen to 5%, and, by *1990*, 64% of all the steel produced worldwide was continuously cast (in Japan, it was more than 90%) [17].

Continuous casting was partly responsible for a new type of steel plant that developed after 1970 – the so-called mini-mill. In this plant, steel was made in an electric arc furnace using an all-scrap charge and was then continuously cast into small-diameter billets for rolling into rods or drawing into wire. Mini-mills were built in industrial regions, where scrap arises, whereas the location of conventional steel plants remained linked to the availability of iron ore and low-cost energy [17].

1.4.1 Forging

Forging is a metalworking process in which metal is shaped by applying compressive forces, typically using a hammer or press. Due to the ever-increasing size of ingots in the late nineteenth century, large hammer forges were developed, simulating the early blacksmiths' hammering action, which were followed by press forges, first built in Britain in *1861* and introduced into the United States by *1877*, in order to accommodate considerably large components. The mechanisms applied therein are quite simple: the upper forging die pressing against the component (ingot) which is situated on the lower anvil by means of a hydraulically operated piston (Steel – Forming and casting, 2020).

1.4.2 Foundry

A foundry is a factory or workshop where metal is melted and cast into specific shapes by pouring it into pre-shaped moulds. In the 1800s, the development of the crucible process meant that steel castings could be produced for the first time at an industrial scale. *From 1824*, steel products were being cast in Germany and Switzerland, and *by 1855*, steel gear wheels were cast in Sheffield, UK, closely followed by the United States, where the production of steel castings started in *1871* in Pittsburgh [17]. The crucible process was eventually replaced as the chief melting method in *1893*, when the Tropenas converter was invented in Sheffield, UK, after which, in *1907*, electric melting in acid-lined furnaces was pioneered in Switzerland. Electric furnace melting is currently used industry-wide for making steel castings [17]. *In the 1920s*, X-ray methods for assessing the soundness of steel castings were first applied in the USSR. During the period 1924–1928, international standards for moulding materials, such as moulding sand, were published in the United States, and magnetic crack detection methods followed *in the 1930s* [17]. In 2019, 1869.9 million tonnes of steel was produced – up from 770 million tonnes in 1989 and starting with a capacity of 30 tonnes per furnace in 1920. A truly extraordinary development! [39].

1.4.3 Who and where are now the biggest steel producers?

The Wirtschaftsvereinigung Stahl [40] is the geopolitical association of the steel industry in Germany, with headquarters in Düsseldorf, and in their publication 'Fakten Stahlindustrie 2014', they have published a series of helpful graphs, courtesy of worldsteel [41], which give an overview of the steel industry's data and developments. The World Steel Association (WSA, worldsteel), a not-for-profit organisation with headquarters in Brussels, represents steel producers, national/regional steel industry associations as well as steel research institutes. Their member-base represents c. 85% of global steel production.

1.4.4 Which industries is steel being produced for?

Allwood et al. [42] produced a magnificent report which gives an overview of the different industries for which steel is being produced as the most used metal on the planet. For each living person, worldwide, over 200 kg of liquid steel is being produced every year. Figure 1.1 demonstrates the year 2008 'flow' of steel within the world economy. The sheer volume of steel production has significantly risen since 2008, but the proportions of the flow have remained similar.

Four key industries, for which steel is being produced, have been identified and shown in this illustration:

A) Over 50% is being used in the construction industry (596 million tonnes/year)
B) The manufacture of metal goods, such as packages, appliances and others (177 million tonnes/year)
C) Industrial equipment (mechanical/electrical) (176 tonnes/year)
D) 139 tonnes/year being used for the production of vehicles

An estimated 67% of liquid steel produced is made from iron ore, and the remainder is made from recycled metal/scrap. Unfortunately, at present, more than half of this scrap results from the manufacturing process itself, rather than from degraded goods.

Every year, 25% of all the finished steel made is being cut off and recycled, because final users want components (such as refrigerators) and not intermediate products (coils of strip steel).

The world population [43] stands at approximately 7.7 billion people today compared with less than 1 billion in c. 1800. This immense increase in human beings populating our planet, alongside the ignorance of how to build a sustainable society, may jeopardise living standards for future generations. The idea of a sustainable society was an unknown concept until its formal definition in 1987 with the Brundtland report, where it was declared to be a concept, instated by an informed society, which was developed in a balanced way, considering both current and future generations' needs [44].

References

1 Griffin, P. and Hammond, G. (2019). Industrial energy use and carbon emissions reduction in the iron and steel sector: a UK perspective. *Applied Energy* 249: 109–125.

2 Spears, D.A. (2007). Aspects of iron incorporation into sediments with special reference to the Yorkshire Ironstones. *Geological Society of London, Special Publication* 46 (1): 19–30.

3 Jankovic, A. (2015). 8 – Developments in iron ore comminution and classification technologies. In: *Iron Ore [Internet]* (ed. L. Lu), 251–282. Woodhead Publishing https://www.sciencedirect.com/science/article/pii/B9781782421566000083 (accessed 18 July 2023).

4 The Welding Institute (TWI) (2023). Austenite Martensite Bainite Pearlite and Ferrite structures [Internet]. https://www.twi-global.com/technical-knowledge/faqs/faq-what-are-the-microstructural-constituents-austenite-martensite-bainite-pearlite-and-ferrite.aspx (accessed 18 July 2023).

5 Acharya, M. (2018). Iron facts, symbol, discovery, properties, uses [Internet]. Chemistry Learner. https://www.chemistrylearner.com/iron.html (accessed 19 July 2023).

6 Senkov, O.N., Miracle, D.B., Chaput, K.J., and Couzinie, J.P. (2018). Development and exploration of refractory high entropy alloys – a review. *Journal of Materials Research* 33 (19): 3092–3128.
7 Güder, Ü., Çeken, M., Yavaş, A. et al. (2022). First evidence of crucible steel production in Medieval Anatolia, Kubadabad: a trace for possible technology exchange between Anatolia and Southern Asia. *Journal of Archaeological Science* 137: 105529.
8 Carr, J. and Taplin, W. (1962). *History Of The British Steel Industry*. Cambridge: Harvard University Press.
9 Russel, B. (2007). *The British Steel Industry*. London: Kluwer Law International.
10 Russel, B. (1968). *The British Steel Industry*. London: Kluwer Law International.
11 Aylen, J. (1980). Innovation in the British Steel Industry. In: *Technical Innovation and British Economic Performance [Internet]* (ed. K. Pavitt), 200–234. London: Palgrave Macmillan UK http://link.springer.com/10.1007/978-1-349-04561-7_12 (accessed 2 November 2022).
12 Birch, A. (2006). *The Economic History of the British Iron and Steel Industry, 1784–1879*. Abingdon: Routledge.
13 Speer, J.G. (2018). 2017 Bessemer lecture: progress in steel product research to advance the industry and society. *Ironmaking & Steelmaking* 45 (10): 858–866.
14 Royal Society of Chemistry (RSC) (2023). Manganese – Element information, properties and uses | Periodic Table [Internet]. https://www.rsc.org/periodic-table/element/25/manganese (accessed 20 July 2023).
15 Royal Society of Chemistry (RSC) (2023). Sulfur – Element information, properties and uses | Periodic Table [Internet]. https://www.rsc.org/periodic-table/element/16/sulfur (accessed 20 July 2023).
16 Royal Society of Chemistry (RSC) (2023). Tungsten – Element information, properties and uses | Periodic Table [Internet]. https://www.rsc.org/periodic-table/element/74/tungsten (accessed 20 July 2023).
17 Guthrie, R.I.L. and Isac, M.M. (2022). Continuous casting practices for steel: past, present and future. *Metals* 12 (5): 862.
18 Royal Society of Chemistry (RSC) (2023). Chromium – Element information, properties and uses | Periodic Table [Internet]. https://www.rsc.org/periodic-table/element/24/chromium (accessed 20 July 2023).
19 Royal Society of Chemistry (RSC) (2023). Molybdenum – Element information, properties and uses | Periodic Table [Internet]. https://www.rsc.org/periodic-table/element/42/molybdenum (accessed 20 July 2023).
20 Ross, R. (2018). Facts about niobium. livescience.com. https://www.livescience.com/34682-niobium.html (accessed 20 July 2023).
21 Royal Society of Chemistry (RSC) (2023). Vanadium – Element information, properties and uses | Periodic Table [Internet]. https://www.rsc.org/periodic-table/element/23/vanadium (accessed 20 July 2023).
22 Gagnon, S. (2023). It's elemental – the element vanadium [Internet]. https://education.jlab.org/itselemental/ele023.html (accessed 20 July 2023).
23 Baker, T.N. (2009). Processes, microstructure and properties of vanadium microalloyed steels. *Materials Science and Technology* 25 (9): 1083–1107.

24 British Stainless-Steel Association (BSSA) (2023). The discovery of stainless steel – British Stainless-Steel Association [Internet]. https://bssa.org.uk/bssa_articles/the-discovery-of-stainless-steel/ (accessed 20 July 2023).

25 tkcopperandbrass (2018). How thyssenkrupp was involved in the development of stainless steel over 100 years ago and is still a leading distributor today [Internet]. Copper and brass sales: a blog about metal and processing. https://tkcopperandbrass.com/2018/06/26/how-thyssenkrupp-was-involved-in-the-development-of-stainless-steel-over-100-years-ago-and-is-still-a-leading-distributor-today/ (accessed 20 July 2023).

26 MAN Mannesmann Salzgitter (2023). Salzgitter AG Geschichte. Kurzüberblick Mannesmann-Geschichte. https://geschichte.salzgitter-ag.com/de/einzelne-geschaeftsbereiche-und-standorte/geschaeftsbereich-stahlverarbeitung/kurzueberblick-mannesmann-geschichte.html (accessed 20 July 2023).

27 English Heritage (2023). HISTORY. Iron Age. https://www.history.com/topics/pre-history/iron-age (accessed 20 July 2023).

28 Warren, K. (1964). The Sheffield Rail Trade, 1861–1930: an Episode in the Locational History of the British Steel Industry. *Transactions and Papers (Institute of British Geographers)* 34: 131.

29 Webb, S.B. (1995). Enterprise and Technology: The German and British Steel Industries, 1865–95. By Ulrich Wengenroth. Cambridge: Cambridge University Press. 1994. Pp. xiv + 293. $69.95. ISBN 0-521-38425-7. *Central European History* 28 (4): 560–561.

30 Gartneer, C.E. (2017). Sciencing. How to convert CWT to pounds. https://sciencing.com/convert-cwt-pounds-8772252.html (accessed 21 July 2023).

31 Zhiming, Y., Zushu, L., Sanjeev, M. et al. (2022). Value in use of lime in BOF steelmaking process. *Ironmaking & Steelmaking* 49 (1): 42–48.

32 Simcoe, C. (2014). The age of steel: Part II: By 1875, the Bessemer steel industry had spread across the country from eastern Pennsylvania to St. Louis, with the epicenter in Pittsburgh – Document – Gale OneFile: Computer Science. Advanced Materials & Processes [Internet]. 172(4). https://go-gale-com.ezproxy.staffs.ac.uk/ps/i.do?p=CDB&u=staffordshire&id=GALE%7CA366864121&v=2.1&it=r&aty=ip (accessed 7 July 2023).

33 Luebering, J.E. (2009). Puddling process | Iron Smelting, Refining & Casting | Britannica [Internet]. https://www.britannica.com/technology/puddling-process (accessed 4 August 2023).

34 Rodriguez, E. (2017). Crucible process | Steelmaking, Refining, Alloying | Britannica [Internet]. https://www.britannica.com/technology/crucible-process (accessed 8 August 2023).

35 Barraclough, K.C. (1981). The development of the early steelmaking processes. https://etheses.whiterose.ac.uk/14433/1/237901_vol.1.pdf (accessed 8 August 2023).

36 Brophy, J.M. (2015). Review of Carl von Siemens, 1829–1906: Ein Leben zwischen Familie und Weltfirma. *The Journal of Modern History* 87 (1): 155–157.

37 Wengenroth, U. (1994). *Enterprise and Technology: The German and British Steel Industries, 1897–1914*. Cambridge: Cambridge University Press.

38 Encyclopedia Britannica (2020). Steel – Basic oxygen process, scrap in blast furnace, oxygen lances | Britannica [Internet]. https://www.britannica.com/technology/steel/Basic-oxygen-steelmaking (accessed 6 July 2023).

39 World Steel Association (WSA) (2019). Worldsteel 2019 [Internet]. https://www.worldsteel.org/en/dam/jcr:96d7a585-e6b2-4d63-b943-4cd9ab621a91/World%2520Steel%2520in%2520Figures%25202019.pdf (accessed 7 July 2023).

40 Wirtschaftsvereinigung Stahl (2023). Rohstahlproduktion in Deutschland: April 2023 | Wirtschaftsvereinigung Stahl [Internet]. Wirtschaftsvereinigung Stahl | Informationen des Verbandes der stahlerzeugenden Unternehmen in Deutschland. https://www.stahl-online.de/medieninformationen/rohstahlproduktion-in-deutschland-april-2023/ (accessed 7 July 2023).

41 World Steel Association (WSA) (2023). Membership [Internet]. https://worldsteel.org/about-us/worldsteel-membership/ (accessed 4 September 2023).

42 Allwood, J.M., Dunant, C., Lupton, R. and Cabrera Serrenho, A. (2019). Steel arising: opportunities for the UK in a transforming global steel industry [Internet]. Apollo – University of Cambridge Repository. https://www.repository.cam.ac.uk/handle/1810/294350 (accessed 3 November 2022).

43 Conejo AN, Birat JP, Dutta A. (2020) A review of the current environmental challenges of the steel industry and its value chain. *Journal of Environmental Management* 2020;259:109782.

44 ARE FO for SD (1987). 1987: Brundtland Report [Internet]. https://www.are.admin.ch/are/en/home/medien-und-publikationen/publikationen/nachhaltige-entwicklung/brundtland-report.html (accessed 24 February 2024).

2

Steelmaking processes

2.1 Introduction

This chapter provides an overview of the different stages of the iron and steelmaking process and provides information on the greenhouse gas (GHG) emissions at every stage. Some cost analysis is being attempted, although the current energy price sector fluctuations are rendering all assumptions made to be taken with caution. The determining factors for quantities and volumes were the blast furnace (BF) and basic oxygen furnace (BOF) capacities, each standing at 330 metric tonnes (equivalent to 330,000 kg). In order to construct process simulations in the various software solutions, a flowchart needs to be compiled with the materials and quantities involved, including the formulae and equations which explain the chemical and metallurgical processes taking place at the different stages of the steelmaking process. The quantities and values in Figure 2.1 display the quantities of iron ore, limestone, lime, coke and 1000 °C oxygen required to be sent to the BF as the first furnace-based process. Following on, the product of the BF, the liquid iron, is sent to the BOF, where the hot metal alongside lime, scrap steel and oxygen at 25 °C is being converted to liquid steel. The steelmaking process is including resource extraction, calcification, coking, sintering, pressurising oxygen and through the BF/BOF route to casting, rolling and finally finish machining. The effects on the total emissions of all the process stages leading to the production of steel will be discussed at a later stage. The simplified flowchart of the steelmaking process is shown in Figure 2.1.

There are more processes and stages involved in steelmaking than one might initially assume. It all starts with the extraction of the resources (iron ore, limestone, coal and oxygen), preparation of these resources (crushing, pelletising, coking, calcination and sintering, alternatively iron ore direct reduction [i.e. with green hydrogen]), until these material streams are fed into the BF and subsequently into the BOF, after which casting, rolling and finishing take place. The following graph will lay out more clearly the stages involved in producing steel.

Circular Steel Production: Pathways to Net-Zero Carbon Emissions, First Edition.
Sandra Kiessling, Hamidreza Gohari Darabkhani, and Abdel-Hamid Soliman.

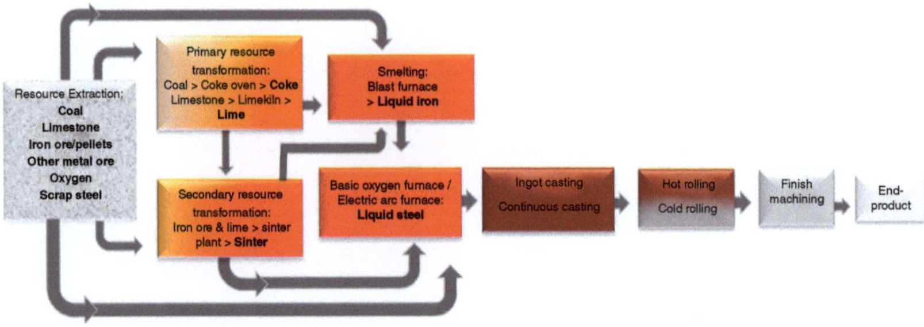

Figure 2.1 Simplified flowchart of the steelmaking process.

2.2 Raw materials

In order to make low-carbon steel, a range of materials are needed: iron ore, coal, lime and oxygen. Later in the process, depending on the quality required, other materials such as chromium, nickel, manganese, molybdenum and tungsten are mixed with iron to produce steel with specific desired qualities. The careful selection and combination of these elements are crucial to achieving the mechanical properties, corrosion resistance and durability required for various applications. Details about these important materials are explained below.

2.2.1 Iron ore and coal extraction

Resource extraction (mining) operations have been in place for thousands of years, and the earliest evidence of mining and quarrying in England occurs during the Early Neolithic period, ~4000 BCE – so, more than 6000 years ago [1]. The processing of mining extract used for steelmaking [2, 3], specifically iron ore, coal and limestone, each undergoes similar production pathways at ambient temperature.

The mined or quarried source needs to be crushed in a dry or wet mill to a workable size, dust and solids need to be separated, and solids and dust are either collected in silos and tanks or transported with conveyor belts to a production site in close proximity or transported via truck [4]. Valuable data of surface coal mining (quarrying) and iron ore mining [2, 4, 5] have been utilised for the compilation of source extraction simulations. Figure 2.2 displays one configuration sample layout of coal mining using a dry mill built in INOSIM software.

The basis for the process simulations was the BF–BOF route steelmaking process, where a 330 metric tonne capacity is assumed. Approximately 2 metric tonnes of coal needs to be mined to have one metric tonne of coke after coking the coal required to be used in the BF. As simulation software works on mass balance, the actual coke quantity had to be used in the input streams: for sintering, 1.6 tonnes is needed, and for the BF, 168 tonnes is required, at 169.6 tonnes in total. For the sintering process, *20.7 tonnes* of iron ore are needed, and for the BF, a total of *97.3 tonnes* of iron ore and iron oxide are needed. Lime and limestone are used at *106.1 tonnes* in sintering, *22.1 tonnes* for the BF and *16.5 tonnes* of lime in the BOF. Therefore, a total of *144.7 tonnes* of lime needs to be added as a source in the input stream.

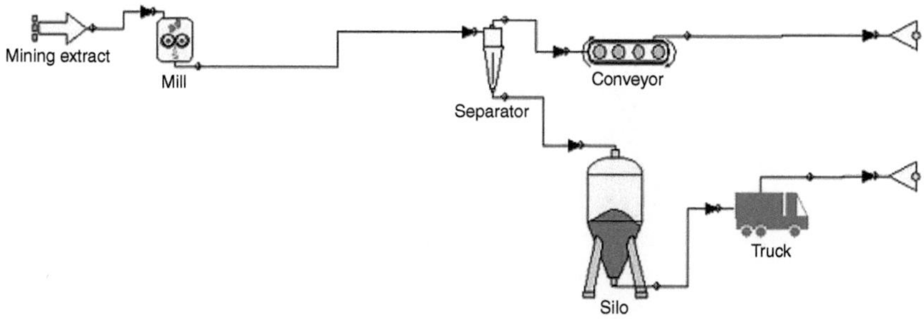

Figure 2.2 INOSIM simplified coal mining process layout.

The compilation in MS Excel was deemed to be the best suited to provide a framework for demonstrating the CO_2 emissions and reduction mechanisms at each stage and the resulting levels of CO_2 emissions by the end of the process. Figure 2.3 demonstrates the flow of material, quantities of the materials involved and results per implementation of the respective steps within the Bio Steel Cycle [6].

2.2.2 Coal

Coal is technically biomass, or at least has biomass as its origin [7]. Millions of years ago, the earth was covered in wetlands, swamps and ancient plant species such as ferns. When the plants died, they sank to the bottom of the swamps. Over the years that followed, the layers of plants were covered by dirt, organic matter from animals and water, compacted by the weight bearing down. Over millions of years, heat and pressure converted the plants into coal, with high carbon content (60% in lignite and 80% in anthracite coal) [7, 8]. The coal seams we excavate today have taken millions of years to form. It is finite – we cannot 'just make' new coal, hence the term 'nonrenewable' [7].

The research used global steel data and literature on sustainability, decarbonisation and CAT, CCS and CCUS technology. The main reason underpinning this choice and course of research is that, as suggested by the WSA Steel Yearbook [9, 10], the global iron and steel industry is still heavily reliant on coal and is responsible for between 7% and 11% of global CO_2 emissions [11–15]. China is thought to be responsible for around 50% of *global* CO_2 emissions [13, 16]. The importance of significantly reducing GHGs and eliminating fossil fuel combustion and usage has never been greater, and fast, practical solutions – on a global scale – are needed. Already in 1912 (Figure 2.4), it was recognised that coal consumption is an environmental hazard and incompatible with keeping global temperatures at a balanced level to sustain life as we know it. Industrial processes have polluted the air we breathe since their conception, and already in 1912, this was recognised, as this Australian newspaper clip demonstrates in Figure 2.4.

It is worth pointing out that in this article, merely the in-furnace coal combustion process is mentioned in connection with carbon emissions. CO_2 emissions from energy consumption, mining, pelletising, coking, sintering, steel smelting, casting,

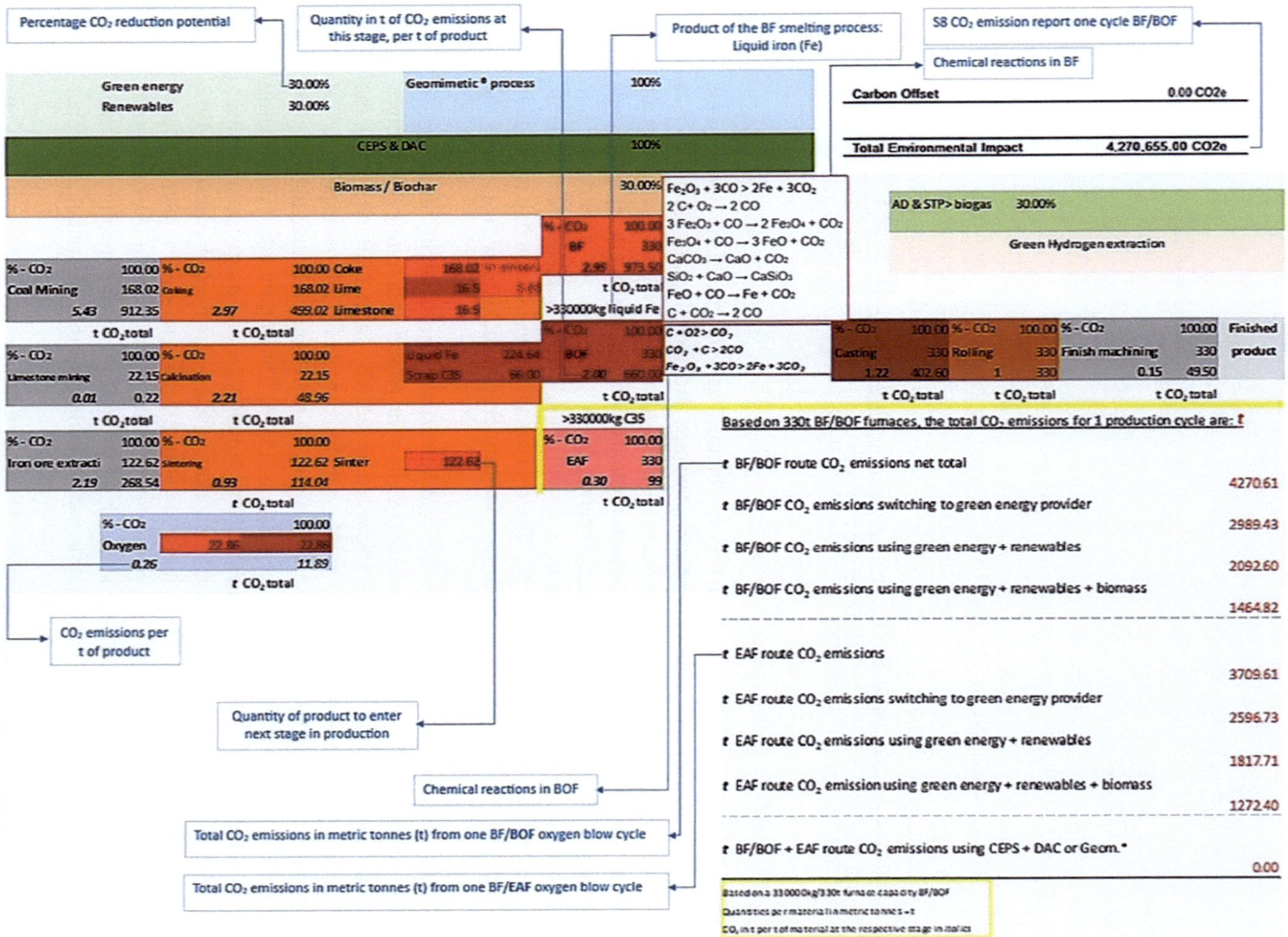

Figure 2.3 Flowchart of the steelmaking process with CO_2 emissions and explanatory comments.

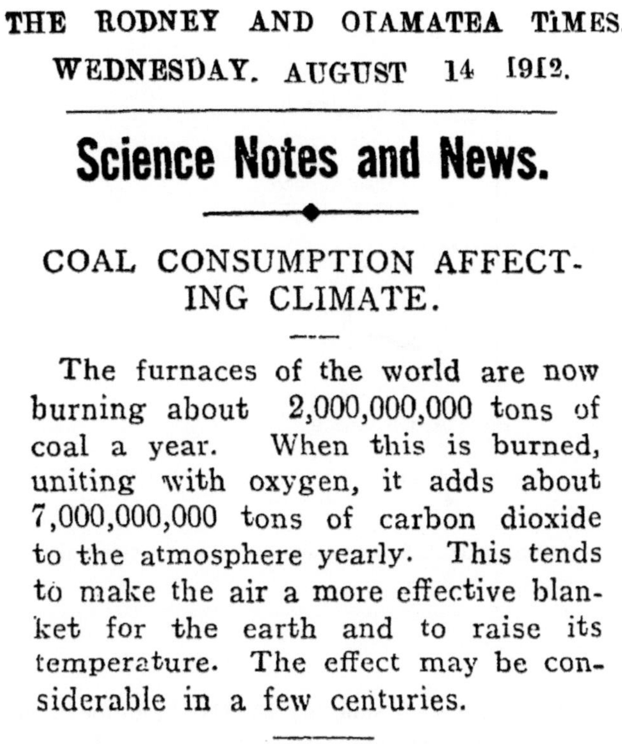
THE RODNEY AND OTAMATEA TIMES,
WEDNESDAY. AUGUST 14 1912.

Science Notes and News.

COAL CONSUMPTION AFFECTING CLIMATE.

The furnaces of the world are now burning about 2,000,000,000 tons of coal a year. When this is burned, uniting with oxygen, it adds about 7,000,000,000 tons of carbon dioxide to the atmosphere yearly. This tends to make the air a more effective blanket for the earth and to raise its temperature. The effect may be considerable in a few centuries.

Figure 2.4 1912 Science Notes and News about Climate Change, The Rodney and Otamatea Times / Public Domain – Waitemata and Kaipara Gazette [17–19].

rolling, annealing, rolling, finish machining and surface treatment and related processes have not been considered at that time. Already more than 100 years ago, in 1912, it was established that BF operations resulted in the release of 3.5 tonnes of CO_2 into the atmosphere for each tonne of steel produced by coal combustion, (7bn t CO_2 emissions per 2bn t, of steel manufactured). It so seems that this figure was possibly based on the carbon content of coal (78–95%).

2.2.2.1 Coal coking

Coking of coal at between 1100 and 2000 °C and calcification of limestone to (burnt) lime at between 780 and 1340 °C required the same considerations as far as mass balance and input streams are concerned. Figure 2.5 shows the simplified coking and calcification route layout.

The average coke oven has a capacity of 36,000 kg, and the lime-kiln has a capacity of 50,000 kg. In order to achieve a mass-balanced simulation, the aforementioned quantities have to be used in the input streams [20–23]: *144,748.62 kg of limestone and 169,665.25 kg of coal.*

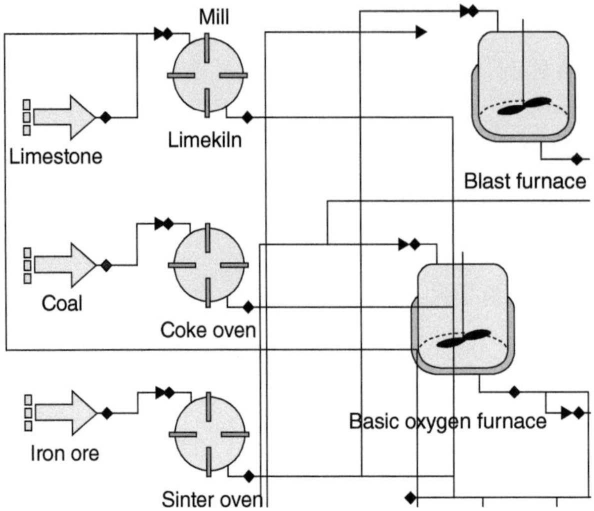

Figure 2.5 INOSIM simplified coking and calcification operation within the steelmaking process.

2.2.3 Lime

Lime and limestone are used at *106,095 kg* in sintering, *22,153.62 kg* for the BF and *16,500 kg* of lime in the BOF. Therefore, a total of *144,748.62 kg* quantity of lime needs to be added as a source in the input stream.

Limestone, as used by the minerals industry, is any rock composed mostly of calcium carbonate ($CaCO_3$) [24]. Limestone is a fundamental raw material used in a range of industrial sectors: metal manufacturing, the building industry, the chemical industry and the agricultural sector. The formation of limestone occurs due to biochemical precipitation of calcium carbonate and resulting continuous compaction over long periods of time. Due to its versatility, there is high demand for limestone and lime products and its use in a growing number of industries has led to its widespread exploration and excavation. Limestone mining mostly takes place in opencast pits (surface mining), with bench formation. Unfortunately, limestone mining is very disruptive and causes a range of disturbances in the natural environment. A myriad of impacts are observed at and surrounding limestone mines, as changes in land-use pattern cause habitat loss, and higher noise levels are disturbing to humans and animals alike, dust emissions diminish the air quality, and changes in aquifer regimes are also observed. This means that where limestone is mined close to or at the site of natural aquifers, the mining operation is disrupting the water table – our fresh water supply. In recent years, environmental concerns and industrial processes accelerating climate change have become increasingly topical, and governments across the globe – at varying levels of effort and efficiency – are trying to create a political framework to mitigate the negative environmental effects [25]. The environmental concerns have triggered the need for a sustainable environment management plan in the mining sector.

- Reducing environmental degradation during operation
- Restoration of degraded lands after final mine closure
- Mitigating the impacts of mining on the environment

The best practices adopted by industries around the world seem suitable to be adapted as per site characteristics, ensuring sustainable mining along with the prevention of environmental degradation [25]. In atmospheric hydrating conditions (ambient temperature), only the calcium oxide fraction of dolomitic quicklime hydrates. This produces a hydrated lime of a specific chemical composition: 46–48% calcium oxide (CaO), 33–34% magnesium oxide (MgO) and 15–17% chemically combined water (calcium hydroxide; inorganic compound: chemical formula $Ca(OH)_2$).

Calcination refers to the thermal treatment of a solid chemical compound, such as limestone. Avoiding the melting of the compound, the lime-kiln charge temperature is raised to ~950 °C under a restricted supply of ambient oxygen (the gaseous O_2 fraction of air) [26], causing the removal of impurities and volatile substances and thermal decomposition [27]. The root of the word calcination can be traced back to its most prominent use with the removal of carbon from limestone (calcium carbonate, $CaCO_3$) through combustion to yield calcium oxide (quicklime). This calcination reaction is $CaCO_3(s) \rightarrow CaO(s) + CO_2(g)$. Calcium oxide is a crucial ingredient in modern cement and is also used as a chemical flux in smelting. Industrial calcination generally emits carbon dioxide (CO_2), making it a major contributor to climate change.

A calciner is a steel cylinder that rotates inside a heated furnace and performs indirect high-temperature processing (550–1150 °C, or 1000–2100 °F) within a controlled atmosphere. The calcination of limestone produces lime, which is 'hungry' for CO_2, as this has been removed during the calcification process. This means that when lime is being exposed to industrial flue stack off-gases, the lime molecules will inevitably try to bind to the CO_2 molecules, which leads to an additional line of thought: CO_2 capture, filtering and cost in various industries: steel, concrete, chemical (solutions, oil and gas) and waste recycling. At present, not many industrial providers seem to be considering this a business proposition, as only businesses in Asia seem to offer solutions in this direction. This could be another opportunity for the UK economy to create employment and training opportunities, as it has yet to be identified which CO_2 capture technology is the most effective for which industry: active coal filters (framework-flexibility-driven CO_2 adsorption on a zeolite), air locks, chemical absorption, phyto-absorption and hydro-absorption (i.e. CO_2 capture in wet and dry superbase ionic liquids) [22, 23].

2.2.4 Oxygen

Steel is an alloy of iron and carbon containing less than 2% carbon and 1% manganese and small amounts of silicon, phosphorus, sulphur and oxygen [9, 28]. Steelmaking starts with mining the resources required to produce steel: iron ore, coal and limestone. Oxygen, used in the BF and BOF operations, needs to be extracted from ambient air and compressed to specifications [9, 28, 29].

Ambient (atmospheric) air contains 21% O_2, 78% N_2 and ~1% is mostly argon and CO_2. With air being the main component required to be present to facilitate combustion, the oxygen in the air is broken down into carbon dioxide, water and energy. As nitrogen in the air has negative impacts on combustion processes, it is being ejected by the reaction to very high combustion temperatures and is expelled through the respective flue stacks [9, 28].

Gasification [30] is the process by which coal or another carbon-based fuel or waste is transformed into a synthesis gas ('syngas'). This syngas is mainly consists of carbon monoxide (CO) and hydrogen (H_2). This gas can be used in a power station to produce electricity or steam. Gasification has many values [31], including the ability to use low British thermal units (BTU) fuels for energy production, easier carbon dioxide capture as compared to traditional sequestration methods, and the ability to produce hydrogen gas. Gasification is carried out in steps. First, the fuel is ground up into a slurry. Next, the fuel is pyrolysed in an oxygen-starved environment. This turns the majority of the carbon into a gas. The next step is to add oxygen or air ('direct gasification') or steam ('indirect gasification') at high temperatures to form carbon monoxide and hydrogen. This is the syngas that can be used in industry for a variety of energy-producing systems [30, 31]. Before it is used, however, the gas may need to undergo more processing, such as removing particulates or cooling, depending on its end-use. Air can be separated into its constituents using a variety of techniques. This paper will address cryogenic, pressure swing adsorption, membrane technology and by-product methods of O_2 generation. Each air separation technology produces oxygen at different purities, pressures and volumetric flow rates. Furthermore, each technology has different running costs. One air separation method is not any better across all industries than any other technique. Depending on the needs of the process, one air separation technique will be more beneficial than others. This paper will look at the glass, gasification and gas-to-liquid industries, with a focus on gasification, their oxygen needs and the effect that oxygen enrichment will have on the process. Current research in metals oxidation will also be addressed. Technologies applying cryogenic air separation units (ASU) can be seen as an 'old' process, which is used to produce high-purity oxygen or nitrogen at high volumes. The process was first developed by Carl Von Linde in 1895 [29] and it remains pretty much the same today. Cryogenics is also the chief method by which liquid oxygen can be produced. The technology is centred on the fact that each of the air's constituents has different boiling points. The idea behind the process is to lower the temperature of the air such that nitrogen and oxygen separate based on their boiling points. This occurs at around − 300 °F. If higher purity oxygen or liquid oxygen is required, further distillation is required. For liquid oxygen, nitrogen is used as the heat transfer fluid to further cool the oxygen. Cryogenic separation is most effective when any of the three criteria need to be met: high-purity oxygen is required (>99.5%), high volumes of oxygen are required (≥102 tonnes of oxygen/day), or high-pressure oxygen is required. Cryogenic air separators take more than an hour to start up. Additionally, since cryogenics can produce such a high purity of oxygen, the waste nitrogen stream is of a usable quality [29]. This can add significant financial benefits to a process integrated with a cryogenic ASU. Pressure swing adsorbers (PSAs) are a much newer technology as compared to cryogenic ASU. PSA devices take atmospheric air into a pressurised tank. Inside the tank are zeolites. Zeolites, under pressure, have the ability to deform and create a dipole. Depending on the zeolite chosen, this dipole allows for the collection of nitrogen but allows oxygen to pass. For oxygen enrichment, the PSA is generally pressurised to a minimum of 1.5 atm. After a certain volume of air has been separated, the zeolite will become saturated with nitrogen. At this point, it needs to be regenerated. This is done by dropping the pressure of the tank back to atmospheric pressure, thus returning the zeolite to its original polarity. This liberates the nitrogen. Vacuum pressure swing adsorbers (VPSA) lower the pressure in the tank to subatmospheric values, thus improving the regeneration process. In both processes, whilst one tank is regenerating, another tank is usually charging. This allows for a continuous production of oxygen. PSA devices are best suited for processes that do not require extremely high purities of oxygen. During the research for establishing the CO_2 emissions along the steelmaking process and all steps involved, the extraction of oxygen was identified as generating 0.26 tonnes of CO_2 emissions per metric tonne of steel produced. The skyrocketing energy prices at the time of writing in 2023/2024 made it difficult to quantify the cost per m^3 of oxygen produced, per tonne of steel.

2.3 Iron ore sintering

Within the sintering process (Figure 2.6), the sinter is made by heating a mix of iron ore powder and fluxes (lime) at ~1300 °C to create an open-grained, consistent substance. The product sinter is then crushed and screened in a centrifuge to separate solids from dust, which at the same time cools the agglomerate [20, 21, 23, 32]. The quantity of 97,367.91 kg of iron ore and iron oxide (sinter) is part of the input stream for the BF.

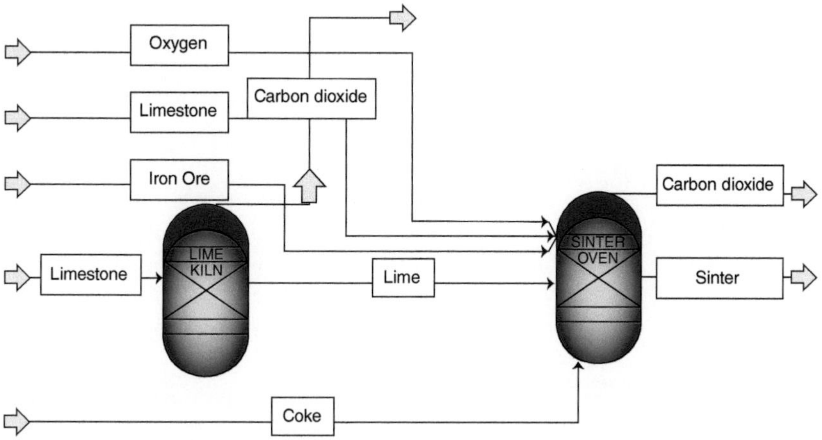

Figure 2.6 The lime-kiln and sinter oven section of the steelmaking process in Aspen + V12.1.

In order to demonstrate proof of concept, process simulation configurations were built in Aspen+V12.1. To establish whether Aspen+ was suitable to trial selected stages of CO_2 emissions reduction measure implementation, the calcination and sintering process were taken as initial trials. RGibbs reactors in steady state were chosen, with an inside vessel pressure of 1.01325 bar. The 50,000 kg capacity was assumed. The section is displayed in Figure 2.6.

Preliminary research findings pointed towards a CO_2 emissions burden of 2.760 and 2.25 tonnes of CO_2/tonne of product, which was reported by the Aspen calculations, as displayed in Figure 2.7.

The chemical reactions taking place are as follows:

Sintering fines at 1250 °C:

$$Fe_2O_3 + C_{135}H_{96}O_9NS + CaCO_3 =$$

$$CaO\ 2\ Fe_2O_3\ (Al_2O_3 + SiO_2)$$

$$2Fe_2O_3 + 3CO \rightarrow 2Fe + 3CO_2$$

Green HDR (hydrogen direct reduction or iron ore) in BF and electric arc furnace (EAF) applications is considered to have a significant impact on reducing CO_2 emissions in steel manufacturing, as this process uses 3.48 MWh of electricity per tonne of steel product and emits only 2.8% of BF CO_2. However, as the prices of fossil-fuel-derived energy have increased significantly, it is imperative to replace fossil-fuel-derived energy with renewable energy technologies and biomass [33–35].

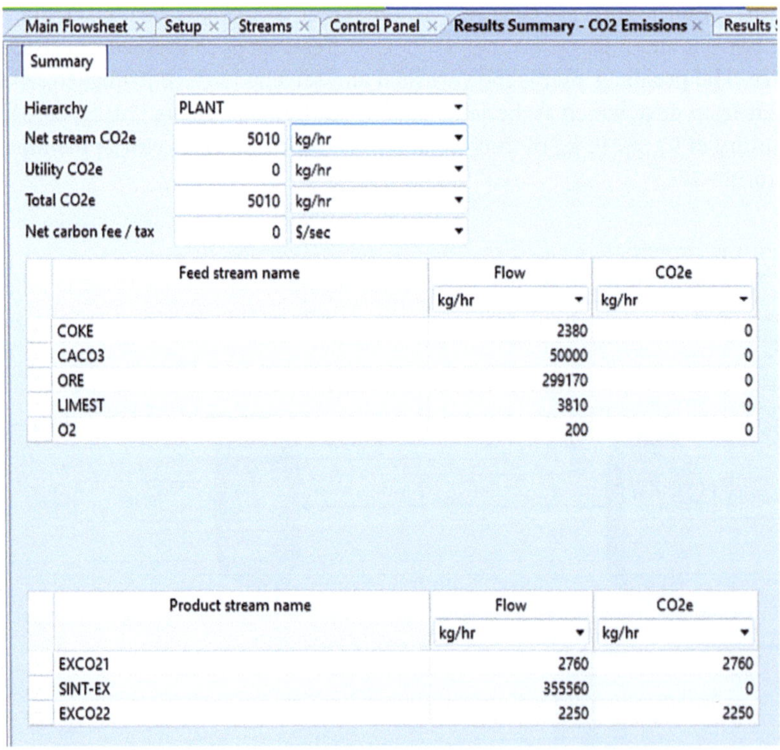

Figure 2.7 Aspen + V12.1 overview material flow and CO_2 emissions.

2.4 Iron ore direct reduction

Conejo, Birat and Dutta [36] and Satyendra [37] described ULCORED as a technology based on iron ore direct reduction, which produces direct reduced iron (DRI) in a shaft furnace. Hereby, natural gas (NG) or gas obtained by gasification of coal, and off-gas are recycled into the furnace after CO_2 has been captured. This is directed into a concentrated stream and stored permanently. The main objective of the ULCORED process was to achieve a significant reduction in NG consumption as a fossil fuel. It was achieved by using partial oxidation (POx) of NG. In tandem with carbon capture and storage (CCS), ULCORED has the potential to reduce 70% of CO_2 emissions, compared with the standard BF route. At its core, CO_2 scrubbing minimises GHG emissions and maintains to reduce energy consumption at the same time. Therefore, ULCORED needs to be seen as an engineered combination of NG oxidisation and carbon capture.

The chemical reactions taking place during the iron's direct reduction are as follows: the iron reduction reactions of the H_2 and CO in a gas-based DRI process:

$$3Fe_2O_3 + H_2 \rightarrow 2Fe_3O_4 + H_2O$$

$$Fe_2O_4 + H_2 \rightarrow 3FeO + H_2$$

$$FeO + H_2 \rightarrow Fe + H_2O$$

2.5 Hydrogen direct reduction (HDR)

Similar to the HDR–EAF process, direct reduction of iron ore using hydrogen has been piloted by a range of entities [38, 39] and the mechanisms and processes involved were clearly defined as follows: direct reduction of iron is the chemical removal of O_2 from solid-state iron ore. At present, the standard procedure to remove carbon from the iron used in the steelmaking process is chemically reduced from iron ore through the use of fossil fuel resources – NG or coke from coal, known as DRI making. The carbon in fossil fuels combines with the oxygen in the iron ore, initially without the addition of heat, producing metallic iron and a CO_2-rich process gas, according to the following simplified chemical reaction:

$$2Fe_2O_3 + 3C \rightarrow 4Fe + 3CO_2$$

(Source: Refs. [38, 40, 41])

Iron ore reduction can also be processed using H_2 instead of carbon, in which case the gas produced is water as per the following reactions:

$$Fe_2O_3 + 3H_2 \rightarrow 2Fe + 3H_2O, FeO + H_2 \rightarrow Fe + H_2O$$

(Source: Refs. [38, 40, 41])

Hydrogen, more specifically green hydrogen, is seen globally as the clean energy source – evidenced by the fact that global industrial giants like Tata Steel, SSAB and ArcelorMittal are investing heavily in developing hydrogen plants with access to seawater, like in Ijmuiden (TATA Steel Hermes Project).

2.6 Blast furnace and the transformation process of ironmaking

The BF is the vessel where smelting of iron ore and fluxes takes place via the addition of 22,864 kg of 1000 °C hot oxygen at 150 psi via tuyeres (lances). A simplified layout can be seen in Figure 2.8.

The injection of the hot oxygen into the charge causes a temperature rise to between 1700 and 2000 °C. The BF product is essentially liquid iron, where a total of *97.4 tonnes* of iron ore and iron oxide, *22.1 tonnes* of lime and limestone and *169.6 tonnes* of coal/coke are used to produce molten iron. This is then transferred to the BOF.

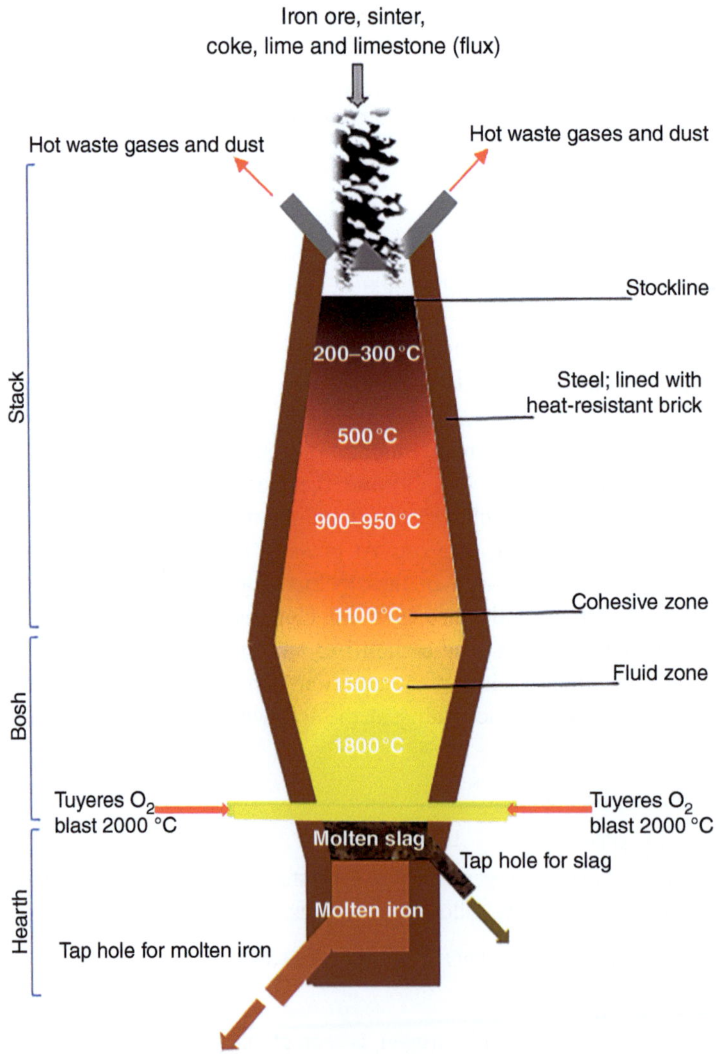

Figure 2.8 Blast furnace simplified side view.

Ironmaking is the last preparatory step before producing the actual steel. Iron is made in the BF, as per Figure 2.8.

The chemical reactions taking place in the BF can be followed in easy steps:

BF Step 1: Bottom blown 1000 °C/O_2 @ 220 psi = $2\,C + O_2 \rightarrow 2CO$ @ 1800 °C

BF Step 2: Furnace top ~ 200 °C– 700 °C = $3\,Fe_2O_3 + CO \rightarrow 2\,Fe_3O_4 + CO_2$

BF Step 4: Air + N_2 (up) flux (down) = $CaCO_3 \rightarrow CaO + CO_2$

BF Step 5: Slag forming = $SiO_2 + CaO \rightarrow CaSiO_3$

BF Step 6: Iron oxide travels to the bottom (1200–2000 °C) = $FeO + CO \rightarrow Fe + CO_2$

BF Step 7: The added coke causes = $C + CO_2 \rightarrow 2CO$

BF gas: Boudouard reaction controlling the gas atmosphere = $2CO \rightleftharpoons CO_2 + C$

After all contents of the BF have been undergoing transformation, the resulting liquid iron is then transferred to the BOF, where the liquid metal mass will be transformed into steel. As Figure 2.9 will show:

Alternatively, steel can also be made by transferring the liquid iron to an EAF instead of the BOF.

2.7 Basic oxygen furnace operation

Steel (C35 as an example) is the product of treating liquid iron from the BF within the BOF with the injection of oxygen at ambient temperature at 220 psi pressure. The chemical reaction of the liquid iron and flux (lime) with the oxygen causes the temperature to rise to between 1700 and 2000 °C. A simplified layout of this configuration can be seen in Figure 2.10.

In the BOF operation, no coal is required within the charge, which can contain up to 20% of scrap steel. The liquid steel is being cast after 40 minutes of blowing oxygen onto the metal mixture.

Steel is an alloy of iron and carbon containing less than 2% carbon and 1% manganese and small amounts of silicon, phosphorus, sulphur and oxygen [9, 28]. Steelmaking starts with mining the resources required to produce steel: iron ore, coal and limestone. Oxygen, used in the BF and BOF operations, needs to be extracted from ambient air and compressed to specifications. Not only during the in-ground

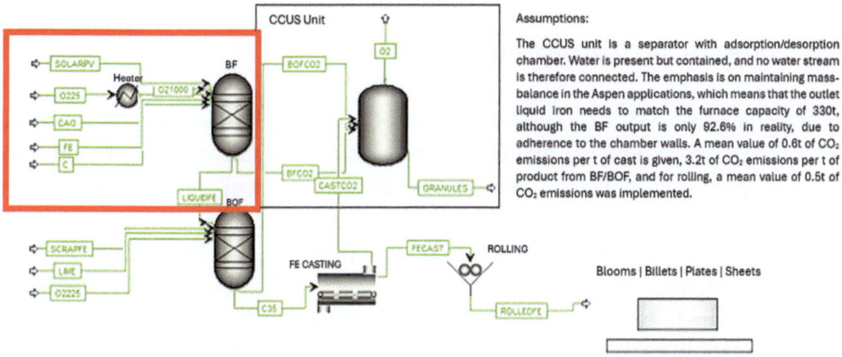

Figure 2.9 Aspen simplified layout of BF–BOF route steelmaking, BF section (red).

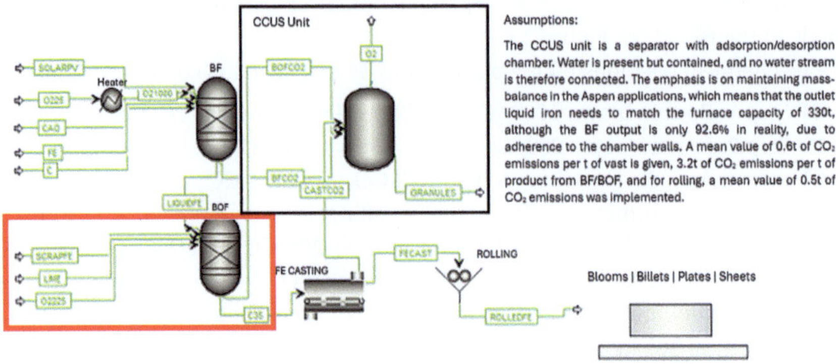

Figure 2.10 Aspen™ simplified layout of BF–BOF route steelmaking, BOF section (red).

or above-ground quarrying of coal, iron ore and limestone can CO_2 emissions savings be made, but also in the following steps of coking, calcination and sintering processes, in the BF (Figure 2.11) and likewise in the BOF operations.

BOF Oxidation at jet impact zone:

$C + O_2 \rightarrow CO_2$

$Si + O_2 \rightarrow SiO_2$

$2Fe + O_2 \rightarrow O_2 \rightarrow 2MnO$

BOF Desulphuration

$(FeS) + CaO \rightarrow CaS + FeO$

$(FeS) + MaO \rightarrow MaS + FeO$

$3Fe_2O_3 + CaS \rightarrow CaO + 6FeO + SO_2$

$2CaS + 3O_2 \rightarrow 2CaO + 2SO_2$

BOF Steelmaking:

$C + O2 \rightarrow CO_2$

$CO_2 + C \rightarrow 2CO$

$Fe_2O_8 + 3CO \rightarrow 2Fe + 3CO_2$

BOF Desilication

$CaCO_8 \rightarrow CaO + CO_2$

$CaO + SiO_2 \rightarrow CaSiO_3$

BOF Dephosphorisation

$4P + 5O_2 \rightarrow 2P_2O_5$

$P_2O_3 + 3CaO \rightarrow Ca3(PO_4)_2$

The stages and chemical reactions during the BF operation are as follows:

The BOF operation produces less CO_2 emissions (on average 2.0 tonnes of CO_2/tonne of cs) than the BF (on average 2.95 tonnes of CO_2/tonne cs).

2.8 Electric arc furnace operation

In order to produce steel, essential resources need to be either mined, produced or manufactured. There are two main process routes to steelmaking: the primary BF–BOF process produces ~79% of the UK crude steel, and the secondary EAF process accounts for the majority of the remaining 21% of the UK crude steel production – the individual steps of which are going to be eluded to in further detail. In order to produce steel, essential resources need to be either mined, produced or manufactured. The CO_2 emissions of all production steps have been investigated throughout [21, 23, 35, 42–57]. The steelmaking table flowchart with BiSC (Figure 2.16) illustrates the individual steps necessary to produce steel, using the

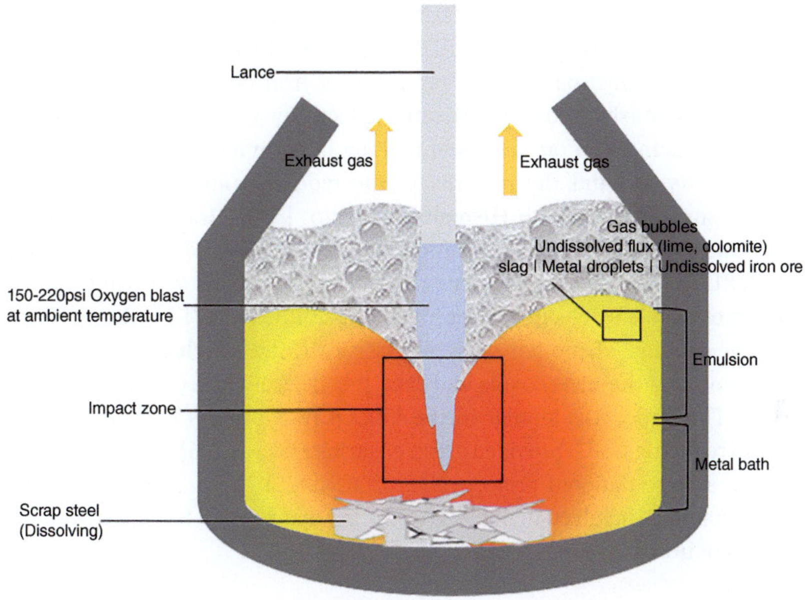

Figure 2.11 Basic oxygen furnace simplified side view.

BF–BOF route. Alternatively, the EAF can be used to make steel in place of the BOF. All required stages in steel production needed to be identified first before an investigation into the individual CO_2 emissions and energy requirements could take place. The flowcharts displayed in this book are congruent with the colour coding of the MS Excel database, which can easily be adapted to accommodate changes within the production process.

2.9 Smelting reduction

Within the smelting processes, two steps are eliminated: coking and pelletising of iron ores. Today, the commercially most successful smelting reduction process is called COREX. During this procedure, pellets and coal are used, but this generates more carbon dioxide than the BF operation. It is therefore a challenge for the iron and steel industry to produce iron directly from iron ore, without using coal/coke and the process of pelletising. Other environmentally detrimental factors are the higher concentrations of sulphur and iron oxide in the metal, as the slag from the COREX process has a higher FeO content in comparison to a BF slag, and also a higher concentration of CO and CO_2. These by-products need higher rates of desulphurisation, in addition to controlling the foaming slags and improving the efficiency of utilising coal particles in order to decrease the carbon footprint [56]. The current energy consumption of the smelting processes can only refer to the COREX process, since this constitutes currently the only commercially viable method. In comparison to the BF, the primary energy consumption of the

COREX process is much higher at 16 GJ/tonne [57]. Efforts to make this process more viable and competitive include recycling the top off-gas, similar to the direct reduction of iron ore/hydrogen direct reduction of iron ore (DRI/HDRI) or BF process. As an established concept, HIsarna effectively eliminates the need for sintering: pre-reducing iron ore fines and melting these in a cyclone are achieved by using pure oxygen, and the majority of the reduction and coal gasification occurs in the smelter (similar to HIsmelt, Rio Tinto) [58]. HIsarna is a smelting technology and the result of a collaboration between ULCOS [59], Tata Steel and Rio Tinto [58]. AIST in the United States supported and sponsored two projects: (i) extraction of metals by electrolysis at high temperatures and (ii) hydrogen flash smelting. The method of molten oxide electrolysis (MOE) includes the process steps of first melting the metal oxide and then the electrolytic separation of metal and oxygen. This process is similar to the ULCOLYSIS process and promises to contribute to reducing GHG, provided green electricity is used [36]. NEDO in Japan is supporting a large national project called COURSE 50 since 2007. One of the multiple mechanisms involved includes introducing hydrogen in the BF to improve the reduction of iron ore [60].

2.10 Electric arc furnace efficiency improvements

Teng et al. [61] Singh and Jaison [62] and Kirschen, Hay and Echterhof [63] have provided specifics of some of the improvements in EAF operation developed in recent years.

A substantial range of innovative technologies has dramatically the role of the EAF operation. The development of the ladle furnace was a milestone development, as it has redefined the EAF role as a melting unit: the processes involved are (a) the development of UHP transformers which are achieving higher melting rates; (b) foamy slag practice which is operating with longer arcs and therefore higher energy density; (c) continuous preheating and charging of scrap metal; (d) higher oxygen injection rates to increase the amount of pig iron in the metallic charge – supplying fossil energy (coal) in addition to electricity; (e) process optimisation of the hot heel practice; (f) eccentric bottom tapping and improved post-combustion treatment and off-gas treatment; (g) using improved oxy-fuel burners; (h) hot charge of DRI and HDRI; and (i) using preheated metal. The level of electrical energy consumed during the EAF operation depends mainly on the metallic charge within. The highest energy consumption levels are in the range between 550 and 700 kWh/tonne, if the metallic charge is either from DRI or from HDRI and can decrease to 300–450 kWh/tonne when the metallic charge is largely based on scrap metal. The lower levels of energy consumption can be achieved when both chemical energy (derived from oxygen injection) and scrap preheating technologies are being implemented in the EAF process. It needs to be emphasised that the scale of typical smelting reduction steelmaking plants is smaller in comparison with standard BF-based plants [64].

Although the COREX process has been the dominating smelting reduction process commercially operating worldwide, Yi et al. [64] found that FINEX can be considered as having more potential for large-scale CO_2 removal, as the systematic development of the FINEX process is showing promise: energy recovery systems, binders for coal briquettes and a steel grid system in FBR (fluidised bed reactor) achieved reductions in economic terms and the overall environmental footprint. One of the biggest advantages of the FINEX over the COREX process is that the FINEX process enables the direct use of sinter feed fine ore [64].

2.11 Improvements in energy consumption and production: energy and emissions

Looking into energy consumed and the resulting GHG emissions in the iron and steel industry, it is estimated that between 2.15 and 2.8 $MtCO_2$/p.a. are energy-related CO_2 emissions [11, 55, 65–70]. As shown in Figure 2.12, the development of electricity usage in steelmaking has been on a downward trajectory since the year 2000, partly due to process and efficiency improvements across most industries.

In stark contrast, quite the opposite observation is made for global electricity consumption [11, 69, 70], as shown in Figure 2.13.

Global electricity consumption has continuously increased during the last 50 years, arriving at an estimated 25,300 terawatt-hours in 2021. Since 1980 and up to 2021, global electricity consumption has increased threefold, and the global population increased by approximately 75%, simultaneously. In line with extended industrialisation and infrastructural improvement, these factors caused a threefold increased electricity demand, with an upward trajectory prognosis, as of the end of 2021.

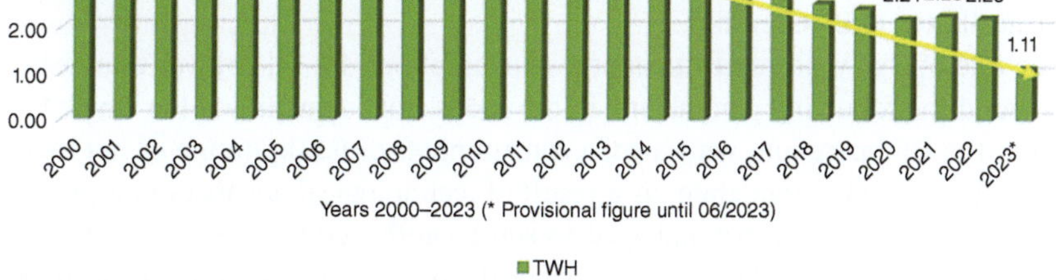

Figure 2.12 Electricity use for iron and steel production in the United Kingdom from 2000 to 2021. Source: Refs. [11, 55, 69–73].

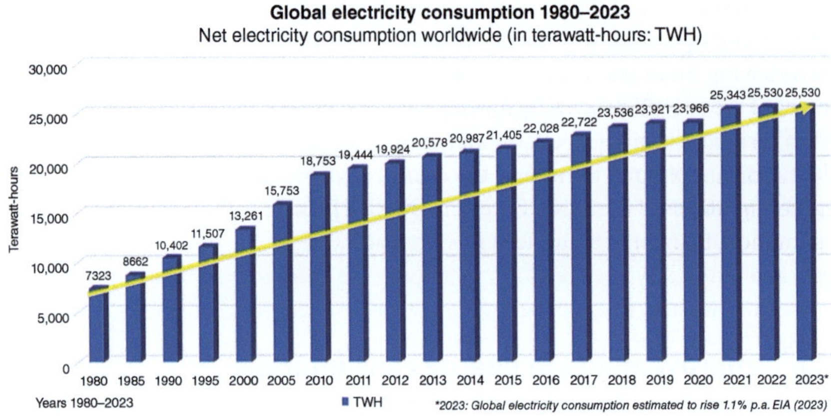

Figure 2.13 Global electricity consumption 1980–2021. Source: Refs. [11, 55, 69–73].

Mandova [48] established that their concept of Bio-CCS could help iron and steel production to become close to carbon neutral, under consideration of the estimated avoidance cost (in the EU) of €80/ tCO_2. They established that the Netherlands, France and Belgium currently have the lowest CCS deployment cost.

It is currently given that [48]:

- 1 Kilowatt hour (kWh) = 3600 kJ (3,600,000 J)
- $1\,J = 2.77777778 \times 10^{-7}$ kWh

Most energy suppliers provide estimated CO_2 emissions per kWh, which state that an estimated 0.46 kg of CO_2 per kWh of energy produced is being emitted by using the current technology mix. Opportunities [74] for both integrated and scrap-based electricity production were identified, with fuel conversion, methane reforming of coke oven gas and cooperation in complex industrial symbioses. A similar approach was taken when the ironmaking and steelmaking process was investigated in detail [75]. It was established that 95% of energy for an integrated steelmaking plant is the result of using coke, 3–4% from gaseous and 1–2% from liquid fuels. Furthermore, it was identified [75] that within the BF/BOF steelmaking route, an energy consumption of 19.8–31.2 GJ/energy/tonne crude steel takes place. In contrast, within the EAF route utilising almost 100% steel scrap, only 9.1–12.5 GJ/energy/tonne crude steel is used. The authors investigating the BF/BOF route of integrated steel attributed ~75% of the fossil-fuel(coal)-derived energy had to be attributed to BF operation.

The role of coal with regard to energy consumption [48, 74, 75] was established and found the following parameters: in the 1950s, coke constituted the single fuel used in the BF operation, with a coke consumption of 950 kg/tonne of hot metal (THM). Fortunately, since then, as a result of technological advances and process improvements, the industry managed to significantly reduce the coke consumption to ~500–550 kg/THM in the 1970s, since the mid-1980s to powdered coal injection (PCI) gradually replacing some of the coke, reducing the coke consumption to ≤400 kg/THM. A vital part of this improvement process was the total injection of

PCI and other fuel of ~100 kg/THM. Today (2023), a PCI rate of ~130–160 kg/THM is quite common. Increasing the PCI rate to ~250 kg/THM would possibly result in reducing the coke consumption down to 250 kg/THM. Allegedly, in Japan, BF is running at ~400 kg/THM. The average PC rate in Japan at the time was considered 120 kg/THM, but individual operations could have operated at 250 kg/THM – such as operation at 266 kg/THM at the Fukuyama No. 3 BF. The authors found, though, that the higher PPCI rate was compromised by the increased total reducing agent consumption, being inextricably tied to the minimum requirement of the reduction of FeO and the energy required for maintaining the optimum thermal state in the BF reactor. Additionally, the optimum temperature of the liquid products of hot metal and slag needs to be maintained.

To date (2023), the energy efficiency of BF–BOF route steelmaking is surpassed by EAF operation. The 15 manufacturing subsectors together consume 95% of all energy used by US manufacturing – although this does not include the energy consumption of fuels which are being used as raw materials [76].

2.12 Exergy analysis

In order to discuss exergy, one must define what components and desired results exergy analysis in the context of steel production entail in detail. Exergy and its effects within the iron and steel industry were thoroughly investigated [64, 70, 73] and confirmed the findings made by mechanical engineer Rant in 1956, who initially coined the term *exergy* as such – based on the first and second laws of thermodynamics, which had already been established in the nineteenth century. In *thermodynamics*, this Slovene mechanical engineer Rant noted for his 1956 introduction of the term '*exergy*' as a new name for 'technically available energy', followed up by establishing the term '*anergy*' as a derivative term [68].

Based on said first and second laws of thermodynamics, approaching the issue from the component point of view, Demirel [74] defines the first law of efficiency or thermal efficiency as the ratio of the work output to the total rate of heat input and the second law of thermodynamics, associated with the definition of entropy S, dS = 6qrev/T – with entropy being a thermodynamic potential and the quantitative measure of irreversibility. Exergy is being described [74] as a unifying concept of various forms of energy – consisting of the physical resource properties and its surrounding environment. Siefert, Narburgh and Chen [67] provided an equation to demonstrate the first law of *thermodynamics* as $\Delta U = Q - W$.

In heavy industry, it means that exergy *loss* means energy *transfer* [77–81]. The work loss $T0_S$ in a continuous process is the difference in the energy (heat) H before and after the process – meaning that available energy $A = H - T0_S$ or exergy is a measure of the departure from the ambient or dead state [77–83].

Several attempts have been made in the past to establish a suitable component-based percentile formula, which would describe overall the exerted/destroyed energy, applicable to any processing in any setting. Already since the 1990s, scientists were convinced [65, 66], that 30% of the heat energy entering any production process is (i) lost and (ii) could be recovered. The technology to do so has experienced a steep learning curve and is now commercially viable and available. Excess heat from any production or manufacturing process can be reused to supply any production site with heat and warm water, partly, due to the simplicity of the technology required. The energy basis and flow have been investigated thoroughly, and an energy industry defining and telling report and graph was produced by Moran and Sciubba [66]. Figure 2.14 displays the exergy analysis power cycle.

The theory of this exergy analysis is based on the fact that if 100% of energy is being inserted into any energy-requiring unit, the amount of 70% will be effectively used for the intended purpose, whilst 30% is being lost due to inefficiencies and deficits within the production and processing infrastructure. It was established that ~70% of energy is being utilised within the production process, as 100% is set as energy input and only 30% could be measured as being available at the end of the production process. This leads to the conclusion that there is the potential for 30% off-heat recovery and utilisation. The exergy analysis, based on this principle, established a 30% loss. But, as every production process is different and works with different parameters, no 'one-size-fits-all' percentile rule of exergy, energy/exergy destruction or heat loss during production could be established. The EERE [76] established an interactive determination process to that effect. Although Moran and Sciubba established some core principles of exergy analysis [65, 66], the practical application has, in fact, been utilised since the beginning of the Industrial Revolution [65]. Establishing exergenetic principles means utilising, monitoring and analysing two systems at different states whenever they are set in direct communication with each other so that these have the opportunity to come into some state of equilibrium. Given that one of the two systems is an adequately idealised system with set parameters called 'the environment' and the other is an 'idea or system of interest', exergy is defined as the maximum *theoretical* shaft work (or electrical work) obtainable as the systems interact to equilibrium. This means that heat transfer only occurs in direct interaction with the environment. Brodyanski, Sorin and Le Goff [65] established very

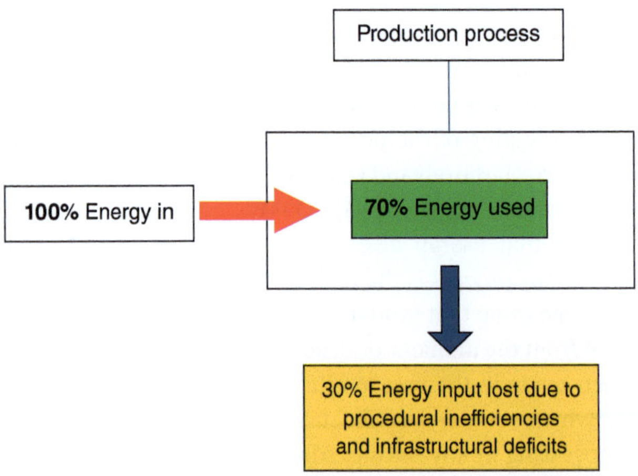

Figure 2.14 Exergy analysis power cycle. Source: Adapted from Ref. [66].

clearly that in order to achieve reliable and replicable data via exergy analysis, one needs to pay attention to the *thermodynamic* base of the concept of *energy efficiency* within *energetics* and *energotechnologies*. The authors found that if the coefficient of energy transformation (CET) and the coefficient of exergy efficiency (CEE) are given, exergy analysis will provide reliable and replicable data, providing the researcher with results showing internal and external exergy losses. Furthermore, by carrying out exergy analysis, the concept of '*transit exergy*', as displayed in Figure 2.15, and its relationship with the CEE of systems and their parameters become clear [65, 67, 68, 85, 86]. The linear flow of energy introduced, the efficacy of usage within the individual system section and relevant resulting losses have been investigated throughout, using the data provided by a coal-fired power plant.

$$\dot{Q}_{env} = \sum_{i=inlet} \dot{n}_i \widehat{h}_i - \sum_{o=outlet} \dot{n}_o \widehat{h}_o - \dot{W} - \sum \dot{Q}_j \tag{2.1}$$

$$\dot{\sigma}_{irr} = \sum_{o=outlet} \dot{n}_o \widehat{s}_o - \sum_{i=inlet} \dot{n}_i \widehat{s}_i + \frac{\dot{Q}_{env}}{T_{env}} + \sum_j \frac{\dot{Q}_j}{T_{j,o} - T_{j,i}} ln\left(\frac{T_{j,o}}{T_{j,i}}\right) \tag{2.2}$$

Siefert, Narburgh and Chen [67] provided equations to demonstrate the first (Eq. 2.1) and second (Eq. 2.2) laws of thermodynamics for a steady-state control volume, and also for solving for the thermal energy transfer into the environment,

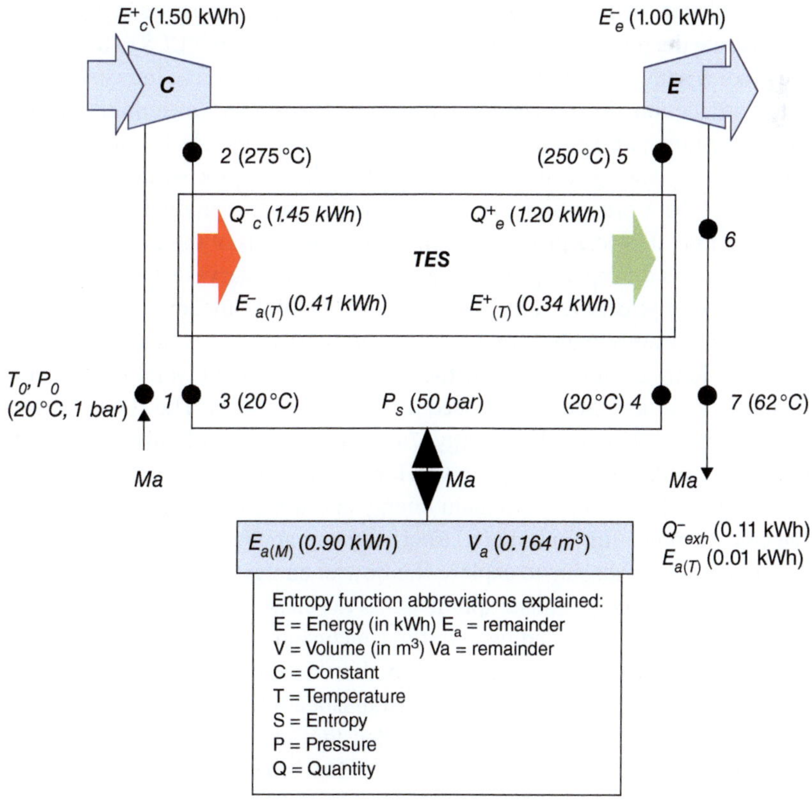

Figure 2.15 Entropy, function.

and exergy (Eq. 2.3) in thermal energy transferred between subsystems, as displayed in the following:

$$\dot{E}_j = \dot{Q}_j \cdot \left[1 - \frac{T_{env}}{T_{j,o} - T_{j,i}} ln\left(\frac{T_{j,o}}{T_{j,i}}\right)\right] \tag{2.3}$$

Furthermore, Demirel [84] describes exergy with – what he classed as – 'a unifying concept of various forms of energy'. Within, heat, mechanical processes, chemical energy and the maximum amount of work theoretically available were thought to bring a resource into equilibrium with its environment through a reversible process – entropy. In Figure 2.15, the function of entropy is explained in detail.

Thus, exergy constitutes a function, consisting of the physical resource properties and its surrounding environment. In practical terms, it means that exergy *loss* means exergy *transfer*. The work loss in a continuous process is the difference in the exergy before and after the process. Available energy or exergy is a measure of the departure from the ambient or dead state. During the practical trials to establish the feasibility and viability of carbon capture and storage (CCS) processes for steel components [66, 67], exergy analysis was considered a useful tool to establish the efficiency of implemented measures. As the environmental and monetary benefits are growing proportionately with input, this effect needs to be demonstrated very clearly to steel-producing businesses and ancillary companies in order to (i) result in behaviour change in the steel component manufacturing companies involved and (ii) demonstrate the potential for remanufacturing processes and CCS to be a transferrable process with applications in a vast variety of metalworking industries.

For steel production, this is an additional element to consider – meaning: researching energy *consumption*. This entails establishing if exergy loss can be minimised or harvested, or if the energy required for the steel production process can be reduced [74, 85, 87–89], as a proportion can be harvested from the production processes within. Therefore, this core principle is one of the elements of the Bio Steel Cycle, where heat emitted during the production process is being harvested and used elsewhere. However, there were some attempts with helpful outcomes, showing the different levels of exergy analysis and results. The approach to establishing a framework applicable to all industries and the energy input for the various industrial sectors had been established and utilised as the starting point. The (US) Office of Energy Efficiency and Renewable Energy [76] created a dynamic Sankey diagram to demonstrate manufacturing energy flows. Herein, the line widths indicate the volume of energy which flows to the main energy end users in manufacturing, with the line colours indicating fuel, steam, electricity and applied or lost energy. Users are able to modify the display to explore the flow of energy use at the macro-scale or even compare the energy consumption across the chosen manufacturing subsectors. This methodology provided us with the conclusion that, in the United States, the best part of 20% of all US energy combined produced is being used to provide energy to the manufacturing sector. They established a new factor within exergy analysis, namely *efficiency loss*. They concluded that of the energy volume being made available to the manufacturing sector, ~20–30% is being lost due to inefficiencies, both on-site and within the infrastructure, with some losses being unavoidable due to technological constraints. The manifold opportunities within exergy analysis to determine energy reclamation potential have been recognised very early, and

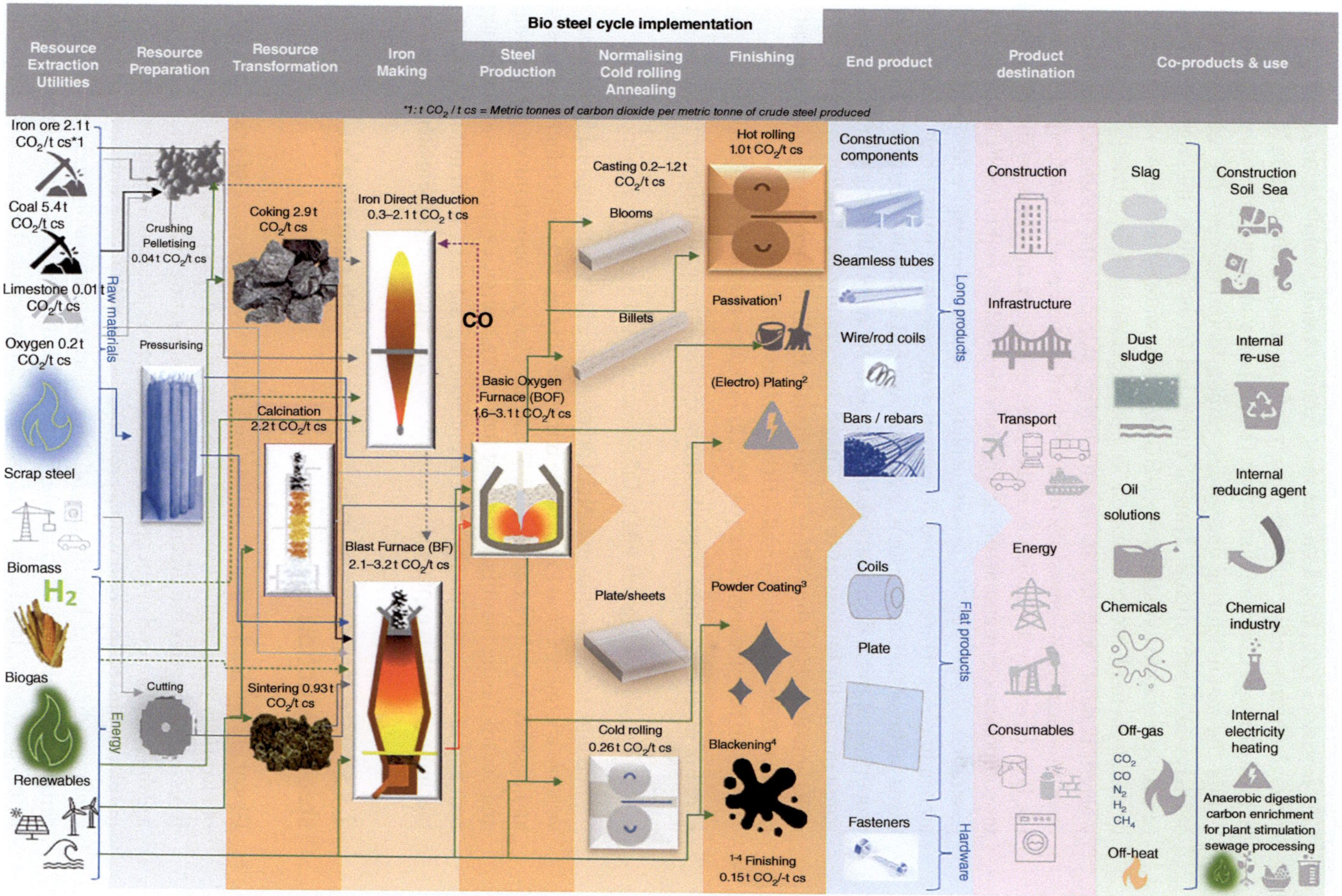

Figure 2.16 Steelmaking table flowchart with BiSC.

simultaneously, the opportunity to determine CO_2 emissions from energy expenditure. In 2018, EERE published a detailed overview of the energy expenditure and CO_2 emissions [90], based on the most recent overall manufacturing energy and carbon footprints, using 2018 US Energy Information Administration (EIA) Manufacturing Energy Consumption Survey (MECS) data and contemporary estimations. Within the manufacturing energy and carbon footprints map, the assumed flow of energy as supply stream, demand curve and losses within the production processes were mapped. Additionally, GHG emissions were estimated, based on energy derived from fossil fuel combustion. The flows demonstrate how energy is used and where it is most likely lost in manufacturing, alongside the estimated simultaneously produced GHG emissions from combustion processes. Each footprint details the flow of energy, described as either fuel, electricity or steam, to the main factor components in manufacturing. This includes boilers, combined heat and power generation, process heaters, process coolers, machine-driven equipment and facility air conditioning. The estimated GHG emissions associated with energy generation, based on fossil fuel combustion, were also mapped. It needs to be said that a great proportion of these figures are estimated, as there are no binding policies or legal requirements for businesses in the United States of any description to report their GHG emissions, and thus, there are no devices in place to record, mitigate or decarbonise production of the factual GHG emissions from industry.

Based on the pioneers' work [64, 66, 72, 74, 86], it can be extrapolated that

- 70% of all energy input is productively applied during manufacturing processes and
- 30% of all energy input is lost to infrastructural and procedural inefficiencies.

Therefore, the individual industrial processes need to be investigated for their harvestable energy and heat losses, to be reinvested in the circular production process, and possibly even for external uses. There are plenty of opportunities for avoiding and decarbonising manufacturing, by harvesting energy, capturing and reutilising heat and off-gases within the steel manufacturing industry, from mining to finish machining, which is explained in some detail in Chapter 4: The Bio Steel Cycle.

2.13 Greater energy independence with sustainable steel production

Global energy market price volatility and an upward trajectory of prices per unit of electricity have sent all industrial sectors and many economies to the brink of recession. Alongside the urgent need for decarbonisation of all industries, achieving a globally higher level of energy independence across all sectors seems imperative. A multidisciplinary approach with a proposed system of CO_2 emissions reduction and capture technologies has the potential for short-term emissions reduction to near zero in the steel industry – although some of the mechanisms can be implemented across most heavy industries. The findings of this research showed a CO_2 emissions reduction of ~30% from 977 tonnes of CO_2 to 684 tonnes in one single BF production cycle (based on 330 tonnes of liquid iron production capacity, with the mean of

2.1–3.2 tonnes CO_2/tonne of steel and chemical reaction emissions applied), by switching the electricity provider for operating the electric heaters to providers generating energy exclusively from renewable sources. Replacing coal with biomass, and adding post-combustion capture units to the BF operation, will add carbon neutrality into the process – resulting in CO_2 emissions reduction to near zero. Carbon capture from biomass utilisation (BECCS – bioenergy from carbon capture and storage) will add the benefit of carbon-negative emissions to the cycle. Simultaneously, energy-saving and process improvement measure implementation (up to 60% efficiency increase), excess heat recovery (<30% of energy savings) and retrofitting renewable energy technology resulted in an energy independence of 88%. Engineering solutions, partly subsidised in the United Kingdom, are readily available for implementation in the iron and steel manufacturing industry.

The volatility of the global energy market and recent price hikes by energy producers have caused never-before-seen levels of profit for the energy companies and untold pressures for businesses and the population in most developed economies. Numerous countries are on the brink of recession across geographical Europe at the time of writing, and energy price increases have made a strong case for the urgent need to achieve greater energy independence – the demonstration of the feasibility being the purpose of this paper. This could be considered one of the foremost important contemporary endeavours. The energy sector and the iron and steel industry, along with heavy industry and petroleum refineries [1], are by far the largest emitters of CO_2 emissions, due to their high fuel and energy demand. The steel industry is accounting for between 7 and 11% [1–7] of global CO_2 emissions as a result of steelmaking, and China is responsible for 50% of these GHGs [7], due to their heavy reliance on coal. The increased use of coal in energy generation, due to imposed oil and gas shortages, was found to be the main factor [1, 8] driving up global energy-related anthropogenic CO_2 emissions by over two billion tonnes, their largest ever rise in absolute terms.

This research focused on the technical solutions currently available for the decarbonisation of the steel industry and how decarbonisation could lead to a higher degree of energy independence, based on the currently known energy consumption trends. The consumption trend evaluation or influence of renewables implementation on consumption trends on a global scale was not the key focus of this project, as these are two completely independent, separate issues altogether but pose an interesting challenge for future research. This research focused on the steel industry and connected subsidiary industries only.

Throughout the research undertaken for this sub-project, global data in connection with renewable energy technology implementation in different settings have been utilised [9–14]. Information on factual CO_2 emissions in steel production [15] and manufacturing [16–18] has been considered, as well as data from other industrial sectors [4, 9, 19, 91, 92]. In order to visualise the opportunities of a circular steel production process, implementation of sustainable elements and opportunities for achieving great energy independence, a comprehensive steel manufacturing overview has been compiled, as displayed in Figure 2.1. The data were accumulated, analysed and used for modelling using MS Excel extrapolations and simultaneously analysed by applying standard mathematical principles and followed for proof of concept with steel production simulations in Simul8, INOSIM and Aspen. Although there are currently several projects in progress concerned with filtering CO_2 emissions from post-combustion off-gas, there is yet a solution to be found which could achieve a 100% filtering of CO_2 emissions.

There are two major hurdles to the implementation of sustainable steel and other production techniques: (i) lack of willingness across all industries to incorporate the required changes in their financial and production planning and managing the installation of suitable technology and (ii) ensuring a viable financing of all efforts using appropriate MCDA and costing and incentives provided by the respective governments. The first hurdle relies on industry leaders to act and drive the decarbonisation of the steel industry to avert climate disaster, as politicians worldwide seem to lack the willingness to implement and enforce adequate policies – the signs are all around us. There are continued research and campaigning efforts taking place in this direction. The second hurdle has been partly addressed by governments across the globe and the UK government with grants, loans and subsidies.

The preceding publication 'The 7 Steps to Net-Zero CO_2 Emissions Steel Production' [15] strategy can be seen as a strategic guidance paper for the decarbonisation of steelmaking, in tandem with this publication as a guideline for achieving higher energy independence. The systems implementation, as described [15], will likely achieve a higher degree of energy independence in the short term (Figure 2.16). It could be achieved in seven easy-to-follow steps, even if only some sections of the following are being applied:

Step 1: switching to a 100% green energy provider
Step 2: installing renewable energy technologies
Step 3: replacing coal and coke with biomass (biochar)
Step 4: installation of post-combustion carbon capture technology
Step 5: utilisation of CO_2 in food and building projects
Step 6: further process improvement in steel manufacturing
Step 7: implementation of AD > biogas > green hydrogen

Conventional energy use and renewable energy component implementation points have been incorporated for highlighting the simplicity of achieving a higher degree of energy independence, whilst simultaneously decarbonising the steelmaking process.

As displayed in Figure 2.1, the same principle applies to the steel (Figure 2.2), in the BF with CCUS unit in the Aspen configuration production process, as far as off-heat is concerned. It was established that a total potential of 425 PJ (1 PJ [petajoule] = 31.6 million m^3 of NG or 278 million kWh of electricity) of excess heat is readily available at a 95 °C temperature, and 960 PJ at ~25 °C [20]. This amount is thought to represent between 4% and 9% of the total industrial final energy demand. Capturing this excess heat means utilising energy potential we have already used in industry, thus reducing the amount of energy to be produced by the same amount. The benefits for agribusinesses utilising off-heat from production and CO_2 in Carbon Enrichment for Plant Stimulation (CEPS) are the subject of ongoing research [20]. Process simulations in Simul8, INOSIM and Aspen+ (BF, Figure 2.2) were used to explain the individual production process implications. The BF/BOF route is the most widespread method of steelmaking, representing ~70% of current global crude steel production [21]. It needs to be emphasised, though, that the displayed CO_2 emissions in metric tonnes are representing the CO_2 emissions at their respective stage, per tonne of product and not per tonne of steel. The emissions burden on each of the input streams in the simulation software systems Simul8 and Aspen is already set within the system parameters and therefore calculated during the process simulations, providing a more detailed set of data, per metric tonne of respective product produced. Additionally, in order to represent the overall mean CO_2 emissions burden in tonnes per tonne of steel, the mean value of 4.95 tonnes CO_2/t cs was allocated for entering in the input stream in S8 as BF 2.95 tonnes of CO_2 per metric tonne of liquid iron produced and BOF 2 tonnes/tonne of CO_2 per metric tonne of steel produced. The implementation is demonstrated in a table flowchart in Figure 2.16 and BF configuration with CCS unit in Figure 2.17.

The BF–EAF process route simulation, representing the second most common method of steelmaking, has a share of ca. 20% of global steel production. Initially, as biomass [58, 92] has the potential to reduce CO_2 emissions by ~30%, the usage of biomass was implemented [58, 92] and filters as well as CCUS units were installed (Figures 2.2 and 2.3), in direct comparison with the MS Excel modelling and mathematical calculations. The calculations are set to work for a 330 tonne BF, similar to the British Steel site in Scunthorpe, and for the operational year of one hour, to reflect the 40-minute average charge processing and discharge time (Figure 2.4). Verifying the results of the preceding evaluations, computations and calculations, initially, 977 tonnes of CO_2 flow was reported by the Aspen system analysis, without any mitigating measures.

The CO_2 reductions in relation to a 330 tonne BF vessel have been modelled in three separate process simulation software packages and have been carried out considering the production parameters of the British Steel Scunthorpe site. Having implemented steps of the Bio Steel Cycle model and strategy, the findings of this research showed a CO_2 emissions reduction of ~30% from 977 tonnes of CO_2 to 684 tonnes in one single BF production cycle (based on 330 tonnes of liquid iron

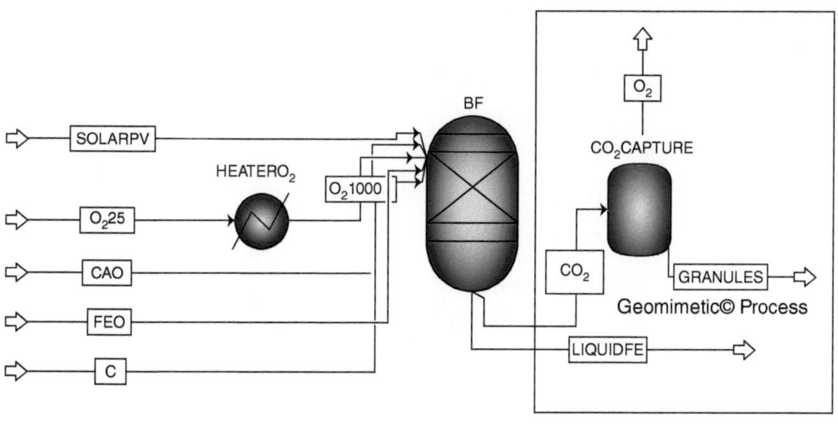

Figure 2.17 Aspen BF configuration with CCUS unit.

production capacity that means 2.1–3.2 tonnes CO_2/tonne of steel), by switching the electricity provider for operating the electric heaters to providers generating energy exclusively from renewable sources. Replacing coal with biomass added carbon neutrality into the process. Adding post-combustion capture units to the BF resulted in CO_2 emissions reduction to near zero. Carbon capture from biomass utilisation also added the benefit of carbon-negative emissions to the cycle (Figure 2.2).

The process simulations and carbon reports in Simul8 demonstrated the effect of implementing biomass and carbon capture technology in the steelmaking process. These simulations were followed by configurations in Aspen and INOSIM, including the implementation of CCUS processes, as the implementation of this process, which includes off-gas flue stack filters, makes it possible to produce steel at almost zero CO_2 emissions – for both, the BF/BOF and BF/EAF process route. Some of the corroborating findings and results from the MS Excel extrapolations, the mathematical formulation and calculations, Simul8 process simulations and configurations in AspenV12.1 and INOSIM are displayed in Figure 2.18.

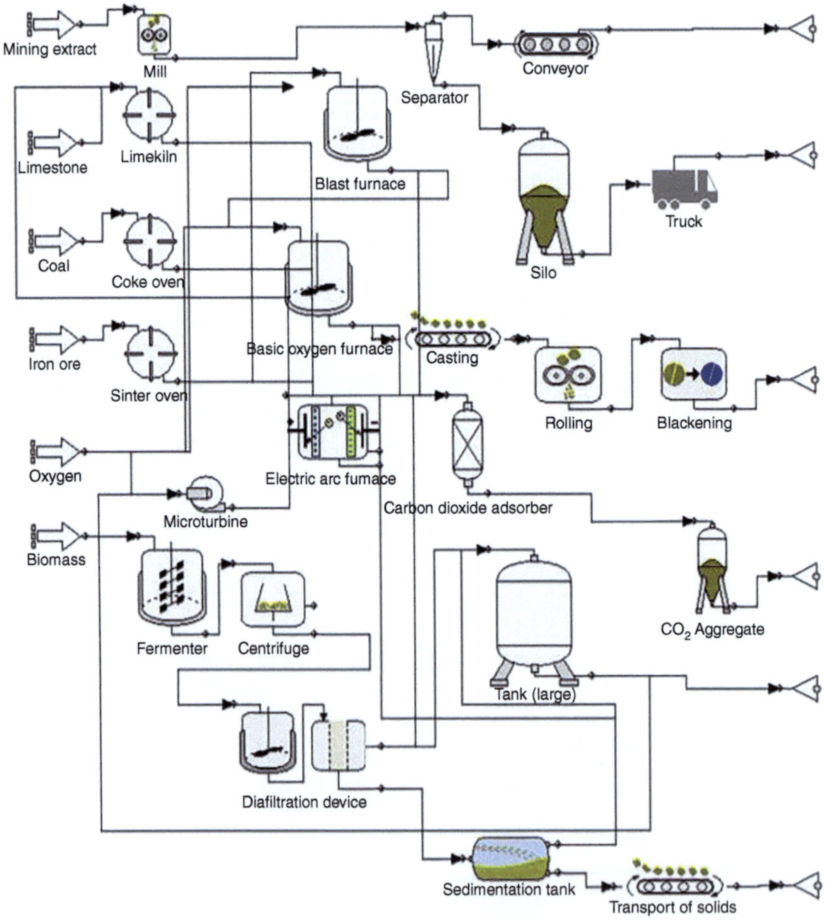

Figure 2.18 INOSIM configuration mining to marketing with CCUS.

The simulation reports provided proof for the concept of the Bio Steel Cycle; insofar as with the implementation of both biomass and the Geomimetic process, the steel production CO_2 emissions could be reduced by 30% and to almost zero, as detailed in Figure 2.4. Therefore, proving the hypothesis that it is possible to produce steel without CO_2 emissions if the novel concept and strategy were being implemented. At the same time, an up to 88% higher degree of energy independence can be achieved, when other components of the Bio Steel Cycle, including renewable energy sources, are being installed which will be further elaborated on in the following text and the summary results are shown in Figure 2.4.

The simulation results are detailed in metric tonnes, occurring during one production cycle, based on 330 metric tonnes of BF and BOF (Figure 2.19).

The first column demonstrates the emissions levels in metric tonnes (t) as per legend underneath from BF/BOF average value in CO_2 emissions, the reduced level of CO_2 emissions after switching to an energy provider deriving their energy at 100% from renewable sources, the reduced level of CO_2 emissions after replacing coal with biomass in the BF, the reduced level of CO_2 emissions after installation of renewable energy technologies (solar, wind and hydro as appropriate) and the reduced level of CO_2 emissions after the installation of post-combustion CO_2 emissions filters. The second column shows the individual percentage proportion in reduction as a result of implementation. The third column shows the successive percentage increase in energy independence. The fourth column shows the successive change of CO_2 emissions in tonnes, based on one cycle of a 330 tonne BF/BOF process. The fifth column shows the remainder of the CO_2 emissions after successive implementation of the suggested technologies – resulting in 0.18 tonnes of CO_2 emissions, but displaying as 0 due to set software parameters, after successful installation of post-combustion CO_2 filters at the individual sites of off-gas outlets.

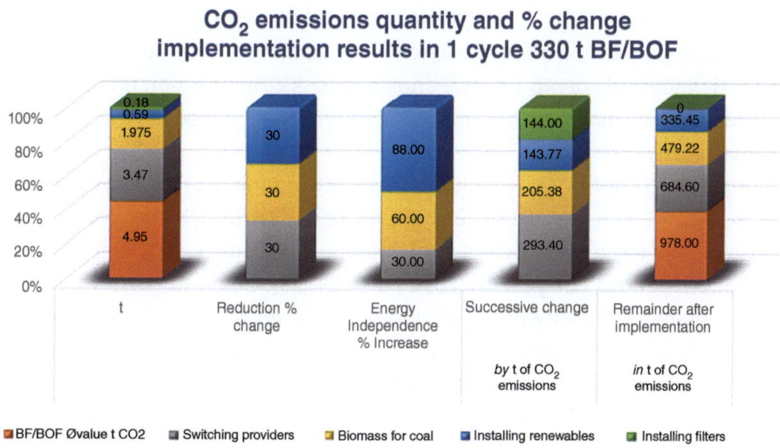

Figure 2.19 CO_2 emissions change in tonnes (t) with CAT and CCUS implementation.

2.14 Heat loss recovery – energy- and CO_2-saving protocols

Already since the 1990s, scientists were convinced [22] that 30% of the heat energy entering any production process is (i) lost and (ii) could be recovered. The required technology has experienced a steep learning curve and is now commercially viable and available. Excess heat from any production or manufacturing process can be reused to supply any production site with heat and warm water, partly due to the simplicity of the technology required. The energy basis and flow have been investigated thoroughly, and an energy industry defining and telling report was produced by Moran and Sciubba [66]. The theory of exergy analysis is based on the fact that if 100% of energy is being inserted into any energy-requiring unit, the amount of 70% will be effectively used for the intended purpose, whilst 30% is being lost due to inefficiencies and deficits within the production and processing infrastructure. The strategy for making existing infrastructure more efficient is to (i) upgrade the existing energy infrastructure, (ii) reduce energy/heat loss by improving energy efficiency with focused energy utilisation and (iii) capture and reutilise lost energy/heat.

The technologies utilised for the strategy points (a)–(c) are as follows:

a) Repair energy/heat escape points and insulate as a preventive measure
b) Using the most focused technology: instead of the BF/BOF route, using the BF/EAF route for steel production
c) Installation of baffles, plate heat exchangers and other energy/heat capturing technology where technically feasible

On a practical level, this means insulation of the heat-bearing infrastructure (furnaces and pipework). Via a connected network of pipes and lines, the energy or heat captured can be exported to nearby agribusinesses or transferred to neighbouring homes and industries through a district energy system. Excess heat is a hidden resource of energy, and it is all around us. Utilising excess heat means enabling greater than 95% energy efficiency. According to the International Energy Agency [1, 5], it is apparent that energy demand is set to grow dramatically in the near future, due to population growth and rising lifestyle energy demands. Without urgent action to tackle the demand side of our lifestyle choices and decarbonisation requirements of the climate crisis, by using energy more efficiently, we will not get on track to meet global climate goals.

Global emissions [23] of CO_2 – including land use and fossil CO_2 – will remain relatively high at 40.5 Gt CO_2 in 2022, but still below their 2019 peak of 40.9 Gt CO_2. A global push for more efficient use of energy can reduce CO_2 emissions by an additional 5 G-tonnes per year by 2030 [1, 8, 24–26], based on current energy demand. The global electricity consumption is displayed in Figure 2.5. However, energy efficiency increases constitute merely 30% of the required CO_2 reduction needed to meet the Net-Zero by 2050 Scenario [1, 8, 24–26]. As far as energy security and greater energy independence are concerned, these energy savings are set to avoid having to produce almost 30 million barrels of oil per day (three times Russia's average annual production, based on 2021 data) and 650 billion m^3 of NG per year – around four times of EU imports from Russia in 2021. Although there has been a steady decline in overall electricity consumption in the iron and steel manufacturing sector (Figure 2.6), due to already implemented structural changes towards more sustainable production processes in the industry, along with the installation of energy efficiency measures, and improved energy efficiency measures [27], there is still a substantial amount of electricity which could be saved (Figure 2.5). Additionally, this would render the existing infrastructure utilisation more efficient. One of the most important factors, the required short-term solutions, is helped by using suitable existing infrastructure as it can be easily retrofitted with technology to prevent heat loss and decarbonise production process technology at the same time. Simultaneously, it could be utilised to (i) capture greater than 98% [28] of CO_2 emissions, use the captured CO_2 in ancillary industries and (ii) installing renewable energy resources (solar, wind and hydro) in suitable locations to increase the level of energy independence. What is being produced on-site does not have to be imported from somewhere else, at fluctuating prices [1, 8, 24]. As shown in Figure 2.6, the development of electricity usage in steelmaking has been on a downward trajectory since the year 2000, partly due to process and efficiency improvements across most industries. In stark contrast, quite the opposite observation was made for global electricity consumption [1, 8, 24] as previously demonstrated in Figure 2.5. Global electricity consumption has continuously increased during the last 50 years, arriving at an estimated 25,530 terawatt-hours in 2023 [25, 29] (Figure 2.6). Since 1980 and up to 2021, global electricity consumption has increased threefold, and the global population increased by ~75%, simultaneously. In line with extended industrialisation and infrastructural improvement, these factors caused a threefold increased electricity demand, with an upward trajectory prognosis, as of the end of 2021. Since the year 2000, China's gross development product (GDP) was recorded as developing a 16-fold increase [7, 25, 26] establishing China as the second largest global economy, after the United States. The development of its billion-strong population and manufacturing industries has caused China to require increased levels of energy, more than any other country. Thus, it has become the largest consumer of electricity worldwide. China and other BRIC countries (Brazil, Russia, India and China) are still vastly outpaced by developed economies with smaller population sizes, in terms of per capita electricity consumption. To place this in context: Iceland, with a population of less than half a million inhabitants, consumes the most electricity, per capita (per person) in the world, followed by Norway, Qatar, Canada, and the United States [7, 25, 26]. Contributing factors such as the existence of power-intensive industries, household sizes, living situations, appliance and efficiency standards, and access to alternative heating fuels have been identified as the determinants of the amount of electricity the average person requires, in the cited countries Given these developments and the looming climate catastrophe, improving energy efficiency, as well as preventing exergy loss, is a technically simple and effective short-term solution. There is vast potential to simultaneously save energy by making existing infrastructure more efficient by reducing the energy/heat loss and therefore saving energy at the same percentage (30%). This means that this is 30% of the energy industry does not have to import and pay for from external sources. At the same time, 30% of

CO_2 emissions for the energy not required, as saved, would not have to be produced. Consequently, improving the existing infrastructure to prevent energy and heat loss would mean a more energy-secure and more sustainable production cycle, in any industry, whilst achieving greater energy independence and reducing the GHG emissions linked to fossil fuel consumption, particularly of energy derived from fossil fuels.

2.15 Retrofitting renewable energy technologies on-site

Renewable energy technologies are an economically viable alternative [12] to combustion processes based on fossil fuels such as coal and gas to produce heat and energy [4, 14, 25–30]. Solar, wind, geothermal and hydropower are well-established technical solutions [4, 14, 25–30], which have already been successfully implemented in a range of countries and settings [31–36]. In the United Kingdom, there are a range of schemes accessible to businesses, such as finance and support from the Department for Business, Energy and Industrial Strategy (2023) and others, as displayed in Table 2.1. Some countries and industries are supplying their entire energy needs via renewable energy solutions [4, 27, 29] – hence why this component is one of the cornerstones of the BiSC, as producing electricity and heating energy accounts for 36% of the UK's CO_2 emissions [14, 24, 25, 27, 37–39]. Besides the emissions savings, using renewable energy technologies exclusively could provide greater independence to businesses across all sectors and increase the UK's energy self-sufficiency. The first step to greener production and greater energy independence is the switch to an energy provider which is deriving their energy at 100% from renewable sources. Utilising existing buildings on industrial and production sites, suitable locations can easily be retrofitted with photovoltaic (PV) solar panels, producing energy from daylight and sunshine. These have the additional benefit of monetary grants (non-repayable), provided by the UK government, and government-backed loans and subsidies [40, 41]. Additionally, the same applies to wind turbines, technology generating electricity using either biomass or hydro-turbines (water-based), and anaerobic digestion (AD) systems. Table 2.1 shows the relevant schemes and their descriptions, mostly detailing how the different technologies could potentially qualify for the individual schemes.

Renewable energy technology is market-ready – now the implementation is key.

The choice and implementation of any of these technologies are entirely dependent on the individual site parameters and need to be thoroughly assessed with regard to their suitability for the identified location and viability with an outlook over the next 30 years. The accurate potential return on investment (ROI) can only be determined for every individual site, as the current contracts with energy providers are one of the determining factors and the authors have not been privy to this kind of information. The MCDA in Table 2.2 includes service and maintenance time and cost, the likelihood of repairs and the availability of suitable service providers to carry out said repairs, servicing and maintenance. Projections for policy-making, return on investment (ROI) or how this could lead to the development of new incentives

Table 2.1 UK support schemes for renewable energy technologies.

Scheme	*Description*
England Woodland Creation Offer	Landowners, land managers and public bodies can apply to the England Woodland Creation Offer (EWCO) for support to create new woodland. Over £10,000 per hectare.
Greening Eden	The CBEN Partnership will complete the calculations using data provided by each company and site visits to provide practical and cost-effective advice on how to reduce emissions. A £400,000 grant fund has been established to help to capital investment projects delivering emissions reductions.
Green Heat Network Fund	Commercialisation and construction of new low- and zero-carbon (LZC) heat networks (including the supply of cooling). Retrofitting and expansion of existing heat networks. Funding will support the uptake of low-carbon technologies like heat pumps, solar and geothermal energy as a central heating source. The GHNF is open to organisations in the public, private, and third sectors in England (no individuals, households and sole traders).
Green Gas Support Scheme	Funding support for biomethane injection to the National Grid.
Clean Heat Grant	Upfront capital funding for households and businesses for the installation of sustainable heating technologies (heat pumps and biomass).
Smart Export Guarantee (SEG)	The SEG funds for the low-carbon electricity exporters, feeding back to the National Grid. Anyone with an installation of one of the following technology types is eligible to apply: solar photovoltaic (solar PV), wind, micro-combined heat and power (micro-CHP), hydro, anaerobic digestion (AD) support and grants for SMEs to help them to reduce carbon emissions.
SMEES	SMEES (SME Energy Efficiency Scheme): guidance and funding for businesses looking to improve their energy efficiency.
Energy for Business	Support and grant funding for SMEs with projects to reduce carbon emissions or save energy.
HNIP	Heat Networks Investment Project (HNIP) government-backed funding.
Low Carbon Dorset	Free support to help businesses in Dorset reduce their carbon emissions, improve energy efficiency and aid the development of new low-carbon products.
Business Energy Efficiency Programme	Energy reviews and grants to help businesses in the West Midlands manage and reduce energy costs.
Low Carbon Workspaces	Offers grants to implement energy efficiency measures, save money and cut waste.
Horizon Europe funding	Funding for research or innovation that is groundbreaking, improves European research standards or responds to challenges like climate change or food security.
Coventry and Warwickshire Green Business Programme	Grants, free energy audits and low-carbon product development support for businesses.

Table 2.2 MCDA excerpt implementation details and cost.

Step 1–7 BiSC	Technique	Process	Company	System/Performance
1	Switching > Green energy	Energy provider	See Appendix 4	100% renewable energy
2	Installing renewables: solar	Solar PV panels	IBC (Internal back contact)	1 kWh/4 panels = 25,667 panels
2	Installing renewables: wind	Horizontal axis wind Turbine	Norvento nED100	100 kW/£317,655.27 × 65
2	Installing renewables: wind	Horizontal axis wind Turbine	Enercon E53	800 kW/£807,581.80 x 8
2	Installing renewables: wind	Horizontal axis wind turbine	EWT DW61	1 MW/£981,368.75 x 6
2	Installing renewables: wind	Horizontal axis wind Turbine	Enercon E82	3 MW/£1,829,271.35 x 3
2	Installing renewables: wind	Horizontal axis wind Turbine	Enercon E126 EP3	3.5 MW/£2,458,302.00 x 2
2	Installing renewables: wind	Vertical axis wind turbine	Patriot Modular	70 kW/£188,196.00/ x 92
2	Installing renewables: hydro	Small closed-loop system	Helios Atlas	6.5 MW
3	Using biomass / green H_2	H_2ermes: H_2 from seawater	HyCC/Tata Steel	15,000 tonnes H_2/p.a.
4	CO_2 filter installation	$CaCO_3$-based CO_2 absorber	Giammarco Vetrocoke	Hot Potassium Carbonate (HPC) solution-based filter
5	Utilisation captured CO_2	Geomimetic: CCUS in aggregate	Blue Planet	100% CCUS
6	Process improvement	HIsarna; ironmaking by simultaneous iron ore reduction and scrap melting combined with biomass and limestone instead of lime	Tata Steel, Horizon 2020, Horizon Europe	3300 tonnes hot metal per day
7	Anaerobic digestion > biogas utilisation in steelmaking	Biogas and H_2 from AD	Biogen	Biogas production

in the realm of renewable energy and sustainable industrial practices were not the substance of this research. Some examples are given in Table 2.2.

Some technical solutions for achieving a higher degree of energy independence have been intensely researched [21, 42–45] and implemented in a range of industrial sectors. Projects containing components such as direct air capture (DAC), redirecting heat and utilising anaerobic digesters to produce biogas (methane and hydrogen), and producing energy on-site with suitably sized turbines [23] can be directly linked to one of the incentives shown in Table 2.1, such as the 'Farming Investment Fund', and includes funding for agricultural businesses such as farmers, foresters, growers and agri-contractors with grants for investing in new technologies, equipment and infrastructure [41]. To make the green industrial revolution happen, the UK government has created new energy efficiency schemes from 2022 to replace the current domestic and non-domestic renewable heat incentives. The cost of manufacturing solar panels and wind turbines has plummeted dramatically in the past decade, making them affordable and often the cheapest form of electricity [14]. As an example, solar module prices fell by up to 93% between 2010 and 2020. During the same period, the global weighted average cost of electricity (LCOE) for utility-scale solar PV projects fell by 85% [14]. Although there is a variety of renewable energy technologies at TRL9 available (Table 2.2), PV technology and on-shore wind generation for independent energy production were established as the mainly utilised renewable energy technology in a competitive market. The second largest dominant technology after solar in the United Kingdom is wind energy. Implementation of sustainable and renewable energy components into any production cycle, such as DAC to (i) capture off-gas carbon and (ii) produce biomass for the production of biogas in AD, CEPS (promoting growth in greenhouses for food production), AD to produce biogas and hydrogen, and capturing and utilising excess heat, can individually make significant contributions to a higher degree of energy independence for the individual commercial entity. The MS Excel extrapolations established how the different components of the '7 Steps to Net-Zero Carbon Emissions Steelmaking' strategy can be implemented and installed. Installing renewable energy technologies can not only help to reduce CO_2 emissions but also supply the production site with renewable energy, where 30% of energy does not have to be imported from third parties and paid for, bringing the reduction in CO_2 emissions down to −49%. At this point in production, where biomass has already replaced the use of coke, and renewable energy technologies (solar, solar PV, wind and hydro) have been installed, an additional 30% of energy can be saved and thus does not have to be bought in, by using biogas, bringing the reduction in CO_2 emissions to −65.7% [46]. This will have been produced in the link-connected anaerobic digester, which produces biogas from connected agribusinesses. The negative percentage reduction in emissions means, in reverse, that at these points in the (steel) production process, a greater degree of energy independence can be achieved at 30%, 49%, 65.7% and 88% [15, 46]. This implies that energy at the same percentage point levels is not required to be imported from external sources, as it is produced either on-site or link connected. Besides energy and heat saving, generating their own energy will inevitably lead to achieve a higher level of independence, at least by 30% and ideally, at 88%; there are savings to be had by not being forced to import and pay for energy from conventional suppliers, energy derived from renewable sources or not. We have been made painfully aware that private and business users of energy and fuel are at the mercy of corporate stakeholder interests and thus vulnerable to high price velocity. Retrofitting the existing industrial building infrastructure with renewable energy technology, with components of the '7 Steps to Net-Zero Carbon Emissions Steelmaking' strategy, can support to: (i) achieve greater energy efficiency and independence, (ii) turbocharge the decarbonisation of energy production, (iii) decarbonise steel and industrial production and (iv) provide savings opportunities via excess heat recovery. The MCDA analysis of different renewable energy systems, taking into consideration the incentives in Table 2.1, leaves much hope for the establishment of the United Kingdom as a green energy hub, as the provided initiatives can enable the stakeholders willing to invest in green energy to not only make the green industrial revolution in the United Kingdom happen but could enable the investors to do so at a minimal cost. Table 2.2 shows an extract of the extensive BiSC implementation MCDA analysis. Observing economic principles and baselines, it can be assumed that, overall, there is a third of the cost involved, with two-thirds of savings on energy cost over a 30-year investment period. Additional, significant positive milestones can be reached, such as investment into a workforce with 'green' skills, future-proofing the business against energy price hikes, besides the positive effects of greater energy independence, decarbonisation of production and rehabilitation of the natural world which has been disrupted beyond recognition by the Anthropocene. The political and legal landscape and the country governmental guidance will have to change dramatically [47, 48] in order to meet the targets, set by the Paris Agreement [49] and COP15 [48] and the dire warnings issued by the recent IPCC reports (2023) [50]. By improving their carbon capture and off-heat utilisation capabilities, and investing in renewable energy technology, businesses are:

- Supporting the decarbonisation of production
- Reaching a higher degree of energy independence
- Achieving a higher level of asset efficiency
- Training workforce in required 'green' skills
- Reducing their energy costs
- Creating a viable additional income stream

Additionally, besides achieving a higher degree of energy independence, economic advantages in monetary terms are almost inevitable when renewable energy systems are being installed. Doing so will also support achieving limiting the global temperature rise to below 2 °C and thus support avoiding climate disaster. The CO_2 emissions by industry sector were analysed [31, 37, 51–54] and besides the iron and steel industry, there are other heavy industries which are CO_2 emitters, and generating copious amounts of off-heat, co-products and therewith resources, which could be harvested and used to power energy-dependent devices: transport, chemical industry, energy supply, residential/commercial buildings sector, agriculture and waste management. There are currently 89 projects globally in process with the aim to achieve climate neutrality at the latest by 2050 [61].

2.16 Precombustion CO_2 avoidance

Replacing precombustion fossil fuels with biomass [35, 38, 48, 93–99] and operating (green) HDR [98, 100–104] as well as capturing post-combustion CO_2 emissions with the Geomimetic® process [105] are efforts which have the potential to reduce the CO_2 emissions in steel production to almost '0'. There is considerable outreach into other industries, such as the application of the Geomimetic® process [105], which produces aggregates from CO_2 emissions and recycled concrete to produce new concrete and utilisation of BOF slacks for road building.

2.17 Steelmaking process improvements

Process improvement technology in steel manufacturing is displayed in Table 2.3 and will give some detailed information about the individual process improvements. Greater material and energy efficiency, and deployment of low-carbon process routes are all critical. The steel production process has been thoroughly investigated in every aspect from mining to recycling, and it can be said that there is currently a global effort underway for developing more environmentally friendly and resource-saving technologies in steel production. Projects such as top gas recycling blast furnace (TGRBF) operation, coal mine methane recovery [23, 35, 82, 106–109, 116, 119] and HIsarna [58, 87, 109, 133] eliminate the need for the sintering process entirely. HIsarna, implemented individually, has the potential to reduce CO_2 emissions from steel production by at least 30%.

2.18 Post-combustion CO_2 capture

Due to power plants being the largest stationary source of anthropogenic CO_2, developing effective and affordable post-combustion CO_2 capture technology has attracted substantial research attention. This study assessed the adsorption of CO_2 onto biochar, a low-cost adsorbent that can be produced from waste biomass through low-temperature pyrolysis. Sugarcane bagasse (BG) and hickory wood (HW) feedstock are converted into biochar at 300, 450, and 600 °C. The sorption of CO_2 on each of the resulting biochar was measured by monitoring its weight changes in CO_2 at 25 or 75 °C. The biochar was found to be effective for CO_2 capture, and the adsorption process could be described by the second-order kinetics model. In general, the biochar produced at higher temperatures had better CO_2 capture performance. The BG biochar produced at 600 °C showed the most adsorption of CO_2 (73.55 mg g^{-1} at 25 °C). However, even when the feedstock was exposed to only 300 °C pyrolysis, the biochar was still able to capture more than 35 mg g^{-1} CO_2 at 25 °C. Experimental results suggest that CO_2 weakly bound to the surface of the biochar through physisorption, so the surface area was a significant determinant of CO_2 adsorption (Pearson's r =0.82); nevertheless, the presence of nitrogenous groups also played a role, when the surface area was sufficient. Biochar's porous structure and unique surface properties enable it to be an efficient CO_2 adsorbent whilst being sustainable and inexpensive. It, therefore, requires

Table 2.3 Steelmaking process improvements.

Production stage	Time frame	Project name	Technology	References
Post-combustion capture:		Athos	Captured carbon usage	[23, 35, 77, 87, 106–110]
		Everest	CCS: from BOG > former gas fields under the North Sea	[23, 35, 77, 87, 106–110]
		Geomimetic®	CO_2 recycling (concrete, flue gases etc.)	[105, 108]
	Short to mid-term	DAC	Land trees (8.3 tonnes/ha/p.a.)3.8 tonnes/tonne of steel	[78, 79, 111–113]
			Land shrubs and herbs	[78, 79, 111–113]
			Water plants	[78, 79, 111–113]
			Organisms (plancton)	[78, 79, 111–113]
		CEPS	Carbon-enriched air 99% utilised in food production	[114, 115]
		Air-blown BF	MEA	[23, 35, 52, 58, 77, 106–108, 110, 116–118]
			MDEA	[23, 35, 52, 58, 77, 106–108, 110, 116–118]
			Selexol	[23, 35, 52, 58, 77, 106–108, 110, 116–118]
			Shift and Selexol	[23, 35, 52, 58, 77, 106–108, 110, 116–118]
		TGRBF	MEA	[23, 35, 52, 58, 77, 106–108, 110, 116–118]
			VPSA	
			Selexol	[23, 35, 52, 58, 77, 106–108, 110, 116–118]
		Corex	MEA	[23, 35, 52, 58, 77, 106–108, 110, 116–118]
			Selexol	[23, 35, 52, 58, 77, 106–108, 110, 116–118]
			Shift and Selexol	[23, 35, 52, 58, 77, 106–108, 110, 116–118]
	Long term	Air-blown BF	Shift membrane reactor + Selexol	
			Selective carbon membrane	
			Hydrate crystallisation	[23, 35, 52, 58, 77, 106–108, 110, 116–118]
		Advanced smelting reduction	Purification only	[23, 35, 52, 58, 77, 106–108, 110, 116–118]
		Mineral carbonation	Two-tiered reaction tower	[25, 37, 54, 60, 79, 105–107, 109, 115–117]
		CCU technology		[25, 37, 54, 60, 79, 105–107, 109, 115–117]

Precombustion process optimisation	Short term	HIsarna	Eliminates sintering, smelting reduction	[8, 9, 58, 108, 113–115, 117]
		ReclaMet	Waste resource recovery	[8, 9, 58, 108, 113–115, 117]
	Long term	Electrolysis	Direct water splitting; biomass > hydrogen	[38, 48, 98, 103]
		Hydrogen EAF	Hydrogen direct reduction in EAF (HDR/EAF)	[38, 48, 98, 103]
		GrInHy	Electrolyser development	[38, 48, 98, 103]
		H2Future	Electrolyser development	[38, 48, 98, 103]
		HYBRIT	Fossil-fuel-free primary steel production	[38, 48, 98, 103]
		HDR	Hydrogen direct reduction (HDR) in BF/EAF/BOF; CO_2 from electricity used	
Research		Renewable energy technology	Solar, wind, hydro and geothermal	[33, 34, 48, 80, 83, 108, 116, 119–122]
Precombustion energy			Wind	[32, 33, 47, 82, 90, 105, 106, 118, 120, 123, 124]
Source replacement			Hydro	
			Geothermal	[32, 33, 47, 82, 90, 105, 106, 118, 120, 123, 124]
Precombustion energy		Industry 4.0	>Optimising energy consumption in production	[23, 75, 106, 107, 125–131]
Consumption optimisation		Industry 4.0	>Model-based steel temperature adjustment	[23, 75, 106, 107, 125–131]
			> Hydrogen control in ladle operations	[23, 75, 106, 107, 125–131]
			> Hydrogen control in casting operations	[23, 75, 106, 107, 125–131]
			> Nitrogen control in ladle/casting operations	[23, 75, 106, 107, 125–131]
Pre-combustion fossil fuel replacement Post-combustion capture:		Coal/Biochar	Biochar/Takachar Waste heat recovery	[44, 75, 126, 130, 132] [44, 75, 126, 130, 132]

careful consideration of the implications and, specifically, the capture of the emitted GHGs, so as not to add to the CO_2 balance sheet.

A further key component of the BiSC is the Geomimetic® process [105], as these units are effectively recycling facilities for the recycling of reclaimed concrete and the reutilisation of CO_2, filters, dust, sludge and slack from (steel) production. These units have the capacity to reduce post-combustion CO_2 emissions to effectively zero and should be on-site of any (steel) production plant. The workings of the Geomimetic process [105] are in its essence carbon utilisation and sequestration processes at the same time, as these recycle CO_2 from flue gases and recycled concrete into synthetic limestone and aggregate in cement production, with the potential of absorbing 100% of the CO_2 emissions produced. This is a technique suitable to be applied in any industrial production setting: energy, steel, concrete, chemical industry, glass industry, paper and transport, to name a few.

Capturing post-combustion CO_2 emissions with the Geomimetic® process [105] is an effort which have the potential to reduce the CO_2 emissions in steel production to almost '0'. There is considerable outreach into other industries, such as the application of the Geomimetic® process [105], which produces aggregates from CO_2 emissions and recycled concrete to produce new concrete and utilisation of BOF slacks for road building.

Blue Planet [103] 2023 is using the so-called Geomimetic process, where recycled concrete and captured carbon are being reformed to make new concrete and aggregate. In combination with manufacturing post-combustion flue stack carbon capture, the utilisation takes place with the subsequent utilisation of the captured carbon in making new concrete (carbon sequestration). This process has the capability to reduce the carbon emissions from steel production by almost 100%. Carbon enrichment for plant stimulation is the ideal growth enhancer for most known plants. Therefore, to utilise the captured off-gas CO_2 and to direct this to greenhouses in close proximity could be the best use of a by-product at present. This system has been tried and tested and enhanced food production significantly.

It has been established that for the majority of land-grown crops, the CO_2 saturation point will be reached at about 1000–1300 ppm in perfect growing conditions. Generally, 800–1000 ppm is recommended for raising tomatoes, cucumbers and pepper seedlings as well as for lettuce production [134].

2.19 Conclusions

Achieving a higher degree of energy independence is within reach of all sectors of society, made possible by technological progress and incentives and grants provided by the respective governments and countries, including the United Kingdom. The opportunities are manifold; they merely require political willingness and implementation across all industries. The process simulations compiled within the systems S8, Aspen+ and INOSIM aligned with the mathematical analysis and modelling in MS Excel and provided almost matching reports, stating a reduction of 30% from 977 tonnes of CO_2 emissions to 684 tonnes of CO_2 flow in one single

BF production cycle (based on 330 tonnes liquid iron production capacity and a mean of 2.1–3.2 tonnes CO_2/tonne of steel and emissions from chemical reactions considered). This is possible by switching the electricity provider for operating the electric heaters to providers generating energy exclusively from renewable sources. Replacing coal with biomass adds further carbon neutrality into the process. Utilising post-combustion capture unit to the BF resulted in a CO_2 emissions reduction to near zero. Carbon capture from biomass utilisation introduces the added benefit of carbon-negative emissions to the cycle. Simultaneously, energy-saving and process improvement measure implementation (up to 60% efficiency increase), excess heat recovery (less than 30% of energy savings) and retrofitting renewable energy technology resulted in an energy independence of 88%. Engineering solutions which are partly subsidised in the United Kingdom – as described in Table 2.1 – are readily available for implementation in the iron and steel manufacturing industry. Furthermore, the application points of renewable energy technology within the steel production process were established where achieving a 30%, 49% and up to 88% higher energy independence is achievable in the short term. The cost risk of installing renewable energy and decarbonisation technology was minimised to less than a third of the initial outlay in individual cases, as the UK government is providing grants, funding and loan schemes to encourage all industrial sectors to work towards Net-Zero. The research preceding the compilation of the current paper has provided answers which are reaching over into multiple other heavy industries, and additionally resulting in further opportunity for a range of research directions, which are listed as follows:

- Utilisation of CO_2 in the building industry (CCUS), agriculture (CEPS), chemical industry, food and drinks industry and pharmaceuticals
- Utilisation of waste products from steelmaking in infrastructural projects
- Other GHG (CO and H_2) captured in TGRBF can also be reused in the chemical industry, thus eliminating the need for waste management

The upgrading and retrofitting of existing and new steel plants and other industrial sites provide an immense opportunity for further research, as there is a vast range of active and abandoned production plants available in the United Kingdom, not limited to:

- Adding solar foil, panels, tiles and shingles to buildings and car parks
- Adding a biogas, replacing fossil fuel, and a hydrogen network from AD for hydrogen direct reduction of iron ore (DRI) in steel production, in addition to reformers and electrolysers
- Installing wind turbines at brownfield sites not suitable for human habitation and
- Developing filters and CO_2 capture units to be retrofitted to existing production plants

Further work is currently underway and will provide more details on the more salient points of this research upon publication.

References

1 Historic England (2018). Pre-industrial mines and quarries | Historic England [Internet]. https://historicengland.org.uk/images-books/publications/iha-preindustrial-mines-quarries (accessed 13 July 2023).

2 Ferreira, H. and Leite, M.G.P. (2015). A life cycle assessment study of iron ore mining. *Journal of Cleaner Production* 108: 1081–1091.

3 McKinsey & Company (2021). Metals & mining practice: the future of the European steel industry. A Road Map toward Economic and Environmental Sustainability. https://www.mckinsey.com/~/media/mckinsey/industries/metals%20and%20mining/our%20insights/the%20future%20of%20the%20european%20steel%20industry/the-future-of-the-european-steel-industry_vf.pdf (accessed 7 July 2023).

4 Bogunovic, D., Kecojevic, V., Lund, V. et al. (2009). Analysis of energy consumption in surface coal mining. *Transactions of the Society for Mining, Metallurgy, and Exploration, Inc.* 326: 79–87.

5 Griffin, P. and Hammond, G. (2021). The prospects for 'green steel' making in a net-zero economy: a UK perspective. *Global Transitions* 1 (3): 72–86.

6 Kiessling, S., Darabkhani, H.G., and Soliman, A.H. (2022). Energies | free full-text | the bio steel cycle: 7 steps to net-zero CO_2 emissions steel production [Internet]. https://www.mdpi.com/1996-1073/15/23/8880 (accessed 11 June 2023).

7 Yang, W., Pudasainee, D., Gupta, R. et al. (2021). An overview of inorganic particulate matter emission from coal/biomass/MSW combustion: sampling and measurement, formation, distribution, inorganic composition and influencing factors. *Fuel Processing Technology* 2013: 106657.

8 EIA US Energy Information Administration (2022). Biomass and the environment – U.S. Energy Information Administration (EIA) [Internet]. https://www.eia.gov/energyexplained/biomass/biomass-and-the-environment.php (accessed 6 July 2023).

9 WSA World Steel Association (2021). Word steel in figures [Internet]. https://www.worldsteel.org/en/dam/jcr:976723ed-74b3-47b4-92f6-81b6a452b86e/World%2520Steel%2520in%2520Figures%25202021.pdf (accessed February 2 2023).

10 WSA World Steel Association (2022). World Steel in figures 2022 a healthy economy needs a healthy steel industry [Internet]. https://worldsteel.org/wp-content/uploads/World-Steel-in-Figures-2022-infographic.pdf?x65430 (accessed 7 July 2023).

11 International Energy Agency (IEA) (2021). Data and statistics [Internet]. https://www.iea.org/data-and-statistics/data-browser?country=WORLD&fuel=Energy%20transition%20indicators&indicator=ETISharesInPowerGen (accessed 11 August 2022).

12 Swalec, C. (2021). Carbon brief. Guest post: these 553 steel plants are responsible for 9% of global CO_2 emissions. https://www.carbonbrief.org/

guest-post-these-553-steel-plants-are-responsible-for-9-of-global-co2-emissions (accessed 7 July 2023).

13 Ren, L., Zhou, S., Peng, T., and Ou, X. (2021). A review of CO_2 emissions reduction technologies and low-carbon development in the iron and steel industry focusing on China. *Renewable and Sustainable Energy Reviews* 143: 110846.

14 COM (European Commission) (2023). Circular economy: Faster progress needed to meet EU resource-efficiency targets, ensure sustainable use of materials and enhance strategic autonomy [Internet]. https://environment.ec.europa.eu/news/circular-economy-faster-progress-needed-meet-eu-resource-efficiency-targets-ensure-sustainable-use-2023-05-15_en#:~:text=In%20recent%20years%2C%20circular%20economy,0.8%25%20of%20the%20EU\stquotes%20GDP (accessed 3 January 2023).

15 International Energy Agency (IEA) (2022). Emissions factors 2024 – data product. [Internet]. https://www.iea.org/data-and-statistics/data-product/emissions-factors-2024 (accessed 6 December 2024).

16 STATISTA (2023). World's biggest CO_2 emitters 2022. https://www.statista.com/statistics/271748/the-largest-emitters-of-co2-in-the-world (accessed 12 December 2023).

17 The Rodney and Otamatea Times (1912).Coal consumption affecting climate. Coal consumption affecting climate. *Rodney and Otamatea Times, Waitemata and Kaipara Gazette* (August 14), p. 7.

18 CARO (2022). News from 110 years ago. https://www.caro.ie/news/news-from-110-year-ago. (accessed 24 January 2024).

19 Foreman, P. (2023). A newspaper published an article about global warming in 1912 [Internet]. Plant Based News. https://plantbasednews.org/news/environment/newspaper-article-printed-warning-global-heating (accessed 24 January 2024).

20 Riesbeck, J., Hooey, L., Kinnunen, K., et al. (2022). Global effects of closing down sinter plant at Ruukki Raahe integrated steel works [Internet]. http://ltu.diva-portal.org/smash/record.jsf?pid=diva2%3A1004013&dswid=-4624 (accessed 7 July 2023).

21 Li, Y. and Zhu, L. (2014). Cost of energy saving and CO_2 emissions reduction in China's iron and steel sector. *Applied Energy* 130: 603–616.

22 Shan, Y., Liu, Z., and Guan, D. (2016). CO_2 emissions from China's lime industry. *Applied Energy* 166: 245–252.

23 Chen, Q., Gu, Y., Tang, Z. et al. (2018). Assessment of low-carbon iron and steel production with CO_2 recycling and utilization technologies: a case study in China. *Applied Energy* 15 (220): 192–207.

24 USGS (2012). Limestone; a crucial and versatile industrial mineral commodity [Internet]. https://pubs.usgs.gov/fs/2008/3089 (accessed 8 September 2023).

25 Ganapathi, H. and Phukan, M. (2020). Environmental hazards of limestone mining and adaptive practices for environment management plan. In: *Environmental Processes and Management: Tools and Practices [Internet]* (ed. R.M. Singh, P. Shukla, and P. Singh), 121–134. Cham: Springer International Publishing (Water Science and Technology Library). https://doi.org/10.1007/978-3-030-38152-3_8 (accessed 8 September 2023).

26 Francey, S., Tran, H., and Berglin, N. (2011). Global survey on lime kiln operation, energy consumption, and alternative fuel usage. *Tappi Journal* 19.

27 Chemistry (IUPAC) TIU of P and A (2014). IUPAC – calcination (C00773) [Internet]. https://goldbook.iupac.org/terms/view/C00773 (accessed 11 September 2023).

28 WSA World Steel Association (2020). Steel statistical yearbook 2020 [Internet]. https://worldsteel.org/wp-content/uploads/Steel-Statistical-Yearbook-2020-concise-version.pdf (accessed 2 February 2021).

29 Linde (2019). Air-separation-plants-history-and-technological-progress-2019_tcm19-457349.pdf [Internet]. Munich: Linde AG. p. 28. https://www.linde-engineering.com/en/images/Air-separation-plants-history-and-technological-progress-2019_tcm19-457349.pdf (accessed 14 December 2023).

30 Chadwick, M.J., Highton, N.H., and Lindman, N. (ed.) (1987). 6 – Coal conversion technologies. In: *Environmental Impacts of Coal Mining & Utilization [Internet]*, 105–155. Amsterdam: Pergamon https://www.sciencedirect.com/science/article/pii/B9780080314273500153 (accessed 14 December 2023).

31 US NETL National Energy Technology Laboratory (2023). http://netl.doe.gov. 5.1. Gasification introduction. https://netl.doe.gov/research/Coal/energy-systems/gasification/gasifipedia/intro-to-gasification (accessed 14 December 2023).

32 Fraunhofer Institute (2009). Methodology for the free allocation of emission allowances in the EU ETS post 2012 Sector report for the iron ore industry by order of the European Commission Study Contract: 07.0307/2008/515770/ETU/C2 Ecofys project number: PECSNL082164 [Internet]. p. 1–15. https://climate.ec.europa.eu/system/files/2016-11/bm_study-iron_ore_en.pdf (accessed 7 June 2023).

33 International Renewable Energy Agency (IRENA) (2012). Renewable energy technologies: cost analysis series [Internet]. https://www.irena.org/-/media/Files/IRENA/Agency/Publication/2012/RE_Technologies_Cost_Analysis-HYDROPOWER.pdf (accessed 7 July 2023).

34 International Renewable Energy Agency (IRENA) (2021). World energy transitions outlook: 1.5°c pathway [Internet]. https://www.irena.org/publications/2021/Jun/World-Energy-Transitions-Outlook (accessed 11 August 2022).

35 Toktarova, A., Karlsson, I., Rootzén, J. et al. (2020). Pathways for low-Carbon transition of the steel industry—a Swedish case study. *Energies* 13 (15): 3840.

36 Conejo, A.N., Birat, J.P., and Dutta, A. (2020). A review of the current environmental challenges of the steel industry and its value chain. *J Environ Manage.* 1 (259): 109782.

37 Satyendra, K.S. (2019). ULCORED process [Internet]. https://www.ispatguru.com/ulcored-process (accessed 9 July 2023).

38 Vogl, V., Åhman, M., and Nilsson, L.J. (2018). Assessment of hydrogen direct reduction for fossil-free steelmaking. *Journal of Cleaner Production* 203: 736–745.

39 WSA World Steel Association. (2022) Hydrogen (H_2)-based ironmaking [Internet]. https://worldsteel.org/wp-content/uploads/Fact-sheet-Hydrogen-H2-based-ironmaking.pdf (accessed 7 July 2023).

40 Rayner-Canham, G. and Overton, T. (2010). *Descriptive Inorganic Chemistry*, 5e. New York: W. H. Freeman and Company.

41 Rice University (2023). Open-stax – chemistry: 19.1 occurrence, preparation, and properties of transition metals and their compounds. https://openstax.org/books/chemistry/pages/19-1-occurrence-preparation-and-properties-of-transition-metals-and-their-compounds (accessed 3 July 2023).

42 Price, L., Sinton, J., Worrell, E. et al. (2002). Energy use and carbon dioxide emissions from steel production in China. *Energy* 27 (5): 429–446.

43 UNIDO United Nations Industrial Development Organization (2010). Carbon capture and storage in industrial applications: technology synthesis report working paper – working paper – [Internet]. Vienna; https://www.unido.org/sites/default/files/2010-12/synthesis_final_0.pdf (accessed 7 July 2023).

44 Worrell, E., Blinde, P., Neelis, M., et al. (2010). Energy efficiency improvement and cost saving opportunities for the U.S. Iron and Steel Industry An ENERGY STAR(R) Guide for Energy and Plant Managers [Internet]. p. LBNL-4779E, 1026806. Report No.: LBNL-4779E, 1026806. http://www.osti.gov/servlets/purl/1026806 (accessed 2 November 2022).

45 Zakkour, P. and Cook, G. (2010). *UNIDO CCS Roadmap for Industry: High-Purity CO_2 Sources. Sectoral Assessment and Final Draft Report*. Vienna, Austria: United Nations industrial development Organization.

46 Griffin, P. and Hammond, G. (2019). Industrial energy use and carbon emissions reduction in the iron and steel sector: a UK perspective. *Journal of Applied Energy* 249: 109–125.

47 Hasanbeigi, A. S C. How clean is the U.S. steel industry? An international benchmarking of energy and CO_2 intensities. *Global Efficiency Intelligence* 2019; https://canadiansteel.ca/files/resources/HowCleanistheU.S.SteelIndustry.pdf (accessed 2 March 2020).

48 Mandova, H., Patrizio, P., Leduc, S. et al. (2019). Achieving carbon-neutral iron and steelmaking in Europe through the deployment of bioenergy with carbon capture and storage. *Journal of Cleaner Production* 218: 118–129.

49 Liu, Z., Ciais, P., Deng, Z. et al. (2020). Near-real-time monitoring of global CO_2 emissions reveals the effects of the COVID-19 pandemic. *Nat Commun* 11 (1): 5172.

50 Liu, X., Peng, R., Bai, C. et al. (2022). Technological roadmap towards optimal decarbonization development of China's iron and steel industry. *Science of The Total Environment* 850: 157701.

51 Mehmood, I., Bari, A., Irshad, S. et al. (2020). Carbon cycle in response to global warming. In: *Environment, Climate, Plant and Vegetation Growth [Internet]* (ed. S. Fahad, M. Hasanuzzaman, M. Alam, et al.), 1–15. Cham: Springer International Publishing http://link.springer.com/10.1007/978-3-030-49732-3_1 (accessed 2 November 2022).

52 Toktarova, A. (2021). *The Low-Carbon Steel Industry-Interactions Between the Hydrogen Direct Reduction of Steel and the Electricity System – ProQuest [Internet]*. Gothenburg, Sweden: Department of Space, Earth and Environment CHALMERS UNIVERSITY OF TECHNOLOGY https://www.proquest.com/openview/c2a4d77d71403fdbffd33f250b13251d/1?pq-origsite=gscholar&cbl=2026366&diss=y (accessed 7 July 2023).

53 Garvey, A., Norman, J.B., and Barrett, J. (2022). Technology and material efficiency scenarios for net zero emissions in the UK steel sector. *Journal of Cleaner Production* 333: 130216.

54 Prakash, R., Muller, M. (2007). Industrial oxygen: its use and generation [Internet]. https://www.eceee.org/library/conference_proceedings/ACEEE_industry/2007/Panel_6/P6_13 (accessed 7 July 2023).

55 OECD 2015 Steel Committee meeting: ETP (Energy Technology Perspective) 2015: iron & steel findings [Internet]. https://www.oecd.org/sti/ind/Item%208b%20-%20IEA_ETP2015_OECD%20Steel%20Committee_final.pdf (accessed 7 July 2023).

56 Manning, C. and Fruehan, R. (2001). Emerging technologies for iron and steelmaking. *JOM* 1 (53): 36–43.

57 Li, H., Zou, Z., Luo, Z. et al. (2019). Model study on burden distribution in COREX melter gasifier. *Processes* 7 (12): 892.

58 Jones, B. (2011). *Tata Steel and Rio Tinto Sign Agreement on HIsarna™ [Internet]*. Ijmuiden: COM, Tata Steel and Rio Tinto https://www.tatasteeleurope.com/corporate/news/tata-steel-and-rio-tinto-hisarna-agreement (accessed 10 July 2023).

59 COM (European Commission), CORDIS | European Commission (2010). Ultra-low CO_2 steelmaking | ULCOS Project | Fact Sheet | FP6. https://cordis.europa.eu/project/id/515960 (accessed 10 July 2023).

60 Gao, X., Zhang, R., You, Z. et al. (2022). use of hydrogen–rich gas in blast furnace ironmaking of V–bearing titanomagnetite: mass and energy balance calculations. *Materials (Basel)* 15 (17): 6078.

61 Teng, L., Ljungqvist, P., Hackl, H., and Andersson, J. (2016). *Process improvement with EMS* 1 (40): 59–62.

62 Singh, J.B. and Jaison, B. (2020). Power factor improvement in EAF using solid state transformer and bidirectional Buck-boost system using battery storage of energy matrix converters. *Solid State Technology* 63 (3): 132–152.

63 Kirschen, M., Hay, T., and Echterhof, T. (2021). Process improvements for direct reduced Iron melting in the electric arc furnace with emphasis on slag operation. *Processes* 9 (2): 402.

64 Yi, S.H., Choi, M.E., Kim, D.H. et al. (2019). FINEX® as an environmentally sustainable ironmaking process. *Ironmaking & Steelmaking* 46 (7): 625–631.

65 Brodyanski, V.M., Sorin, M.V., and Le Goff, P. (1994). The efficiency of industrial processes: exergy analysis and optimization. https://www.osti.gov/etdeweb/biblio/6901683 (accessed 6 July 2023).

66 Moran, M.J. and Sciubba, E. (1994). Exergy analysis: principles and practice. *Journal of Engineering for Gas Turbines and Power.* 116 (2): 285–290.

67 Siefert, N.S., Narburgh, S., and Chen, Y. (2016). Comprehensive exergy analysis of three IGCC power plant configurations with CO_2 capture. *Energies* 9 (9): 669.

68 Brockett, R.W. (2017). Thermodynamics with time: exergy and passivity. *Systems & Control Letters* 1 (101): 44–49.

69 International Energy Agency (IEA) (2018). Global energy demand grew by 2.1% in 2017, and carbon emissions rose for the first time since 2014 – News. https://www.iea.org/news/global-energy-demand-grew-by-21-in-2017-and-carbon-emissions-rose-for-the-first-time-since-2014 (accessed 7 July 2023).

70 International Energy Agency (IEA) (2021). Global CO_2 emissions rebounded to their highest level in history in 2021 – News. https://www.iea.org/news/global-co2-emissions-rebounded-to-their-highest-level-in-history-in-2021 (accessed 7 July 2023).

71 International Energy Agency (IEA) (2000). IEA Greenhouse gas R&D programme (IEA GHG) [Internet]. https://ieaghg.org/docs/General_Docs/Reports/PH3-30%20iron-steel.pdf (accessed 7 July 2023).

72 International Energy Agency (IEA) (2023) [IEA data and statistics. https://www.iea.org/data-and-statistics (accessed 7 July 2023).

73 STATISTA (2023). Global electricity consumption 2022. https://www.statista.com/statistics/280704/world-power-consumption (accessed 28 November 2023).

74 Johansson, M.T. and Söderström, M. (2011). Options for the Swedish steel industry – energy efficiency measures and fuel conversion. *Energy* 36 (1): 191–198.

75 Yang, Y., Raipala, K., and Holappa, L. (2014). Ironmaking. In: *Treatise on Process Metallurgy [Internet]*, 2–88. Elsevier https://linkinghub.elsevier.com/retrieve/pii/B9780080969886000171 (accessed 2 November 2022).

76 Office of Energy Efficiency & Renewable Energy (EERE) (US) (2010). http://Energy.gov. Dynamic manufacturing energy sankey tool (2010, units: trillion BTU). https://www.energy.gov/eere/iedo/dynamic-manufacturing-energy-sankey-tool-2010-units-trillion-btu-0 (accessed 6 July 2023).

77 TATA Steel UK (2020). Tata steel in the UK fact sheet [Internet]. https://www.tatasteeleurope.com/sites/default/files/Tata%20Steel%20UK%20Factsheet%202020%20(1).pdf (accessed 7 July 2023).

78 Nix, J. (2020). *John Nix Pocketbook – For Farm Management*, 50e. Leicestershire: The Anderson Centre.

79 Suopajärvi, H., Umeki, K., Mousa, E. et al. (2018). Use of biomass in integrated steelmaking – status quo, future needs and comparison to other low-CO_2 steel production technologies. *Applied Energy* 213 (C): 384–407.

80 Intergovernmental Panel for Climate Change (IPCC) (2022a). Climate change 2022: mitigation of climate change [Internet]. https://www.un.org/en/climatechange/reports?gclid=EAIaIQobChMIxsyF3ZSl-AIVoo1oCR19Dw3IEAAYAiAAEgItyfD_BwE (accessed 30 April 2022).

81 Cornelissen, R. (1997). Thermodynamics and sustainable development. https://research.utwente.nl/en/publications/thermodynamics-and-sustainable-development (accessed 6 July 2023).

82 Aylen, J. (1980). Innovation in the British steel industry. In: *Technical Innovation and British Economic Performance [Internet]* (ed. K. Pavitt), 200–234. London: Palgrave Macmillan UK http://link.springer.com/10.1007/978-1-349-04561-7_12 (accessed 2 November 2022).

83 IPCC (2023). Intergovernmental panel for climate change. AR6 synthesis report climate change 2023 [internet]. Report No.: AR6. https://www.ipcc.ch/report/ar6/syr/downloads/report/IPCC_AR6_SYR_SPM.pdf (accessed 5 January 2023).

84 Demirel, Y. (2007). Nonequilibrium thermodynamics. In: *Transport and Rate Processes in Physical, Chemical and Biological Systems*, 2e. London: Elsevier.

85 Waugh, R., Pauliuk, S., Allwood, J., and Müller, D. (2013). The roles of energy and material efficiency in meeting steel industry CO_2 targets. *Environmental Science and Technology* 47: 3455–3462.

86 Li, J., Li, C., Zhang, W. et al. (2022). Material, energy and exergy flows of the oxygen blast furnace process with sintering flue gas injection. *Journal of Cleaner Production* 371: 133294.

87 Tata Steel Europe (2020). Tata steel in Europe sustainability report 2019/2020 [Internet]. https://www.tatasteeleurope.com/sites/default/files/TSE%20Sustainability%20report%202019-20%20(EN).pdf (accessed 7 July 2023).

88 Schmitz, N., Sankowski, L., Kaiser, F. et al. (2021). Towards CO_2-neutral process heat generation for continuous reheating furnaces in steel hot rolling mills – A case study. *Energy* 224 (15): 120155.

89 United States Environmental Protection Agency (USEPA) (2023). Greenhouse gas (GHG) emissions and removals [Internet]. https://www.epa.gov/ghgemissions (accessed 2 January 2023).

90 Office of Energy Efficiency & Renewable Energy (EERE) (US) (2018). http://Energy.gov. Manufacturing energy and carbon footprints (2018 MECS). https://www.energy.gov/eere/iedo/manufacturing-energy-and-carbon-footprints-2018-mecs (accessed 6 July 2023).

91 International Energy Agency (IEA) (2023a). Coal_documentation.pdf [Internet]. http://wds.iea.org/wds/pdf/coal_documentation.pdf (accessed 12 July 2023).

92 Bliss, J.D., Hayes, T.S., and Orris, G.J. (2024). U.S. Geological Survey Fact Sheet 2008–3089, version 1.1 [Internet]. Limestone; A Crucial and Versatile Industrial Mineral Commodity. https://pubs.usgs.gov/fs/2008/3089 (accessed 24 January 2024).

93 Wilkerson, J. (2018). HARVARD UNIVERSITY BLOG biomass over coal: burning different Carbon to mitigate climate change [internet]. Science in the News. https://sitn.hms.harvard.edu/flash/2018/biomass-over-coal-burning-different-carbon-to-mitigate-climate-change (accessed 7 July 2023).

94 Albanito, F., Hastings, A., Fitton, N. et al. (2019). Mitigation potential and environmental impact of centralized versus distributed BECCS with domestic biomass production in Great Britain. *GCB Bioenergy* 11 (10): 1234–1252.

95 COM (European Commission). (2019). BIONICO: a pilot plant for turning biomass directly into hydrogen | BIONICO Project | Results in Brief | H2020 | CORDIS | European Commission [Internet]. https://cordis.europa.eu/article/id/394984-bionico-a-pilot-plant-for-turning-biomass-directly-into-hydrogen (accessed 6 July 2023).

96 Cumicheo, C., Mac Dowell, N., and Shah, N. (2019). Natural gas and BECCS: a comparative analysis of alternative configurations for negative emissions power generation. *International Journal of Greenhouse Gas Control* 90: 102798.

97 Wang, R., Liang, L., Wang, Y., and Roskilly, A.P. (2019). Process simulation of blast furnace operation with biomass syngas injection for clean production | Energy Proceedings. https://www.energy-proceedings.org/process-simulation-of-blast-furnace-operation-with-biomass-syngas-injection-for-clean-production (accessed 12 July 2023).

98 US Department of Energy Efficiency & Renewable Energy (EERE). (2021). Hydrogen production: natural gas reforming [Internet]. https://www.energy.gov/eere/fuelcells/hydrogen-production-natural-gas-reforming (accessed 8 November 2022).

99 Tanzer, S.E., Blok, K., and Ramírez, A. (2021). Decarbonising industry via BECCS: promising sectors, challenges, and techno-economic limits of negative emissions. *Curr Sustainable Renewable Energy Rep* 8 (4): 253–262.

100 Blume, L. (2021). GrInHy2.0 – wasserstoff-niedersachsen.de [Internet]. http://hydrogen-lower-saxony.com. https://www.wasserstoff-niedersachsen.de/en/grinhy2-0 (accessed 7 July 2023).

101 Garcia-Herrero, A., Tagliapietra, S., Vorsatz, V. (2021). Hydrogen development strategies: a global perspective. – document – Gale Academic OneFile [Internet]. https://go-gale-com.ezproxy.staffs.ac.uk/ps/i.do?id=GALE%7CA673932617&sid=googleScholar&v=2.1&it=r&linkaccess=abs&issn=&p=AONE&sw=w&userGroupName=anon%7E4246e29b&aty=open+web+entry (accessed 6 July 2023).

102 Bhaskar, A., Abhishek, R., Assadi, M., and Somehesaraei, H.N. (2022). Decarbonizing primary steel production : techno-economic assessment of a hydrogen based green steel production plant in Norway. *Journal of Cleaner Production* 20 (350): 131339.

103 COM (European Commission). CORDIS | European Commission. (2022). Green industrial hydrogen via reversible high-temperature electrolysis | GrInHy Project | Fact Sheet | H2020. https://cordis.europa.eu/project/id/700300 (accessed 6 July 2023).

104 Öhman, A., Karakaya, E., and Urban, F. (2022). Enabling the transition to a fossil-free steel sector: the conditions for technology transfer for hydrogen-based steelmaking in Europe. *Energy Research & Social Science* 84: 102384.

105 Blue Planet. (2023). Blue Planet Systems. Technology. https://www.blueplanetsystems.com/technology (accessed 6 July 2023).

106 Kuramochi, T., Ramírez, A., Turkenburg, W., and Faaij, A. (2011). Techno-economic assessment and comparison of CO_2 capture technologies for industrial processes: preliminary results for the iron and steel sector. *Energy Procedia* 4: 1981–1988.

107 Kuramochi, T., Höhne, N., Schaeffer, M. et al. (2018). Ten key short-term sectoral benchmarks to limit warming to 1.5 °C. *Climate Policy* 18 (3): 287–305.

108 Santos, M.P.S. and Hanak, D.P. (2022). Carbon capture for decarbonisation of energy-intensive industries: a comparative review of techno-economic feasibility of solid looping cycles. *Front Chem Sci Eng* 16 (9): 1291–1317.

109 Tata Steel Europe. (2023). Tata steel in Europe. Sustainability home page. https://www.tatasteeleurope.com/sustainability (accessed 11 July 2023).

110 British Steel. (2023). How we make steel. https://britishsteel.co.uk/what-we-do/how-we-make-steel (accessed 6 July 2023).

111 Nix, J. (2023). *John Nix Pocketbook – For Farm Management*, 53e. Leicestershire: The Anderson Centre.

112 Seethalakshmi, K.K., Jijeesh, C.M., and Balagopalan, M. (2016). *Bamboo Plantations: An Approach to Carbon Sequestration [Internet]*. Kerala/Trichur: Kerala Forest Research Institute https://www.researchgate.net/publication/215475397_Bamboo_plantations_An_approach_to_Carbon_sequestration.

113 Forestry Commission (2003). *Forests, Carbon and Climate Change: The UK Contribution*. Edinburgh.

114 Bao, J., Lu, W.H., Zhao, J., and Bi, X.T. (2018). Greenhouses for CO_2 sequestration from atmosphere. *Carbon Resources Conversion* 1 (2): 183–190.

115 Farla, J.C.M. and Hendriks, C.A. (1995). Carbon dioxide recovery from industrial processes. *Climatic Change* 29: 439–461.

116 Bataille, C., Åhman, M., Neuhoff, K. et al. (2018). A review of technology and policy deep decarbonization pathway options for making energy-intensive industry production consistent with the Paris agreement. *Journal of Cleaner Production* 187: 960–973.

117 Tata Steel. (2023). Tata steel in Europe. Sustainability performance at our sites Port Talbot. https://www.tatasteeleurope.com/construction/sustainability/performance-at-our-sites/port-talbot (accessed 13 July 2023).

118 Tata Steel Europe. (2023). Tata steel in Europe. Sustainability performance at our sites IJmuiden. https://www.tatasteeleurope.com/construction/sustainability/performance-at-our-sites/ijmuiden (accessed 7 July 2023).

119 Bataille, C., Nilsson, L.J., and Jotzo, F. (2021). Industry in a net-zero emissions world: new mitigation pathways, new supply chains, modelling needs and policy implications. *Energy and Climate Change* 2: 100059.

120 Intergovernmental Panel for Climate Change (IPCC). (2022). Climate change 2022: impacts, adaptation and vulnerability. [Internet]. https://www.ipcc.ch/report/ar6/wg2 (accessed 3 January 2023).

121 National Renewable Energy Laboratory (NREL). (2019). 2019 year in review – NREL's top 20 stories [Internet]. https://www.nrel.gov/news/program/2019/2019-year-in-review-nrel-top-20-stories.html (accessed 11 July 2023).

122 Waisman, H., Bataille, C., Winkler, H. et al. (2019). A pathway design framework for national low greenhouse gas emission development strategies. *Nature Climate Change* 9: 261.

123 Kittipongvises, S. (2017). Assessment of environmental impacts of limestone quarrying operations in Thailand. *Environmental and Climate Technologies* 20 (1): 67–83.

124 European Lime Association (EULA). (2012). How much lime per tonne of steel? Summary of the technical report a competitive and efficient lime industry – cornerstone for a sustainable Europe [Internet]. https://www.eula.eu/wp-content/uploads/2019/02/A-Competitive-and-Efficient-Lime-Industry-Summary_0.pdf (accessed 8 November 2022).

125 Jahani, N., Sepehri, A., Vandchali, H.R., and Tirkolaee, E.B. (2021). Application of industry 4.0 in the procurement processes of supply chains: a systematic literature review. *Sustainability* 13 (14): 7520.

126 Franciosi, C., Voisin, A., Miranda, S., and Iung, B. (2020). Integration of I4.0 technologies with maintenance processes: what are the effects on sustainable manufacturing? *IFAC-PapersOnLine* 53 (3): 1–6.

127 Wei, R., Li, H., Chen, Y. et al. (2022). 9.07 – environmental issues related to bioenergy. In: *Comprehensive Renewable Energy [Internet]*, 2e (ed. T.M. Letcher), 92–106. Oxford: Elsevier https://www.sciencedirect.com/science/article/pii/B978012819727100011X (accessed 7 July 2023).

128 Babich, A. and Senk, D. (2015). Recent developments in blast furnace iron-making technology. In: *Iron Ore [Internet]*, 505–547. Elsevier https://linkinghub.elsevier.com/retrieve/pii/B9781782421566000174 (accessed 2 November 2022).

129 Calvo, L.M. and Domingo, R. (2015). Influence of process operating parameters on CO_2 emissions in continuous industrial plants. *Journal of Cleaner Production* 1 (96): 253–262.

130 Gupta, R.C. (2014). *Treatise on Process Metallurgy Volume 3: Industrial Processes [Internet]*, vol. Vol. 3. London: Elsevier https://www.sciencedirect.com/topics/engineering/coke-rate (accessed 7 July 2023).

131 Copeland, C. (2013). A practical engineering approach to improving the reliability of blast furnace tuyeres. https://www.semanticscholar.org/paper/A-Practical-Engineering-Approach-to-Improving-the-Copeland/310bee8b0bec31a17dcf88d158469e46ccbef783 (accessed 11 July 2023).

132 Keung, K. (2021). TAKACHAR Email from Kevin Keung, Research and Development Lead, 18 October.

133 Tata Steel Nederland. Tata Steel in Europe. (2023). Sustainability performance at our sites IJmuiden. https://www.tatasteeleurope.com/construction/sustainability/performance-at-our-sites/ijmuiden (accessed 10 July 2023).

134 Ministry of Agriculture, Food and Rural Affairs Canada. (2023). Published plans and annual reports 2022–2023: Ministry of Agriculture, Food and Rural Affairs | Ontario.CA [Internet]. http://www.ontario.ca/page/published-plans-and-annual-reports-2022-2023-ministry-agriculture-food-and-rural-affairs (accessed 13 July 2023).

3

Introduction to the Bio Steel Cycle

3.1 Introduction

The requirement to drastically reduce GHG emissions, and particularly, carbon dioxide (CO_2) emissions, has never been greater than today. The Kyoto Protocol (KP), the Paris Agreement and recent 2023 reports from the IPCC and EIA have clearly set out the impact, which the highest ever recorded anthropogenic CO_2 emissions are having on our environment and climate.

The steel industry is thought to be responsible for between 7 and 11% of global CO_2 emissions, but some scientists are of the opinion it could be significantly higher [1–5], due to high fossil fuel and energy consumption and the fact that there is no unified way of recording or reporting CO_2 emissions. The wide-ranging difference in percentage shares can largely be attributed to the different approaches to (i) which level of carbon emissions has been observed, (ii) how many different stages and processes in (steel) production have been included and measured, (iii) which methods have been used to measure the quantity of CO_2 emissions and (iv) the different geographical physical locations they have been observed at. There is currently no existing globally applicable legal framework with guidance to report greenhouse gas (GHG) emissions or guidance for any industry to pinpoint the emissions data extraction points for emissions level determination. The methods of data collection and collation vary wildly, and there is no unified method. Therefore, these percentage-level statements need to be taken cautiously, as the GHG emissions could be even higher than the already identified levels [2, 4, 6–10].

The factual level of CO_2 emissions for every tonne of steel produced currently stands at more than 4.6 tonnes of CO_2 emissions [11] for the BF/BOF route alone and 12.94 tonnes of CO_2/tonne of CS from mining to marketing; the onus is on industry to remedy the environmental damage caused in the past two centuries. The anthropogenic carbon emissions are at an all-time high with reported ~65.6 Gt CO_2 emissions in 2019 [12]. To remedy future environmental damage caused by CO_2 emissions, the innovative concept, model and strategy 'The Bio Steel Cycle' was created, as shown in Figure 3.1.

Circular Steel Production: Pathways to Net-Zero Carbon Emissions, First Edition.
Sandra Kiessling, Hamidreza Gohari Darabkhani, and Abdel-Hamid Soliman.

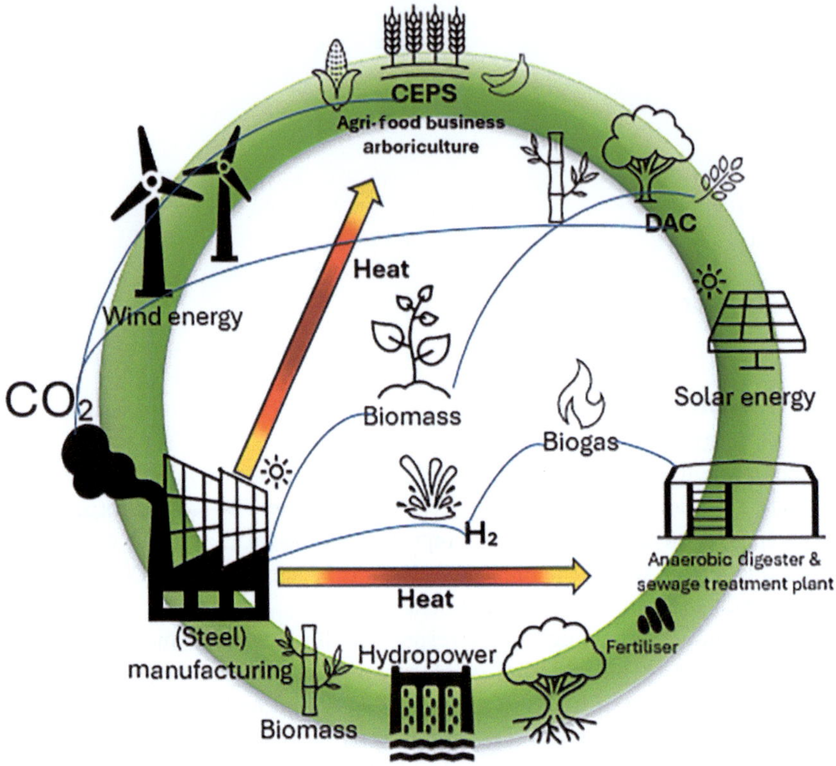

Figure 3.1 The Bio Steel Cycle.

The 64 steel-producing countries produced between January and December 2021 [13] 17 Gt CO_2 equivalent of CO_2 emissions as a result of the current linear steel manufacturing. The importance of significantly reducing GHGs and eliminating fossil fuel combustion and usage has never been greater, and fast, practical solutions – on a global scale – are needed.

Already in 1912, it was recognised that coal consumption is an environmental hazard and incompatible with keeping global temperatures at a balanced level to sustain life. Industrial processes have for more than two hundred years polluted the air we breathe, and already in 1912, this was recognised as displayed in an Australian newspaper clip [14]. It is worth pointing out that, in this mentioned article, merely the in-furnace coal combustion process is mentioned in connection with carbon emissions. CO_2 emissions from energy consumption, mining, pelletising, coking, sintering, steel smelting, casting, rolling, annealing, rolling, finish machining and surface treatment and related processes have not been considered at that time. Already more than 100 years ago, in 1912, it was established that blast furnace operations resulted in the release of 3.5 tonnes of CO_2 into the atmosphere by combustion the factor applied was 3.5 (7bn t of CO_2 emissions with 2bn t of steel produced). It so seems that this figure was possibly based on the carbon content of coal (78–95%) [15].

Although the steelmaking process has undergone significant improvements and the BOF operation has allegedly reduced the CO_2 emissions to between 3.2 and 1.6 tonnes/CO_2/tonne of steel produced [1, 16–21], it begs the question of why this knowledge has not been used to establish the true CO_2 emissions over the past two-centuries – including all necessary processes required to produce steel. The objectives of this multidisciplinary and multi-industry overarching study are to identify the most efficient implementation opportunities of the chosen processes and technologies to reduce the current BF/BOF route 4.6 tonnes CO_2 emissions/tonne of steel produced to factual '0', ordered in seven easy steps, from short-term to long-term solution implementation.

3.2 Importance of demonstrating all carbon emissions along the steelmaking process

The devastating effects of climate change, such as wildfires increasing in numbers and frequency, hurricanes and earthquakes, have sparked the desire and motivation for wanting to make a difference in the iron and steel industry. The Bio Steel Cycle is aiming to work within the principles of the circular (tech-) economy. Waste and CO_2 emissions avoidance and reduction, and pollution reduction are the key elements of this model. This is an industry-overreaching, multidisciplinary concept and systematic approach to steel manufacturing, considering all aspects along the production process and emphasising the need for production process improvement. This approach will benefit society, businesses and our environment, as well as reducing production cost in heavy industry. It needs to be emphasised that the Bio Steel Cycle is an adaptable model and strategy with the level of implementation based on the individual site parameters. Every industrial production site requires to be assessed for the most suitable components of the BiSC to be planned and implemented within infrastructural and budgetary constraints. Based on the literature reviewed, a range of process components was researched and approaches to CCS were considered, such as a three-phase model [18], the techno-economic pathway concept, where biomass [22], hydrogen and electrification of steel production were identified as possible factors to reduce CO_2 emissions – although they are largely making the case for electrification of steel production.

Reading inspiring publications by the Ellen McArthur Foundation [23], the desire and motivation for wanting to make a difference were born. The core message and principles which [18] have developed are the circular economy with the basis of three principles, driven by design:

- Eliminate waste and pollution
- Circulate products and materials (at their highest value)
- Regenerate nature.

Essentially underpinned by a societal transition to renewable energy, materials and circular economic principles, a resilient system can be created which is positive for the environment. With habitat recreation and reversing biodiversity loss, it is much healthier for the planet's human population; as the living conditions are improved, the effects of climate change such as habitat and harvest loss due to extended and prolonged wildfires and unseasonal flooding can be halted or even reversed. Additionally, it is beneficial for businesses, as the usage of virgin material and its destructive extraction from natural sources can be reduced to a minimum – if not entirely avoided.

The developed concept 'Bio Steel Cycle' and the '7 Steps to Net-Zero CO_2 Emissions Steelmaking' have been published in November 2022 [11] and outlined how decarbonisation of the steel industry could be achieved in the very near future. For the benefit of the reader, it should be mentioned that in the previously published paper, not the whole scale of the identified CO_2 emissions in the steelmaking process has been mentioned, but the focus was rather on the BF/BOF process route.

The impact of different technologies [18] on the processes at all stages in the steelmaking process [1], and iron and steel manufacturing [13] and related databases and corresponding literature was investigated. Additionally, literature with regard to CAT, CCS and CCUS and circular economic economy, sustainability and decarbonisation of the steel industry was categorised for ease of implementation, from easy to more challenging and depending on duration. An Excel database was used for data collection, modelling and key calculations. The parameters were defined as tonne of CO_2/tonne of steel produced, although further parameters for future research are being allowed. The research works with metric tonnes only. Additionally, engineering simulation software has been used, such as Simul8, Aspen Plus V11 and INOSIM.

The flowchart in Figure 3.2 provides some detail about the different stages involved in making steel and the possible CO_2 and heat transfer flow when the various components of the Bio Steel Cycle have been schematically integrated. This overview provides some detail about the theoretical framework and some guidance on implementation:

3.3 BF/BOF route carbon capture

3.3.1 CO_2 capture mechanisms

The standard steel production (SSP) process in combination with the currently operational newly developed technologies can achieve a reduction of more than 50% [24] with successive implementation to less than 8 metric tonnes/tonne of steel produced. By incorporating the BiSC components of CAT, CCS and CCUS into existing steel production sites, an almost 100% CO_2 emissions reduction can be achieved immediately.

The post-combustion capture of CO_2 (CCS) and other GHGs and the exploration of carbon scrubbing of flue gases have been explored. There are several possible technologies and processes to be considered for post-combustion carbon capture:

- Mechanical capture
- Compression and dehydration

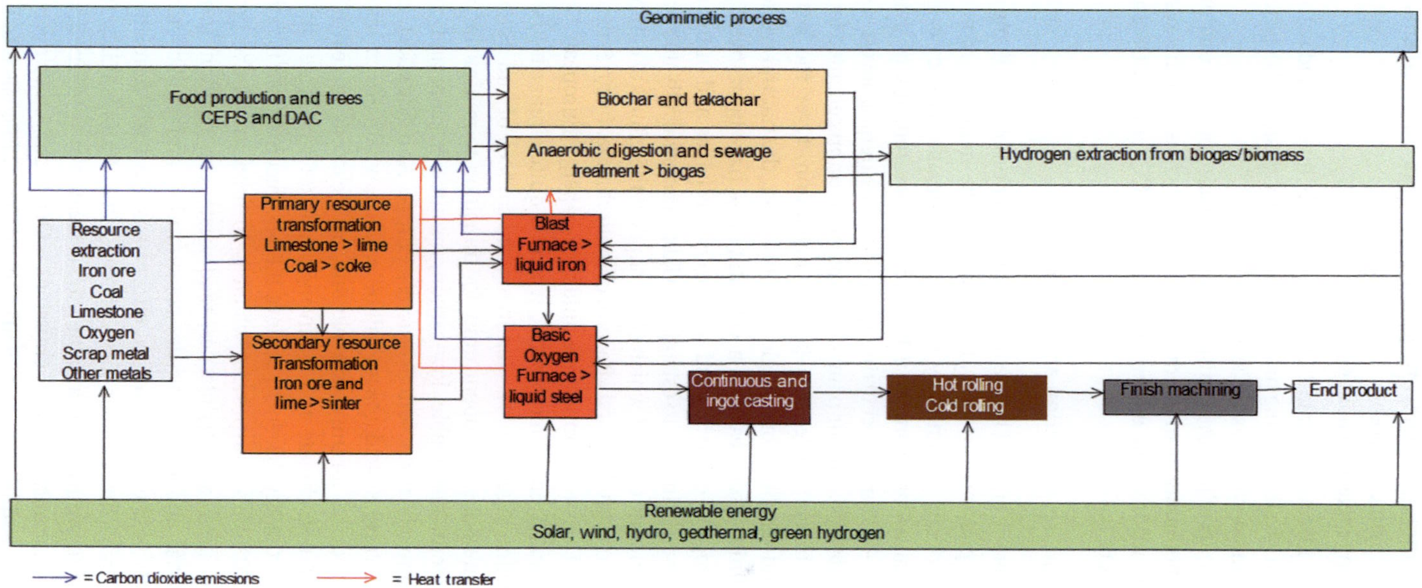

Figure 3.2 Simul8 steel linear production configuration.

- Membrane installation
- Guiding off-gas through troughs of physical solvents/solid sorbents (such as Zeolite13X) and chemical solvents
- As well as utilising metal/organic frameworks

[1, 25–30]. Suitable literature (Table 2.3) was thoroughly investigated with regard to applied and innovative steelmaking procedures, CAT, CCS and CCUS processes and improved management systems such as I4.0 [31, 32].

The key components within the Bio Steel Cycle are based on a circular production process and are functioning in an interactive manner. The basis for this system is the BF/BOF route and involves the aforementioned CAT, CCS and CCUS and process improvements where possible.

3.3.2 Carbon capture, utilisation and storage

Carbon dioxide is being emitted by the respiratory function of most living creatures, for a reason: most plants require CO_2 to grow [33]. The natural carbon emissions cycle has been perfectly designed by nature, but humans have caused the production of CO_2 emissions to increase beyond the requirements of living things for this – initially – naturally occurring gas to life-threatening levels. One of the results is the climate crisis, as our planet is eating up to unprecedented temperature levels, which are incompatible with human habitation. Over the last 20 years, climate change and its effects on the global population, the natural world and our way of using up finite resources at an accelerating rate have been investigated and researched by a range of countries and numerous specialists and in the respective fields. The researchers' aim was to establish a viable, economically and ecologically sound circular economic concept which could support the efforts of carbon capture and storage (CCS) to avert the worst effects of climate change. In 2015, the European Commission (COM) issued a milestone guidance paper, a European Union (EU) action plan for circular economic process, which is quite similar to the US Environmental Protection Agency [34], identifying the sources of and potential ways to reduce GHG emissions. The opportunity for transition to minimising waste and developing a resilient, sustainable, low-carbon, resource-efficient and competitive economic model must be taken seriously across Europe. As a side effect, energy will be saved and also irreversible damages will be avoided, which are currently caused by using up resources at an unsustainable rate, as it exceeds the earth's capacity for renewal – as far as climate and biodiversity, air, soil and water pollution are concerned. One of the key drivers to achieve the set objectives is a major initiative called 'Horizon 2020', the purpose of which is to fund innovative projects and targeted action in areas such as plastics, food waste, construction, (critical) raw materials, industrial and mining waste, consumption and public procurement. One of the results of the EU countries' efforts was the Paris Agreement, which was signed in 2016 by the member states of the United Nations (UN). It binds all signing countries to secure a sustainable global economic recovery and net-zero emissions future (not only zero CO_2 emissions – GHGs and other toxic emissions are also ring-fenced within this agreement). One of the main goals is to limit global temperature rise to no more than 1.5 °C, as achieving this would avert the worst ecological, financial and economic impacts of climate change.

HM Government issued 'The Ten Point Plan for a Green Industrial Revolution' in 2020, covering the issue that most emissions are being produced through the burning of fossil fuels for electricity, heat and transportation. The six sectors identified for being the most polluting are as follows: electricity/power generation, transportation, industry, commercial/residential building, agriculture and land use/forestry.

The efforts and plans currently in place amount to a total of achievable CO_2 reduction of 40.76%, with the potential to increase these efforts, based on the road map for further policies, to achieve a net-zero carbon economy by 2050.

The European Environment Agency Report 03/2020 [35] constitutes the official inventory submission of the EU for 2020 under the UNFCCC [36], as well as the KP [37], covering the years between 1990 and 2018. The EU, party to the UNFCCC, issues these reports on GHG inventories annually, for emissions and removals within the area taking place within its territory.

Due to the nature of this lag in reporting, the 2020 inventory report cannot yet reflect the effects which the COVID-19 pandemic might have had. Although the United Kingdom left the EU on 24 February 2020, EU law still applies until the end of the transition period. Moreover, key provisions of Regulation (EU) No 525/2013 ('Mechanism for Monitoring and Reporting GHG') and of Decision No 406/2009/EC ('Effort Sharing') apply to the United Kingdom also in respect of GHGs emitted during 2019 and 2020. The EU and its member states – including Iceland – agreed to a quantified emissions reduction commitment, which will limit their average annual GHG emissions to 80% of the sum of their base year emissions, which is reflected in the Doha Amendment [38].

GHG emissions decreased in the majority of sectors between 1990 and 2018, with the notable exception of transport, including international transport, refrigeration and air conditioning. Considering the aggregate level, the highest emissions reductions were achieved in manufacturing industries, construction, electricity/heat production, iron/steel production and domestic combustion.

Sausen and Schumann [39] and Lee et al. [40] had already thoroughly investigated GHG in 2000 and – specifically – aviation CO_2 emissions, using data from the Global Carbon Project [40] and went on to research issues revolving around these urgent matters and established how significant the impact of global aviation factually is in terms of CO_2 emissions produced, radiative forcing (RF) and effective radiative forcing (ERF) and its impact on climate change. The authors looked into both RF and ERF (and their effects on increasing the air temperature), with sums being calculated for the years 2000–2018.

Allegedly, contrail cirrus (the condensed clouds one can observe as an air signature of a plane's flight path), consisting of linear contrails and the cirrus cloudiness arising from them, yields the most significant warming ERF, closely followed by CO_2 and NO_x emissions. Interestingly, emissions of sulphate aerosol formation result in a negative (cooling) term.

In conclusion, the mean contrail cirrus ERF/RF ratio of 0.42 indicated that contrail cirrus is less effective in surface and atmospheric warming than other terms. It was established that by 2018, the net aviation ERF is +100.9 milliwatts (mW)/m^2, meaning a 5–95% likelihood range (of 55, 145) with significant contributions from contrail cirrus (57.4 mW/m^2), CO_2 (34.3 mW/m^2) and also NO_x (17.5 mW/m^2).

Already in 2007 [41], the opportunities which lie within hydrogen production from solar energy were researched, with hydrogen being a credible alternative to using fossil fuels in industrial applications. Effectively, CO_2 emissions were removed from hydrogen production, establishing that it is both economically and practically viable to produce hydrogen using solar energy – making hydrogen a renewable energy source.

House et al. [42] followed similar lines of thought in 2011, when they investigated the opportunities of economic and energetic analysis of capturing CO_2 from ambient air. In conclusion, they established that the total levelised cost of $80.-/MWh natural gas advanced combined cycle is the lowest total levelised energy cost, whereas, with regard to the $/(t$CO_2$ avoided), hydropower is by far the most cost-efficient and bears the lowest risk for humans and the environment, at $748/tonnes of CO_2 captured.

Further critical and thorough literature reviews of concepts concerning circular economic processes were carried out [23, 43], covering the following issues, presented in order of conceptual scale (from more to less encompassing):

- Sustainable development
- Ecological transition
- Green economy
- Functional economy
- Life cycle thinking
- Cradle-to-cradle thinking
- Shared value
- Industrial ecology
- Extended producer responsibility
- Eco-design

The study of these concepts has led the researcher to develop the Bio Steel Cycle [11].

O'Callaghan [44] wrote for the European Commission about 'green' hydrogen – produced utilising exclusively renewable energy technologies – as an alternative to fossil fuels. Possibly inspired by this, Sunny, Mac Dowell and Shah [45] followed on from this by investigating the issue of what might be needed to deliver carbon-neutral heat using hydrogen and CCS, which led to the discovery that it is entirely possible to produce hydrogen based on using renewable energy technologies.

The European Commission established that, currently, around 20% of all GHG emissions are produced by industries such as steel and cement – it is, therefore, vital to develop promising technologies and concepts in these areas. Siemens [46–50] was referred to as claiming that '… Hydrogen and fuels derived … capable of reducing the CO_2 from fossil fuels … down to zero'. Hydrogen technology was further referred to [44, 45] as a concept capable of generating energy exclusively utilising renewable energy technologies.

To this effect, a thorough literature review has been carried out to determine the various possibilities of steelmaking process CAT, CCS and CCUS technologies, with key literature as the basis for the development of the Bio Steel Cycle in Table 3.1 and decarbonisation projects in the iron and steel industry in Table 3.2.

Table 3.1 Literature review list fundamental information for the development of the Bio Steel Cycle concept, model and strategy.

Title	Description	References
How we make steel	An overview of the steelmaking process in the United Kingdom	[51]
Carbon capture for decarbonisation of energy-intensive industries: a comparative review of techno-economic feasibility of solid looping cycles	CCS of energy-intensive industries (steel, cement, paper and chemical) techno-economic feasibility and implementation opportunities	[52]
Hydrogen production: natural gas reforming	Endothermic – heat dependent. Steam–methane reforming; Steam (700–1000 °C) to produce hydrogen from natural gas. Methane reacts with steam under 3–25 bar pressure (1 bar = 14.5 psi) in the presence of a catalyst to produce hydrogen, carbon monoxide and some CO_2.	[53]
Energy Outlook 2022 – Energy Outlook 2022 – Table 18. Energy-related carbon dioxide emissions by sector and source	2022 overview: energy-related CO_2 emissions by sector	[54]
Energy Outlook 2021 – Energy Outlook 2021 – Table 18. Energy-related carbon dioxide emissions by sector and source	2022 overview: energy-related CO_2 emissions by sector	[55]
Forests, carbon and climate change: the UK contribution	UK tree cover and DAC	[56]
Data and statistics	Global energy and fuel data	[57]
Basic oxygen furnace	Process data	[58]
Electric arc furnace	Process data	[59]
Application of industry 4.0 in the procurement processes of supply chains: a systematic literature Rv.	I4.0 digitisation and CO_2-saving opportunities	[32]
Takachar mobile units	Biochar units scalable and deployable internationally	[60]
December 2021 crude steel production and 2021 global crude steel production totals	World crude steel production for the 64 countries reporting to the WSA was 158.7 million tonnes (Mt) in December 2021, a 3.0% decrease compared to December 2020	[61]
Vegetation uptake of mercury and impacts on global cycling	GHG vegetation uptake mechanism, vegetation cycling and model representation of vegetation Hg (mercury) cycling	[62]

(Continued)

Table 3.1 (Continued)

Title	Description	References
Integration of I4.0 technologies with maintenance processes: what are the effects on sustainable manufacturing?	Digitisation of manufacturing and possible CO_2-reducing effects	[31]
Near-real-time monitoring of global CO_2 emissions reveals the effects of the COVID-19 pandemic	Global CO_2 emissions development affected by COVID-19	[63]
John Nix Pocketbook – for Farm Management (50th edn)	Guidance book for agri- and agri-tech businesses and management	[64]
Pathways for low-carbon transition of the steel industry – a Swedish case study	Applicable to global steel manufacture, providing real-time data on technology supporting transitioning to Net-Zero	[18]
Sustainability report	Fact Sheet EU sustainable and zero-carbon steel manufacturing data	[65]
BIONICO: a pilot plant for turning biomass directly into hydrogen'	Project within Horizon 2020 – Biogas membrane reformer for decentralised hydrogen production	[66]
A pathway design framework for national low greenhouse gas emissions development strategies	Range of specialists in sustainability; Prof Fischedick/IPCC; provided some key data to build the BiSC from	[67]
Industrial energy use and carbon emissions reduction in the iron and steel sector: a UK perspective	Opportunities and challenges to reducing industrial energy demand and (CO_2) emissions in the iron and steel sector are evaluated	[1]
Development of optimal modes and mathematical models of energy performance of electric steelmaking production	Development of the underlying mathematical formulae in energy performance of electric steelmaking	[68]
Achieving carbon-neutral iron and steelmaking in Europe through the deployment of bioenergy with carbon capture and storage	Carbon-neutral steelmaking guidance	[27]
2019 year in review – NREL's top 20 stories	National Renewable Energy Laboratory (NREL) Key data for BiSC Renewable Energy Technology. US Department of Energy	[69]
2019 UK greenhouse gas emissions	Overview: UK CO_2 emissions by sector and industry	[70]

Sustainability. Theory and Methods for Engaged Inquiry. 2nd ed.	Tata Steel sustainable process improvement solutions	[71]
Steel Statistical Yearbook 2019	WSA – almost complete set of global data iron and steel industry	[72]
A review of technology and policy deep decarbonisation pathway options for making energy-intensive industry production consistent with the Paris Agreement	CCS, CAT and CCUS review and policy review including decarbonisation of energy-intense industrial production	[26]
Assessment of low-carbon iron and steel production with CO_2 recycling and utilisation technologies: a case study in China	Low-carbon iron and steel production process data	[73]
Techno-economic assessment and comparison of CO_2 capture technologies for industrial processes: preliminary results for the iron and steel sector	Techno-economic assessment of CCS iron and steel sector	[74]
Use of biomass in integrated steelmaking—status quo, future needs, and comparison to other low-CO_2 steel production technologies	Biomass in steelmaking and CAT/CCS/CCUS analysis	[75]
Assessment of hydrogen direct reduction for fossil-free steelmaking	HDR	[76]
Assessment of environmental impacts of limestone quarrying operations in Thailand	Limestone and Lime	[77]
Bamboo plantations: an approach to carbon sequestration	Plant carbon sequestration	[78]
Substance flow analysis of aluminum industry in mainland China	Analysis of process improvement in metal manufacturing	[79]
Recent developments in blast furnace iron-making technology	Blast furnace processing data	[80]
Influence of process operating parameters on CO_2 emissions in continuous industrial plants	Process changes effects on carbon emissions	[81]

(Continued)

Table 3.1 (Continued)

Title	Description	References
Treatise on process metallurgy Chapter 4.2 – Energy resources, its role and use in metallurgical industries	Energy analysis and application in multi-metal manufacturing	[82]
Cost of energy saving and CO_2 emissions reduction in China's iron and steel sector	Techno-economic assessment of energy and carbon emissions savings in iron/steel in China	[17]
Chapter 1.1 – Ironmaking', treatise on process metallurgy	Process details and benchmarks	[83]
A practical engineering approach to improving the reliability of blast furnace tuyeres	Process improvement	[84]
Metallurgical coke	Coke/coal data	[85]
How much lime per tonne of steel? Summary of the technical report a competitive and efficient lime industry – cornerstone for a sustainable Europe	Analysis of lime in steel manufacturing	[86]
Process integration of steelmaking and methanol production for suppressing CO_2 emissions – a study of different auxiliary fuels	Overview of technological advances and efficiency of methanol in CO_2 emissions	[87]
Assessment of energy efficiency improvement and CO_2 emissions reduction potentials in India's cement and iron and steel industries	Overview of cement and steel CO_2 emissions reduction technology potential (India)	[88]
Climate policy – Ten key short-term sectoral benchmarks to limit	Analysis of key benchmarks per industry sector to avert climate change	[74]
Energy efficiency improvement and cost saving opportunities for the U.S. Iron and Steel Industry	Techno-economic assessment of US iron and steel industry	[16]
Industrial oxygen: its generation and use	Oxygen generation	[89]

Table 3.2 A selection of decarbonisation projects and endeavours.

Project name	Description	References
The ULCOS PROJECTS – ULCOS/ ULCOS-BF	The Ultra-Low CO_2 steelmaking (ULCOS) program: ULCOS is a project series involving 48 companies in the EU and to drastically decrease CO_2 emissions, of the order of at least 50% over the whole steel mill. Phase I, with a budget larger than 50 million euros, started in 2004 and Phase II in 2010. ULCOS eventually proposed several breakthrough technologies: top gas recycling blast furnace (TGRBF) or ULCOS-BF, ULCORED, ULCOWIN and ULCOLYSIS.	[90–99]
ULCORED	ULCORED is being described as a technology based on iron ore direct reduction, which produces direct reduced iron (DRI) in a shaft furnace. Hereby, natural gas (NG) or gas obtained by gasification of coal, and off-gas are recycled into the furnace after CO_2 has been captured. This is directed into a concentrated stream and stored permanently. The main objective of the ULCORED process was to achieve a significant reduction in natural gas consumption as a fossil fuel. It was achieved by using partial oxidation (PO_x) of NG. In tandem with CCS, ULCORED has the potential to reduce 70% of CO_2 emissions, compared with the standard BF route. At its core, CO_2 scrubbing minimises GHG emissions and maintains to reduce energy consumption at the same time. Therefore, ULCORED needs to be seen as an engineered combination of natural gas oxidisation and carbon capture.	[99, 100]
ULCOWIN	ULCOWIN and Siderwin are engineering processes which aim to produce iron by electrolysis at low temperature and are researching to develop a radically new steel production process without CO_2 emissions, based on electrolysis technology. It follows an initial project launched in the early 2000s, one of whose missions was to lay the foundations for electrolytic steel production at the laboratory stage. The carbon is replaced by electricity and requires neither agglomeration nor coke. The operation of the blast furnace could be entirely electric, with oxygen being the only off-gas emitted. With no need to treat the hot metal, the adjustment to grade – say C35 – with the addition of carbon would be performed during the melting of iron metal plates via electrolysis. One of the key benefits of this process is the easily implementable incorporation of renewable energy technologies into the process.	[95, 101]
ULCOLYSIS	With ULCOLYSIS, an electrolysis-based process is being described which operates at liquid steel temperature, where molten iron oxide (iron ore = Fe_2O_3) is transformed into liquid metal and oxygen during an electrolytic smelting process. This method can be applied with other oxides too.	[92, 99]

(Continued)

Table 3.2 (Continued)

Project name	Description	References
HIsarna	HIsarna is a smelting technology resulting from a collaboration between ULCOS/European Commission, Tata Steel and Rio Tinto. To reach the 50%+ GHG emissions reduction target, all of the aforementioned technologies except for electrolysis can be associated with CCS. Smelting processes in general eliminate two steps: coking and pelletising of iron ores. HIsarna slightly pre-reduces iron ore fines and melts them in a cyclone by using pure oxygen, whilst most of the reduction and coal gasification takes place in the smelter (similar to the HIsmelt technology (Rio Tinto)). Coal powder is injected into the smelter. Encouraging pilot plant test results demonstrated 20% fewer CO_2 emissions in comparison to the conventional BF operation. A higher percentage of CO_2 avoidance was achieved when CCUS of off-gas was integrated (Birat [92]). An additional 10–20% of CO_2 emissions savings was thought to be attributed to the elimination of the requirement for coking and pelletising.	[102–104]
NER-300/400	The European Commission is involved in the NER-400 program, similar to the previous NER-300 program which was centred on low-carbon technologies, but this did not lead to any CCS project. In the NER-400 endeavours, TGRBF CO_2 is separated from the top gas. The blast furnace uses pure oxygen instead of air at the tuyeres, which generates a (mostly) nitrogen-free gas. With this concept, using 600 m^3/tonne of reducing gases can decrease the amount of fuel to 190–200 kg/tonne maintaining a pulverised carbon injection (PCI) rate of 170–180 kg/tonne. Model calculations of energy consumption in the TGR-BF are 20–25% lower than the conventional blast furnace. In a downsized oxygen BF, maximum shear stresses can be 20–30% lower compared to a conventional BF of the same output, thus making it in principle possible to charge coke with lower mechanical strength. Another innovation would be replacing pulverised coal injection (PCI) with powdered biomass, particularly charcoal. Replacement of PCI with heavy oil or natural gas is also possible, but it does not contribute to reducing CO_2. Coal might be replaced by charcoal to a large extent. High-quality charcoal with a carbon content higher than coal can be produced by controlling pyrolysis temperature and using a wood-based feedstock. Furthermore, charcoal has lower concentrations of sulphur and acid oxides in addition to higher concentrations of basic oxides.	[75, 92, 99, 105, 106]
AIST MOE/ COURSE 50	In addition to ULCOS, AIST in the United States sponsored two projects: one is related to the extraction of metals by electrolysis at high temperatures and another to hydrogen flash smelting. The concept of molten oxide electrolysis (MOE) involves the melting of metal oxide and then the electrolytic separation of metal and oxygen. This process can contribute to decreasing GHG if lean electricity is employed and is somewhat similar to the ULCOLYSIS process. In Japan, there is a large National project called COURSE 50 supported by NEDO which was launched in 2007. One of its many dimensions is to introduce hydrogen in the blast furnace to reinforce the reduction of iron ore.	[99, 107]

COREX	The only commercially successful smelting reduction process today is COREX. The process uses pellets and coal but generates even more CO_2 than the BF. Direct ironmaking from iron ore without coke and pelletising presents important technical challenges. Among them are the higher concentrations of both sulphur and iron oxide in the metal (COREX slag has an unusual FeO content compared to a BF slag) as well as higher concentrations of CO and CO_2. These require higher rates of desulphurisation, control of foaming slags and improved utilisation of coal particles to decrease the carbon footprint. Actual energy consumption by smelting processes can only refer to the COREX process since it is the only commercial one. Primary consumption of energy of COREX is higher in comparison to the BF, of the order 16 GJ/tonne. In order to make it competitive, this process requires recycling the top off-gas in another plant like the DRI process or BF process.	[17, 108, 109]
EAF process improvement	A large number of innovations have drastically changed the shape, size and role of the EAF, and some of the innovations include the following: the development of the ladle furnace, redefining the EAF role as a melting unit; the development of UHP transformers achieving higher melting rates; foamy slag practice to operate with longer arcs and therefore higher energy density; continuous scrap preheating and charging; higher oxygen injection rates to increase the amount of pig iron in the metallic charge and to supply fossil energy (coal) in addition to electricity; optimisation of the hot heel practice; eccentric bottom tapping, improved post-combustion; oxy-fuel burners; hot charge of DRI; and use of preheated metal, to name a few. Electrical energy consumption in the EAF varies depending on the metallic charge: the highest values range from 550 to 700 kWh/tonne, when the metallic charge is DRI and decrease to 300–450 kWh/tonne when the metallic charge is scrap. The lower values are achieved using both chemical energy (derived from oxygen injection) and scrap preheating technologies.	[30, 110–113]
Continuous casting/thin slab casting	Until the 1950s, ingot casting was the main technology in operation to produce steel. Whilst the amount of ingot casting decreased, continuous casting (CC) incrementally increased in volume and eventually largely replaced ingot casting. Currently, it is estimated that more than 96% of steel is produced by CC. Using this technology also reduced the volume of metal-containing ores, from 1.45 to 1.27 tonnes of liquid steel to produce 1 tonne of cold sheet. Continuous casting pours liquid steel into a turn dish and then into an oscillating mould. Thin slab casting (TSC) is a more recent version of CC; the post-cast-processing is almost the same, as some hot rolling is still required. The casting speed is increased in TSC in comparison with CC. During near-net-shape casting (NNSC), the resulting product exits the production line with a thickness of a few millimetres, and the majority of downstream processes (reheating, hot rolling, scarfing, pickling and annealing) become unnecessary. In CC, a typical slab thickness is 250 mm, whilst, in TSC, the thickness is reduced to ca. 50–65 mm, and in NNSC, the thickness is 2–20 mm. In addition to higher productivity and lower investment, the steel quality of NNSC products is free from oscillation marks observed in the material after CC. The simplified processing of NNSC has a ca. 80% lower energy consumption than conventional CC. The resulting emissions are reduced by ca. 60% and currently estimated to be 170 kg CO_2/tonne hot strip produced using NNSC.	[99, 114–118]

(Continued)

Table 3.2 (Continued)

Project name	Description	References
CASTRIP® TSC/ EUROSTRIP	The CASTRIP® facility at Nucor Steel's Crawfordsville started commercial operations in 2002, including a range of steel strips at thickness from 0.9 to 1.5 mm. EUROSTRIP ran a commercial facility in Krupp Thyssen Nirosta, following up on demonstrators in Isbergues, France, and Terni, Italy.	[99, 114, 116, 119]
Athos	The Athos Project (Amsterdam-IJmuiden CO_2 Transport Hub and Offshore Storage) aims at developing a CO_2 transport and storage network in the North Sea Canal area to enable CCUS. Athos is beside Porthos in Rotterdam, the second initiative in the Netherlands to build an infrastructure for CCUS. The Athos phase I ended on 20 September 2021 as a direct result of Tata Steel's decision to switch to hydrogen-based DRI steelmaking technology. The Athos partners continue to engage with Tata Steel in phase II to assess potential support and to contribute to regional CO_2 reduction targets. In the North Sea Canal area, Athos is part of a broader package of initiatives to drive sustainable development in the region.	[120, 121]
Everest	The Everest (Enhancing Value by Emissions Reuse and Emissions Storage) project consists of two phases: in the first phase, a carbon capture (from steelmaking process gases) facility has been installed. In phase two, the captured CO_2 is then delivered to the Athos (Amsterdam-IJmuiden CO_2 Transport Hub and Offshore Storage) infrastructure, aiming to convert the remaining gases into usable raw materials, with implementation in IJmuiden planned for 2027. The expected CO_2 savings results have been estimated for the first phase at approximately 3 Mt of CO_2 p.a. and phase two (reuse of process gases) at approximately 4 Mt of CO_2 p. a.	[120, 121]
Hermes	The Tata Steel/Hermes project is at its core the reduction of iron ore with H_2. It is estimated that enough supplies of green hydrogen and the steel production process technology to produce steel with hydrogen will be available by 2040. Tata Steel is collaborating with the chemical company Nouryon (at the port of Amsterdam) to develop the largest green hydrogen 100-megawatt renewable-energy-powered water electrolysis facility in Europe (at Ijmuiden).	[120, 121]
Electrolysis	Siemens Energy is aiming to create the hydrogen economy of the future. They are working closely with partners from government, academia and industry to create a sustainable and profitable hydrogen economy. The technologies in development embrace the whole hydrogen value chain: wind energy, infrastructure for power transmission and seawater electrolysis (similar to Tata Steel), in addition to hydrogen compression, transport, storage and hydrogen-based power and heat generation.	[24, 46, 47, 121]

Hydrogen EAF	Using hydrogen in the direct reduction–EAF route (DR–EAF) means reducing iron ore with hydrogen whilst in a solid state, to produce DRI (sponge iron). This sponge iron is then fed into the EAF, where inserted electrodes are generating a current, which melts the sponge iron, and finally producing steel. However, some carbon is needed to produce steel, which is added in the form of pulverised coal or coal dust, biomethane or other biogenic carbon sources. The electrode energy consumption and added carbon equate to ±53 kg/CO_2 per tonne of steel, and where the electricity used is not from fully renewable energy sources, this carbon intensity adds to that figure. ArcelorMittal is developing a new innovation project in Hamburg (Germany), the first industrial-scale production and process integration of DRI with 100% hydrogen as the reductant. The annual production capacity is estimated at 100,000 tonnes of steel.	[122–124]
GrInHy	The European Commission's GrInHy project (Green Industrial Hydrogen via Reversible High-Temperature Electrolysis) was carried out between 2016 and 2019. The technical and economic opportunities of high-temperature electrolysis (HT electrolysis) and the higher electrical system efficiency compared to low-temperature electrolysis technologies led to establishing the following main objectives: • delivering proof of achieving electrical efficiency of at least 80% LHV • increasing the scale of the SOEC core to a DC power input of 120 kW electricity • producing a unit with a lifetime of ≥10,000 h with a degradation rate below 1%/1000 h; • operation for ≥7000 h whilst simultaneously adhering to the hydrogen quality standards of the steel industry. The electrical efficiency of 80% has been largely achieved by operating the HT electrolyser in close proximity to the thermal-neutral operation point. The installation consisted of an optimised multi-stack module design with a total capacity of 120 kW electricity.	[125, 126]
H2Future	Another project under the Horizon 2020 umbrella and managed by a cooperation of VERBUND, VOESTALPINE and SIEMENS, a 26-month project of a 6-MW electrolysis power plant, was installed at the VOESTALPINE LINZ plant in Austria. Replicability of the experimental results at larger scales in EU27 for the steel industry is currently being studied and involves the technical, economic and environmental assessment of the experimental results utilising the CertifHY tools.	[24]
HYBRIT	HYBRIT is a technology which is replacing CO_2 emissions with H_2O – water. Up to now, steelmaking involved coal to remove oxygen from iron ore, emitting copious amounts of CO_2 in the process. SSAB is set to revolutionise the industry with the innovative HYBRIT® technology, using hydrogen instead of coal in the ore reduction process and emitting H_2O instead of CO_2.	[127, 128]

(Continued)

Table 3.2 (Continued)

Project name	Description	References
Hydrogen direct reduction (HDR)	Similar to the HDR–EAF process, direct reduction of iron ore using hydrogen has been piloted by a range of entities and the mechanisms and processes involved were clearly defined: direct reduction of iron is the chemical removal of O_2 from solid-state iron ore. At present, the standard procedure to remove carbon from the iron used in the steelmaking process is chemically reduced from iron ore through the use of fossil fuel resources – natural gas or coke from coal, known as direct reduced ironmaking (DRI). The carbon in fossil fuels combines with the oxygen in the iron ore, initially without the addition of heat, producing metallic iron and a CO_2-rich process gas. Hydrogen, and more specifically green hydrogen, is seen globally as the clean energy source – evidenced by the fact that global industrial giants like Tata Steel, SSAB and ArcelorMittal are investing heavily in developing hydrogen plants with access to seawater, like in Ijmuiden (Tata Steel/Hermes).	[13, 76, 129–131]
Horizon 2020	Horizon 2020 is the EU's flagship research and development funding programme, which was active from 2014 until 2020. It had a sizeable budget of ~€80 billion. Horizon 2020 has been succeeded by Horizon EuropeEN. Some of the aforementioned projects have received funding from the Horizon 2020 finance provision as declared. Horizon EuropeEN is the follow-on project after conclusion of Horizon 2020.	[91, 96, 97, 104]
Horizon EuropeEN	Horizon EuropeEN is the follow-on project of Horizon 2020 and is a research and innovation funding programme, active until 2027, with a budget of €95.5 billion. The aims and objectives are tackling climate change, helping to achieve the UN's Sustainable Development Goals and boosting the EU's competitiveness and growth. Legal entities from within the EU and associated countries can participate.	[96–98]
HM Government 'The Ten Point Plan for a Green Industrial Revolution'	HM Government issued 'The Ten Point Plan for a Green Industrial Revolution' in 2020 covering the issue that most emissions are being produced through the burning of fossil fuels for electricity, heat and transportation. The six sectors identified for being the most polluting are as follows: • electricity. • transportation. • industry. • commercial/residential. • agriculture. • land use/forestry. The efforts and plans currently in place amount to a total of achievable CO_2 reduction of 40.76%, with the potential to increase these efforts, based on the road map for further policies, to achieve a net-zero carbon economy by 2050.	[132]

3.4 BF/BOF off-heat utilisation

Food production – in greenhouses – anaerobic digestion and sewage treatment plants require a stable ambient temperature throughout the year, which requires a conventional source of heating. Reusing heat from the steelmaking process, the expenditure of installing heating systems and using fuel and energy is simply not required, as the air has already been heated. Although the installation of a suitable infrastructure would have to be considered, the economical case has to be further investigated, and the financial viability – cost of ducts and pipework against the installation of a heating and cooling system including energy requirements – needs to be solid to make a case for reutilising flue stack off-heat. The flue stack heat can be as high as 1650 °C and would have to cool down, i.e. via travelling through adequately sized pipework, ducts and possibly turbines (which in their own right would possibly be able to generate electricity, via convection, baffle systems or plate heat exchangers). The energy saved on heating food production facilities is deserving of further investigation, as this will effectively contribute to achieving net-zero carbon emissions in steel production.

3.5 Renewable energy technologies

Renewable energy technologies are one of the key components within the Bio Steel Cycle, as CO_2 emissions in steelmaking can be reduced by more than 30% [26, 73, 133–137], if commercial entities in iron and steel production [26, 67] were to simply switch their energy provider [27, 69] to those which supply energy which was derived using 100% renewable energy technology and producing their own energy by retrofitting their plants with renewable energy technologies (wind and solar PV). The same applies to greenhouses, as there is a vast amount of roof space available, which has to date not been utilised. The static requirements would obviously have to be considered, but, as the cost of solar energy and solar PV has decreased significantly over recent years [138] to less than £3/m^2, it can be considered an unmissable opportunity.

To increase energy efficiency makes economic sense, too: it is possible to use fewer natural resources whilst simultaneously achieving the same economic output and – at the same time – reducing the environmental impact and – as an added bonus – saving money [139].

Energy productivity is assessed on the basis of final energy consumption rather than primary energy consumption. This enhances the validity of the indicator because losses in the energy supply system through energy conversion and transport do not appear in the balance. Final energy consumption includes electricity as well as heat; therefore, weather conditions and stockpiling of fuels influence the indicator.

The Bio Steel Cycle model and strategy could be seen as the engineering solution to the urgent issue of CO_2 emissions reduction requirements. The technology to avoid and reduce CO_2 emissions is available – all it takes now is the willingness of industry and governments to implement emissions reduction technologies and the provision of political frameworks and monetary incentives. One could argue that heavy industries have taken the natural environment for granted and have taken all the profits with total disregard for the consequences, for all of us, and the onus is therefore on them to remedy the wrongs they have committed. However, now is not the time for splitting hairs – now is the time for fast and effective action. CO_2 emissions and GHG emissions on the whole need to be drastically reduced, to ensure our planet will be a habitable place for generations to come.

The Bio Steel Cycle implementation can deliver this in the short term. Other industries could utilise this system, without much need for adaptation.

3.6 DAC woodlands

In the spirit of innovative, multidisciplinary approaches to solving contemporary CO_2 emissions issues, the positive effects of direct air capture (DAC) [140] and utilisation of woodlands for carbon capture cannot be emphasised enough. As one of the critical components of the BiSC, woodlands/trees for DAC would even be a profitable sideline for steel producers, as illustrated in Chapter 4, Section 4.5 DAC woodlands and plant matter: after five years, a 10ha woodland can generate more than £130,000 profit.

Trees and vegetation as natural carbon sinks [56, 64, 140] should ideally be planted around steel production plants to absorb the remaining CO_2 emissions via DAC. Whilst at the same time, the plant matter could feed the anaerobic digester (AD), biochar plants, or directly used at selected quality in ironmaking and steelmaking as readily available biomass. In this respect, bamboo beats deciduous native plants with its carbon sequestration capacities: on average, one hectare of bamboo stand absorbs ~17 tonnes of carbon/year [78]. Native deciduous and non-deciduous trees have a carbon sequestration capacity of on average 9 tonnes of CO_2/ha of tree plantation [56, 64, 140, 141]. Planting a sufficient number of trees should be considered in the planning for the updating of existing steel production plants and for any new development in order to meet the UK government's zero emissions targets. The UK tree cover stands at 13% (3.25 million hectares), which is the lowest in the Northern Hemisphere. In comparison, forests and wooded land cover over 182 million hectares in the EU, which is about 42% of the EU's total land area. This equates to 0.36 hectares of forest per capita in the EU in comparison [142]. Woodlands not only capture post-combustion CO_2 and create biomass for anaerobic digestion, but they also may create recreation and employment opportunities and additional sources of commercial activity. Additionally, they offer a low-cost opportunity for carbon offsetting, which could be seen as a commercial opportunity in itself.

3.7 Biogas, biomass and hydrogen

Coal is technically biomass or at least has biomass as its origin [143] millions of years ago; the earth was covered in wetlands, swamps and ancient plant species such as ferns. When the plants died, they sank to the bottom of the swamps. Over the years that followed, the layers of plants were covered by dirt, organic matter from animals and water, compacted by the weight bearing down. Over millions of years, heat and pressure converted the plants into coal, with a high carbon content (60% in lignite and 80% in anthracite coal; [144]). The coal seams we excavate today have taken millions of years to form. It is finite – we cannot just 'make' new coal, hence the term 'nonrenewable' [143].

Innovative technologies such as HIsarna [102] and GrynHy [52, 125, 126] and HDR have CO_2-saving potential in their own right, as explained in more detail as follows. Removing coal as primary energy source or using (green) hydrogen (essentially biomass) direct reduction, an immediate 30% CO_2 emissions reduction is possible [16, 18, 63, 73, 75, 77, 145, 146], and therefore, replacing coal with green hydrogen or other biomass would reduce the CO_2 emissions from steelmaking potentially by the same percentage. According to Siemens [24, 46, 47], a 50% carbon emissions reduction is immediately possible via utilisation of green HDR. The efforts for conversion of biomass from plants into renewable energy and reducing the resulting CO_2 emissions have seen extensive efforts across a variety of industries, using thermochemical and biochemical approaches [60, 73, 147]. Converting biomass thermochemically, nitrogen oxides (NO_x), particulate matter (PM) and tar are the environmentally hazardous by-products, which can fortunately be recycled using existing technology [60, 73, 147]. During bioconversion of biomass, namely anaerobic digestion as an example, volatile organic compounds (VOCs), such as sulphur-containing gases (hydrogen sulphide/H_2S), wastewater and biogas slurry [148–152], are being produced. These can be utilised as a value-added resource, and wastewater and VOCs can be treated via physical–chemical or biological methods in sewage treatment plants [141, 150]. Developing new industrial technologies or processes that use biomass and its by-products as a source of energy should be seen as the future direction of clean production [145]. Fortunately, there are multiple suitable varieties of plants, which could be exploited for the production of biomass. Mainly, the dry matter content is an important factor, as this determines the species' value in carbon-reliant combustion processes. The dry matter content of *Chusquea culeou* (Chile), as an example, can be determined at 156–162 tonnes/ha, whilst that of *Phyllostachys pubescence* (Japan) and *Gigantochloa alter* (Indonesia) levels can be determined at approximately 138 and 45 tonnes/ha [78]. The lowest level of dry matter was recorded for *Bashania fangiana* (China) at 0.35 tonnes/ha. Bamboo species in years 4, 6 and 8 have been reported to have 122, 225 and 286 tonnes/ha of dry matter. They are therefore to be seen as performing as well as a plantation of 10-year-old, fast-growing *Causarina equisetifolia* (292.68 tonnes/ha) or *Eucalyptus tereticornis* (254.97 tonnes/ha) [66, 75, 78, 141, 153].

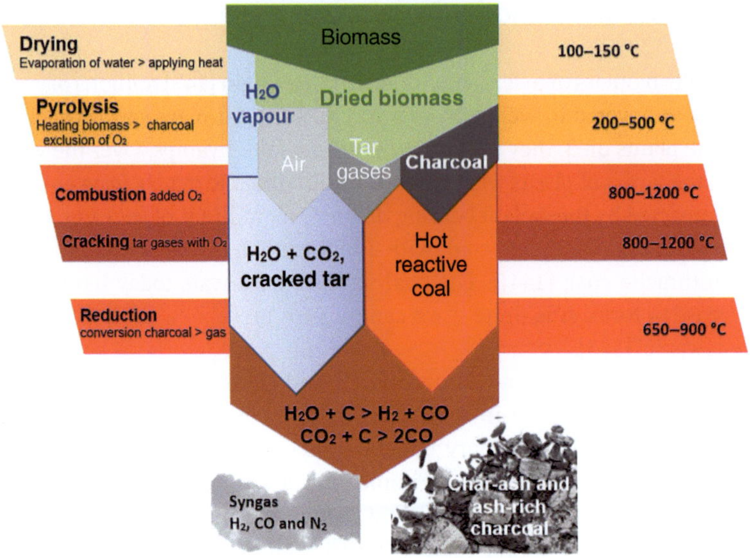

Figure 3.3 Biomass gasification process flowchart.

UK native, deciduous plants (trees) are able to sequester approximately 9 tonnes/ha/year [64], whereas bamboo species are able to store almost double this amount at 17 tonnes/ha/year [78]. The gasification of biomass with high dry matter content undergoes several steps, which, unfortunately, themselves are emitters of CO_2 and other GHGs, as displayed in Figure 3.3.

In order to determine the validity of current findings established during the literature review process, the gasification of biomass has been simulated using software packages from S8, Aspen+ and INOSIM V13. More to this is in Chapter 7: Results.

An additional type of biomass is biochar – it is biomass which has undergone extensive transformation.

Due to power plants being the largest stationary source of anthropogenic CO_2, developing effective and affordable post-combustion CO_2 capture technology has attracted substantial research attention. This study assessed the adsorption of CO_2 onto biochar, a low-cost adsorbent that can be produced from waste biomass through low-temperature pyrolysis. Sugarcane bagasse (BG) and hickory wood (HW) feedstock were converted into biochar at 300, 450 and 600 °C. The sorption of CO_2 on each of the resulting biochar was measured by monitoring its weight changes in CO_2 at 25 or 75 °C. The biochar was found to be effective for CO_2 capture, and the adsorption process could be described by the second-order kinetics model [66, 75, 78, 141, 153]. In general, the biochar produced at higher temperatures had better CO_2 capture performance. The BG biochar produced at 600 °C showed the most adsorption of CO_2 (73.55 mg/g at 25 °C). However, even when the feedstock was exposed to only 300 °C pyrolysis, the biochar was still able to capture more than 35 mg/g CO_2 at 25 °C [66, 75, 78, 141, 153]. Experimental results suggest that CO_2 weakly bound to the surface of the biochar through physisorption, so the surface

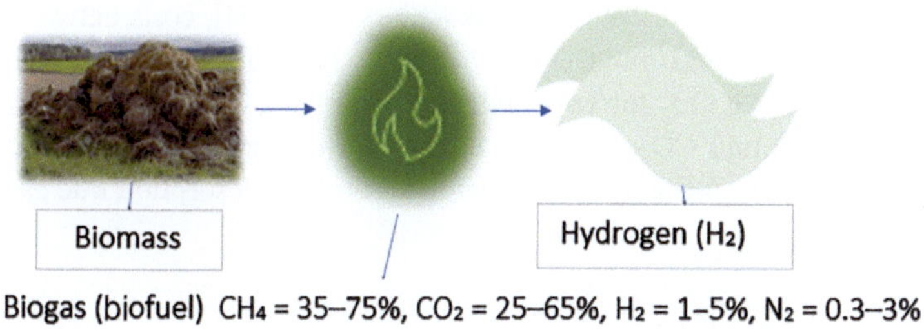

Figure 3.4 Pathway of biomass conversion via biogas to hydrogen.

area was a significant determinant of CO_2 adsorption (Pearson's $r = 0.82$); nevertheless, the presence of nitrogenous groups also played a role, when the surface area was sufficient. Biochar's porous structure and unique surface properties enable it to be an efficient CO_2 adsorbent whilst being sustainable and inexpensive. It, therefore, requires careful consideration of the implications and, specifically, the capture of the emitted GHGs, so as not to add to the CO_2 balance sheet [66, 75, 78, 141, 153].

An additional source of biomass to be utilised in steelmaking and other production as an alternative to imported energy is biogas. Biogas can be made in anaerobic digestion from waste products such as cow manure: 1 kg will result in ~40 l of biogas, and 1 kg of chicken dung will result in ~70 l. Biogas has an energy content equivalent to that of fossil fuel. The calorific value of 1 m^3 biogas is equivalent to 0.6–0.8 l of kerosene [151]. Figure 3.4 shows the transformation pathway of biomass to hydrogen.

The resources for making energy without producing a large amount of GHG emissions are basically being produced on a daily basis, namely manure and other organic matter. This requires recognition as an energy resource and to be treated as such – energy independence can be achieved in the short term, with the existing technology for conversion.

3.8 CEPS

Incorporating the Carbon Enrichment for Plant Stimulation (CEPS) [25] process into the BiSC-based circular steelmaking process involves CCUS by means of driving CO_2-enriched flue stack off-gas from combustion processes into CEPS units. Subsequently, this at almost 100% carbon-enriched air is then directed into greenhouses to stimulate plant growth. The chosen CEPS model is deemed scalable and has the capacity to provide 85 tonnes of CO_2 p.a. in its current configuration: concentrating CO_2 from ambient air (400 ppm) into an enriched product stream at 1000 ppm CO_2. Locating the source of CO_2 (i.e. flue gases from production in steel, cement or energy production) and greenhouses in close proximity of each other eliminates costs associated with filtering, deactivation, compression, transportation, handling, distribution and storage entirely. Every successfully installed CEPS [25] unit/greenhouse infrastructure is effectively a CO_2 sequestration station, and the economic feasibility is based on 1 kWh = 3600 kWs = 3.6 MJ [154], costs between £0.11 and £0.21/kWh and 17.8 kJ/mol CO_2 = 17.8 kJ/44 g CO_2 for pure CO_2 versus 8.5 kJ/mol CO_2 = 8.5 kJ/44 g CO_2 for enriched, 1000 ppm CO_2, costing effectively between \$15 and \$309/tonne CO_2 [55]. The efficiency, technical and economic viability are making the case for temperature swing absorption/desorption flue gas carbon capture.

3.9 Industrial filters: carbon capture and sequestration

Installation options of carbon capture flue stack filter technologies are wide ranging, and every production site has their own parameters and challenges to overcome. Thorough surveying of the sites and (GHG) emissions points needs to be identified, and depending on the situation, a suitable GHG capturing system can be installed. The utilisation of captured carbon in concrete and food production is the next step within a circular system, and there is a range of systems available. Blue Planet [155] is using the so-called Geomimetic® process, where recycled concrete and captured carbon are being reformed to make new concrete and aggregate. In combination with manufacturing, post-combustion flue stack carbon capture, such as provided by Giammarco Vetrocoke and Yurcent, makes the utilisation of the thus captured carbon and subsequent utilisation in making new concrete (carbon sequestration) easily implementable. These processes have the capability to reduce the carbon emissions from steel production by almost 100%.

3.10 Anaerobic digestion, sewage treatment and geothermal units

3.10.1 Anaerobic digestion

Projects containing components such as DAC [156], redirecting heat and utilising ADs [157–160] to produce biogas (methane and hydrogen) and producing energy on-site with turbines [149, 151, 161] can be directly linked to one of the incentives currently (2023/2024) offered by the UK government, such as the 'Farming Investment Fund', and include funding for agricultural businesses such as farmers, foresters, growers and agri-contractors with grants for investing in new technologies, equipment and infrastructure. The AD is touted in agribusiness circles as the 'unsung hero of waste management' [157–160]. These have the ability to convert almost any organic (non-contaminated) matter into biogas and substrate, at relatively low running and maintenance cost, in relation to the initial installation outlay. ADs are an essential part of the BiSC in a drive to generate biogas and potentially hydrogen at a later stage, to be recirculated into the steelmaking process.

3.10.2 Sewage treatment and steel slag reutilisation

The AD and sewage treatment facility are vital components within the BiSC and would be ideally integrated into the steel mill whilst quite possibly be independent businesses in their own right, conveniently located on-site of the steel production facility. These units would be able to accommodate debris from nearby woodland management and additional biomass from surrounding residential and commercial entities. Steelmaking by-products, such as brown water, can be treated at the sewage treatment facility [102, 105, 112–115, 117–119]; the resulting clean water is reused in steel production and the remaining sediment as additional aggregate in building and construction [119–135]. Although the process of steelmaking and all the stages involved preceding the actual conversion of liquid iron into steel use large quantities of water, only a fractional amount is actually consumed, as the majority share of brown water is either treated and reused or recycled and returned to the source. Steelmking involves water also being used in cooling operations, descaling and dust scrubbing at various stages, such as post-combustion off-gas situations, whereas seawater is widely used for once-through cooling after an antifouling pretreatment has taken place [138]. Approximately 61 Mt of steel slag is globally generated on an annual basis, with more than 50% (33 Mt/annum) produced by the United States and EU [162, 163], 27 Mt generated by China and Japan [164] and 1.2 Mt from Russia [165]. Although there are various applications for using steel waste and by-products in construction [162, 166], as beneficial soil additives [167–169] and in geo- and environmental engineering applications [165, 170, 171], an estimated 35% [119] of steel slags is generated in Europe. Given the varied after-uses [172, 173], interested stakeholders directed their focus towards any potential negative environmental impacts that may arise during weathering [174, 175] of exposed deposits, either reused in construction or final disposal in landfill sites [176–178]. As a result, a unique 36-year dataset has been used to carry out laboratory leaching tests to arrive at large datasets concerning aqueous and solid phases within steel slag leachates [179, 180]. The findings suggested that the long-term development of leachates under ambient environmental conditions at large disposal sites, in contained isolated and exposed samples, temporal patterns of Ca concentration, alkalinity and effects on measured pH were detected, and confirmed the longevity of pollution generated by legacy steelworks' disposal sites. These findings supported the conclusion that recycling of steel slag in building and construction was the environmentally less harmful solution to final steel slag disposal [119, 123].

3.10.3 Geothermal energy and ground-source heat pumps

Geothermal energy is an additional renewable energy source, working with heat pump units in tandem with water-filled boreholes, which is then naturally heated by the hot rock. The hot water produces vapour that can be used to drive turbines and create electricity. There are now some developed countries [138] which are discovering the potential of geothermal energy (GTE) as an innovative and sustainable energy source. During an investigation of the complex geological structures in Serbia and Turkey, it was found that due to the considerable number of thermo-mineral springs and geothermal wells, the geothermal heat flow density ranges from 80 to 120 mW/m^2 – which is above most European countries' average of 60 mW/m^2. The conclusions were that geothermal energy produced by enhanced geothermal systems (EGS) and hybridised geothermal systems (HGS) for domestic and commercial use, as well as power generation in a variety of power plant set-ups, has the potential to significantly reduce harmful GHGs [139]. Particularly, the geothermal situation in Turkey [181] provides an immense potential for the production of green hydrogen with the total hydrogen production potential of using geothermal resources estimated to be ~560 ktonnes.

One technology to extract geothermal energy is ground-source heat pumps (GSHP), which function in a subsurface situation and operate with extracting heat via a system of antifreeze fluid-filled pipes buried underground from a maximum depth of 200 metres. In comparison to the simplicity of solar or wind energy systems, the installation of geothermal or GSHP systems not only seems time-consuming but also costly and has a greater environmental impact during installation – as huge mounds of soil require to be disturbed. Additionally, once installed, the pipes are difficult to access, and potential antifreeze leakage issues cannot easily be detected or resolved.

3.11 CAT, CCS and CCUS

Carbon dioxide is being emitted by the respiratory function of most living creatures, for a reason: most plants require CO_2 to grow [33]. The natural carbon emissions cycle has been perfectly designed by nature, but humans have caused the production of CO_2 emissions to increase beyond the requirements of living things for this – initially – naturally occurring gas to life-threatening levels. One of the results is the climate crisis, as our planet is heating up to unprecedented temperature levels, which are incompatible with human habitation.

Over the last 20 years, climate change and its effects on the global population, the natural world and our way of using up finite resources at an accelerating rate have been investigated and researched by a range of countries and numerous specialists and in the respective fields. The researchers' aim was to establish a viable, economically and ecologically sound circular economic concept which could support the efforts of CCS to avert the worst effects of climate change. In 2015, the European Commission [182] issued a milestone guidance paper, an EU action plan for circular economic process, which is quite similar to the paper issued by the US Environmental Protection Agency [34], identifying the sources of and potential ways to reduce GHG emissions.

The opportunity for transition to minimising waste and developing a resilient, sustainable, low-carbon, resource-efficient and competitive economic model must be taken seriously across Europe. As a side effect, energy will be saved and also irreversible damages will be avoided, which are currently caused by using up resources at an unsustainable rate, as it exceeds the earth's capacity for renewal – as far as climate and biodiversity, air, soil and water pollution are concerned. One of the key drivers to achieve the set objectives is a major initiative called 'Horizon 2020' [91], the purpose of which is to fund innovative projects and targeted action in areas such as plastics, food waste, construction, (critical) raw materials, industrial and mining waste, consumption and public procurement. One of the results of the EU countries' and UN efforts was the Paris Agreement [183], which was signed in 2016 by the UN member states. It binds all signing countries to secure a sustainable global economic recovery and net-zero emissions future (not only zero CO_2 emissions – GHGs and other toxic emissions are also ring-fenced within this agreement). One of the main goals is to limit global temperature rise to no more than 1.5 °C, as achieving this would avert the worst ecological, financial and economic impacts of climate change. The UK government issued 'The Ten Point Plan for a Green Industrial Revolution' in 2020 [132], covering the issue that most emissions are being produced through the burning of fossil fuels for electricity, heat and transportation. The six sectors identified for being the most polluting are as follows: electricity, transportation, industry, commercial/residential buildings, agriculture and land use/forestry.

The efforts and plans currently in place amount to a total of achievable CO_2 reduction of only 40.76%, with the potential to increase these efforts, based on the road map for further policies, to achieve a net-zero carbon economy by 2050. The European Environment Agency Report [35] constitutes the official inventory submission of the EU for 2020 under the UNFCCC [36], as well as the Kyoto Protocol KP [37], covering the years between 1990 and 2018. The EU, party to the UNFCCC, issues these reports on GHG inventories annually, for emissions and removals within the area taking place within its territory. Due to the nature of this lag in reporting, the 2020 inventory report cannot yet reflect the effects which the COVID-19 pandemic might have had. Although the United Kingdom left the EU on 1 February 2020, EU law still applies until the end of the transition period. Moreover, key provisions of Regulation (EU) No 525/2013 ('Mechanism for Monitoring and Reporting GHG') and of Decision No 406/2009/EC ('Effort Sharing') apply to the United Kingdom also in respect of GHGs emitted during 2019 and 2020. The EU and its member states – including Iceland – agreed to a quantified emissions reduction commitment, which will limit their average annual GHG emissions to 80% of the sum of their base year emissions, which is reflected in the Doha Amendment [38].

GHG emissions decreased in the majority of sectors between 1990 and 2018, with the notable exception of transport, including international transport, refrigeration and air conditioning. Considering the aggregate level, the highest emissions reductions were achieved in manufacturing industries, construction, electricity/heat production, iron/steel production and domestic combustion.

Greenhouse gas and – specifically – aviation CO_2 emissions had been already thoroughly investigated [39] using data from the Global Carbon Project [40], whereas research seemingly following on from these findings established [34] how severe the impact of global aviation factually is in terms of CO_2 emissions produced, RF (radiative forcing) and ERF (effective radiative forcing) and its impact on climate change. The findings reflected the research into RF and ERF (and their effects on increasing the air temperature), with sums being calculated for the years 2000–2018. Allegedly, contrail cirrus (the condensate clouds one can observe as evidence of a plane's taken flight path), consisting of linear contrails and the cirrus clouds arising from them, results in the most significant warming ERF, closely followed by CO_2 and NO_x emissions. However, emissions of sulphate aerosol formation were reported as having a negative – meaning cooling – effect. In conclusion, the mean contrail cirrus ERF/RF ratio of 0.42 indicated that contrail cirrus is less effective in surface and atmospheric warming than other terms. It was established that by 2018, the net aviation ERF is +100.9 mW/m^2, meaning a 5–95% likelihood range (of 55, 145) with significant contributions from contrail cirrus (57.4 mW/m^2), CO_2 (34.3 mW/m^2) and also NO_x (17.5 mW/m^2) [34, 35].

The opportunities in connection with hydrogen production from solar energy were also researched [53, 184–189], as hydrogen is a credible alternative to using fossil fuels in industrial applications. Effectively, CO_2 emissions can be removed from hydrogen production, establishing that it is both economically and practically viable to produce hydrogen using solar energy – making hydrogen a renewable (green) energy source.

After a thorough analysis of the economic and energetic aspects of capturing CO_2 from ambient air [42], it was concluded that the total levelised cost of \$80.-/MWh natural gas advanced combined cycle is the lowest total levelised energy cost, whereas – with regard to the \$/tCO_2 avoided – hydropower is by far the most cost-efficient and bears the lowest risk for humans and the environment, at \$748/tonnes of CO_2 captured [37].

CIRAIG and EMAF [23, 43] carried out thorough critical literature reviews of concepts concerning circular economic processes, covering the following issues, presented in order of conceptual scale (from more to less encompassing):

- Sustainable development
- Ecological transition
- Green economy
- Functional economy
- Life cycle thinking
- Cradle-to-cradle thinking
- Shared value
- Industrial ecology
- Extended producer responsibility
- Eco-design

The study of these concepts has led the researcher to develop the Bio Steel Cycle [11].

The European Commission investigated 'green' hydrogen technologies [44], where exclusively renewable energy technologies were considered as an alternative to fossil fuels, which are vital components of the Bio Steel Cycle and the implementation thereof. The issue of what might be needed to deliver carbon-neutral heat using hydrogen and CCS was further investigated [45], which led to the discovery that it is entirely possible to produce hydrogen based on using renewable energy technologies exclusively.

As currently, around between 20 and 35% of global GHG emissions are produced by industries such as steel and cement [44]; it is therefore vital to develop promising technologies and concepts in these areas. Siemens was referred to as claiming that '... Hydrogen and fuels derived ... capable of reducing the CO_2 from fossil fuels ... down to zero' [39]. To this effect, a thorough literature review has been carried out to determine the various possibilities of steelmaking process improvement technologies and policy guidance, informed by peer-reviewed scientific material, as demonstrated in Table 3.3.

3.11.1 CO_2 avoidance with production process improvement

The conventional linear steel manufacturing process, including all processes such as coal, iron ore and limestone extraction, crushing, pelletising, sintering, smelting, casting, rolling and finish machining, has been investigated and resulted in establishing that the BF/BOF route alone produces ~4.6 tonnes of CO_2 emissions [11, 208], which are one of the undesired by-products of every metric tonne of steel produced. The results to date of research into the CO_2 emissions at the various stages of the steelmaking process and related subprocesses have been established, including the findings for the emissions burden generated during the BF/BOF operation. CO_2 emissions as a direct result of steel manufacture increased by 1.6% or 8.7 Gt CO_2 in 2020, although in the Net-Zero Emissions by 2050 Scenario, industry emissions are estimated to decrease by 2.3% annually at 6.9 Gt CO_2 by 2030 – despite expected industrial production growth [57]. Improved material and energy efficiency, the increased utilisation of renewable energy technologies, and the development and deployment of low-carbon process routes (including CCS and hydrogen) are all considered to have an emissions reduction effect of individually between 12 and 30% [208]. Fossil-fuel-derived energy generation and combustion of fossil fuels for industrial activity, construction and transport – including aviation – are the biggest sources of CO_2 emissions – worldwide [18, 27, 55, 209–212]. Steel production is in close third place, followed by cement production and chemical and glass industry.

One possible solution to combat CO_2 emissions in the steel industry is the Bio Steel Cycle (BiSC) [11] as a model, based on circular tech economic principles. Steel scrap is considered a resource and, similar to recycling waste and by-products, is an integral part of the circular production process. Off-gases (CO_2 and other GHG) are being captured and reutilised, and alongside implementation of steelmaking process improvements, furnace heat capture and utilisation, CAT, CCS and CCUS technologies and processes, and multidisciplinary external components are closing the circle [26, 27, 52, 67, 69, 135, 212, 213]. Technologies such as ReclaMet (waste resource recovery, post-combustion) [65, 103] and electrolysis projects, i.e. GrInHy and H2Future [125, 126] (direct water splitting: biomass > hydrogen, precombustion), all have an impact in the magnitude of between 12 and 25%, although further research is required to establish not only the most effective technology in terms of environmental impact but also which technology can be deployed the fastest and the most cost-efficient.

Table 3.3 Steelmaking process improvement technologies and policy guidance.

Title	Description	References
From Australian iron ore to green steel: the opportunity for technology-driven decarbonisation	Green steel for Australia – economic modelling of a green steel value chain	[190]
Technological roadmap towards optimal decarbonisation development of China's iron and steel industry	Policy guidance exploring the deep decarbonisation pathways	[19]
Blast furnace gas decarbonisation through calcium assisted steel-mill off-gas hydrogen production. Experimental and modelling approach	The calcium-assisted steel mill off-gas H_2 production process (CASOH) in integrated steelmaking plants	[191]
Policy and pricing barriers to steel industry decarbonisation: a UK case study	Policy guidance exploring the decarbonisation pathways in the UK iron and steel industry	[192]
Decarbonisation options of the iron and steelmaking industry based on a three-dimensional analysis	Decarbonisation options – China's iron and steel industry as a case study	[29]
Energy saving technologies and optimisation of energy use for decarbonised iron and steel industry	Suitable decarbonisation technologies are categorised	[5]
Decarbonisation of steel mill gases in an energy-neutral chemical looping process	Strategy for carbon capture from steel mill gases	[30]
Regional supply chains for decarbonising steel: Energy efficiency and green premium mitigation	Green hydrogen-based iron ore reduction and renewable electricity-based steelmaking	[193]
Decarbonising the iron and steel sector for a 2 °C target using inherent waste streams	Global steel demand and limitations of techno-economically feasible options for low-carbon steelmaking …	[194]
Impact of the carbon border adjustment mechanism: an economic and geopolitical assessment of the German-Chinese aluminium trade flows	Assessment on how incentivising non-EU trade partners to adopt measures comparable to the EU's and yielding revenue to reuse in accelerating the decarbonisation of steelmaking	[195]

Technology and material efficiency scenarios for net zero emissions in the UK steel sector	Assessment including steel plant retrofit and grid electricity decarbonisation	[196]
A highly granular model of China's coal production, transport and consumption system shows how its decarbonisation and energy security plans will affect coal …	The effect of decarbonisation on coal imports	[197]
Valorisation of waste materials for high-temperature thermal storage: a review	Decarbonisation of the energy system and decarbonisation of both the electricity and steelmaking industry	[198]
Enabling the transition to a fossil-free steel sector: the conditions for technology transfer for hydrogen-based steelmaking in Europe	Decarbonisation of steelmaking via hydrogen	[199]
Value in the use of lime in the BOF steelmaking process	Lime quality and its impact on metallurgical steel quality	[200]
The low-carbon steel industry-interactions between the hydrogen direct reduction of steel and the electricity system	A case study in Sweden; cost-optimal design of the steelmaking industry and electricity system with close to '0' CO_2	[167]
Editorial for the special issue: overview, state of the art, recent developments and future trends regarding Hydrogen route for a green steelmaking process	Steel production based on hydrogen is one of the key factors to improve the green footprint of the steel industry	[189]
Industry in a net-zero emissions world: new mitigation pathways, new supply chains, modelling needs and policy implications	Mitigation pathway investigation towards decarbonisation of steelmaking; analysis of supply chains and policy needs	[135]
Collection of possible decarbonisation barriers (Deliverable 1.5)	Identification of major barriers to the decarbonisation process of the EU iron and steel industry	[201]

(Continued)

Table 3.3 (Continued)

Title	Description	References
The prospects for 'green steel' making in a net-zero economy: a UK perspective	Investigation into making UK steel production more environmentally benign and advancing decarbonisation	[202]
Preparing for a just transition: meeting green skills needs for a sustainable steel industry	Identification of skills required for the steel industry in the transition to sustainability	[203]
Hydrogen development strategies: a global perspective.	Hydrogen is a candidate to fully decarbonise European steelmaking, global aviation and maritime transport	[204]
Which countries are prepared to green their coal-based steel industry with electricity? Reviewing climate and energy policy	Policy and guidance by country for 'Green' steelmaking	[205]
Decarbonising steel production using CO_2 Capture and Storage (CCS): results of focus group discussions in a Welsh steel-making community	Decarbonisation of steel production by switching to renewable sources – local focus group discussions	[28]
Decarbonising industry via BECCS: promising sectors, challenges, and techno-economic limits of negative emissions	Decarbonisation of BECCS in the iron and steel industry	[136]
Opportunities and challenges for decarbonising steel production by creating markets for 'green steel' products	Investigation into potential markets for green steel products	[206]
CO_2 abatement in the steel industry through carbon recycle and electrification by means of advanced polymer membranes	… For this, a conceptual process design and assessment were performed for a process that is a combination of carbon recycling and electrification of the steelmaking process ….	[207]

3.11.2 CO_2 and off-heat utilisation in food production

Food production – in greenhouses – anaerobic digestion and sewage treatment plants require a stable ambient temperature throughout the year, which requires a conventional source of heating. Reusing heat from the steelmaking process, the expenditure of installing heating systems and using fuel and energy is simply not required, as the air has already been heated. Although the installation of a suitable infrastructure would have to be considered, the economical case has to be further investigated, and the financial viability – cost of ducts and pipework against the installation of a heating and cooling system including energy requirements – needs to be solid to make a case for reutilising flue stack off-heat. The flue stack heat can be as high as 1650 °C and would have to cool down, i.e. via travelling through adequately sized pipework, ducts and possibly turbines (which in their own right would possibly be able to generate electricity, via convection, baffle systems or plate heat exchangers). The energy saved on heating food production facilities is deserving of further investigation, as this will effectively contribute to achieving net-zero carbon emissions in steel production. Additionally, incorporating the CEPS [25] process into the BiSC-based circular steelmaking process involves CCUS by means of driving CO_2-enriched flue stack off-gas from combustion processes into CEPS units. Subsequently, this at almost 100% carbon-enriched air is then directed into greenhouses to stimulate plant growth. The chosen CEPS model is deemed scalable and has the capacity to provide 85 tonnes of CO_2 p.a. in its current configuration: concentrating CO_2 from ambient air (400 ppm) into an enriched product stream at 1000 ppm CO_2. Locating the source of CO_2, i.e. flue gases from production in steel, cement or energy production) and greenhouses in close proximity of each other, eliminates costs associated with filtering, deactivation, compression, transportation, handling, distribution and storage entirely. Every successfully installed CEPS unit/greenhouse infrastructure is effectively a CO_2 sequestration station, and the economic feasibility is based on 1 kWh = 3600 kWs = 3.6 MJ [154], costs between £0.11 and £0.21/kWh and 17.8 kJ/mol CO_2 = 17.8 kJ/44 g CO_2 [214] for pure CO_2 versus 8.5 kJ/mol CO_2 = 8.5 kJ/44 g CO_2 for enriched, 1000 ppm CO_2, costing effectively between \$15 and \$309/tonne CO_2 [159]. The efficiency, technical and economic viability are making the case for temperature swing absorption/desorption flue gas carbon capture [25].

3.11.3 CO_2 utilisation in the building industry

A further key component of the BiSC is the Geomimetic® process [155], as these units are effectively recycling facilities for the recycling of reclaimed concrete and the reutilisation of CO_2, filters, dust, sludge and slack from (steel) production. These units have the capacity to reduce post-combustion CO_2 emissions to effectively zero and should be on-site of any (steel) production plant. The workings of the Geomimetic® process [155] are in its essence carbon utilisation and sequestration processes at the same time, as these recycle CO_2 from flue gases and recycled concrete into synthetic limestone and aggregate in cement production, with the potential of absorbing 100% of the CO_2 emissions produced. This is a technique suitable to be applied in any industrial production setting: energy, steel, concrete, chemical industry, glass industry, paper and transport, to name a few.

3.11.4 DAC and agricultural waste for anaerobic digestion

In the spirit of innovative, multidisciplinary approaches to solving contemporary CO_2 emissions issues, the positive effects of DAC [140] and utilisation of woodlands for carbon capture cannot be emphasised enough. As one of the critical components of the BiSC, woodlands/trees for DAC would even be a profitable sideline for steel producers. Trees and vegetation as natural carbon sinks [56, 64] should ideally be planted around steel production plants to absorb the remaining CO_2 emissions via DAC. Whilst at the same time, the plant matter could feed the AD, biochar plants, or directly used at selected quality in ironmaking and steelmaking as readily available biomass. In this respect, bamboo beats deciduous native plants with its carbon sequestration capacities: on average, one hectare of bamboo stand absorbs ~17 tonnes of carbon/year. [78]. Native deciduous and non-deciduous trees have a carbon sequestration capacity of on average 9 tonnes of CO_2/ha of tree plantation [56, 64, 140, 141]. Planting a sufficient number of trees should be considered in the planning for the updating of existing steel production plants and for any new development in order to meet the UK government's zero emissions targets. The UK tree cover stands at 13% (3.25 million hectares), which is the lowest in the Northern Hemisphere. In comparison, forests and wooded land cover over 182 million hectares in the EU, which is about 42% of the EU's total land area. This equates to 0.36 hectares of forest per capita in the EU in comparison [142]. Woodlands not only capture post-combustion CO_2 and create biomass for anaerobic digestion, but they also may create recreation and employment opportunities and additional sources of commercial activity. Additionally, they offer a low-cost opportunity for carbon offsetting, which could be seen as a commercial opportunity in itself.

3.11.5 Anaerobic digestion, sewage treatment and hydrogen from biogas

The AD and sewage treatment facility are vital components within the BiSC and would be ideally integrated into the steel mill or quite possibly be independent businesses in their own right, conveniently located on-site of the steel production facility. These units would be able to accommodate debris from nearby woodland management and additional biomass from surrounding residential and commercial entities. Steelmaking by-products, such as brown water, can be treated at the sewage treatment facility. The cleared sewage can subsequently be utilised to fertilise the food production units [215]. The anaerobic digestion process in itself produces biogas, which can be used in steel production, but it also provides the base for the extraction of hydrogen. The green hydrogen produced at or nearby the anaerobic digestion facility can then be used in (steel) production within the HDR process. As this has been derived from biogas as a result of anaerobic digestions, this can therefore be considered green hydrogen. Hydrogen direct reduction has been piloted over recent years and has shown to have great CO_2 avoidance potential, and green hydrogen technologies are currently developed by a number of significant industry leaders, such as Mannesmann Salzgitter, in cooperation with the European Commission and Tata Steel [125, 126]. Green HDR (or iron ore) in blast furnace and electric arc furnace application is considered as having a significant impact on reducing CO_2 emissions in steel manufacturing, as this process uses 3.48 MWh of electricity per tonne of steel product and emits only 2.8% of blast furnace CO_2. However, as the prices of fossil-fuel-derived energy have increased significantly, it is imperative to replace fossil-fuel-derived energy with renewable energy technologies and biomass [14, 85, 157].

3.11.6 Renewable energy technologies' utilisation

Renewable energy technologies are one of the key components within the Bio Steel Cycle, as CO_2 emissions in steelmaking can be reduced by more than 30% [26, 73, 133–137], if commercial entities in iron and steel production [26, 67] were to simply switch their energy provider [27, 69] to those which supply energy which was derived using 100% renewable energy technology and producing their own energy by retrofitting their plants with renewable energy technologies (wind and solar PV). The same applies to greenhouses, as there is a vast amount of roof space available, which has to date not been utilised. The static requirements would obviously have to be considered, but, as the cost of solar energy and solar PV has decreased significantly over recent years [138] to less than £3/m^2, it can be considered an unmissable opportunity.

To increase energy efficiency makes economic sense too: it is possible to use fewer natural resources whilst simultaneously achieving the same economic output and – at the same time – reducing the environmental impact and – as an added bonus – saving money [139].

Energy productivity is assessed on the basis of final energy consumption rather than primary energy consumption. This enhances the validity of the indicator because losses in the energy supply system through energy conversion and transport do not appear in the balance. Final energy consumption includes electricity as well as heat; therefore, weather conditions and stockpiling of fuels influence the indicator.

The Bio Steel Cycle model and strategy could be seen as the engineering solution to the urgent issue of CO_2 emissions reduction requirements. The technology to avoid and reduce CO_2 emissions is available – all it takes now is the willingness of industry and governments to implement emissions reduction technologies and the provision of political frameworks and monetary incentives. One could argue that heavy industries have taken the natural environment for granted and have taken all the profits with total disregard for the consequences, for all of us, and the onus is therefore on them to remedy the wrongs they have committed. However, now is not the time for splitting hairs – now is the time for fast and effective action. CO_2 emissions and GHG emissions on the whole need to be drastically reduced, to ensure our planet will be a habitable place for generations to come.

The Bio Steel Cycle implementation can deliver this in the short term. Other industries could utilise this system, without much need for costly adaptation.

References

1 Griffin, P. and Hammond, G. (2019). Industrial energy use and carbon emissions reduction in the iron and steel sector: a UK perspective. *Journal of Applied Energy* 249: 109–125.

2 Hasanbeigi, A. and Springer, C. (2019). How clean is the U.S. Steel Industry? An international benchmarking of energy and CO_2 intensities. *Global Efficiency Intelligence* 1–42.

3 Baena-Moreno, F., Cid-Castello, N., Arellano-García, H., and Reina, T. (2021). Towards emissions free steel manufacturing – Exploring the advantages of a CO_2 methanation unit to minimize CO_2 emissions. *Science of The Total Environment* 781 (7): 146776.

4 Ren, L., Zhou, S., Peng, T., and Ou, X. (2021). A review of CO_2 emissions reduction technologies and low-carbon development in the iron and steel industry focusing on China. *Renewable and Sustainable Energy Reviews* 143: 110846.

5 Wang, R. (2022). Energy saving technologies and optimisation of energy use for decarbonised iron and steel industry [Internet] [Doctoral]. Durham University. http://etheses.dur.ac.uk/14289/ (accessed 7 July 2023).

6 Zhang, S., Yi, B., Guo, F., and Zhu, P. (2022). Exploring selected pathways to low and zero CO_2 emissions in China's iron and steel industry and their impacts on resources and energy. *Journal of Cleaner Production* 340: 130813.

7 International Energy Agency (IEA) (2023). Net zero roadmap: a global pathway to keep the 1.5 °C goal in reach – analysis. https://www.iea.org/reports/net-zero-roadmap-a-global-pathway-to-keep-the-15-0c-goal-in-reach (accessed 18 October 2023).

8 Zhang, H., Sun, W., Li, W., and Wang, Y. (2022). Physical and chemical characterization of fugitive particulate matter emissions of the iron and steel industry. *Atmospheric Pollution Research* 13 (1): 101272.

9 Wang, C., Walsh, S.D.C., Haynes, M.W. et al. (2022). From Australian iron ore to green steel: the opportunity for technology-driven decarbonisation. Geoscience Australia [Internet]. https://www.researchgate.net/profile/Israel-Lutalo/publication/362546377_From_Australian_iron_ore_to_green_steel_the_opportunity_for_technology-driven_decarbonisation/links/62f0e52e0b37cc34477d9141/From-Australian-iron-ore-to-green-steel-the-opportunity-for-technology-driven-decarbonisation.pdf (accessed 15 May 2023).

10 Swalec, C. (2021) Clear on climate: The iron and steel industry is responsible for 11% of global carbon dioxide (CO_2) emissions and will need to change rapidly to align with the world's climate goals. Carbon Brief 2021 [Internet].

https://www.carbonbrief.org/guest-post-these-553-steel-plants-are-responsible-for-9-of-global-co2-emissions/ (accessed 9 March 2022).
11 Kiessling, S., Darabkhani, H.G. and Soliman, A.H. (2022). Energies | Free Full-Text | The bio steel cycle: 7 steps to net-zero CO_2 emissions steel production [Internet]. https://www.mdpi.com/1996-1073/15/23/8880 (accessed 11 July 2023).
12 IPCC Intergovernmental Panel for Climate Change (2023). Summary for policymakers. Climate Change 2021 – The Physical Science Basis: Working Group I Contribution to the Sixth Assessment Report of the Intergovernmental Panel on Climate Change [Internet]. Cambridge University Press, p. 3–32. https://www.cambridge.org/core/books/climate-change-2021-the-physical-science-basis/summary-for-policymakers/8E7A4E3AE6C364220F3B76A189CC4D4C (accessed 12 July 2023).
13 World Steel Association (WSA) (2022). World Steel in Figures 2022 A healthy economy needs a healthy steel industry [Internet]. https://worldsteel.org/wp-content/uploads/World-Steel-in-Figures-2022-infographic.pdf?x65430 (accessed 7 July 2023).
14 The Rodney and Otamatea Times (1912). Coal consumption affecting climate. Rodney and Otamatea Times, Waitemata and Kaipara Gazette, 14 August 1912; p. 7.
15 International Energy Agency (IEA) (2023). coal_documentation.pdf [Internet]. http://wds.iea.org/wds/pdf/coal_documentation.pdf (accessed 12 July 2023).
16 Worrell, E., Blinde, P., Neelis, M. et al. (2010). Energy efficiency improvement and cost saving opportunities for the U.S. Iron and Steel Industry An energy star(R) guide for energy and plant managers [Internet]. Report No.: LBNL-4779E, 1026806. http://www.osti.gov/servlets/purl/1026806/ (accessed 2 November 2022).
17 Li, Y. and Zhu, L. (2014). Cost of energy saving and CO_2 emissions reduction in China's iron and steel sector. *Applied Energy* 130: 603–616.
18 Toktarova, A., Karlsson, I., Rootzén, J. et al. (2020). Pathways for low-carbon transition of the steel industry – a Swedish case study. *Energies* 13 (15): 3840.
19 Liu, X., Peng, R., Bai, C. et al. (2022). Technological roadmap towards optimal decarbonization development of China's iron and steel industry. *Science of the Total Environment* 850: 157701.
20 Suer, J., Ahrenhold, F., and Traverso, M. (2022). Carbon footprint and energy transformation analysis of steel produced via a direct reduction plant with an integrated electric melting unit. *Journal of Sustainable Metallurgy* 8 (4): 1532–1545.
21 Tian, B., Zhu, R., Zhou, Y. et al. (2022). A static balance model and analysis of 430 stainless steel produced by basic oxygen furnace–argon oxygen decarburization furnace process – Tian – 2022 – steel Research International – Wiley Online Library. Steel Research International [Internet]. 93(7). https://onlinelibrary.wiley.com/doi/abs/10.1002/srin.202100852 (accessed 7 July 2023).
22 Wang, R., Liang, L., Wang, Y., Roskilly, A.P. (2019). Process simulation of blast furnace operation with biomass syngas injection for clean production | Energy Proceedings. https://www.energy-proceedings.org/process-simulation-

of-blast-furnace-operation-with-biomass-syngas-injection-for-clean-production/ (accessed 12 July 2023).
23 EMAF Ellen MacArthur Foundation (2019). Completing the picture – how the circular economy tackles climate change [Internet]. https://www.ellenmacarthurfoundation.org/assets/downloads/Completing_The_Picture_How_The_Circular_Economy-_Tackles_Climate_Change_V3_26_September.pdf (accessed 9 October 2022).
24 Voestalpine (2023). Voestalpine, Siemens and VERBUND are building a pilot facility for green hydrogen at the Linz location. voestalpine, Siemens and VERBUND are building a pilot facility for green hydrogen at the Linz location – voestalpine. https://www.voestalpine.com/group/en/media/press-releases/2017-02-07-voestalpine-siemens-and-verbund-are-building-a-pilot-facility-for-green-hydrogen-at-the-linz-location/ (accessed 10 July 2023).
25 Bao, J., Lu, W.H., Zhao, J., and Bi, X.T. (2018). Greenhouses for CO_2 sequestration from atmosphere. *Carbon Resources Conversion* 1 (2): 183–190.
26 Bataille, C., Åhman, M., Neuhoff, K. et al. (2018). A review of technology and policy deep decarbonization pathway options for making energy-intensive industry production consistent with the Paris Agreement. *Journal of Cleaner Production* 187: 960–973.
27 Mandova, H., Patrizio, P., Leduc, S. et al. (2019). Achieving carbon-neutral iron and steelmaking in Europe through the deployment of bioenergy with carbon capture and storage. *Journal of Cleaner Production* 218: 118–129.
28 Williams, R., Jack, C., Gamboa, D., and Shackley, S. (2021). Decarbonising steel production using CO_2 Capture and Storage (CCS): results of focus group discussions in a Welsh steel-making community. *International Journal of Greenhouse Gas Control* 104: 103218.
29 Lu, X., Tian, W., Li, H. et al. (2022). Decarbonization options of the iron and steelmaking industry based on a three-dimensional analysis [Internet]. http://ijmmm.ustb.edu.cn/article/doi/10.1007/s12613-022-2475-7 (accessed 7 July 2023).
30 Singh, V., Buelens, L.C., Poelman, H. et al. (2022). Decarbonisation of steel mill gases in an energy-neutral chemical looping process. *Energy Conversion and Management* 254: 115248.
31 Franciosi, C., Voisin, A., Miranda, S., and Iung, B. (2020). Integration of I4.0 technologies with maintenance processes: what are the effects on sustainable manufacturing? *IFAC-PapersOnLine* 53 (3): 1–6.
32 Jahani, N., Sepehri, A., Vandchali, H.R., and Tirkolaee, E.B. (2021). Application of industry 4.0 in the procurement processes of supply chains: a systematic literature review. *Sustainability* 13 (14): 7520.
33 National Aeronautics and Space Administration (NASA) (2023). Global climate change – vital signs of the planet [Internet]. https://climate.nasa.gov/news/2436/co2-is-making-earth-greenerfor-now/#:~:text=Studies%20have%20shown%20that%20increased,chief%20culprit%20of%20climate%20change (accessed 7 April 2023).

34 United States Environmental Protection Agency (USEPA) (2023). Greenhouse Gas (GHG) emissions and removals [Internet]. https://www.epa.gov/ghgemissions (accessed 2 January 2023).

35 European Environment Agency (EEA) (2020). The European environment – state and outlook 2020: knowledge for transition to a sustainable Europe – European Environment Agency [Internet]. https://www.eea.europa.eu/soer/2020 (accessed 11 July 2023).

36 United Nations Framework Convention on Climate Change (2015). About the secretariat | UNFCCC [Internet]. https://unfccc.int/about-us/about-the-secretariat (accessed 11 July 2023).

37 UNFCCC (2005). What is the Kyoto Protocol? | UNFCCC [Internet]. https://unfccc.int/kyoto_protocol (accessed 11 July 2023).

38 UNFCCC (2012). The Doha Amendment | UNFCCC [Internet]. https://unfccc.int/process/the-kyoto-protocol/the-doha-amendment?gclid=EAIaIQobChMI8sDOgKiGgAMVtYZoCR3uuwx4EAAYASAAEgIhzPD_BwE (accessed 11 July 2023).

39 Sausen, R. and Schumann, U. (2000). Estimates of the climate response to aircraft CO_2 and NO_x-emission scenarios. *Climatic Change* 44: 27–58.

40 Lee DS, Fahey DW, Skowron A, Allen MR, Burkhardt U, Chen Q, et al. The contribution of global aviation to anthropogenic climate forcing for 2000 to 2018. *Atmospheric Environment* (1994). 2021 Jan 1;244:117834.

41 Ture, E. (2007). Hydrogen production from solar energy. In: *Assessment of Hydrogen Energy for Sustainable Development* (ed. J.W. Sheffield and Ç. Sheffield), 135–146. Dordrecht: Springer Netherlands (NATO Science for Peace and Security Series C: Environmental Security).

42 House, K.Z., Baclig, A.C., Ranjan, M. et al. (2011). Economic and energetic analysis of capturing CO_2 from ambient air. *Proceedings of the National Academy of Sciences of the United States of America* 108 (51): 20428–20433.

43 International Reference Centre for the Life Cycle of Products, Processes and Services (CIRAIG) (2015). *Circular Economy: A Critical Literature Review of Concepts*. Canada: Polytechnique Montréal.

44 O'Callaghan, J. (2018). Can we produce enough green hydrogen to save the world? | Research and Innovation [Internet]. https://ec.europa.eu/research-and-innovation/en/horizon-magazine/can-we-produce-enough-green-hydrogen-save-world (accessed 11 July 2023).

45 Sunny, N., Mac Dowell, N., and Shah, N. (2020). What is needed to deliver carbon-neutral heat using hydrogen and CCS? *Energy & Environmental Science* 13 (11): 4204–4224.

46 SIEMENS (2022). siemens-energy.com Global Website. Hydrogen solutions. https://www.siemens-energy.com/global/en/offerings/renewable-energy/hydrogen-solutions.html (accessed 7 July 2023).

47 SIEMENS (2023). siemens-energy.com Global Website. Hydrogen for energy. https://www.siemens-energy.com/global/en/priorities/future-technologies/hydrogen.html (accessed 7 July 2023).

48 SIEMENS (2023). siemens.com Global Website. Siemens & Halske AG, 1897–1966. https://www.siemens.com/global/en/company/about/history/people/managing-board-chairmen-up-to-1966/siemens-and-halske.html (accessed 8 August 2023).
49 Siemens, F. (1885). Scientific American. New Method of Heating the Regenerative Gas Furnace. https://www.scientificamerican.com/article/new-method-of-heating-the-regenerative-gas-furnace/ (accessed 8 August 2023).
50 Brophy, J.M. (2015). Review of Carl von Siemens, 1829–1906: Ein Leben zwischen Familie und Weltfirma. *The Journal of Modern History* 87 (1): 155–157.
51 British Steel (2023). British Steel. How we make steel. https://britishsteel.co.uk/what-we-do/how-we-make-steel/ (accessed 6 July 2023).
52 Santos, M.P.S. and Hanak, D.P. (2022). Carbon capture for decarbonisation of energy-intensive industries: a comparative review of techno-economic feasibility of solid looping cycles. *Frontiers of Chemical Science and Engineering* 16 (9): 1291–1317.
53 US Department of Energy Efficiency & Renewable Energy (EERE) (2021). Hydrogen production: natural gas reforming [Internet]. https://www.energy.gov/eere/fuelcells/hydrogen-production-natural-gas-reforming (accessed 6 July 2023).
54 IEA (2023). World Energy Outlook 2023 – Analysis [Internet]. https://www.iea.org/reports/world-energy-outlook-2023 (accessed 25 January 2024).
55 US Energy Information Administration (EIA) (2021). U.S. Energy Information Administration – EIA – Independent Statistics and Analysis – International Energy Outlook 2021 [Internet]. https://www.eia.gov/outlooks/aeo/data/browser/#/?id=1-IEO2021&sourcekey=0 (accessed 11 July 2023).
56 FC Forestry Commission (2003). *Forests, Carbon and Climate Change: the UK Contribution*. Edinburgh: Forestry Commission.
57 International Energy Agency (IEA) (2021). Data and statistics [Internet]. https://www.iea.org/data-and-statistics/data-browser?country=WORLD&fuel=Energy%20transition%20indicators&indicator=ETISharesInPowerGen (accessed 8 November 2022).
58 The Institute for Industrial Productivity (IETD) (2021). Basic oxygen furnace [Internet]. http://www.iipinetwork.org/wp-content/Ietd/content/basic-oxygen-furnace.html (accessed 8 November 2022).
59 The Institute for Industrial Productivity (IETD) (2021). Electric arc furnace [Internet]. http://www.iipinetwork.org/wp-content/Ietd/content/electric-arc-furnace.html (accessed 8 November 2022).
60 Keung, K. (2021). TAKACHAR Email from Kevin Keung, Research and Development lead.
61 WSA (2021). December 2021 crude steel production and 2021 global crude steel production totals. worldsteel.org. https://worldsteel.org/media-centre/press-releases/2022/december-2021-crude-steel-production-and-2021-global-totals/ (accessed 5 February 2024).
62 Zhou, J., Obrist, D., Dastoor, A. et al. (2021). Vegetation uptake of mercury and impacts on global cycling. *Nature Reviews Earth and Environment* 2: 269–284.

63 Liu, Z., Ciais, P., Deng, Z. et al. (2020). Near-real-time monitoring of global CO_2 emissions reveals the effects of the COVID-19 pandemic. *Nature Communications* 11 (1): 5172.

64 Nix, J. (2020). *John Nix Pocketbook – For Farm Management*, 50e. Leicestershire: The Anderson Centre.

65 Tata Steel Europe (2020). Tata Steel in Europe Sustainability Report 2019/2020 [Internet]. https://www.tatasteeleurope.com/sites/default/files/TSE%20Sustainability%20report%202019-20%20(EN).pdf (accessed 7 July 2023).

66 European Commission (COM) (2019). BIONICO: A pilot plant for turning biomass directly into hydrogen | BIONICO Project | Results in brief | H2020 | CORDIS | European Commission [Internet]. https://cordis.europa.eu/article/id/394984-bionico-a-pilot-plant-for-turning-biomass-directly-into-hydrogen (accessed 6 July 2023).

67 Waisman, H., Bataille, C., Winkler, H. et al. (2019 Mar). A pathway design framework for national low greenhouse gas emission development strategies. *Nature Climate Change* 1 (9): 261.

68 Khakimov, H.T., Shayumova, Z.M., Kurbanbaeva, Z.K., and Khusanov, B.M. (2019). Development of optimal modes and mathematical models of energy performance of electric steelmaking production. Voropai N, Senderov S, Michalevich A, Guliev H, editors. *E3S Web Conferences* 139: 01076.

69 National Renewable Energy Laboratory (NREL) (2019). 2019 year in review – NREL's top 20 stories [Internet]. https://www.nrel.gov/news/program/2019/2019-year-in-review-nrel-top-20-stories.html (accessed 11 July 2023).

70 Office for National Statistics (ONS) (2019). Provisional UK greenhouse gas emissions national statistics 2019. GOV.UK. https://www.gov.uk/government/statistics/provisional-uk-greenhouse-gas-emissions-national-statistics-2019 (accessed 11 July 2023).

71 TATA Steel UK (2019). Sustainability [Internet]. Theory and Methods for Engaged Inquiry. 2nd ed. https://www.tatasteeleurope.com/en/sustainability/sustainability%E2%80%93overview (accessed 8 November 2022).

72 World Steel Association (WSA) (2019). Steel Statistical Yearbook 2019 [Internet]. https://worldsteel.org/wp-content/uploads/Steel-Statistical-Yearbook-2018.pdf (accessed 11 July 2023).

73 Chen, Q., Gu, Y., Tang, Z. et al. (2018). Assessment of low-carbon iron and steel production with CO_2 recycling and utilization technologies: a case study in China. *Applied Energy* 220: 192–207.

74 Kuramochi, T., Höhne, N., Schaeffer, M. et al. (2018). Ten key short-term sectoral benchmarks to limit warming to 1.5 °C. *Climate Policy* 18 (3): 287–305.

75 Suopajärvi, H., Umeki, K., Mousa, E. et al. (2018). Use of biomass in integrated steelmaking – status quo, future needs and comparison to other low-CO_2 steel production technologies. *Applied Energy* 213 (C): 384–407.

76 Vogl, V., Åhman, M., and Nilsson, L.J. (2018). Assessment of hydrogen direct reduction for fossil-free steelmaking. *Journal of Cleaner Production* 203: 736–745.

77 Kittipongvises, S. (2017). Assessment of Environmental Impacts of Limestone Quarrying Operations in Thailand. *Environmental and Climate Technologies* 20 (1): 67–83.

78 Seethalakshmi, K.K., Jijeesh, C.M., and Balagopalan, M. (2016). Bamboo plantations: an approach to Carbon sequestration [Internet]. Kerala/Trichur: Kerala Forest Research Institute. https://www.researchgate.net/publication/215475397_Bamboo_plantations_An_approach_to_Carbon_sequestration (accessed 5 March 2020).

79 Chen, W., Shi, L., and Qian, Y. (2010). Substance flow analysis of aluminium in mainland China for 2001, 2004 and 2007: exploring its initial sources, eventual sinks and the pathways linking them. *Resources, Conservation and Recycling* 54 (9): 557–570.

80 Babich, A. and Senk, D. (2015). Recent developments in blast furnace iron-making technology. In: *Iron Ore [Internet]*, 505–547. Elsevier https://linkinghub.elsevier.com/retrieve/pii/B9781782421566000174 (accessed 2 November 2022).

81 Calvo, L.M. and Domingo, R. (2015). Influence of process operating parameters on CO_2 emissions in continuous industrial plants. *Journal of Cleaner Production* 96: 253–262.

82 Gupta, R.C. (2013). Treatise on process metallurgy Chapter 4.2 – Energy resources, its role and use in metallurgical industries. In: *Treatise on Process Metallurgy, 3: Industrial Processes* (ed. S. Seetharaman), 1425–1458. Elsevier.

83 Yang, Y., Raipala, K., and Holappa, L. (2014). Chapter 1.1 – Ironmaking. In: *Treatise on Process Metallurgy [Internet]* (ed. S. Seetharaman), 2–88. Boston: Elsevier https://www.sciencedirect.com/science/article/pii/B9780080969886000171 (accessed 7 July 2023).

84 Copeland, C. (2013). A Practical Engineering Approach to Improving the Reliability of Blast Furnace Tuyeres. https://www.semanticscholar.org/paper/A-Practical-Engineering-Approach-to-Improving-the-Copeland/310bee8b0bec31a17dcf88d158469e46ccbef783 (accessed 11 July 2023).

85 Satyendra, K.S. (2013). Metallurgical coke – IspatGuru [Internet]. https://www.ispatguru.com/metallurgical-coke/ (accessed 7 July 2023).

86 European Lime Association (EULA) (2012). How much lime per tonne of steel? Summary of the technical report. A competitive and efficient lime industry – cornerstone for a sustainable Europe [Internet]. https://www.eula.eu/wp-content/uploads/2019/02/A-Competitive-and-Efficient-Lime-Industry-Summary_0.pdf.

87 Ghanbari, H., Helle, M., and Saxén, H. (2012). Process integration of steelmaking and methanol production for suppressing CO_2 emissions – a study of different auxiliary fuels. *Chemical Engineering and Processing Process Intensification* 61: 58–68.

88 Morrow, W.R., Hasanbeigi, A., Sathaye, J., and Xu, T. (2014). Assessment of energy efficiency improvement and CO_2 emission reduction potentials in India's cement and iron & steel industries. *Journal of Cleaner Production* 65: 131–141.

89 Prakash, R. and Muller, M. (2007) Industrial oxygen: its use and generation [Internet]. https://www.eceee.org/library/conference_proceedings/ACEEE_industry/2007/Panel_6/P6_13/ (accessed 7 July 2023).

90 European Commission Horizon 2020 (2019). BIONICO: a pilot plant for turning biomass directly into hydrogen [Internet]. European Commission. https://cordis.europa.eu/article/id/394984-bionico-a-pilot-plant-for-turning-biomass-directly-into-hydrogen (accessed 8 November 2022).

91 COM (European Commission) (2023). Horizon 2020 – Work Programme 2014–2015. G. Technology readiness levels (TRL). https://ec.europa.eu/research/participants/data/ref/h2020/wp/2014_2015/annexes/h2020-wp1415-annex-g-trl_en.pdf (accessed 6 July 2023).

92 Birat, J.P. (2010). ULCOS program: status & progress. In Brussels: European Commission. p. 1–30. https://www.eesc.europa.eu/sites/default/files/resources/docs/estep_ulcos_nov_2010.pdf (accessed 4 April 2020).

93 COM (European Commission). CORDIS | European Commission. 2010 [cited 2023 Jul 10]. Ultra-Low CO2 steelmaking | ULCOS Project | Fact Sheet | FP6. https://cordis.europa.eu/project/id/515960

94 Directorate-General for Research and Innovation (European Commission), Feiterna, A., Zagaria, A.M. et al. (2014). ULCOS top gas recycling blast furnace process (ULCOS TGRBF): Final report [Internet]. LU: Publications Office of the European Union. https://data.europa.eu/doi/10.2777/59481 (accessed 12 February 2024).

95 Quader, A.M., Ahmed, S., Dawal, S.Z., and Nukman, Y. (2016). Present needs, recent progress and future trends of energy-efficient Ultra-Low Carbon Dioxide (CO_2) Steelmaking (ULCOS) program. *Renewable and Sustainable Energy Reviews* 55: 537–549.

96 COM (European Commission). How Horizon Europe was developed [Internet]. https://research-and-innovation.ec.europa.eu/funding/funding-opportunities/funding-programmes-and-open-calls/horizon-europe/how-horizon-europe-was-developed_en (accessed 6 July 2023).

97 COM (European Commission) (2023). Horizon Europe [Internet]. https://research-and-innovation.ec.europa.eu/funding/funding-opportunities/funding-programmes-and-open-calls/horizon-europe_en (accessed 6 July 2023).

98 European Commission (2024). Horizon Europe – European Commission [Internet]. https://research-and-innovation.ec.europa.eu/funding/funding-opportunities/funding-programmes-and-open-calls/horizon-europe_en (accessed 7 Mar 2023).

99 Conejo, A.N., Birat, J.P., and Dutta, A. (2020). A review of the current environmental challenges of the steel industry and its value chain. *Journal of Environmental Management* 259: 109782.

100 Satyendra, K.S. (2019). ULCORED process [Internet]. https://www.ispatguru.com/ulcored-process/ (accessed 9 July 2023).

101 ArcelorMittal (2023). Siderwin: reducing iron ore via electrolysis [Internet]. https://corporate.arcelormittal.com/corporate-library/reporting-hub/siderwin-reducing-iron-ore-via-electrolysis (accessed 9 July 2023).

102 Jones, B. (2011). Tata Steel and Rio Tinto sign agreement on HIsarnaTM [Internet]. Ijmuiden: COM, Tata Steel and Rio Tinto. https://www.tatasteeleurope.com/corporate/news/tata-steel-and-rio-tinto-hisarna-agreement (accessed 10 July 2023).

103 Tata Steel Europe (2023). Sustainability Home Page. https://www.tatasteeleurope.com/sustainability (accessed 11 July 2023).

104 COM (European Commission) (2023). CORDIS | European Commission. Steel industry boost research into cleaner technologies. https://cordis.europa.eu/article/id/29184-steel-industry-boost-research-into-cleaner-technologies (accessed 6 July 2023).

105 Cañete, M.A. (2014). European Union: a Global Leader in Climate Action. In: Climate Action [Internet]. https://climate.ec.europa.eu/news-your-voice/news/european-union-global-leader-climate-action-2014-11-12_en (accessed 10 July 2023).

106 Ariyama, T., Sato, M., Nouchi, T., and Takahashi, K. (2016). Evolution of blast furnace process toward reductant flexibility and carbon dioxide mitigation in steel works. *ISIJ International* 56 (10): 1681–1696.

107 Gao, X., Zhang, R., You, Z. et al. (2022). Use of hydrogen–rich gas in blast furnace ironmaking of V–bearing titanomagnetite: mass and energy balance calculations. *Materials (Basel)* 15 (17): 6078.

108 Manning, C. and Fruehan, R. (2001). Emerging technologies for iron and steelmaking. *Journal of Metals* 53: 36–43.

109 Li, J., Li, C., Zhang, W. et al. (2022). Material, energy and exergy flows of the oxygen blast furnace process with sintering flue gas injection. *Journal of Cleaner Production* 371: 133294.

110 Teng, L., Ljungqvist, P., Hackl, H., and Andersson, J. (2016). Process improvement with EMS. *Steel Times International* 40: 59–62.

111 Teng, L, Ljungqvist, P., Hackl, H., and Andersson, J. (2016). Process improvement with EMS in the electric arc furnace (EAF) process. Steel Times International. https://www.semanticscholar.org/paper/Process-improvement-with-EMS-In-the-electric-arc-(-Teng-Ljungqvist/f20b76fb304b35e129651a138d94bc3d3055d163 (accessed 10 July 2023).

112 Singh, J.B. and Jaison, B. (2020). Power factor improvement in EAF using solid state transformer and bidirectional buck-boost system using battery storage of energy matrix converters. *Solid State Technology* 63 (3): 132–152.

113 Kirschen, M., Hay, T., and Echterhof, T. (2021). Process improvements for direct reduced iron melting in the electric arc furnace with emphasis on slag operation. *Processes* 9 (2): 402.

114 Thyssen Krupp (2002). ThyssenKrupp Nirosta wins “Milestones” business award Strip casting revolutionizes stainless steel production [Internet]. https://www.thyssenkrupp.com/en/newsroom/press-releases/thyssenkrupp-nirosta-wins--milestones--business-award-2219.html (accessed 10 July 2023).

115 Luebering, J.E. (2009). Puddling process | Iron Smelting, Refining & Casting | Britannica [Internet]. https://www.britannica.com/technology/puddling-process (accessed 4 August 2023).

116 Sosinsky, D., Campbell, P., Mahapatra, R. et al. (2009). The CASTRIP® process – recent developments at Nucor steel's commercial strip casting plant. *Metallurgist* 52: 691–699.

117 Wänerholm, M. (2016). Rapport nr 2016-008 – Climate impact of metal-casting [Internet]. p. 1–26. https://www.diva-portal.org/smash/get/diva2:1140576/FULLTEXT01.pdf (accessed 7 August 2023).

118 Guthrie, R.I.L. and Isac, M.M. (2022). Continuous casting practices for steel: past, present and future. *Metals.* 12 (5): 862.

119 Syre, R. (2023). thyssenkrupp. thyssenkrupp nucera Supplies the Electrolyzers for H_2 Green Steel to Build One of the Largest Integrated Green Steel Plants in Europe. https://www.thyssenkrupp.com/en/newsroom/press-releases/pressdetailpage/thyssenkrupp-nucera-supplies-the-electrolyzers-for-h2-green-steel-to-build-one-of-the-largest-integrated-green-steel-plants-in-europe-224050 (accessed 24 July 2023).

120 SCCS (2023). ATHOS project details [Internet]. https://www.geos.ed.ac.uk/sccs/project-info/2507 (accessed 10 July 2023).

121 Tata Steel Nederland (2023). Sustainability performance at our sites IJmuiden. Tata Steel in Europe. https://www.tatasteeleurope.com/construction/sustainability/performance-at-our-sites/ijmuiden (accessed 10 July 2023).

122 Bellona, E. (2021). Hydrogen in steel production: what is happening in Europe – part two. Bellona.org. https://bellona.org/news/eu/2021-05-hydrogen-in-steel-production-what-is-happening-in-europe-part-two (accessed 6 July 2023).

123 ArcelorMittal (2023). Hydrogen-based steelmaking to begin in Hamburg. https://corporate.arcelormittal.com/media/case-studies/hydrogen-based-steelmaking-to-begin-in-hamburg (accessed 5 March 2023).

124 ArcelorMittal (2023). Manufacturing and forming processes [Internet]. https://constructalia.arcelormittal.com/en/sica-steel-academy/properties-of-steel/manufacturing-and-forming-processes (accessed 9 August 2023).

125 Blume, L. (2021). GrInHy2.0 – wasserstoff-niedersachsen.de [Internet]. hydrogen-lower-saxony.com. https://www.wasserstoff-niedersachsen.de/en/grinhy2-0/ (accessed 7 July 2023).

126 COM (European Commission) (2022). CORDIS | European Commission. Green industrial hydrogen via reversible high-temperature electrolysis | GrInHy Project | Fact Sheet | H2020. https://cordis.europa.eu/project/id/700300 (accessed 6 July 2023).

127 SSAB Svenskt Stål AB (2022) Sustainability at SSAB. https://www.ssab.com/en/company/sustainability/first-in-fossil-free-steel/environmental-benefits (accessed 7 July 2023).

128 SSAB (2023). HYBRIT. A new revolutionary steelmaking technology [Internet]. https://www.ssab.com/en-gb/fossil-free-steel/insights/hybrit-a-new-revolutionary-steelmaking-technology (accessed 10 July 2023).

129 Rayner-Canham, G. and Overton, T. (2010). *Descriptive Inorganic Chemistry*, 5e. New York: W. H. Freeman and Company.

130 World Steel Association (WSA) (2023). Carbon capture and storage (CCS) [Internet]. Brussels: WSA; p. 2. https://worldsteel.org/wp-content/uploads/Carbon-Capture-Storage_2023.pdf (accessed 15 February 2024).

131 Rice University (2023). CHEMISTRY 19.1 occurrence, preparation, and properties of transition metals and their compounds - chemistry | OpenStax [Internet]. https://openstax.org/books/chemistry/pages/19-1-occurrence-preparation-and-properties-of-transition-metals-and-their-compounds (accessed 7 July 2023).

132 HM Government UK (2020). The ten point plan for a green industrial revolution – building back better, supporting green jobs, and accelerating our path to net zero [Internet]. https://www.legislation.gov.uk/ukpga/2008/27/contents (accessed 7 July 2021).

133 International Renewable Energy Agency (IRENA) (2012). Renewable energy technologies: cost analysis series [Internet]. https://www.irena.org/-/media/Files/IRENA/Agency/Publication/2012/RE_Technologies_Cost_Analysis-HYDROPOWER.pdf (accessed 7 July 2023).

134 Albanito, F., Hastings, A., Fitton, N. et al. (2019Oct). Mitigation potential and environmental impact of centralized versus distributed BECCS with domestic biomass production in Great Britain. *GCB Bioenergy* 11 (10): 1234–1252.

135 Bataille, C., Nilsson, L.J., and Jotzo, F. (2021). Industry in a net-zero emissions world: new mitigation pathways, new supply chains, modelling needs and policy implications. *Energy and Climate Change* 2: 100059.

136 Tanzer, S.E., Blok, K., and Ramírez, A. (2021). Decarbonising industry via BECCS: promising sectors, challenges, and techno-economic limits of negative emissions. *Current Sustainable/Renewable Energy Reports* 8 (4): 253–262.

137 Toktarova, A. (2021) The low-carbon steel industry-interactions between the hydrogen direct reduction of steel and the electricity system – ProQuest [Internet]. Gothenburg, Sweden: Department of Space, Earth and Environment Chalmers University of Technology. https://www.proquest.com/openview/c2a4d77d71403fdbffd33f250b13251d/1?pq-origsite=gscholar&cbl=2026366&diss=y (accessed 7 July 2023).

138 Dastoor, P. (2021). Solar foil. Email 18/10/2021.

139 Wilke, S. and Umweltbundesamt (2017). Indicator: Greenhouse gas emissions avoided by renewable energies. https://www.umweltbundesamt.de/en/data/environmental-indicators/indicator-ghg-emissions-avoided-through-the-use-of (accessed 6 July 2023).

140 Nix, J. (2023). *John Nix Pocketbook – For Farm Management*, 53e. Leicestershire: The Anderson Centre.

141 Yi, Q., Zhao, Y., Huang, Y. et al. (2018). Life cycle energy-economic-CO_2 emissions evaluation of biomass/coal, with and without CO_2 capture and storage, in a pulverized fuel combustion power plant in the United Kingdom. *Applied Energy* 225: 258–272.

142 EUROSTAT (2023). Forests, forestry and logging [Internet]. https://ec.europa.eu/eurostat/statistics-explained/index.php?title=Forests,_forestry_and_logging (accessed 12 July 2023).

143 Yang, W., Pudasainee, D., Gupta, R. et al. (2021). An overview of inorganic particulate matter emission from coal/biomass/MSW combustion: sampling and measurement, formation, distribution, inorganic composition and influencing factors. *Fuel Processing Technology* 2013: 106657.

144 US Energy Information Administration (EIA) (1994). Carbon dioxide emission factors for coal [Internet]. https://www.eia.gov/coal/production/quarterly/co2_article/co2.html (accessed 6 July 2023).

145 Benetto, E., Popovici, E.C., Rousseaux, P., and Blondin, J. (2004). Life cycle assessment of fossil CO_2 emissions reduction scenarios in coal-biomass based electricity production. *Energy Conversion and Management* 45 (18): 3053–3074.

146 Wilkerson, J. (2018). Harvard University Blog biomass over coal: burning different carbon to mitigate climate change [Internet]. Science News. https://sitn.hms.harvard.edu/flash/2018/biomass-over-coal-burning-different-carbon-to-mitigate-climate-change/ (accessed 7 July 2023).

147 Wei, R., Li, H., Chen, Y. et al. (2022). 9.07 – Environmental issues related to bioenergy. In: *Comprehensive Renewable Energy*, 2e (ed. T.M. Letcher), 92–106. Oxford: Elsevier https://www.sciencedirect.com/science/article/pii/B978012819727100011X (accessed 7 July 2023).

148 Lacy, P. and Rutqvist, J. (2015). *Waste to Wealth [Internet]*. London: Palgrave Macmillan UK http://link.springer.com/10.1057/9781137530707 (accessed 22 November 2022).

149 Mosayeb-Nezhad, M., Mehr, A.S., Lanzini, A. et al. (2019). Technology review and thermodynamic performance study of a biogas-fed micro humid air turbine. *Renewable Energy* 140: 407–418.

150 Zhao, Q., Chu, X., Mei, X. et al. (2020). Co-treatment of waste from steelmaking processes: steel slag-based carbon capture and storage by mineralization. *Frontiers in Chemistry* 8: https://www.frontiersin.org/articles/10.3389/fchem.2020.571504/full (accessed 5 March 2021).

151 Kabeyi, M.J. and Olanrewaju, O.A. (2022). Biogas production and applications in the sustainable energy transition. *Journal of Energy* 8750221.

152 US Energy Information Administration (EIA) (2023). U.S. Energy Information Administration – EIA – Independent Statistics and Analysis Annual Energy Outlook 2022 [Internet]. https://www.eia.gov/outlooks/aeo/data/browser/#/?id=1-AEO2022&cases=ref2022&sourcekey=0 (accessed 6 July 2023).

153 Seethalakshmi, K.K., Balagopalan, M., Jijeesh, C.M. (2009). Bamboo plantations: An approach to Carbon sequestration. https://forest.kerala.gov.in/images/gwarm/17bmbopacs.pdf (accessed 5 March 2020).

154 Universität Leipzig (2021). Units of Energy | Energy Fundamentals [Internet]. 2021 [cited 2023 Jul 12]. https://home.uni-leipzig.de/energy/energy-fundamentals/03.htm (accessed 5 March 2020).

155 Blue Planet (2023). Blue Planet Systems. Technology. https://www.blueplanetsystems.com/technology (accessed 6 July 2023).

156 Sendi, M., Bui, M., Mac Dowell, N., and Fennell, P. (2022). Geospatial analysis of regional climate impacts to accelerate cost-efficient direct air capture deployment. *One Earth* 5 (10): 1153–1164.

157 Green City Times (2015). What is anaerobic digestion? | Green City Times [Internet]. [cited 2023 Jul 19]. https://www.greencitytimes.com/anaerobic-digestion-a-proven-solution-to-our-waste-problem/ (accessed 19 July 2023).

158 HM Government UK (2023). Section 3: Anaerobic digestion. GOV.UK. https://www.gov.uk/government/statistics/area-of-crops-grown-for-bioenergy-in-england-and-the-uk-2008-2020/section-3-anaerobic-digestion (accessed 7 July 2023).

159 Nag, R., Auer, A., Markey, B.K. et al. (2019). Anaerobic digestion of agricultural manure and biomass – critical indicators of risk and knowledge gaps. *Science of The Total Environment* 690: 460–479.

160 US EPA O (2019). How does anaerobic digestion work? [Internet]. https://www.epa.gov/agstar/how-does-anaerobic-digestion-work (accessed 2 August 2023).

161 Bazooyar, B. and Darabkhani, H.G. (2020). Design, manufacture and test of a micro-turbine renewable energy combustor. *Energy Conversion and Management* 213: 112782.

162 Motz, H. and Geiseler, J. (2001). Products of steel slags an opportunity to save natural resources. *Waste Management* 21 (3): 285–293.

163 Proctor, D.M., Fehling, K.A., Shay, E.C. et al. (2000). Physical and chemical characteristics of blast furnace, basic oxygen furnace, and electric arc furnace steel industry slags. *Environmental Science & Technology* 34 (8): 1576–1582.

164 Shen, H. and Forssberg, E. (2003). An overview of recovery of metals from slags. *Waste Management* 23 (10): 933–949.

165 Riley, A.L. and Mayes, W.M. (2015). Long-term evolution of highly alkaline steel slag drainage waters. *Environmental Monitoring and Assessment* 187 (7): 463.

166 Shi, C. (2004). Steel slag – its production, processing, characteristics, and cementitious properties. *Journal of Materials in Civil Engineering* 16 (3): 230–236.

167 Zhou, R., Liu, X., Luo, L. et al. (2017). Remediation of Cu, Pb, Zn and Cd-contaminated agricultural soil using a combined red mud and compost amendment. *International Biodeterioration and Biodegradation* 118: 73–81.

168 Li, Z., Shen, J., Wang, F. et al. (2023). Impacts of organic materials amendment on the soil antibiotic resistome in subtropical paddy fields. *Frontiers in Microbiology* 13: 1075234.

169 Rubin, R., Oldfield, E., Lavallee, J. et al. (2023). Climate mitigation through soil amendments: quantification, evidence, and uncertainty. *Carbon Management* 14 (1): 2217785.

170 Goetz, E.R. and Riefler, R.G. (2014). Performance of steel slag leach beds in acid mine drainage treatment. *Chemical Engineering Journal* 240: 579–588.

171 Munn, D. (2005). Steel industry slags compared with calcium carbonate in neutralizing acid mine soil. https://www.semanticscholar.org/paper/Steel-Industry-Slags-Compared-with-Calcium-in-Acid-Munn/0c308b6b2f7c1478362bb494e5c7a9cf960744f8 (accessed 5 February 2024).

172 SciELO (2024). SciELO – Brazil – Analysis of carbonation and hydration of BOF steel slag for application as railway ballast [Internet]. https://www.scielo.br/j/remi/a/gmjgPj8J6Gsjy6PCFGCFxjv/?lang=en (accessed 5 February 2024).

173 Chiricuţă, I. and Rusu, I. Tests carried out on roads containing steel mill slag. *Current Trends in Civil & Structural Engineering* 10 (2): 000733.

174 Navarro, C., Díaz, M., and Villa-García, M.A. (2010). Physico-chemical characterization of steel slag. study of its behavior under simulated environmental conditions. *Environmental Science & Technology* 44 (14): 5383–5388.

175 Scott, I.S.P.C. and Penn, C.J. (2021). Estimating the variability of steel slag properties and their influence in phosphorus removal ability. *Chemosphere* 276: 130205.

176 Song, Q., Guo, M.Z., Wang, L., and Ling, T.C. (2021). Use of steel slag as sustainable construction materials: a review of accelerated carbonation treatment. *Resources, Conservation and Recycling* 173: 105740.

177 Fisher, L.V. and Barron, A.R. (2019). The recycling and reuse of steelmaking slags – a review. *Resources, Conservation and Recycling* 146: 244–255.

178 Ren, Z. and Li, D. (2023). Application of steel slag as an aggregate in concrete production: a review. *Materials (Basel)* 16 (17): 5841.

179 De Windt, L., Chaurand, P., and Rose, J. (2011). Kinetics of steel slag leaching: batch tests and modeling. *Waste Management* 31 (2): 225–235.

180 Fällman, A.M. (2000). Leaching of chromium and barium from steel slag in laboratory and field tests – a solubility controlled process? *Waste Management* 20: 149–154.

181 Karayel, G.K., Javani, N., and Dincer, I. (2022). Effective use of geothermal energy for hydrogen production: a comprehensive application. *Energy* 249: 123597.

182 COM (European Commission) (2015). First circular economy action plan – European Commission [Internet]. https://environment.ec.europa.eu/topics/circular-economy/first-circular-economy-action-plan_en (accessed 5 February 2024).

183 UN United Nations (2016). Paris Agreement 2016 [Internet]. United Nations. https://www.un.org/en/climatechange/paris-agreement#:~:text=To%20tackle%20climate%20change%20and,2015%3A%20the%20historic%20Paris%20Agreement (accessed 2 March 2020).

184 Al-Alawi, M. (2007). Biohydrogen production by anaerobic biological fermentation of agriculture waste. In: *Assessment of Hydrogen Energy for Sustainable Development* (ed. J.W. Sheffield and Ç. Sheffield), 177–185. Dordrecht: Springer Netherlands (NATO Science for Peace and Security Series C: Environmental Security).

185 Barbir, F. (2007). Fuel cells for clean power generation: status and perspectives. In: *Assessment of Hydrogen Energy for Sustainable Development* (ed. J.W. Sheffield and Ç. Sheffield), 113–121. Dordrecht: Springer Netherlands (NATO Science for Peace and Security Series C: Environmental Security).

186 Benhamou, K. (2007). Hydrogen fueling sustainability of energy systems, regional integration, and development: the Sahara wind project. In: *Assessment of Hydrogen Energy for Sustainable Development* (ed. J.W. Sheffield and Ç. Sheffield), 59–73. Dordrecht: Springer Netherlands (NATO Science for Peace and Security Series C: Environmental Security).

187 Bhaskar, A., Abhishek, R., Assadi, M., and Somehesaraei, H.N. (2022). Decarbonizing primary steel production : Techno-economic assessment of a hydrogen based green steel production plant in Norway. *Journal of Cleaner Production* 350: 131339.

188 Bose, T.K., Hourri, A., Champagne, G.Y., and Fournier, R.P. (2007). Pathway for hydrogen in urban transit system. In: *Assessment of Hydrogen Energy for Sustainable Development* (ed. J.W. Sheffield and Ç. Sheffield), 75–82. Dordrecht: Springer Netherlands (NATO Science for Peace and Security Series C: Environmental Security).

189 Colla, V. and Matino, I. (2021). Editorial for the Special Issue: Overview, state of the art, recent developments and future trends regarding Hydrogen route for a green steel making process. *Matériaux & Techniques* 109 (3–4): E301.

190 Wang, C., Walsh, S., Haynes, M.W. et al. (2022). From Australian iron ore to green steel: the opportunity for technology-driven decarbonisation [Internet]. https://ecat.ga.gov.au/geonetwork/srv/api/records/5377741d-ece0-4e01-b87e-6bf49da11c03 (accessed 7 July 2023).

191 Grasa, G., Díaz, M., Fernández, J.R. et al. (2022). Blast furnace gas decarbonisation through calcium assisted steel-mill off-gas hydrogen production. Experimental and modelling approach. *SSRN Journal* 191: 507–522. https://www.ssrn.com/abstract=4171691 (accessed 2 November 2022).

192 Richardson-Barlow, C., Pimm, A.J., Taylor, P.G., and Gale, W.F. (2022). Policy and pricing barriers to steel industry decarbonisation: a UK case study. *Energy Policy* 168: 113100.

193 Devlin, A. and Yang, A. (2022). Regional supply chains for decarbonising steel: energy efficiency and green premium mitigation. *Energy Conversion and Management* 254: 115268.

194 Sun, Y., Tian, S., Ciais, P. et al. (2022 Dec). Decarbonising the iron and steel sector for a 2 °C target using inherent waste streams. *Nature Communications* 13 (1): 297.

195 Steenbrink, F. (2022). Impact of the carbon border adjustment mechanism: an economic and geopolitical assessment of the German-Chinese aluminium trade flows. https://repository.tudelft.nl/islandora/object/uuid%3A1a176161-908d-41b0-a9b0-d7b961744a9d (accessed 7 July 2023).

196 Garvey, A., Norman, J.B., and Barrett, J. (2022). Technology and material efficiency scenarios for net zero emissions in the UK steel sector. *Journal of Cleaner Production* 333: 130216.

197 Gosens, J., Turnbull, A., and Jotzo, F. (2021). A highly granular model of China's coal production, transport and consumption system shows how its decarbonization and energy security plans will affect coal imports.

198 Jacob, R., Sergeev, D., and Müller, M. (2022). Valorisation of waste materials for high temperature thermal storage: a review. *Journal of Energy Storage* 47: 103645.

199 Öhman, A., Karakaya, E., and Urban, F. (2022). Enabling the transition to a fossil-free steel sector: the conditions for technology transfer for hydrogen-based steelmaking in Europe. *Energy Research and Social Science* 84: 102384.

200 Zhiming, Y., Zushu, L., Sanjeev, M. et al. (2022). Value in use of lime in BOF steelmaking process. *Ironmaking and Steelmaking* 49 (1): 42–48.

201 Kempken, T., Hauck, T., DeSantis, M. et al. (2021) Collection of possible decarbonisation barriers (Deliverable D1.5). [Internet]. Centre for European Policy Studies (CEPS), Brussels (Belgium): Green Steel for Europe Consortium. https://www.estep.eu/assets/Uploads/D1.5-Collection-of-possible-decarbonisation-barriers.pdf (accessed 2 March 2022).

202 Griffin, P.W. and Hammond, G.P. (2021). The prospects for 'green steel' making in a net-zero economy: a UK perspective. *Global Transitions* 3: 72–86.

203 Antonazzo, L., Stroud, D., Weinel, M. et al. (2021). Preparing for a just transition—Meeting green skill needs for a sustainable steel industry [Internet]. Cardiff: Cardiff University. p. 1–20. https://orca.cardiff.ac.uk/id/eprint/145353/1/RGB_Meeting_Green_Skills_Needs_A4_booklet_singlepages.pdf (accessed 8 September 2022).

204 Garcia-Herrero, A., Tagliapietra, S., and Vorsatz, V. (2021). Hydrogen development strategies: a global perspective. Document – Gale Academic OneFile [Internet]. https://go-gale-com.ezproxy.staffs.ac.uk/ps/i.do?id=GALE%7CA673932617&sid=googleScholar&v=2.1&it=r&linkaccess=abs&issn=&p=AONE&sw=w&userGroupName=anon%7E4246e29b&aty=open+web+entry (accessed 6 July 2023).

205 Arens, M., Ahman, M., and Vogl, V. (2021). Which countries are prepared to green their coal-based steel industry with electricity? – Reviewing climate and energy policy as well as the implementation of renewable electricity? *Renewable and Sustainable Energy Reviews* 143: 110938.

206 Muslemani, H., Liang, X., Kaesehage, K. et al. (2021). Opportunities and challenges for decarbonizing steel production by creating markets for 'green steel' products. *Journal of Cleaner Production* 315: 128127.

207 Sarić, M., Dijkstra, J.W., and van Delft, Y.C. (2021). CO_2 abatement in the steel industry through carbon recycle and electrification by means of advanced polymer membranes. *Membranes* 11 (11): 856.

208 Kiessling, S., Gohari Darabkhani, H., and Soliman, A.H. (2024). Greater energy independence with sustainable steel production. *Sustainability* 16 (3): 1174.

209 International Energy Agency (IEA) (2021). Global CO_2 emissions rebounded to their highest level in history in 2021 – News. https://www.iea.org/news/global-co2-emissions-rebounded-to-their-highest-level-in-history-in-2021 (acccessed 7 July 2023).

210 International Energy Agency (IEA) (2023). IEA data and statistics. https://www.iea.org/data-and-statistics (accessed 7 July 2023).

211 IPCC Intergovernmental Panel for Climate Change (2021). *Climate Change 2021 – The Physical Science Basis Summary for Policymakers [Internet]*. Cambridge: Cambridge University Press https://www.ipcc.ch/report/ar6/wg1/downloads/report/IPCC_AR6_WGI_SPM.pdf (acccessed 7 July 2023).

212 International Renewable Energy Agency (IRENA) (2021). World energy transitions outlook: 1.5 °c pathway [Internet]. https://www.irena.org/publications/2021/Jun/World-Energy-Transitions-Outlook (accessed 8 November 2022).

213 IPCC Intergovernmental Panel for Climate Change (2022). Climate change 2022: impacts, adaptation and vulnerability [Internet]. https://www.ipcc.ch/report/ar6/wg2/ (acccessed 3 January 2023).

214 Energy Information Administration (EIA) (2021). CO_2 per kWh [Internet].: https://www.eia.gov/tools/faqs/faq.php?id=74&t=11 (accessed 8 November 2022).

215 Firth, A.E.J., Mac Dowell, N., Fennell, P.S., and Hallett, J.P. (2020). Assessing the economic viability of wetland remediation of wastewater, and the potential for parallel biomass valorisation. *Environmental Science: Water Research & Technology* 6 (8): 2103–2121.

4

Key components of the Bio Steel Cycle

4.1 Introducing the BiSC key components for net-zero carbon steel manufacturing

In this chapter, the individual steps within the Bio Steel Cycle (BiSC) are being explored, based on the concept that nothing is going to waste, and waste and by-products are seen as resources, which are either reused or recycled: an example of circular economic principles in application. It provides in-depth information on the basic circular concept of the BiSC model, from carbon avoidance and carbon utilisation to carbon sequestration processes, including the implementation of advanced renewable energy technologies and biomass utilisation.

In order to prove the 1 hypothesis, the most likely implementable scenarios have been considered and simulated using standard mathematical principles, MS Excel calculations and process engineering software. The resulting reports are encouraging as they provide strong evidence for the technical viability of the principles of the BiSC model and that it can possibly be applicable to most heavy industries, such as cement, chemical, glass, paper and transport. It includes using renewable energy technologies, avoiding carbon dioxide (CO_2) emissions by incorporating process improvement technologies, recycling waste and by-products and capturing post-combustion emissions where possible [1–8] as displayed in Figure 4.1.

Analyses of the technical and economical long-term potential of novel steel production technologies carbon avoidance technology (CAT), carbon capture and storage (CCS) and carbon capture, utilisation and storage (CCUS) in the United Kingdom [9], Germany [10] and beyond used techno-economic models to model three research-stage [4, 11–17], ore-based steelmaking routes versus blast furnace (BF)–basic oxygen furnace (BOF) route [18–21]. It was concluded that in comparison with the BF with CCS1 (BFCCS) [10, 12, 22–24], hydrogen direct reduction (HDR) [25–29] and iron ore electrolysis (EW) [30–32], energy and raw material efficiency is significantly higher for HDR and EW [10, 20, 25, 33–35], and the 80% reduction target by 2050 [36] was thought to be perfectly achievable in the scenario, as per Tata Steel Europe's vision map 2050 [34, 35, 37]. As shown in Table 4.1, it was found that there are a sufficient number of viable CAT, CCS and CCUS technologies, methods and strategies at technical readiness levels (TRLs) 6–9 available and already installed, which are suitable for immediate BiSC

Circular Steel Production: Pathways to Net-Zero Carbon Emissions, First Edition.
Sandra Kiessling, Hamidreza Gohari Darabkhani, and Abdel-Hamid Soliman.

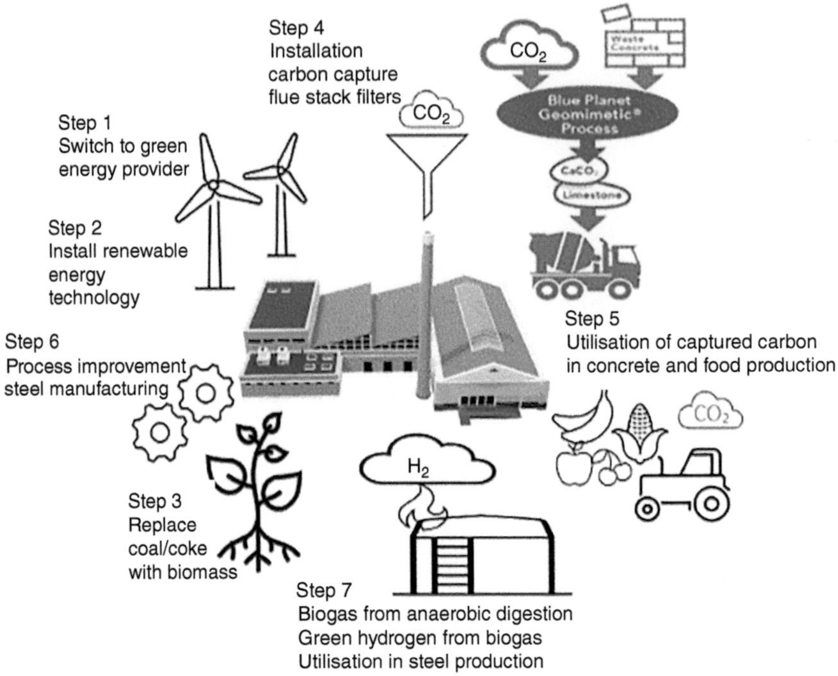

Figure 4.1 The Bio Steel Cycle (BiSC) resource utilisation flow in simple steps.

implementation and achieving short- to medium-term significant reduction of CO_2 emissions in steelmaking. The urgency for sufficient prioritisation throughout all industries and political willingness (subsidies) cannot be emphasised enough [36] – and with it, the need to create a viable commercial environment, due to the required high capital investment [2, 16, 72, 75–81] and a significant dependency on electricity prices.

4.2 BF/BOF route carbon capture

The bulk of emissions are being produced during the coking process, energy generation from fossil fuel and the BF/BOF process route of steelmaking. However, there are various opportunities to decarbonise this process. The standard steel production (SSP) process in combination with the currently operational newly developed technologies can achieve a reduction of more than 50% [42, 82]) with successive implementation to less than 8 metric tonnes per tonne of steel produced. By incorporating the BiSC components of CAT, CCS and CCUS into existing steel production sites, an almost 100% CO_2 emission reduction can be achieved immediately.

The post-combustion capture of CO_2 (CCS) and other greenhouse gases (GHGs) and the exploration of carbon scrubbing of flue gases have been explored. There are

Table 4.1 Process improvement technologies in development and installed.

Stage	Time frame	Name	Technology	References
Post-combustion capture	Short- to mid-term	Athos	Captured carbon usage	[14, 16, 34, 35, 37–41]
		Everest	CCS: from BOG > former gas fields under the North Sea	[14, 16, 34, 35, 37–41]
		Geomimetic	CO_2 recycling (concrete, flue gases, etc.)	[42–44]
		DAC	Land trees (8.3t/ha/p.a.) 3.8 t/tonne of steel	[45–52]
			Land shrubs and herbs	[45–52]
			Water plants	[45–52]
			Organisms (plancton)	[45–52]
		CEPS	Carbon-enriched air 99% utilized in food production	[53, 54]
		Air-blown BF	MEA	[3, 6, 14, 16, 34, 35, 37–41]
			MDEA	[3, 6, 14, 16, 34, 35, 37–41]
			Selexol	[3, 6, 14, 16, 34, 35, 37–41]
			Shift and Selexol	[3, 6, 14, 16, 34, 35, 37–41]
		TGRBF	MEA	[3, 6, 14, 16, 34, 35, 37–41]
			VPSA	[3, 6, 14, 16, 34, 35, 37–41]
			Selexol	[3, 6, 14, 16, 34, 35, 37–41]
		COREX	MEA	[3, 6, 14, 16, 34, 35, 37–41]
			Selexol	[3, 6, 14, 16, 34, 35, 37–41]
			Shift and Selexol	[3, 6, 14, 16, 34, 35, 37–41]
	Long term	Air-blown BF	➢ Shift membrane reactor + Selexol	[3, 6, 14, 16, 34, 35, 37–41]
			➢ Selective carbon membrane	[3, 6, 14, 16, 34, 35, 37–41]
			➢ Hydrate crystallisation	[3, 6, 14, 16, 34, 35, 37–41]
		Advanced smelting reduction	Purification only	[3, 6, 14, 16, 34, 35, 37–41]
		Mineral carbonation	Two-tiered reaction tower	[3, 6, 14, 16, 34, 35, 37–41]
		CCU technology		[3, 6, 14, 16, 34, 35, 37–41]

(Continued)

Table 4.1 (Continued)

Stage	Time frame	Name	Technology	References
Pre-combustion process optimization	Short term	HIsarna	Eliminates sintering and smelting reduction	[34, 35, 37, 55–62]
		ReclaMet	Waste resource recovery	[33, 34, 37]
	Long term	Electrolysis	Direct water splitting; biomass > hydrogen	[4, 10, 25, 58]
		Hydrogen EAF	Hydrogen direct reduction in EAF (H-DR/EAF)	[4, 10, 25, 58]
		GrInHy	Electrolyser development	[4, 10, 25, 58]
		H2Future	Electrolyser development	[4, 10, 25, 58
		HYBRIT	Fossil-fuel-free primary steel production	[4, 10, 25, 58]
		Hydrogen direct reduction (HDR)	HDR in BF/EAF/BOF; CO_2 from electricity used	[4, 10, 25, 58]
Research (TRLs 6–9)				[4, 10, 25, 58]
Pre-combustion energy source replacement		Renewable energy technology	Solar ǀ Wind ǀ Hydro ǀ Geothermal	[1–8, 63, 64]
			Wind	[1–8, 63, 64]
			Hydro	[1–8, 63, 64]
			Geothermal	[1–8, 63, 64]
Pre-combustion energy consumption optimization		Industry 4.0:	➢ Optimising energy consumption in production	[14, 19, 20, 41, 65–69]
			➢ Model-based steel temperature adjustment	[14, 19, 20, 41, 65–69]
			➢ Hydrogen control in ladle operations	[14, 19, 20, 41, 65–69]
			➢ Hydrogen control in casting operations	[14, 19, 20, 41, 65–69]
			➢ Nitrogen control in ladle operations	[14, 19, 20, 41, 65–69]
			➢ Nitrogen control in casting operations	[14, 19, 20, 41, 65–69]
Pre-combustion fossil fuel replacement		Coal/Biochar	Biochar/Takachar	[22, 23, 70, 71]
Post-combustion capture			Waste heat recovery	[19, 65, 68, 72–74]

several possible technologies and processes to be considered for post-combustion carbon capture:

- mechanical capture
- compression and dehydration
- membrane installation
- guiding off-gas through troughs of physical solvents/solid sorbents (such as Zeolite 13X) and chemical solvents
- as well as utilising metal–organic frameworks

[3, 4, 15, 24, 53, 83, 84]. Suitable literature (Appendix 1) was thoroughly investigated with regard to applied and innovative steelmaking procedures, CAT, CCS and CCUS processes (Appendix 2) and improved management systems such as I4.0 [65, 66].

The key components within the BiSC are based on a circular production process and are functioning in an interactive manner. The basis for this system is the BF/BOF route and involves the aforementioned CAT, CCS and CCUS and process improvements where possible.

Table 4.1 provides an overview of the process improvement technologies which are in development at TRLs 6–9 or are already in industrial application.

4.3 BF/BOF off-heat utilisation in iron and steelmaking

4.3.1 Off-heat utilisation in anaerobic digestion and sewage treatment

Anaerobic digestion (AD) and sewage treatment plants require a stable ambient temperature throughout the year, which requires a conventional source of heating. In order to achieve the maximum biogas output, the AD chamber requires a constant equilibrium. In countries of temperate environmental conditions, sufficient insulation and – depending on the ambient temperature – an adequate heating system is vitally important to ensure a healthy environment for the bacteria, which are the workers who break down organic matter under the exclusion of oxygen. The hot air from ironmaking in the BF and steelmaking in the BOF can be utilised to achieve this stable temperature, which can also be used in food production. For this purpose, the top gas recycling BF (TGRBF) system is required to be furnished with baffle segments or insulated water-filled pipes (back-boiler principle), which lead the hot air past the inserts and transfer the heat onto the metal and further along in the network. At the same time, the rising off-gas cools down by travelling through this structure and can be directed through a network of pipes and ducts, either filled with water loops to generate warm water or an air circulation system, and filters through to greenhouses for food production, to keep a constant temperature for enhanced plant growth, as displayed in Figure 4.2.

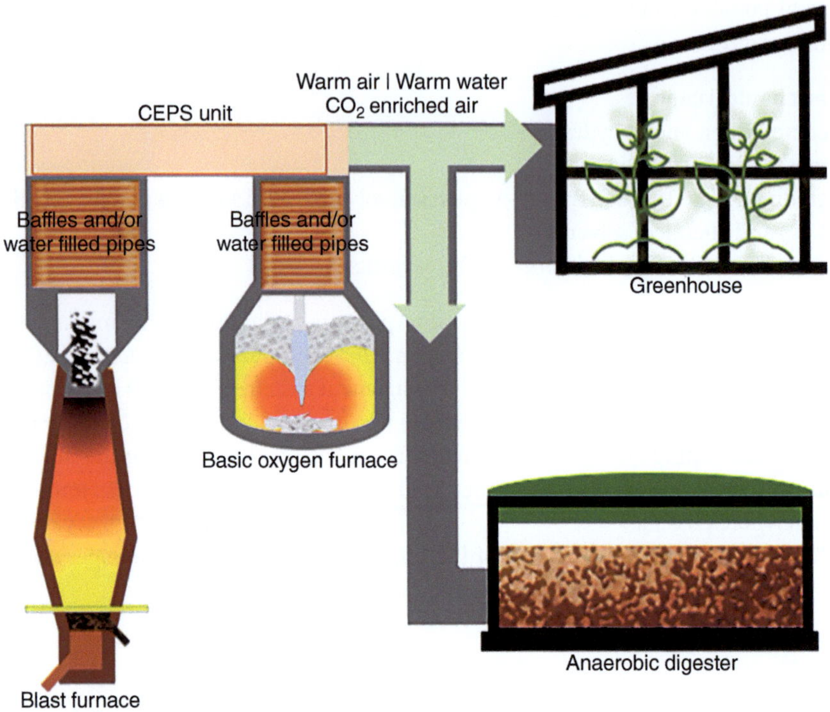

Figure 4.2 Theoretical BF and BOF off-gas recycling layout.

Additionally, the off-gas could be sent through a carbon enrichment for plant stimulation (CEPS) bed, where the off-heat is utilised for the adsorption/desorption process, where the temperature in the water bath is raised to motivate the CO_2 molecules to detach from the Zeolite 13X anodes and increase the CO_2 density to 99%.

4.3.2 CO_2 and off-heat utilisation in food production

Food production in greenhouses requires a stable temperature throughout the year, which requires a permanently available and constant source of heating. Reusing heat from the steelmaking process, the expenditure of installing heating systems and using fuel and energy is simply not required, as the air has already been heated. Although the installation of a suitable infrastructure would have to be considered, the economical case has to be further investigated, and the financial viability – the cost of ducts and pipework against the installation of a heating and cooling system including energy requirements – needs to be solid to make a case for reutilising flue stack off-heat. The flue stack heat can be as high as 1650 °C and would have to cool down, i.e. via travelling through adequately sized pipework, ducts and possibly turbines (which in their own right would possibly be able to generate electricity, via convection, baffle systems or plate heat exchangers). The energy saved on heating food production facilities is deserving of further investigation, as this will effectively contribute to achieving net-zero carbon emissions in steel production. Additionally, incorporating the CEPS [53] process into the BiSC-based circular steelmaking process involves CCUS by means of driving CO_2-enriched flue stack off-gas from combustion processes into CEPS units. Subsequently, this at almost 100% carbon-enriched air is then directed into greenhouses to stimulate plant growth. The chosen CEPS model is deemed scalable and has the capacity to provide 85 tonnes of CO_2 p.a. in its current configuration: concentrating CO_2 from ambient air (400 ppm) into an enriched product stream at 1000 ppm CO_2. Locating the source of CO_2, i.e. flue gases from production in steel, cement or energy production and greenhouses in close proximity to each other, eliminates costs associated with filtering, deactivation, compression, transportation, handling, distribution and storage entirely. Every successfully installed CEPS unit per greenhouse infrastructure is effectively a CO_2 sequestration station, and the economic feasibility is based on 1 kWh = 3600 kWs = 3.6 MJ [85], costing between £0.11 and £0.21/kWh, and 17.8 kJ/mol CO_2 = 17.8 kJ/44 g CO_2 (eia, 2021) for pure CO_2 versus 8.5 kJ/mol CO_2 = 8.5 kJ/44 g CO_2 for enriched, 1000 ppm CO_2, costing effectively between \$15 and \$309/t CO_2. The efficiency, technical and economic viability are making the case for temperature swing absorption/desorption flue gas carbon capture.

4.4 Renewable energy technologies

4.4.1 Renewable energy technologies utilisation

There is a wide range of available installation points for renewable energy technologies at existing (steel) production sites, depending on the site parameters: solar PV and thermal panels, horizontal and vertical wind turbines, hydropower and wave power systems and geothermal units. The solutions are manifold, and the best-case scenario depends on the structure and soundness of existing production facilities and the location of the premises. Any customer in the United Kingdom, whether commercial or residential, can already make the step to reduce their carbon footprint by 30% – by switching their electricity and gas provision to a company, which derives their electricity and gas at 100% from renewable sources.

4.4.2 Wind energy

The second largest dominant technology after solar energy is onshore wind energy [86]. Wind energy pylons are – besides solar – another effective way to produce electricity from a natural source (wind). This technology is mature and widely used; there is a wide range of products on the market, and the site parameters will determine which system would be suitable for the location in question (VWT Power, 2023). Utilising the vast UK coastline, large offshore wind parks have been established, producing renewable energy reliably for some time now. Given that the United Kingdom is an island, technically, one could argue that it only makes logical sense – since land in the United Kingdom is scarcer than water – that offshore wind

Figure 4.3 133 11-kW wind turbine. Source: Earthmill [157] / https://www.earthmill.co.uk/wind-turbines-for-sale/small-wind-turbines/gaia-133-11kw-wind-turbine / / last accessed 7 February 2024.

park development should be given priority to onshore wind farms. The actual wind turbine designs are manifold, and information on new interpretations of the basic principles and their expression of innovative designs are being published on an almost monthly basis. Horizontal axis turbines are the gold standard (Figure 4.3), but the newest development is vertical axis wind turbines, where contrary to the traditional ‘windmill’ design (Figure 4.3), the blades are mounted on two star-shaped frames and mounted through the pillar, as shown in Figure 4.4, similar to caged vertical wind turbines [87].

There are currently four types of wind turbine energy generators (WTGs), which are as follows:

1) Direct current generators (DC)
2) Alternating current synchronous generators (AC)
3) AC asynchronous generators
4) Switched reluctance generators

This design is much more suitable in commercial or residential settings in comparison to traditional wind turbine designs not only due to the smaller dimension of the mounted multiple blades but also due to their much smaller size. Although it has become apparent that the number of blades increases not only efficiency but also noise levels, the speed at which these technologies are developing gives hope that a solution will soon be found to combat these issues. Additionally, the installed renewable energy technologies (solar, wind, geothermal or hydropower) and the back-feeding of biogas from AD all provide the industrial processes with renewable energy, from their own infrastructure. Thus, the energy produced on site will make most industries inevitably more energy independent.

Figure 4.4 Vertical axis wind turbine (Qr6 VAWT). Source: Author Photo.

4.4.3 Solar energy

Renewable energy technologies are an essential part of the BiSC, as CO_2 emissions in steelmaking can be reduced by more than 60% [2–6, 14, 23, 64, 71, 88, 89] if companies' energy and gas supply was a) switched to a provider using 100% renewable energy technology and producing their own energy by retrofitting their plants with renewable energy technologies. The global weighted average cost of electricity (LCOE) for solar PV projects fell by 85% [86]. There are a variety of renewable energy technologies at TRL 9 available (Table 4.1), but solar PV technology and onshore wind generation for independent energy production were established as the mainly utilised renewable energy technology in a competitive market [86]. A brief overview of currently available solar technology systems is shown in Table 4.2.

Table 4.2 Examples of solar technologies.

Solar technology	Manufacturer research facility	Product name	Reference
Solar foil	University of Newcastle Centre for Organic Electronics	Organic solar cells	[90]
Solar film	Power Roll	Power Roll solar film	[91]
Solar PV	UKSOL	UKSOL polycrystalline solar technology	[92]
Flexible solar panels	UKSOL	UKSOL flexible solar	[93]
Bifacial solar panels	UKSOL	UKSOL bifacial modules	[94]
Solar thermal	Grant UK	Grant Sahara collectors	[95]
Tracker solar panels	EcoFlow	EcoFlow single-axis solar tracker	[96]
Solar window panels	Spirit Energy	KB solar glass PV windows	[97]
Integrated solar panels	Wienerberger UK	Sandtoft in-roof solar	[98]
Solar shingles	Hanergy	Hantile	[99]

The newest development in solar PV panels is the next-generation perovskite solar cells which allegedly can be manufactured at 50% cost of traditional silicon cells, at 50% greater efficiency, according to researchers from Nanjing University (China) [100].

4.4.4 Hydropower

Technology generating electricity using either biomass or hydro turbines (water-based), and AD systems [47] are also included in some of the schemes currently supported by the 2023/2024 UK governments' incentives to promote sustainable energy generation, as displayed in Chapter 6: Greater Energy Independence with Sustainable Steel Production [42]. Currently, there are closed-loop and open

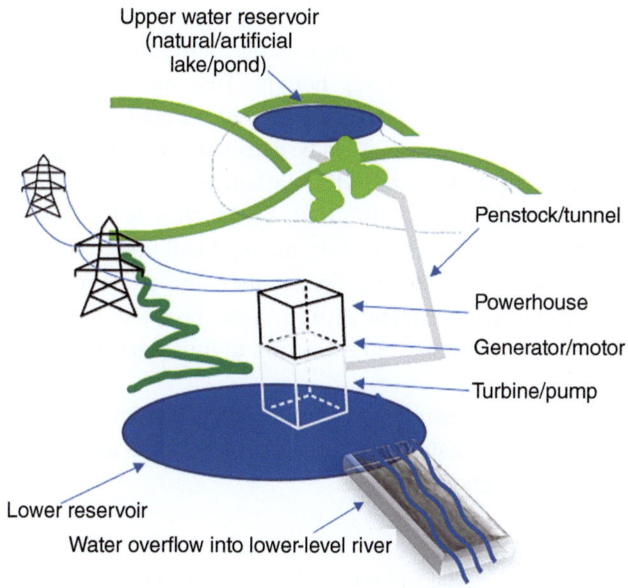

Figure 4.5 Open-loop system configuration.

hydropower systems available. Given the volatility of UK river water levels and contamination with faeces released by water companies, closed-loop hydropower systems might be the technology of choice. The following image Figure 4.5 demonstrates the layout of an open-loop system.

Additionally, the installed renewable energy technologies (solar, wind, geothermal and hydropower) and the back-feeding of biogas from AD all provide the industrial processes with renewable energy, from their own infrastructure. Thus, the energy produced on site will make most industries inevitably more energy independent.

4.5 DAC woodlands and plant matter

The BiSC includes CO_2 capture and utilisation, implementing renewable energy sources (solar, wind, microturbines and green H_2), plantation of biomass for direct air capture (DAC), CO_2 absorption and provision of biomass. Additionally, the unsung hero of waste management, AD, will feature as the source of green hydrogen and biogas, working in tandem with turbines [101–103], for utilisation in the steelmaking process. The 7-step implementation strategy starts with replacing energy sources, develops over process improvement and installation of flue gas carbon capture and concludes with utilising biogas-derived hydrogen, as a product from AD of the grown agri-food in the cycle. In order to understand the steel production process and to identify CO_2 emissions and energy usage, the carbon life cycle within requires focused attention [104, 105]. Our atmosphere consists of not only oxygen but also a range of natural gases: The greatest proportion of naturally occurring gas is nitrogen (N_2) at ca. 78% of air, followed by oxygen (O_2) at ca. 21%. The air also contains inert gases, such as argon (Ar), as the third most abundant gas at 0.93%. Trace amounts of CO_2, neon (Ne), helium (He), methane (CH_4), krypton (Kr), hydrogen (H_2), nitrous oxide (NO), xenon (Xe), ozone (O_3), iodine (I2), carbon monoxide (CO) and ammonia (NH_3) can be detected in the atmosphere [104–106]. To start, one needs to understand what carbon actually represents. Carbon is one name of a GHG called CO_2 [107, 108], by which it is commonly referred to. CO_2 is the result of combustion processes, mainly fossil fuels (coal, gas and oil) [104–106, 109], but also solid waste and plant matter, and also a result of industrial processes, i.e. manufacturing cement [110]. In steel production, CO_2 is being emitted at every stage of production, from extraction to finish machining, as shown in Figure 4.6.

CO_2 is sequestered from the atmosphere when it is absorbed by terrestrial and aqueous plants via DAC as part of the natural carbon cycle [52]. Plants utilise CO_2 during the biological process called photosynthesis,

$$6CO_2 + 6H_2O \rightarrow C_6H_{12}O_6 + 6O_2$$

where CO_2 is being absorbed and oxygen is released [52]. The CO_2 being emitted during industrial processes is then to a certain extent sequestered by plant matter and converted to O_2; broad-leaved tree planting can sequester up to 500 t/CO_2/ha/60 years [47, 48], and bamboo species are able to sequester almost 2 ½ times this amount [18, 50, 51].

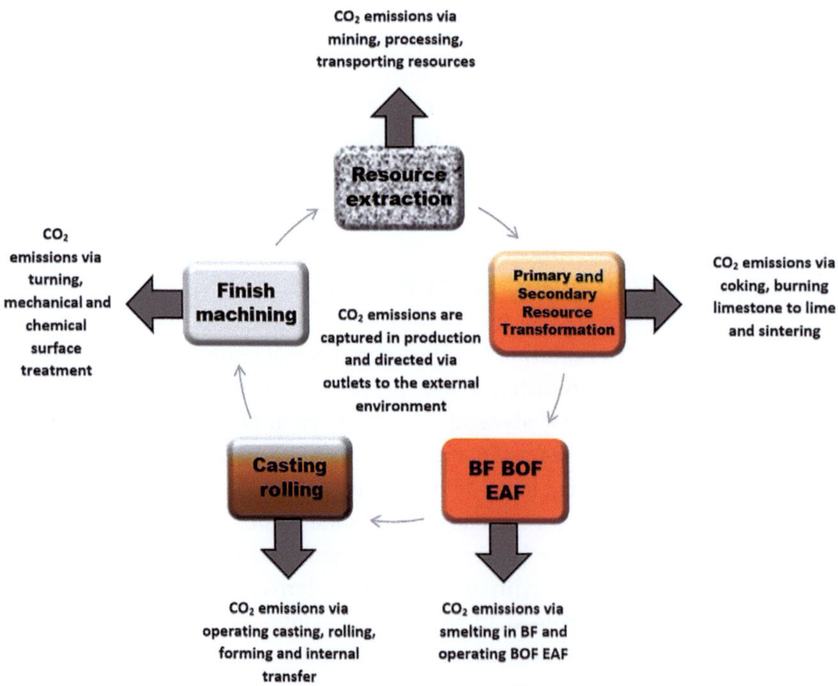

Figure 4.6 CO_2 life cycle in steel production.

In the spirit of innovative, multidisciplinary approaches to solving contemporary CO_2 emission issues, the positive effects of DAC [47] and utilisation of woodlands for carbon capture cannot be emphasised enough. In the United Kingdom, there is now a grant offering of income for planning and establishing woodlands, per hectare (ha, 1,0000 m^2). The Forestry Commission and connected parties provide a range of funding to suit different woodland creation projects and are offering farmers and landowners the opportunity with financial support to plan, create and maintain commercially viable new woodlands in England. For example, schemes covering approximately 10 ha have already received £163,900 (£12,000 per ha) from woodland creation grants. The total consists of financial incentives paid via the Woodland Creation Planning Grant (WCPG) and the England Woodland Creation Offer (EWCO) [111–116]. As one of the critical components of the BiSC, woodlands/trees for DAC would even be a profitable sideline for (steel) manufacturers in England, as illustrated in the calculations, as follows:

Cost for the first five years:	£ 27,500
EWCO maintenance payment	- £ 41,000
EWCO standard costs	- £ 81,500
WCPG	- £ 2100
EWCO additional contributions:	
Recreational access	- £ 17,300
Nature recovery low	- £ 15,100
Close to settlements	- £ 6800
Profit after first five years	£ 136,300

Follow-on applications for further EWCO maintenance payments can be made, and follow-up grants for additional spaces or wetland and other habitat creation are also open. Trees and vegetation as natural carbon sinks [46–48, 52] should ideally be planted around steel production plants to absorb the remaining CO_2 emissions via DAC. Whilst, at the same time, the plant matter could feed the anaerobic digester, biochar plants are directly used at selected quality in ironmaking and steelmaking as readily available biomass. In this respect, bamboo beats deciduous native plants with its carbon sequestration capacities: on average, one hectare of bamboo stand absorbs ~17 tonnes of carbon per year [50, 51]. Native deciduous and non-deciduous trees have a carbon sequestration capacity of on average 9 t of CO_2/ha of tree plantation [47, 48, 117–119]. Planting a sufficient number of trees should be considered in the planning for the updating of existing steel production plants and for any new development in order to meet the UK government's zero emissions targets. The UK tree cover stands at 13% (3.25 million hectares), which is the lowest in the Northern Hemisphere. In comparison, forests and wooded land cover over 182 million hectares in the European Union (EU), which is about 42% of the EU's total land area. This equates to 0.36 hectares of forest per capita in the EU in comparison [120]. Woodlands not only capture post-combustion CO_2 and create biomass for AD but also may create recreation and employment opportunities and additional sources of commercial activity. Additionally, they offer a low-cost opportunity for carbon offsetting, which could be seen as a commercial opportunity in itself. Trees are natural carbon sinks [46–48, 52], and ideally, woodlands would be planted around steel production plants to absorb the remaining CO_2 emissions via DAC – whilst simultaneously, the trees would provide some of the material for producing biochar and organic matter to be fed into the anaerobic digester, alongside agricultural businesses. Planting a sufficient number of trees [46–48, 52] and feeding both anaerobic digester and biochar plants [25, 121, 122] are vital components within the BiSC and instrumental to meet the UK government's zero emissions target. They should be considered in the planning for the updating of existing steel production plants and for any new steel plant development or refurbishment. As the UK tree cover stands at 13.2% per 2020 (3.2 million ha, 66.65 m people = 0.048 ha per capita) [123]), it is fair to say that this is the lowest percentage in the Northern Hemisphere. EU forests and wooded land cover over 182 million hectares (42%) of the EU's total land area. Biochar [22, 23, 70] can easily be used as a direct replacement for coke or coal. Biogas and biomass are alternatives to commercial gases and fossil fuels [1, 89, 101, 103, 124], as their properties allow for 1:1 replacement. Using biochar instead of coke in (steel) production could reduce CO_2 emissions by 30% [21, 64, 121, 122]. DAC also provides the potential for restoring former wasteland or brownfield sites to commercially viable woodlands, the opportunity for habitat creation and increased biodiversity and an opening for an integrated approach to – not only – steel production. Meaning: by having a great proportion of the resources produced on site (biomass and biogas) and recycling plants (anaerobic digester, biochar plant, recycling plant and hydrogen extraction site) required for steel production on site or close by, greater independence from international imports (coal, coke, wood and gas) would be achieved. Additionally, it was found that the creation of woodlands could produce sizeable profits and create upskilling, training and employment opportunities – as would the integration of anaerobic digester, biochar plant, recycling plant and the hydrogen extraction site. Woodland management, soil management, plant operations and maintenance are all opportunities to create employment in the so-called green economy.

4.6 CEPS

CEPS [53] is the ideal growth enhancer for most known plants. To utilise the captured off-gas CO_2 and direct this to greenhouses in close proximity could be the best use of a by-product yet. This system has been tried and tested (Monbiot, 2023) and enhanced food production yield significantly. It has been established that for the majority of land-grown crops, the CO_2 saturation point will be reached at about 1000–1300 ppm in perfect growing conditions. Generally, 800–1000 ppm is recommended for raising tomatoes, cucumbers and pepper seedlings as well as for lettuce production (Ministry of Agriculture, Food and Rural Affairs Canada, 2023). A simple overview of the CEPS mechanisms involved can be seen, as shown in Figure 4.7.

CEPS [53] units connected with a greenhouse infrastructure are effectively a CO_2 capture and sequestration station, and the economic feasibility is based on

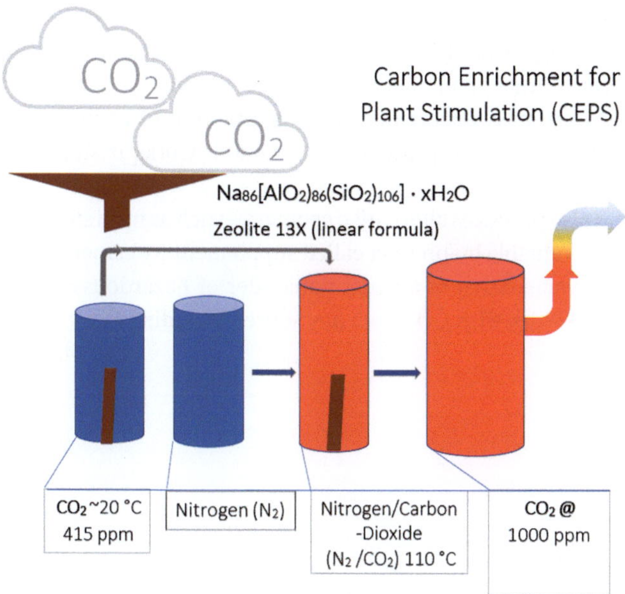

Figure 4.7 Carbon enrichment for plant stimulation (CEPS) system.

1 kWh = 3600 kWs = 3.6 MJ [110]), costing between £0.11 and £0.21/kWh, and 17.8 kJ/mol CO_2 = 17.8 kJ/44 g CO_2 for pure CO_2 versus 8.5 kJ/mol CO_2 = 8.5 kJ/44 g CO_2 for enriched, 1000 ppm CO_2, costing between £12 and £240/t CO_2 [84]. The efficiency, technical and economic viability are making the case for temperature swing absorption/desorption flue gas carbon capture.

4.7 Geomimetic process

The paradigm-shifting Geomimetic mineralisation technology uses CO_2 from any source as input to create carbon-negative aggregate for the construction industry. Blue Planet's [43] technology combines waste CO_2 with calcium (Ca) sourced from waste to produce synthetic limestone aggregates, effectively sequestering the waste CO_2 safely and permanently. The core process of the Geomimetic system is the mimicking of the Earth's natural process of creating geomass (carbonate rocks or limestone/$CaCO_3$). The calcium for effectively recycling material is sourced from:

- Demolished/returned concrete
- Cement kiln dust
- Steel slag
- Fly ash
- Air pollution control residues (APCr)
- Bauxite residue
- Silicate rocks

Waste concrete is used as a source of alkalinity and calcium in industry, and the resulting recycled concrete aggregate (RCA) can be used in structural concrete and at a higher replacement rate than that of aggregate produced from virgin material. The exposure to geomass means that the consumption of chemicals is of a negligible amount and there is no process waste. How it works is shown in Figures 4.8 and 4.9.

The Blue Planet process also produces alternative geomass – such as from steel slag and cement kiln dust – in a valuable by-product called supplementary cementitious materials (SCM) and uses valorisation of a small remainder of hazardous wastes. The Blue Planet process can work with CO_2 from any source including:

- Cement plants
- Iron and steel plants
- Natural gas power stations
- Refineries
- DAC
- Hydrogen/SMR

The beauty of this process is the fact that the CO_2 does not need to be purified, liquified or concentrated in a CEPS process, as Blue Planet is able to utilise flue gas directly as well as pure CO_2 from DAC or process streams and has been successfully demonstrated to work well on gas streams containing as little as 5% CO_2.

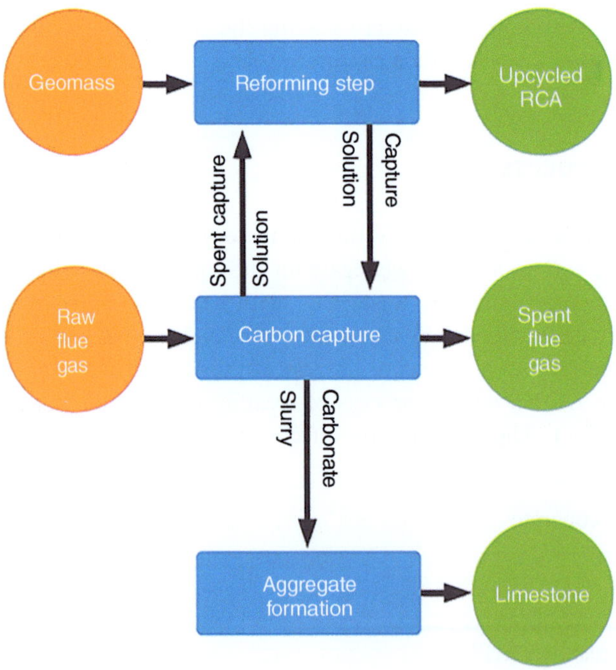

Figure 4.8 A circular solution.

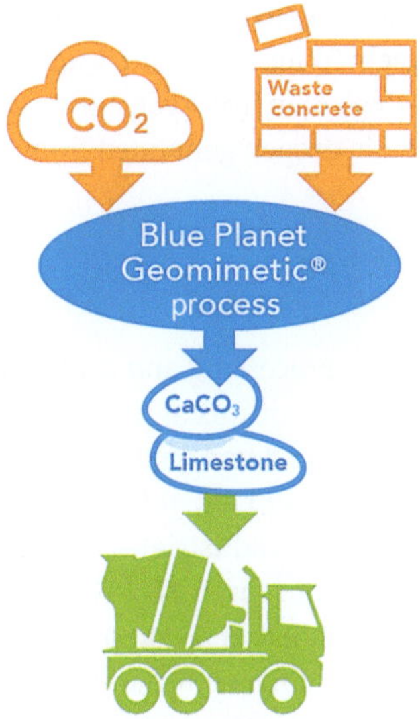

Figure 4.9 The Geomimetic process.

4.8 Anaerobic digestion

Anaerobic digestion can surely be called the 'unsung hero of waste recycling'. Organic matter, organic waste and purpose-grown crops are used to generate bioenergy via AD. In order to produce biogas, organic matter such as food waste, slurry and manure, crops and crop residues are fed into the AD chambers and converted into biogas [47, 48, 115].

The natural process of AD uses organic matter which is converted into biogas and digestate by microorganisms, such as bacteria, under the exclusion of oxygen or air. There are no costly processes involved: the bacteria just busy themselves 24/7 consuming their feed (organic matter), and the digestion of the feed by the bacteria generates biogas. The biogas contains ca. 60% methane and ca. 40% CO_2, on average, which can be used to provide heat, power or transport fuel. By removing the CO_2 proportion, biomethane is created, which is suitable to be fed directly into the public natural gas grid – just like natural gas – or it can be used as a vehicle fuel.

As shown in Table 4.3, the uptake of AD in farming as a means of recycling manure and slurry has been slow, partly due to the high capital expenditure (CAPEX) [47, 48, 115, 125–128].

Table 4.3 Percentile usage of different organic matters in AD [47, 48, 115, 125–128].

Year	Manure/ slurry	Crops	Food waste	Crop waste	Other waste
2018/19	16%	31%	31%	4%	19%
2019/20	17%	30%	29%	4%	20%

AD is a vital component in the BiSC, not only to recycle waste but also to generate biogas from by-products as a result of steelmaking (slag, slurry and dust). The biogas can not only provide a source for generating green hydrogen but also can be sent back into the steelmaking cycle, as well as the hydrogen generated from it.

4.9 Sewage treatment

During steelmaking, fresh water is used for processing and cooling purposes and is consequently discharged, whilst containing a toxic mix of recalcitrant contaminants. The advanced oxidation processes (AOPs) exhibited higher efficacy in removing recalcitrant contaminants (heavy metals >95% and phenol >99%), but the economic viability is the main limiting factor restricting industrial-scale application. As an alternative, and hence the implementation into the BiSC [44], the combination of a series of biological reactors, in cascade installed in AD and sewage treatment stations, can enable better control over the microbial community, thus improving the operational parameters whilst providing an economical and sustainable solution for effluent treatment and water reuse [129]. As the water consumption is ca. 10 m^3/t of crude steel and contains up to 5% grease and oil, treatments of these effluents are necessary and the following processes have been used in the industry: (i) addition of sodium chloride to break up the emulsion bond, (ii) addition of calcium hydroxide to achieve saponification of the grease and oil and (iii) addition of aluminium sulphate as flocculation agent, and calcium carbonate is used for neutralisation. The Ruthner process [130] has been developed for treating pickling wastewater [131, 132].

4.10 Biogas, biomass and hydrogen

4.10.1 Biogas

Biogas – as mentioned in Section 4.8 Anaerobic digestion – is the result of processes taking place during AD. Biogas as the result of recycled waste products is a vital component of green steel, produced using the BiSC model and strategy. Biogas is being fed back into the steelmaking process, with the raw materials for its production resulting from debris from reforested areas utilising DAC, treating slags and waste water and organic matter from external sources. Particularly, the organic matter sourced from external third parties could potentially combine three or more industries and solve multiple issues in one sweeping step: the steel industry, agriculture and agri-waste and waste-water treatment industries. By motivating landowners and farmers alike to dispose of their wastes at the nearby anaerobic digester facility, less nitrogen is being deposited on arable land and in bodies of water.

4.10.2 Biomass and green hydrogen utilisation

Innovative technologies such as HIsarna [133], GrInHy [30, 134] and HDR have CO_2-saving potential in their own right, as explained in more detail, as follows. Removing coal as a primary energy source or using (green) HDR (essentially biomass), an immediate 30% CO_2 emissions reduction is possible [11, 14, 16, 49, 56, 72, 121, 135], and therefore, replacing coal with green hydrogen or other biomass would reduce the CO_2 emissions from steelmaking potentially by the same percentage. According to Siemens [136], even a drastic 50% carbon emissions reduction is immediately possible via the utilisation of green HDR.

4.11 Multi-criteria decision analysis (MCDA) of BiSC

4.11.1 Introduction

In order to make decisions for any business to commit to capital expenditure (CAPEX), the multi-criteria decision analysis (MCDA) is used. Particularly for the steel industry, where most assets are calculated to have a three-year amortisation time, detailed analysis is vital. The first exhaustive publication on MCDA was the pioneer [137], whose work is still the benchmark for the creation of meaningful data to support decision-making in a political or industrial context. They built their practice on decision theory: (i) decision trees, (ii) modelling of uncertainty and (iii) the expected utility rule. With the application of decision factor identification theory to accommodate multifaceted future outcomes, Keeney and Raiffa created a theoretically sound multilevel integration model of the uncertainty associated with future events and the varied objectives those criteria might influence. This was followed by publications of governmental guidelines [138] which provided further guidance on the practical application of MCDA.

4.11.2 Basics of MCDA of BiSC

The first steps to take on the journey of MCDA are as follows:

Steps in multi-criteria analysis of technologies used in steel manufacturing and improved steelmaking processes:

1) Context: (i) aims of the MCDA and (ii) identify decision makers.
2) Identifying suitable projects.
3) Identifying (i) objectives, (ii) criteria and (iii) value associated with the consequence of each option.
4) Describing the expected percentage reduction in CO_2 emissions, cost and time of installation and project lifetime.
5) Attributing weighting to each criterion relative to their importance to the decision-making process.
6) Combining calculated weights and scores for each of the options to determine an overall value.
7) Sensitivity analysis to variation in scores or weighting.
8) Examination and analysis of the derived results.

It should be emphasised that the application of MCDA is not limited to scenarios with the aim to determine only one most appropriate option for implementation.

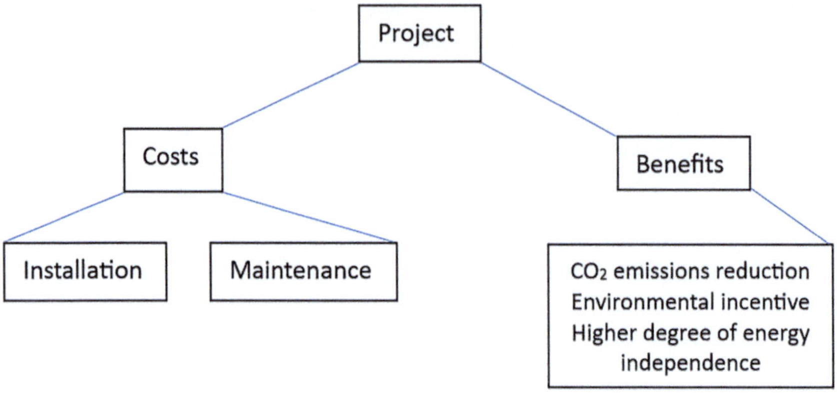

Figure 4.10 Value tree for objective determination.

Using the different steps of the MCDA process can be especially helpful when the requirement is either to short-list a set of options for further analysis or to compile a meaningful grouping of options.

In order to determine the options, the value tree for objective determination needs to be established, as shown in Figure 4.10.

Based on the value tree for objectives in Figure 4.10, the categories most suitable for the current setting of CO_2 emissions reduction and decarbonisation of the steel industry are TRL determination, percentage efficiency, installation cost, CO_2 reduction percentage, implementation project lifetime and level of environmental intrusion. Based on [137] work, the steps for the MCDA in the efforts of decarbonising the steel industry are therefore:

1) Established context: (i) the aim of the MCDA is to provide scores and weighting for the chosen projects to support the selection of the most suitable process improvement and decarbonisation strategy and (ii) identify decision makers.
2) Identifying suitable projects.
3) Identifying (i) objectives, (ii) criteria and (iii) value associated with the consequence of each option.
4) Describing the expected percentage reduction in CO_2 emissions, cost and time of installation and project lifetime.
5) Attributing weighting to each criterion relative to their importance to the decision-making process.
6) Combining calculated weights/scores for each option to determine an overall value.
7) Sensitivity analysis to variation in scores or weighting.
8) Examination and analysis of the derived results.

Having established a performance matrix, the scores for the projects need to be established.

This is the point where things become complicated: it is not normally possible to combine TRL determination, percentage efficiency, installation cost, CO_2 reduction percentage, implementation project lifetime and level of environmental intrusion.

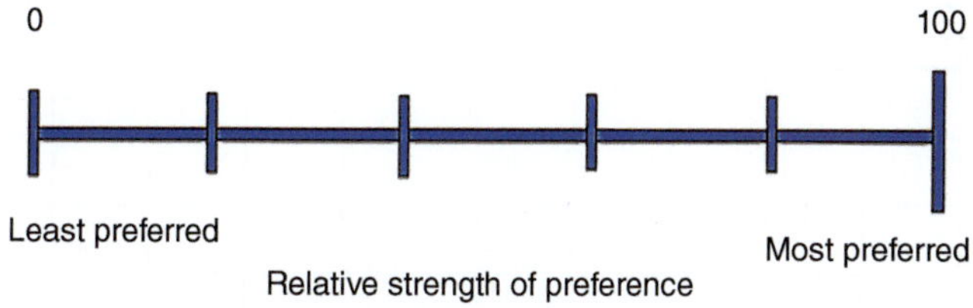

Figure 4.11 Relative strength of preference scales example.

However, with an MCDA, it can be done. The core idea is to construct scales representing preferences for the consequences (i.e. from low to high or high to low), to weight the scales for their relative importance (share out of 100) and then to calculate weighted averages across the preference scales, to obtain a combined rating as a result. For the purpose of this exercise, relative preference scales will be illustrated. The base example image is shown in Figure 4.11.

Relative strength preference scales are anchored at their left ends by the least, valued at '0', and at their right ends by the most preferred options of a criterion, valued at '100'. It so follows that the most preferred option is assigned a preference score of 100, and the least preferred option is assigned a score of 0, similar to a temperature °Celsius scale. Scores are assigned to all options, resulting in the numbers representing differences in strength of preference. It needs to be emphasised that these are relative judgements, and a certain bias can be assumed, as is human nature. However, these can be seen as sound judgements, based on a solid foundation of researched facts and datasets.

Having established the required steps to carry out a meaningful MCDA and having chosen the projects to analyse, a project duration and lifetime Gantt chart and performance matrix have been established, using MS Excel, in Tables 4.4 and 4.5.

The weighting of preferences and objectives scales contains the values and weightings at set stages, which are demonstrated in Table 4.6.

The overall weighted scores at each stage, for each objective, are then calculated, resulting in an overall score for each option by using the following formula [137]:

$$Si = \sum W_i W_i \tag{4.1}$$

4.11.3 MCDA of BiSC

There are multiple factors and outcomes to consider, which require analysis of the results and possible effects on different scenarios. The weighting and scores have been attributed to the author's best knowledge and are not influenced by any bias.

For every objective, the weighted scores were added: $S = W_1S_1 + W_2S_2 + W_3S_3 + W_4S_4 + W_5S_5 + W_6S_6$, resulting in the weighted scores on the right-hand side of the MCDA matrix in Table 4.7.

4.11.4 MCDA sensitivity analysis to variation in scores or weighting

The scores and weighting were given thorough thought, and it was determined that the best approach was to establish weighting in line with this work's aims and

Table 4.4 Gantt chart project duration and lifetime.

BiSC steps	Technique	Project or process	Company	Implementation System/performance	Project lifetime	Year 1 Month	1	2	3	4	5	6	7	8	9	10	11	12
1	Switching to green energy	Green Energy Providers	See Appendix	100% renewable energy	1 d													
2	Installing solar panels	Solar PV panels	IBC (Internal back contact)	1 kWh/4 panels = 25,667panels	2 yr													
2	Installing wind turbines	Horizontal axis w. turbine	Norvento NED100	100 kW/ £317,655.27 × 65	6 mo													
2	Installing wind turbines	Horizontal axis w. turbine	Enercon E53	800 kW/ £807,581.80 ×8	6 mo													
2	Installing wind turbines	Horizontal axis w. turbine	EWT DW61	1 MW/ £981,368.75 ×6	6 mo													
2	Installing wind turbines	Horizontal axis w. turbine	Enercon E82	3 MW/ £1,829,271.35 ×3	6 mo													
2	Installing wind turbines	Horizontal axis w. turbine	Enercon E126 EP3	3.5 MW/ £2,458,302.00 ×2	6 mo													
2	Installing wind turbines	Vertical axis wind turbine	Patriot Modular	70 kW/ £188,196.00/ × 92	6 mo													
2	Installing hydro energy sys.	Small closed loop system	Helios Atlas	6.5 MW	10 mo													
3	Using biomass/green H_2	H_2ermes: H_2 from seawater	HyCC/Tata Steel	15,000 H_2/p.a.	2 yr													
4	CO_2 filters installation	$CaCO_3$ based CO_2 absorber	Giammarco Vetrocoke	Hot potassium Carbonate	1 mo													
5	Utilisation captured CO_2	Geomimetic process	Blue Planet	100% CCUS	1 mo													
6	Process improvement	Hisarna	Tata Steel	3300 t hot metal per day	3 yr													
7	Anaerobic digestion → biogas utilisation in steelmaking	Biogas and H_2 from AD	Biogen	Biogas production	3 mo													

Table 4.5 Performance matrix.

Steps 1–7 in BiSC	Technique	Project or process	Company	System/ performance	TRLs 6-9	% Efficiency	Installation cost £	CO_2 reduction percentage	Implementation project lifetime	Level 1–5 environmental intrusion
1	Switching > green energy	Energy provider	See Appendix 4	100% renewable energy	√[a]	√	√	√	√	√
2	Installing renewables: solar	Solar PV panels	Internal back contact (IBC)	1kWh/4 panels = 25,667 panels	√	√	√	√	√	√
2	Installing renewables: wind	Horizontal axis wind turbine	Norvento nED100	100 kW/ £317,655.27 x 65	√	√	√	√	√	√
2	Installing renewables: wind	Horizontal axis wind turbine	Enercon E53	800 kW/ £807,581.80 x 8	√	√	√	√	√	√
2	Installing renewables: wind	Horizontal axis wind turbine	EWT DW61	1 MW/ £981,368.75 x 6	√	√	√	√	√	√
2	Installing renewables: wind	Horizontal axis wind turbine	Enercon E82	3 MW/ £1,829,271.35 x 3	√	√	√	√	√	√
2	Installing renewables: wind	Horizontal axis wind turbine	Enercon E126 EP3	3.5 MW/ £2,458,302.00 x 2	√	√	√	√	√	√
2	Installing renewables: wind	Vertical axis wind turbine	Patriot Modular	70kW/ £188,196.00 x 92	√	√	√	√	√	√

(Continued)

Table 4.5 (Continued)

Steps 1–7 in BiSC	Technique	Project or process	Company	System/ performance	TRLs 6-9	% Efficiency	Installation cost £	CO_2 reduction percentage	Implemen-tation project lifetime	Level 1–5 environ-mental intrusion
2	Installing renewables: hydro	Small closed-loop system	Helios Atlas	6.5 MW	√	√	√	√	√	√
3	Using biomass/ green H_2	H_2Ermes: H_2 from seawater	HyCC/Tata Steel	15,000 t H_2/p.a.	√	√	√	√	√	√
4	CO_2 filter installation	$CaCO_3$-based CO_2 absorber	Giammarco Vetrocoke	Hot potassium carbonate (HPC) solution	√	√	√	√	√	√
5	Utilisation of captured CO_2	Geomimetic: CCUS in aggregate	Blue Planet	100% CCUS	√	√	√	√	√	√
6	Process improvement	HIsarna; ironmaking by simultaneous iron ore reduction and scrap melting combined with biomass and limestone instead of lime	Tata Steel, Horizon 2020, Horizon Europe	3300 t hot-metal per day	√	√	√	√	√	√
7	Anaerobic digestion > biogas utilisation in steelmaking	Biogas and H_2 from anaerobic digestion	Biogen	Biogas production	√	√	√	√	√	√

a) The presence of a tick in the data field indicates the presence of relevant data.

Table 4.6 Weighting of preferences and objectives.

TRL determination/ 16 points weighting	Percentage efficiency/ 10 points weighting
Low > high	Low > high
1–25 = 4	0–25 = 3
26–50 = 8	26–50 = 5
51–75 = 12	51–75 = 12
76–100 = 16	76–100 = 16
Installation cost/ 25 points weighting	CO_2 reduction percentage/ 30 points weighting
High > low	Low > high
0–10 m£ = 25	0–50 = 15
11–20 m£ = 20	51–100 = 30
21–30 m£ = 15	
31–40 m£ = 10	
41-> = 5	
Implementation project lifetime/ 14 points weighting	Level of environmental intrusion/ 5 points weighting
High > low	Low > high
1 d–6 mo = 14	Level 0 = 5
7 mo–2 yr = 10	Level 1 = 4
2 yr and longer = 6	Level 2 = 3
	Level 3 = 2
	Level 4 = 1
	Level 5 = 0

objectives: establishing ways and means for decarbonisation of the iron and steel industry and achieving a higher degree of energy independence. In the twenty-first century capitalist Britain, the negative effects of cost implications preventing implementation of decarbonisation technology have been colossal, at great cost to the natural environment. But even before they [7, 8, 63, 150, 151] issued a stark warning, industry leaders have realised their responsibilities and have started globally 89 projects [152]) with the aim to establish technologies which have the potential to decarbonise heavy industry, and not only the steel industry.

Therefore, the weighting has been attributed as follows:

As identifying suitable CO_2 reduction technologies is imperative, the weighting has been attributed to be 30 points out of 100, as this is the most crucial objective. The levels have been split into two categories, 0–50% CO_2 emissions reduction potential at 15 points and 51–100% CO_2 emissions reduction potential at 30 points. Next in line is the installation cost, as the prevention of implementation of suitable

Table 4.7 MCDA of Bio Steel Cycle implementation steps.

BiSC steps 1–7	Technique	Project or process	Company	System/ performance	Technical readiness level (TRL)	S_1 score	% Efficiency	S_2 score	Installation cost £	S_3 score	Reduction % CO_2	S_4 score	Project lifetime	S_5 score	Levels 1–5 environ-mental intrusion	100 best–0 worst score	Ranking	References
1	Switching> green energy	Green energy provider	See Appendix	100% renewable energy	9	16	100	10	0.00	25	30	15	1 d	14	0	85	1	App.
2	Installation: solar panels	Solar PV panels	Internal back contact	1kWh/4 panels = 25,667 panels	9	16	23	3	51,334,000	5	30	15	2 yr	10	0	54	2	[47, 139–142]
2	Installation: wind turbine	Horizontal axis turbine	Norvento NED100	100 kW/ £317,655.27 x 65	9	16	40	5	20,647,593	15	30	15	6 mo	14	3	67	2	[47, 139–143]
2	Installation: wind turbine	Horizontal axis turbine	Enercon E53	800 kW/£807,581.80 x 8	9	16	40	5	6,460,654	25	30	15	6 mo	14	3	77	2	[47, 139–143]
2	Installation: wind turbine	Horizontal axis turbine	EWT DW61	1 MW/£981,368.75 x 6	9	16	40	5	5,888,213	25	30	15	6 mo	14	3	77	2	[47, 139–143]
2	Installation: wind turbine	Horizontal axis turbine	Enercon E82	3 MW/£1,829,271.35 x 3	9	16	40	5	5,487,814	25	30	15	6 mo	14	3	77	2	[47, 139–143]
2	Installation: wind turbine	Horizontal axis turbine	Enercon E126 EP3	3.5 MW/£2,458,302.00 x2	9	16	40	5	4,916,604	25	30	15	6 mo	14	3	77	2	[47, 139–143]
2	Installation: wind turbine	Vertical axis turbine	Patriot Modular	70 kW/£188,196.00 x 92	8	12	40	5	17,314,032	20	30	15	6 mo	14	3	68	2	[47, 139–143]
2	Installation: hydro system	Closed-loop system	Helios Atlas	6.5 MW	8	12	80	10	23,488,991	15	30	15	10 mo	10	4	63	2	[2, 86, 144]
3	Using biomass/ green H_2	H_2 from seawater	Hermes/ HyCC/Tata Steel	15,000 t H_2/p.a.	7	8	80	10	56,212,441	5	30	15	2 yr	10	1	52	2	[2, 35, 86, 133, 144, 145]
4	CO_2 filter installation	$CaCO_3$-based CO_2 absorber	Giammarco Vetrocoke	Hot potassium carbonate solution	9	16	99	10	68,145	25	100	30	1 mo	14	0	100	1	[146–148]
5	Utilisation of captured CO_2	Geomimetic process	Blue Planet	100% CCUS	9	16	100	10	0.00	25	100	30	1 mo	14	0	100	1	[43]
6	Process improvement	HIsarna	Tata Steel	3300 t hot-metal per day	6	4	80	10	60,469,288	5	55	30	3 yr	6	2	58	2	[35, 133, 149]
7	AD → biogas in steel-making	Biogas and H_2 from AD	Biogen	Biogas production	9	16	90	10	19,500,000	25	60	30	3 mo	14	5	95	2	[47, 115, 128]
		Weighting			W_1		W_2		W_3		W_4		W_5		W_6	—		
					16	—	10	—	25	—	30	—	14	—	5	—		
					Low > high	—	Low > high	—	High > low	—	Low > high	—	High > low	—	High > low	—		

decarbonisation technology due to high cost is one of the main factors preventing timely CO_2 emissions reduction in heavy industry. Costs of 0-10 m£ have been weighted at 25 points, 11–20 m£ at 20, 21–30 m£ at 15, 31–40 m£ at 10 and 41 m£ + at 5 points. It is quite obviously crucial that the identified technology is available to industry, and therefore, the TRLs have been weighted at 16 points, with TRL 6 weighted at 4 points, TRL 7 weighted at 8 points, TRL 8 weighted at 12 points and TRL 9 weighted at 16 points. The project lifetime factor is feeding into the requirement for a higher degree of energy independence and has therefore been attributed to 14 points. From fastest to slowest, the weighting has been determined to be for one day – six months at 14 points, seven months – two years at 10 points and any project taking two years and longer from start to completion at 6 points. The technology efficiency has been weighted at 10 points, with 0–25% efficiency at 3 points, 26–50% at 5 points, 51–75% at 8 points and 76–100% at 10 points. As one of this work's driving factors is preventing further damage to our natural environment and halting or at best reversing environmental damage, the environmental intrusion of the chosen technology needs to be taken into consideration too. The least environmentally intrusive technology has been set at Level 0 with the highest score of 5 points, Level 1 at 4 points, Level 2 at 3 points, Level 3 at 2 points, Level 4 at 1 point and Level 5 with a high impact factor of 0 points.

4.11.5 MCDA examination and analysis of the derived results

As cost and time as time is money seem to be the most crucial factors deciding the successful technology implementation, the initial installation cost has been plotted against the expected project lifetime. Figures 4.12, 4.13 and 4.14 demonstrate the relation.

As cost and time as time is money seem to be the most crucial factors deciding the successful technology implementation, the initial installation cost has been plotted against the expected project lifetime. Figure 4.12 demonstrates the relation.

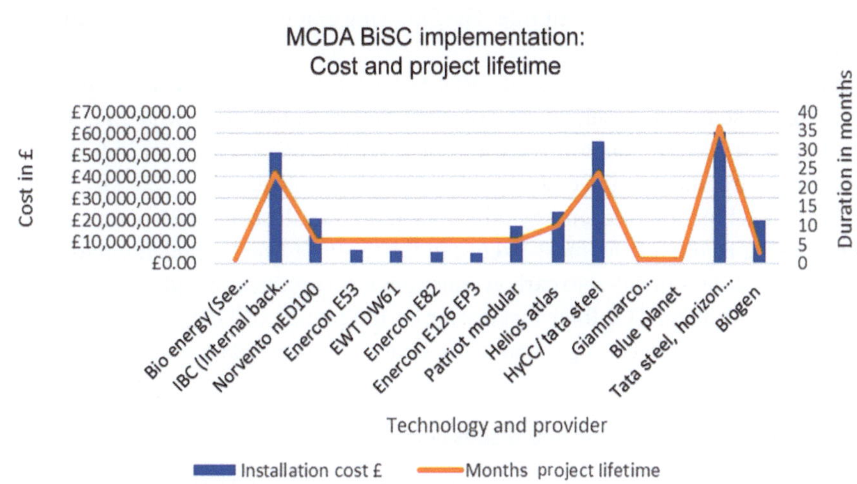

Figure 4.12 MCDA of BiSC cost and project duration.

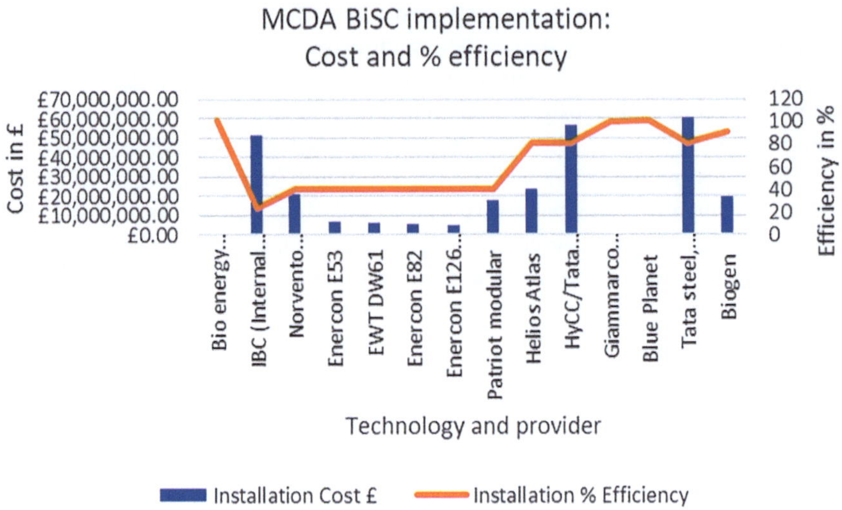

Figure 4.13 MCDA of BiSC cost and % efficiency.

Based on the graph and comparison of cost and project lifetime, the technologies standing out with relatively low cost and installation time are switching to a bioenergy provider and wind energy for energy procurement, installing CO_2 filter mechanisms and using Blue Planet's Geomimetic process for CCUS.

But this is not the only deciding factor, the technology efficiency also plays an important role, as displayed in Figure 4.13.

Comparing cost and technology efficiency, the best choices from this viewpoint are switching to a bioenergy provider and wind energy for energy procurement, installing CO_2 filter mechanisms, using Blue Planet's Geomimetic process for CCUS and installing an anaerobic digester to produce biogas.

As industry leaders have a responsibility for their company staff and stakeholders, the TRLs also play an important role. The following graph in Figure 4.14 reveals the connection.

As all chosen technologies are at TRLs 6–9, each of the options is viable as far as company staff and stakeholder responsibilities are concerned. Even Tata Steel's HIsarna is at TRL 6, which means it is now ready to be used in an industrially relevant environment in the case of key enabling technologies.

With Switching to green energy providers at an overall score of 85 points, it has been ranked as one of the most likely candidates for a timely implementation of the suggested steps to net-zero carbon emissions steel production and higher energy independence, along CO_2 filter installation at 100 points, CCUS at 100 points and AD for the production of biogas and use in heavy industry at 95 points.

4.11.6 MCDA conclusions and discussion

The MCDA has provided in-depth detail for decision makers, supporting the planning and implementation of suitable projects to achieve decarbonisation of the

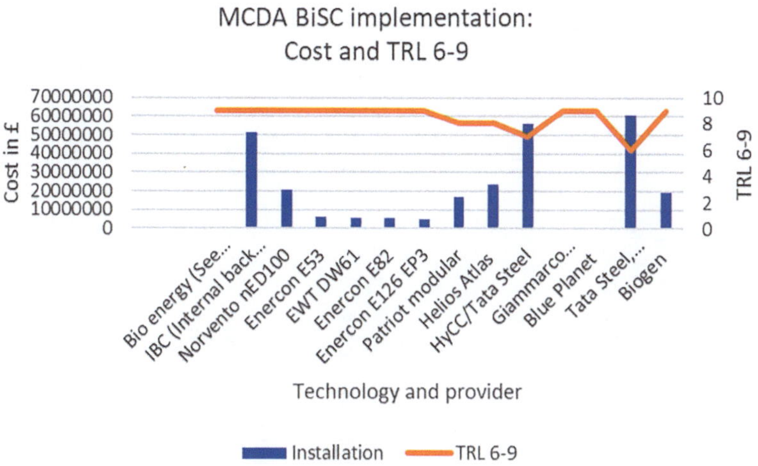

Figure 4.14 MCDA of BiSC cost and TRLs 6–9.

iron and steel industry and achieve a higher degree of energy independence. Saving on CO_2 emissions by switching to green energy providers now seems not only the easiest but also the most fiscally responsible course of action. Although solar PV panels are one of the easiest technologies to install, they are almost prohibitively costly – but the alternative as far as renewables are concerned, is readily available in the form of wind turbine technology. These steps, combined with CO_2 filter installation, carbon utilisation as aggregates and anaerobic digester installation to produce biogas seem the most sensible paths to timely and efficient decarbonisation of the iron and steel industry and for achieving a higher degree of energy independence.

4.11.7 The Bio Steel Cycle implementation stages, support schemes and cost

The BiSC implementation of stage-relevant components with the aim to decarbonise the steelmaking process requires significant upfront capital investment. The current retail price for a metric tonne of steel is currently, at the time of writing (08/2023), at £1000/ t cs, and the cost of producing 1 t of cs (08/2023) stands at £433.13. This shows gross profit margins of currently 66.87%, despite the retail price slip from £1645 in April 2022 to £1000 in August 2023. In theory, this should provide steel manufacturers with the means to invest in decarbonisation of the steelmaking process, without affecting business operations or efficiency. Additionally, there are extraordinary efforts by most governments in the developed world underway (United Kingdom/EU) to increase the momentum of decarbonisation of heavy industry.

Table 4.8 demonstrates how the implementation of the BiSC components can achieve an almost zero CO_2 emissions steel production process, whilst not financially impacting business performance, and, additionally, can even provide an added source of income (DAC).

Table 4.8 BiSC implementation, financial support and cost.

	10,000t/day × 360 days = annual capacity steel cs production per BF/BOF	*4.6 t CO_2 /t cs via BF/BOF = p.a.*	*Cost of producing one t of cs via BF/BOF, in UK, in £, 08/2023*	*UK retail price per t of cs, in £, 08/2023*			
Baseline	3,600,000.00	16,560,000.00	**433.13**	**1,000.00**			
Stage BiSC	*Technology*	*Installation cost in £*	*Maintenance cost ~5% p.a., in £*	*Amortisation over 20 yr, incl. maintenance p.a., in £*	*Potential UK/EU financial support scheme, no. as per Table 5.1*	*Theoretical UK/EU financial support scheme value in %/£*	*Added cost or -profit/t cs*
CO_2 capture post combustion	Filter insertion	68,145.00	3,407.25	3,577.61	2, 8, 12, 14, 15	100%	0.00
	CEPS $35–300/T-$CO_2$ = mean £132/t; daily capacity 10,000 t	132,000.00	6,600.00	6,930.00	2, 8, 12, 14, 15	100%	0.00
	DAC 10 ha, 5 yr	43,600.00	2,180.00	2,289.00	1	10,000 per ha	−56,400.00
Installation renewable energy technologies	Wind	4,916,604.00	245,830.20	258,121.71	6, 7, 8, 11, 14, 15	100%	0.00
	Solar	51,334,000.00	2,566,700.00	2,695,035.00	6, 7, 8, 11, 14, 15	100%	0.00
	Hydro	23,488,990.54	1,174,449.53	1,233,172.00	6, 7, 8, 11, 14, 15	100%	0.00
	Anaerobic Digestion	19,500,000.00	975,000.00	1,023,750.00	From 2 to 15	100%	0.00
Steel production process improvement	Hisarna	60,469,288.43	3,023,464.42	3,174,637.64	From 2 to 15	100%	0.00

The technology and financial support are available in developed countries to decarbonise heavy industry, particularly the steel industry, in the short term. All that is required now is the willingness to immediate implementation.

4.11.8 Conclusions and discussion

Implementation of sustainable and renewable energy components into any production cycle, such as DAC to (i) capture off-gas carbon and (ii) produce biomass for the production of biogas in AD, and CEPS (promoting growth in greenhouses for food production) and AD to produce biogas and hydrogen, and capture and utilise excess heat, can individually make significant contributions to a higher degree of energy independence for the individual commercial entity. Figure 4.15 provides more clarity by displaying the relevant percentage changes.

The MS Excel extrapolations established how the different components of the '7 steps to net-zero carbon emissions steelmaking' strategy can be implemented and installed. Installing renewable energy technologies can not only help to reduce CO_2 emissions but also supply the production site with renewable energy, where 30% of energy does not have to be imported from third parties and paid for, bringing the reduction in CO_2 emissions down to −49%. At this point in production, where biomass has already replaced the use of coke, and renewable energy technologies (solar, solar PV, wind and hydro) have been installed, an additional 30% of energy can be saved and thus does not have to be bought in, using biogas, bringing the reduction in CO_2 emissions to −65.7% [153]. This will have been produced in the link-connected anaerobic digester, which produces biogas from connected agribusinesses.

The negative percentage reduction in emissions means, in reverse, that at these points in the (steel) production process, a greater degree of energy independence can be achieved at 30%, 49% and 65.7% [44, 153]. This implies that energy at the same percentage point levels is not required to be imported from external sources, as it is produced either on site or link connected.

Besides energy and heat saving, generating their own energy will inevitably lead to achieving a higher level of independence, at least by 30%, and ideally, at 65.7%, there are savings to be had by not being forced to import and pay for energy from conventional suppliers, energy derived from renewable sources or not. We have been made painfully aware that private and business users of energy and fuel are at the mercy of corporate stakeholder interests and thus vulnerable to high price velocity. Retrofitting the existing industrial building infrastructure with renewable energy technology, with components of the '7 step to net-zero carbon emissions steelmaking'strategy, can support to (i) achieve greater energy efficiency and independence, (ii) turbo-charge the decarbonisation of energy production, (iii) decarbonise steel and industrial production and (iv) provide savings opportunities via excess heat recovery.

Observing economic principles, it can be assumed that, overall, there is a third of the cost involved, with two-thirds of savings over a 30-year investment period. Additional, significant positive milestones can be reached, such as investment into

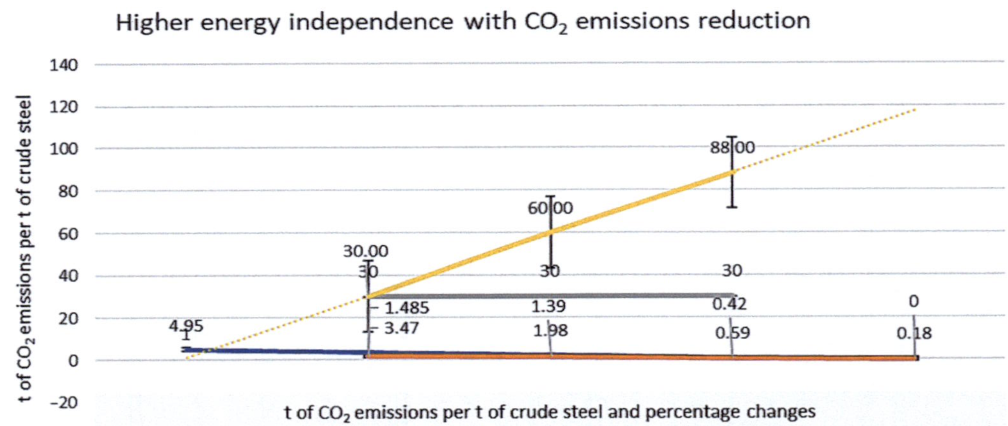

t CO2 emissions	4.95	3.47	1.98	0.59	0.18
t t reduction		1.485	1.39	0.42	0
Reduction % change		30	30	30	
Energy Independence Increase % change		30.00	60.00	88.00	
#REF!	1				

Figure 4.15 CO_2 emissions levels and reduction of the BF/BOF and BF/EAF route.

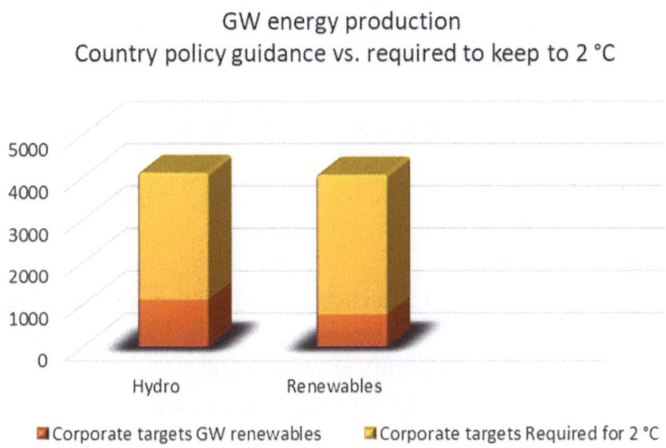

Figure 4.16 Country policy guidance versus renewables implementation required to keep the temperature rise below 2°C. Source: Adapted from UNEP [154].

a workforce with 'green' skills, future-proofing the business against energy price hikes, besides the positive effects of greater energy independence, decarbonisation of production and rehabilitation of the natural world which has been disrupted beyond recognition by the Anthropocene. Figure 4.16 shows the current level of country policy guidance versus actual requirements [154].

The legal landscape and the country's governmental guidance will have to change dramatically, in order to meet the targets, set by the Paris Agreement [36] and COP15 [155] and the dire warnings issued by the recent IPCC reports [7, 8, 63, 150, 151].

By improving their carbon capture and off-heat utilisation capabilities and investing in renewable energy technology, businesses are:

a) Supporting achieving the decarbonisation of production
b) Reaching a higher degree of energy independence
c) Achieving a higher level of asset efficiency
d) Training workforce in required "green" skills
e) Reducing their energy costs
f) Creating a viable additional income stream for their present and future operations

Not only will a higher degree of energy independence achieve an economic advantage in monetary terms, but also any installation of renewable energy systems will support limiting the global temperature rise to below 2 °C and thus support avoiding climate disaster. The CO_2 emissions by industry sector were analysed [4, 14, 66, 156], and besides the iron and steel industry, there are other industries which are CO_2 emitters, which are also generating copious amounts of off-heat, co-products and therewith resources, which could be harvested and used to power energy-dependent devices: transport, energy supply, business (including heavy industry), residential sector, agriculture and waste management.

Achieving a higher degree of energy independence is within reach of all sectors of society, made possible by technological progress and incentives and grants provided by the respective governments and countries, besides the United Kingdom. The opportunities are manifold; they merely require political willingness and implementation across all industries.

The research preceding the compilation of the current paper has provided answers which are reaching over into multiple other heavy industries, and additionally resulting in a further opportunity for a range of research directions, which are listed as follows:

1) The CO_2 captured in the iron and steel industry can be used in the building industry (concrete – Geomimetic process), agriculture (CEPS), food and drinks industry (CO_2 for beverages) and pharmaceutical applications.
2) Waste products from steelmaking, i.e. sludge and slag, can be used in infrastructural projects, such as road and bridge-building.
3) Other GHGs captured from post-combustion processes can be used in the chemical industry and be reused or recycled, eliminating the need for waste management.

The upgrading and retrofitting of existing and new steel plants and other industrial production sites provide an immense opportunity for further research, as there is a vast range of implementation points at active and abandoned production plants, not limited to:

1) Adding solar foil, panels, tiles and shingles to buildings and car parks structurally sound and capable of taking additional load.
2) Adding a hydrogen transfer network for HDR in steel production from a network of anaerobic digesters.
3) Adding wind turbines on suitable onshore brownfield sites which are not suitable for human habitation.
4) Further work is currently underway and will provide more detail on the more salient points of this paper.

References

1 International Renewable Energy Agency (IRENA) (2021). World energy transitions outlook: 1.5°c pathway [Internet]. https://www.irena.org/publications/2021/Jun/World-Energy-Transitions-Outlook (accessed 11 August 2022).

2 International Renewable Energy Agency (IRENA) (2012). Renewable energy technologies: cost analysis series [Internet]. https://www.irena.org/-/media/Files/IRENA/Agency/Publication/2012/RE_Technologies_Cost_Analysis-HYDROPOWER.pdf (accessed 7 July 2023).

3 Bataille, C., Åhman, M., Neuhoff, K. et al. (2018). A review of technology and policy deep decarbonization pathway options for making energy-intensive industry production consistent with the Paris agreement. *Journal of Cleaner Production* 187: 960–973.

4 Mandova, H., Patrizio, P., Leduc, S. et al. (2019). Achieving carbon-neutral iron and steelmaking in Europe through the deployment of bioenergy with carbon capture and storage. *Journal of Cleaner Production* 218: 118–129.

5 National Renewable Energy Laboratory (NREL) (2019). 2019 year in review – NREL's top 20 stories [Internet]. https://www.nrel.gov/news/program/2019/2019-year-in-review-nrel-top-20-stories.html (accessed 11 July 2023).

6 Waisman, H., Bataille, C., Winkler, H. et al. (2019). A pathway design framework for national low greenhouse gas emission development strategies. *Nature Climate Change* 9: 261.

7 Intergovernmental Panel for Climate Change (IPCC I) (2022). Climate change 2022: impacts, adaptation and vulnerability. [Internet]. https://www.ipcc.ch/report/ar6/wg2 (accessed 3 January 2022).

8 Intergovernmental Panel for Climate Change (IPCC) (2022a). Climate change 2022: mitigation of climate change [Internet]. https://www.un.org/en/climatechange/reports?gclid=EAIaIQobChMIxsyF3ZSl-AIVoo1oCR19Dw3IEAAYAiAAEgItyfD_BwE (accessed 30 April 2022).

9 Office for National Statistics (ONS) (2023). Foreign direct investment, experimental UK subnational estimates: 2021 – Office for National Statistics [Internet]. www.ons.gov.uk/releases/foreigndirectinvestmentexperimentaluksubnationalestimatesmarch2023 (accessed 14 July 2023).

10 Vogl, V., Åhman, M., and Nilsson, L.J. (2018). Assessment of hydrogen direct reduction for fossil-free steelmaking. *Journal of Cleaner Production* 203: 736–745.

11 Benetto, E., Popovici, E.C., Rousseaux, P., and Blondin, J. (2004). Life cycle assessment of fossil CO_2 emissions reduction scenarios in coal-biomass based electricity production. *Energy Conversion and Management.* 45 (18): 3053–3074.

12 Zakkour, P. and Cook, G. (2010). *UNIDO CCS Roadmap for Industry: High-Purity CO_2 Sources. Sectoral Assessment and Final Draft Report.* Vienna, Austria: United Nations industrial development Organization.

13 Morrow, W.R., Hasanbeigi, A., Sathaye, J., and Xu, T. (2014). Assessment of energy efficiency improvement and CO_2 emission reduction potentials in India's cement and iron & steel industries. *Journal of Cleaner Production* 65: 131–141.

14 Chen, Q., Gu, Y., Tang, Z. et al. (2018). Assessment of low-carbon iron and steel production with CO_2 recycling and utilization technologies: a case study in China. *Applied Energy* 220: 192–207.

15 Griffin, P. and Hammond, G. (2019). Industrial energy use and carbon emissions reduction in the iron and steel sector: a UK perspective. *Applied Energy* 249: 109–125.

16 Toktarova, A., Karlsson, I., Rootzén, J. et al. (2020). Pathways for low-carbon transition of the steel industry — a Swedish case study. *Energies* 13 (15): 3840.

17 Bhaskar, A., Abhishek, R., Assadi, M., and Somehesaraei, H.N. (2022). Decarbonizing primary steel production: techno-economic assessment of a hydrogen based green steel production plant in Norway. *Journal of Cleaner Production* 350: 131339.

18 Ananthanarayanan, P.S., Balachandran, G., Balasubramanian, N. et al. (2014). Contributors to volume 3. In: *Treatise on Process Metallurgy [Internet]* (ed. S. Seetharaman), xxvii–xxx. Boston: Elsevier https://www.sciencedirect.com/science/article/pii/B9780080969886099879 (accessed 5 April 2023).

19 Yang, Y., Raipala, K., and Holappa, L. (2014). Chapter 1.1 – ironmaking. In: *Treatise on Process Metallurgy [Internet]* (ed. S. Seetharaman), 2–88. Boston: Elsevier https://www.sciencedirect.com/science/article/pii/B9780080969886000171 (accessed 7 July 2023).

20 Babich, A. and Senk, D. (2015). Recent developments in blast furnace iron-making technology. In: *Iron Ore [Internet]*, 505–547. Elsevier https://linkinghub.elsevier.com/retrieve/pii/B9781782421566000174 (accessed 2 November 2022).

21 Wang, R. (2022). Energy saving technologies and optimisation of energy use for decarbonised iron and steel industry [Internet] [Doctoral]. Durham University. http://etheses.dur.ac.uk/14289 (accessed 7 July 2023).

22 Cumicheo, C., Mac Dowell, N., and Shah, N. (2019). Natural gas and BECCS: a comparative analysis of alternative configurations for negative emissions power generation. *International Journal of Greenhouse Gas Control* 90: 102798.

23 Tanzer, S.E., Blok, K., and Ramírez, A. (2021). Decarbonising industry via BECCS: promising sectors, challenges, and techno-economic limits of negative emissions. *Current Sustainable Renewable Energy Reports* 8 (4): 253–262.

24 Williams, R., Jack, C., Gamboa, D., and Shackley, S. (2021). Decarbonising steel production using CO_2 capture and storage (CCS): results of focus group discussions in a welsh steel-making community. *International Journal of Greenhouse Gas Control* 104: 103218.

25 COM (European Commission) (2019). BIONICO: a pilot plant for turning biomass directly into hydrogen | BIONICO Project | Results in brief | H2020 | CORDIS | European Commission [Internet]. https://cordis.europa.eu/article/id/394984-bionico-a-pilot-plant-for-turning-biomass-directly-into-hydrogen (accessed 6 July 2023).

26 Colla, V. and Matino, I. (2021). Editorial for the special issue: overview, state of the art, recent developments and future trends regarding hydrogen route for a green steel making process. *Matériaux & Techniques* 109 (3–4): E301.

27 Garcia-Herrero, A., Tagliapietra, S., and Vorsatz, V. (2021). Hydrogen development strategies: a global perspective. – Document – Gale Academic OneFile [Internet]. https://go-gale-com.ezproxy.staffs.ac.uk/ps/i.do?id=GALE%7CA673932617&sid=googleScholar&v=2.1&it=r&linkaccess=abs&issn=&p=AONE&sw=w&userGroupName=anon%7E4246e29b&aty=open+web+entry (accessed 6 July 2023).

28 Grasa, G., Díaz, M., Fernandez, J.R., et al. (2022). Blast furnace gas decarbonisation through calcium assisted steel-mill off-gas hydrogen production. Experimental and modelling approach. SSRN Journal [Internet]. https://www.ssrn.com/abstract=4171691 (accessed 2 November 2022).

29 Öhman, A., Karakaya, E., and Urban, F. (2022). Enabling the transition to a fossil-free steel sector: the conditions for technology transfer for hydrogen-based steelmaking in Europe. *Energy Research & Social Science* 84: 102384.

30 COM (European Commission). CORDIS | European Commission (2022). Green industrial hydrogen via reversible high-temperature electrolysis | GrInHy Project | Fact Sheet | H2020. https://cordis.europa.eu/project/id/700300 (accessed 6 July 2023).

31 Galitskaya, E. and Zhdaneev, O. (2022). Development of electrolysis technologies for hydrogen production: a case study of green steel manufacturing in the Russian Federation. *Environmental Technology & Innovation* 27: 102517.

32 ArcelorMittal (2023). Siderwin: reducing iron ore via electrolysis [Internet]. https://corporate.arcelormittal.com/corporate-library/reporting-hub/siderwin-reducing-iron-ore-via-electrolysis (accessed 9 July 2023).

33 TATA Steel UK (2020). Tata steel in the UK fact sheet [Internet]. https://www.tatasteeleurope.com/sites/default/files/Tata%20Steel%20UK%20Factsheet%202020%20(1).pdf (accessed 7 July 2023).

34 Tata Steel Europe (2020). Tata steel in Europe sustainability report 2019/2020 [Internet]. https://www.tatasteeleurope.com/sites/default/files/TSE%20Sustainability%20report%202019-20%20(EN).pdf (accessed 7 July 2023).

35 Tata Steel Europe (2023). Tata steel in Europe. Sustainability home page. https://www.tatasteeleurope.com/sustainability (accessed 11 July 2023).

36 UN United Nations (2016). Paris Agreement 2016 [Internet]. https://www.un.org/en/climatechange/paris-agreement#:~:text=To%20tackle%20climate%20change%20and,2015%3A%20the%20historic%20Paris%20Agreement (accessed 2 march 2020).

37 Tata Steel (2024). Tata steel in Europe. Zeremis. https://www.tatasteeleurope.com/sustainability/green-steel-solutions/zeremis (accessed 9 February 2024).

38 British Steel (2023). How we make steel. https://britishsteel.co.uk/what-we-do/how-we-make-steel (accessed 6 July 2023).

39 Santos, M.P.S. and Hanak, D.P. (2022). Carbon capture for decarbonisation of energy-intensive industries: a comparative review of techno-economic feasibility of solid looping cycles. *Front Chem Sci Eng* 16 (9): 1291–1317.

40 Kuramochi, T., Ramírez, A., Turkenburg, W., and Faaij, A. (2011). Techno-economic assessment and comparison of CO_2 capture technologies for industrial processes: preliminary results for the iron and steel sector. *Energy Procedia* 4: 1981–1988.

41 Kuramochi, T., Höhne, N., Schaeffer, M. et al. (2018). Ten key short-term sectoral benchmarks to limit warming to 1.5°C. *Climate Policy* 18 (3): 287–305.

42 Kiessling, S., Gohari Darabkhani, H., and Soliman, A.H. (2024). Greater energy independence with sustainable steel production. *Sustainability* 16 (3): 1174.

43 Blue Planet Systems (2023). Blue Planet. Technology. https://www.blueplanetsystems.com/technology (accessed 6 July 2023).

44 Kiessling, S., Darabkhani, H.G., and Soliman, A.H. (2022). Energies | free full-text | the bio steel cycle: 7 steps to net-zero co_2 emissions steel production

[Internet]. https://www.mdpi.com/1996-1073/15/23/8880 (accessed 11 July 2023).

45 Zhou, J., Obrist, D., Dastoor, A. et al. (2021). Vegetation uptake of mercury and impacts on global cycling. *Nature Reviews Earth and Environment* 2: 269–284.

46 Forestry Commission (2023). *Forests, Carbon and Climate Change: The UK Contribution [Internet]*. Edinburgh: Forestry Commission; p. 1–60. https://assets.publishing.service.gov.uk/government/uploads/system/uploads/attachment_data/file/1162830/Forestry-Commission-Key-Performance-Indicators-Report-2022-23.pdf (accessed 7 June 2023).

47 Nix, J. (2023). *John Nix Pocketbook – For Farm Management*, 53rde. Leicestershire: The Anderson Centre.

48 Nix, J. (2020). *John Nix Pocketbook – For Farm Management*, 50the. Leicestershire: The Anderson Centre.

49 Suopajärvi, H., Umeki, K., Mousa, E. et al. (2018). Use of biomass in integrated steelmaking – status quo, future needs and comparison to other low-CO_2 steel production technologies. *Applied Energy* 213 (C): 384–407.

50 Seethalakshmi, K.K., Jijeesh, C.M., and Balagopalan, M. (2016). *Bamboo Plantations: An Approach to Carbon Sequestration [Internet]*. Kerala/Trichur: Kerala Forest Research Institute https://www.researchgate.net/publication/215475397_Bamboo_plantations_An_approach_to_Carbon_sequestration.

51 Seethalakshmi, K.K., Balagopalan, M., and Jijeesh, C.M. (2009). Bamboo plantations: an approach to carbon sequestration. https://forest.kerala.gov.in/images/gwarm/17bmbopacs.pdf (accessed 5 march 2020).

52 Forestry Commission (2003). *Forests, Carbon and Climate Change: The UK Contribution*. Edinburgh.

53 Bao, J., Lu, W.H., Zhao, J., and Bi, X.T. (2018). Greenhouses for CO_2 sequestration from atmosphere. *Carbon Resources Conversion* 1 (2): 183–190.

54 Farla, J.C.M., Hendriks, C.A., and Blok, K. (1995). Carbon dioxide recovery from industrial processes. *Climatic Change* 29 (4): 439–461.

55 Energy Information Administration (EIA) (2021). CO_2 per kWh [Internet]. https://www.eia.gov/tools/faqs/faq.php?id=74&t=11 (accessed 11 August 2022).

56 Kittipongvises, S. (2017). Assessment of environmental impacts of limestone quarrying operations in Thailand. *Environmental and Climate Technologies* 20 (1): 67–83.

57 Office of Energy Efficiency & Renewable Energy (EERE) (US) (2018). http://Energy.gov. Manufacturing energy and carbon footprints (2018 MECS). https://www.energy.gov/eere/iedo/manufacturing-energy-and-carbon-footprints-2018-mecs (accessed 6 July 2023).

58 US Department of Energy Efficiency & Renewable Energy (EERE) (2021). Hydrogen production: natural gas reforming [Internet]. https://www.energy.gov/eere/fuelcells/hydrogen-production-natural-gas-reforming (accessed 6 July 2023).

59 Energy Information Administration (EIA) (2023). Net zero roadmap: a global pathway to keep the 1.5 °c goal in reach – analysis. https://www.iea.org/reports/

net-zero-roadmap-a-global-pathway-to-keep-the-15-0c-goal-in-reach (accessed 18 October 2023).

60 World Steel Association (WSA) (2021). Word Steel in figures [Internet]. https://www.worldsteel.org/en/dam/jcr:976723ed-74b3-47b4-92f6-81b6a452b86e/World%2520Steel%2520in%2520Figures%25202021.pdf (accessed 2 February 2022).

61 European Lime Association (EULA) (2012). How much lime per tonne of steel? Summary of the technical report. A competitive and efficient lime Industry – cornerstone for a sustainable Europe [Internet]. https://www.eula.eu/wp-content/uploads/2019/02/A-Competitive-and-Efficient-Lime-Industry-Summary_0.pdf (accessed 8 November 2022).

62 European Lime Association (EULA) (2012). *A Competitive and Efficient Lime Industry [Internet]*. Brussels: EULA (European Lime Association; p. 1–28. https://www.eula.eu/wp-content/uploads/2019/02/A-Competitive-and-Efficient-Lime-Industry-Summary_0.pdf (accessed 7 June 2023).

63 Intergovernmental Panel for Climate Change (IPCC) (2023). AR6 synthesis report climate change 2023 [Internet]. Report No.: AR6. https://www.ipcc.ch/report/ar6/syr/downloads/report/IPCC_AR6_SYR_SPM.pdf (accessed 5 January 2023).

64 Mandova, H., Leduc, S., Wang, C. et al. (2018). Possibilities for CO_2 emission reduction using biomass in European integrated steel plants. *Biomass and Bioenergy* 115: 231–243.

65 Franciosi, C., Voisin, A., Miranda, S., and Iung, B. (2020). Integration of I4.0 technologies with maintenance processes: what are the effects on sustainable manufacturing? *IFAC-PapersOnLine* 53 (3): 1–6.

66 Jahani, N., Sepehri, A., Vandchali, H.R., and Tirkolaee, E.B. (2021). Application of industry 4.0 in the procurement processes of supply chains: a systematic literature review. *Sustainability* 13 (14): 7520.

67 Calvo, L.M. and Domingo, R. (2015). Influence of process operating parameters on CO_2 emissions in continuous industrial plants. *Journal of Cleaner Production* 96: 253–262.

68 Gupta RC. *Treatise on Process Metallurgy Volume 3: Industrial Processes [Internet]*. Vol. 3. London: Elsevier; 2014. https://www.sciencedirect.com/topics/engineering/coke-rate (accessed 7 July 2023).

69 Copeland, C. (2013). A practical engineering approach to improving the reliability of blast furnace Tuyeres. https://www.semanticscholar.org/paper/A-Practical-Engineering-Approach-to-Improving-the-Copeland/310bee8b0bec31a17dcf88d158469e46ccbef783 (accessed 11 July 2023).

70 Keung, K. (2021). TAKACHAR Email from Kevin Keung, Research and Development lead. (18 October 2021).

71 Albanito, F., Hastings, A., Fitton, N. et al. (2019). Mitigation potential and environmental impact of centralized versus distributed BECCS with domestic biomass production in Great Britain. *GCB Bioenergy* 11 (10): 1234–1252.

72 Worrell, E., Blinde, P., Neelis, M., et al. (2010). Energy efficiency improvement and cost saving opportunities for the U.S. iron and steel industry. An energy star(R) guide for energy and plant managers [Internet]. p. LBNL-4779E,

1026806. Report No.: LBNL-4779E, 1026806. http://www.osti.gov/servlets/purl/1026806 (accessed 2 November 2022).

73 Jafari, M., Khan, M.I., Al-Ghamdi, S.G. et al. (2023). Waste heat recovery in iron and steel industry using organic Rankine cycles. *Chemical Engineering Journal* 477: 146925.

74 Inayat, A. (2023). Current progress of process integration for waste heat recovery in steel and iron industries. *Fuel* 338: 127237.

75 Li, Y. and Zhu, L. (2014). Cost of energy saving and CO_2 emissions reduction in China's iron and steel sector. *Applied Energy* 130: 603–616.

76 IETD (The Institute for Industrial Productivity) (2021). Basic oxygen furnace [Internet]. http://www.iipinetwork.org/wp-content/Ietd/content/basic-oxygen-furnace.html (accessed 8 November 2022).

77 IETD (The Institute for Industrial Productivity) (2021). Electric arc furnace [Internet]. http://www.iipinetwork.org/wp-content/Ietd/content/electric-arc-furnace.html (accessed 8 November 2022).

78 Sendi, M., Bui, M., Mac Dowell, N., and Fennell, P. (2022). Geospatial analysis of regional climate impacts to accelerate cost-efficient direct air capture deployment. *One Earth* 5 (10): 1153–1164.

79 International Energy Agency (IEA) (2000). IEA Greenhouse gas R&D programme (IEA GHG) [Internet]. https://ieaghg.org/docs/General_Docs/Reports/PH3-30%20iron-steel.pdf (accessed 7 July 2000).

80 International Energy Agency (IEA) (2023). IEA data and statistics. https://www.iea.org/data-and-statistics (accessed 7 July 2023).

81 International Energy Agency (IEA) (2018). Global energy demand grew by 2.1% in 2017, and carbon emissions rose for the first time since 2014 – News. https://www.iea.org/news/global-energy-demand-grew-by-21-in-2017-and-carbon-emissions-rose-for-the-first-time-since-2014 (accessed 7 July 2023).

82 Voestalpine (2023). voestalpine, Siemens and VERBUND are building a pilot facility for green hydrogen at the Linz location. https://www.voestalpine.com/group/en/media/press-releases/2017-02-07-voestalpine-siemens-and-verbund-are-building-a-pilot-facility-for-green-hydrogen-at-the-linz-location (accessed 10 July 2023).

83 Lu, X., Tian, W., Li, H., et al. (2022). Decarbonization options of the iron and steelmaking industry based on a three-dimensional analysis [Internet]. http://ijmmm.ustb.edu.cn/article/doi/10.1007/s12613-022-2475-7 (accessed 7 July 2023).

84 Singh, V., Buelens, L.C., Poelman, H. et al. (2022). Decarbonisation of steel mill gases in an energy-neutral chemical looping process. *Energy Conversion and Management* 254: 115248.

85 Universität Leipzig (2021). Units of energy | energy fundamentals [Internet]. https://home.uni-leipzig.de/energy/energy-fundamentals/03.htm (accessed 12 July 2023).

86 IRENA (2022). Renewable power remains cost-competitive amid fossil fuel crisis [Internet]. https://www.irena.org/news/pressreleases/2022/Jul/Renewable-

Power-Remains-Cost-Competitive-amid-Fossil-Fuel-Crisis (accessed 13 July 2023).

87 Shubham, S., Wright, N., Avallone, F., and Ianakiev A. (2023). AIAA AVIATION Forum. Aerodynamic and aeroacoustic investigation of vertical axis wind turbines with different number of blades using mid-fidelity and high-fidelity methods. https://arc.aiaa.org/doi/10.2514/6.2023-3642 (accessed 7 July 2023).

88 Bataille, C., Nilsson, L.J., and Jotzo, F. (2021). Industry in a net-zero emissions world: new mitigation pathways, new supply chains, modelling needs and policy implications. *Energy and Climate Change* 2: 100059.

89 Toktarova, A. *The Low-Carbon Steel Industry-Interactions between the Hydrogen Direct Reduction of Steel and the Electricity System – ProQuest [Internet]*. Gothenburg, Sweden Department of Space, Earth and Environment CHALMERS UNIVERSITY OF TECHNOLOGY; 2021. https://www.proquest.com/openview/c2a4d77d71403fdbffd33f250b13251d/1?pq-origsite=gscholar&cbl=2026366&diss=y (accessed 7 July 2023).

90 Peacock, B. and PV Magazine Australia (2022). Printed solar panels developed in Newcastle to power Tesla's journey around Australia. https://www.pv-magazine-australia.com/2022/04/26/printed-solar-panels-developed-in-newcastle-to-power-teslas-journey-around-australia (accessed 7 February 2024).

91 Power Roll (2023). Power roll – unique architecture for thin film electronics. https://powerroll.solar/unique-solar-film/microgrooves-and-perovskites (accessed 7 February 2024).

92 UKSOL (2023). UKSOL | The British solar PV module producer [Internet]. 260-340W poly solar modules. https://www.uksol.uk/poly-modules (accessed 7 February 2024).

93 UKSOL (2023). UKSOL | The British solar PV module producer. Flexible solar solutions & modules. https://www.uksol.uk/flexible-solar (accessed 7 February 2024).

94 UKSOL (2023). UKSOL | The British solar PV module producer. 310-700W bifacial solar modules. https://www.uksol.uk/bifacial-modules (accessed 8 February 2024).

95 Grant UK Ltd (2023). Solar thermal collectors – solar thermal panels hot water systems. https://www.grantuk.com/products/solar-thermal/grant-solar-collectors (accessed 7 February 2024).

96 ECOFLOW (2023). EcoFlow single axis solar tracker – EcoFlow UK [Internet]. https://uk.ecoflow.com/products/single-axis-solar-tracker (accessed 7 February 2024).

97 Spirit Energy (2023). Solar glass – spirit energy [Internet] www.spiritenergy.co.uk/kb-solar-glass-photovoltaic-windows (accessed 7 February 2024).

98 Wienerberger UK (2023). United Kingdom. In roof solar. www.wienerberger.co.uk/products/roof/in-roof-solar.html (accessed 7 February 2024).

99 Hanenergy (2023). Hantile solar roof tiles. https://www.hanergy.eu/hantile-solar-roof-tiles (accessed 8 February 2024).

100 The Independent (2023). 'Miracle material' solar panels to finally enter production. www.independent.co.uk/tech/perovskite-solar-panels-cells-china-b2361555.html (accessed 7 July 2023).

101 Kabeyi, M.J.B.O. and Olanrewaju, O.A. (2022). Biogas production and applications in the sustainable energy transition. *Journal of Energy* 8750221.

102 Bazooyar, B. and Darabkhani, H.G. (2020). Design, manufacture and test of a micro-turbine renewable energy combustor. *Energy Conversion and Management* 213: 112782.

103 Nezhad, M.M., Mehr, A.S., Lanzini, A. et al. (2019). Technology review and thermodynamic peformance study of a biogas-fed micro humid air turbine. *Renewable Energy* 140: 407–418.

104 National Aeronautics and Space Administration (NASA) (2023). Global climate change – vital signs of the planet. [Internet]. https://climate.nasa.gov/news/2436/co2-is-making-earth-greenerfor-now/#:~:text=Studies%20have%20shown%20that%20increased,chief%20culprit%20of%20climate%20change (accessed 7 April 2023).

105 National Aeronautics and Space Administration (NASA), Reiny, S. Climate change: vital signs of the planet. CO_2 is making earth greener—for now. https://climate.nasa.gov/news/2436/co2-is-making-earth-greenerfor-now (accessed 7 July 2023).

106 UCAR C for SE (2023). What's in the Air? | Center for Science Education [Internet]. https://scied.ucar.edu/learning-zone/air-quality/whats-in-the-air (accessed 8 February 2024).

107 United States Environmental Protection Agency (USEPA) (2023). Greenhouse Gas (GHG) emissions and removals [Internet]. https://www.epa.gov/ghgemissions (accessed 2 January 2023).

108 United States Environmental Protection Agency (USEPA) (2015). Sources of Greenhouse gas emissions [Internet]. https://http://www.epa.gov/ghgemissions/sources-greenhouse-gas-emissions (accessed 6 July 2023).

109 Rudd, L., Kulshreshtha, S., Belcher, K., and Amichev, B. (2021). Carbon life cycle assessment of shelterbelts in Saskatchewan Canada. *Journal of Environmental Management* 297: 113400.

110 Global Cement and Concrete Association (GCCA) (2021). Global cement and concrete industry announces roadmap to achieve groundbreaking 'net zero' co_2 emissions by 2050 – includes milestone commitment to cut co_2 emissions by a quarter by 2030.

111 HM Government UK GOV.UK. Finance and support for your business. https://www.gov.uk/business-finance-support (accessed 7 July 2023).

112 HM Government UK (2023) Low carbon workspaces [Internet]. www.lowcarbonworkspaces.co.uk (accessed 7 July 2023).

113 HM Government UK (2022). Green heat network fund – GOV-UK find a grant [Internet]. https://www.find-government-grants.service.gov.uk/grants/green-heat-network-fund (accessed 13 July 2023).

114 HM Government UK (2020). GOV.UK. Future support for low carbon heat. https://www.gov.uk/government/consultations/future-support-for-low-carbon-heat (accessed 13 July 2023).

115 HM Government UK (2021). GOV.UK. Section 3: anaerobic digestion. https://www.gov.uk/government/statistics/area-of-crops-grown-for-bioenergy-in-england-and-the-uk-2008-2020/section-3-anaerobic-digestion (accessed 8 February 2024).

116 HM Revenue and Customs (2022). GOV.UK. Corporation tax rates and allowances. https://www.gov.uk/government/publications/rates-and-allowances-corporation-tax/rates-and-allowances-corporation-tax (accessed 14 July 2023).

117 Forestry Commission (FC) (2003). *Forests, Carbon and Climate Change: The UK Contribution*. Edinburgh: Forestry Commission.

118 Yi, Q., Zhao, Y., Huang, Y. et al. (2018). Life cycle energy-economic-CO_2 emissions evaluation of biomass/coal, with and without CO_2 capture and storage, in a pulverized fuel combustion power plant in the United Kingdom. *Applied Energy* 225: 258–272.

119 Yi, S.H., Choi, M.E., Kim, D.H. et al. (2019). FINEX® as an environmentally sustainable ironmaking process. *Ironmaking & Steelmaking* 46 (7): 625–631.

120 EUROSTAT (2023). Forests, forestry and logging [Internet]. https://ec.europa.eu/eurostat/statistics-explained/index.php?title=Forests,_forestry_and_logging (accessed 12 July 2023).

121 Wilkerson, J. (2018). HARVARD UNIVERSITY blog biomass over coal: burning different carbon to mitigate climate change [Internet]. Science in the News. https://sitn.hms.harvard.edu/flash/2018/biomass-over-coal-burning-different-carbon-to-mitigate-climate-change (accessed 7 July 2023).

122 Wang, R., Liang, L., Wang, Y., and Roskilly, A.P. (2019). *ID: 9. Process Simulation of Blast Furnace Operation with Biomass Syngas Injection for Clean Production [Internet]*. Newcastle, UK: Sir Joseph Swan Centre for Energy Research, Newcastle University. https://www.energy-proceedings.org/wp-content/uploads/2020/02/9_Paper_0705055709.pdf (accessed 10 January 2020).

123 Forest Research (2023). Tools and resources woodland statistics [Internet]. www.forestresearch.gov.uk/tools-and-resources/statistics/statistics-by-topic/woodland-statistics (accessed 7 December 2023).

124 Mosayeb-Nezhad, M., Mehr, A.S., Lanzini, A. et al. (2019). Technology review and thermodynamic performance study of a biogas-fed micro humid air turbine. *Renewable Energy* 140: 407–418.

125 Nag, R., Auer, A., Markey, B.K. et al. (2019). Anaerobic digestion of agricultural manure and biomass – critical indicators of risk and knowledge gaps. *Science of the Total Environment* 690: 460–479.

126 Al-Alawi, M. (2007). Biohydrogen production by anaerobic biological fermentation of agriculture waste. In: *Assessment of Hydrogen Energy for Sustainable Development* (ed. J.W. Sheffield and Ç. Sheffield), 177–185. Dordrecht: Springer Netherlands (NATO Science for Peace and Security Series C: Environmental Security).

127 Green City Times (2015). What is anaerobic digestion? | Green City Times [Internet]. https://http://www.greencitytimes.com/anaerobic-digestion-a-proven-solution-to-our-waste-problem/ (accessed July 19 2023).

128 US EPA (2019). How does anaerobic digestion work? [Internet]. https://www.epa.gov/agstar/how-does-anaerobic-digestion-work (accessed 2 August 2023).

129 Rawat, A., Srivastava, A., Bhatnagar, A., and Gupta, A.K. (2023). Technological advancements for the treatment of steel industry wastewater: effluent management and sustainable treatment strategies. *Journal of Cleaner Production* 383: 135382.

130 Ghorbanzadeh, S., Hosseini, S.A., Taghdiri, A., and Alishahi, M. (2022). Water oxidation electrocatalyst: a new application area for Ruthner powder waste material. *Boletín de la Sociedad Española de Cerámica y Vidrio* 61 (4): 336–346.

131 ChemREADY (2023). Steel production. https://www.getchemready.com/steel-production/ (accessed 8 February 2024).

132 Jørgensen, S.E. (ed.) (1979). Chapter 14: waste water from the iron and steel industry and mining. In: *Studies in Environmental Science [Internet]*, 217–227. Elsevier (Industrial Waste Water Management; Vol. 5). https://www.sciencedirect.com/science/article/pii/S0166111608716054 (accessed 8 February 2024).

133 Jones, B. (2011). *Tata Steel and Rio Tinto Sign Agreement on HIsarna™ [Internet]*. Ijmuiden: COM, Tata Steel and Rio Tinto. https://www.tatasteeleurope.com/corporate/news/tata-steel-and-rio-tinto-hisarna-agreement (accessed 10 July 2023).

134 Blume, L. (2021). GrInHy2.0 – wasserstoff-niedersachsen.de [Internet]. http://hydrogen-lower-saxony.com. https://http://www.wasserstoff-niedersachsen.de/en/grinhy2-0/ (accessed 7 July 2023).

135 Liu, Z., Ciais, P., Deng, Z. et al. (2020). Near-real-time monitoring of global CO_2 emissions reveals the effects of the COVID-19 pandemic. *Nature Communications* 11 (1): 5172.

136 SIEMENS http://siemens-energy.com Global Website. (2022). Hydrogen solutions. https://www.siemens-energy.com/global/en/offerings/renewable-energy/hydrogen-solutions.html (accessed 7 July 2023).

137 Keeney, R., Raiffa, H., and Rajala, D. (1979). Decisions with multiple objectives: preferences and value trade-offs systems, man and cybernetics. *IEEE Transactions on Systems* 9: 403–403.

138 HM Government UK (2009). GOV.UK. Multi-criteria analysis manual for making government policy. https://www.gov.uk/government/publications/multi-criteria-analysis-manual-for-making-government-policy (accessed 13 July 2023).

139 Clean Energy Reviews (2023). Most efficient solar panels 2023. https://www.cleanenergyreviews.info/blog/most-efficient-solar-panels (accessed 6 July 2023).

140 The Renewable Energy Hub UK How much do solar panel systems cost? [Internet]. www.renewableenergyhub.co.uk/main/solar-panels/how-much-do-solar-panel-systems-cost (accessed 7 July 2023).

141 The Renewable Energy Hub UK (2023). How do solar thermal panels work [Internet]. www.renewableenergyhub.co.uk/main/solar-thermal-information/how-do-solar-thermal-panels-work (accessed 3 April 2024).

142 Renewables First – The Renewable Energy Company How much does a wind turbine cost?. www.renewablesfirst.co.uk/home/renewable-energy-technologies/windpower/windpower-learning-centre/how-much-does-a-wind-turbine-cost/ (accessed 7 July 2023).

143 Renewables First – The Renewable Energy Company (2015) Wind turbine planning permission – renewables first [Internet]. https://renewablesfirst.co.uk/renewable-energy-technologies/windpower/wind-turbine-consenting/wind-turbine-planning-permission/ (accessed 3 April 2024).

144 Helios Atlas (2019). How facilities with closed-loop water systems can cut costs by capturing energy from falling water [Internet]. https://www.heliosaltas.com/how-facilities-with-closed-loop-water-systems-can-cut-costs-by-capturing-energy-from-falling-water/ (accessed 7 July 2023).

145 Shumkov, I. (2022). *HyCC Partners with Port of Amsterdam for 500-MW Green Hydrogen Project [Internet]*. Amsterdam: HyCC and Tata Steel Europe. https://renewablesnow.com/news/hycc-partners-with-port-of-amsterdam-for-500-mw-green-hydrogen-project-791360/ (accessed 7 July 2023).

146 International Energy Agency (IEA) (2021a). Is carbon capture too expensive? – analysis. https://www.iea.org/commentaries/is-carbon-capture-too-expensive (accessed 7 July 2023).

147 International Energy Agency (IEA) (2021). Global CO_2 emissions rebounded to their highest level in history in 2021 – News. https://www.iea.org/news/global-co2-emissions-rebounded-to-their-highest-level-in-history-in-2021 (accessed 7 July 2023).

148 Giammarco Vetrocoke (2023). HPC – VoidCapture™. https://www.giammarco-vetrocoke.com/services/low-energy/ (accessed 28 November 2023).

149 TATA Steel UK (2020). Fact sheet [Internet]. https://www.tatasteeleurope.com/ts/sites/default/files/Tata%20Steel%20UK%20Factsheet%202020%20%281%29.pdf

150 Intergovernmental Panel for Climate Change (IPCC) (2021). *Climate Change 2021 – the Physical Science Basis Summary for Policymakers [Internet]*. Cambridge: Cambridge University Press. https://www.ipcc.ch/report/ar6/wg1/downloads/report/IPCC_AR6_WGI_SPM.pdf (accessed 7 July 2023).

151 Intergovernmental Panel for Climate Change (IPCC) (2023). Summary for policymakers. In: *Climate Change 2021 – The Physical Science Basis: Working Group I Contribution to the Sixth Assessment Report of the Intergovernmental Panel on Climate Change [Internet]*, 3–32. Cambridge: Cambridge University Press https://www.cambridge.org/core/books/climate-change-2021-the-physical-science-basis/summary-for-policymakers/8E7A4E3AE6C364220F3B76A189CC4D4C (accessed 12 July 2023).

152 Leadership Group for Industry Transition (LEADIT) (2023). Green steel tracker. https://www.industrytransition.org/green-steel-tracker/ (accessed 18 September 2023).

153 Devlin, A. and Yang, A. (2022). Regional supply chains for decarbonising steel: energy efficiency and green premium mitigation. *Energy Conversion and Management* 254: 115268.

154 United Nations EnvironmentProgram (UNEP) (2020). UN Environment. The six-sector solution to the climate crisis. https://www.unep.org/interactive/six-sector-solution-climate-change/ (accessed 13 July 2023).

155 United Nations EnvironmentProgram (UNEP) (2022). UN Environment Programme. UN Biodiversity Conference (COP 15). http://www.unep.org/un-biodiversity-conference-cop-15 (accessed 13 July 2023).

156 Office for National Statistics (ONS) (2019). GOV.UK. Provisional UK Greenhouse gas emissions national statistics 2019. https://www.gov.uk/government/statistics/provisional-uk-greenhouse-gas-emissions-national-statistics-2019 (accessed 11 July 2023).

157 Earthmill (2023). Gaia 133 11kW wind turbine | earthmill | sustainable energy specialists [Internet]. www.earthmill.co.uk/wind-turbines-for-sale/small-wind-turbines/gaia-133-11kw-wind-turbine (accessed 7 February 2024).

5

The CAT and reducing effects of the BiSC key components implementation

5.1 Introduction

This chapter focuses on the impact of the Bio Steel Cycle (BiSC) key elements in percentage changes of the individual components, with details on how the implementation will either avoid or reduce carbon emissions. Emphasis is on the overall end effect and by which percentile the applied technology or process can reduce the carbon emissions per metric tonne (t) of steel produced, which has been established, as shown in Table 5.1.

Table 5.1 CO_2 emissions at the individual stages of steelmaking and subsidiary processes.

Step in production/resource/process	t CO_2/t product	References
Extraction resources		
Coal	5.43	[1–6]
Iron ore	0.3–2.19	[7, 8]
Oxygen	0.2–0.26	[9, 10]
Limestone	0.01–0.04	[3, 7, 11–13]
Primary resource transformation		
Coal > coke oven > coke	0.38–2.59	[3, 7, 14]
Limestone > lime-kiln > lime	0.4–0.93	[3, 15, 16]
Secondary resource transformation		
Coke + iron ore > sinter	0.34–2.21	[3, 7, 17]
Iron production		
Blast furnace iron ore smelting	2.1–3.2	[3, 8, 18–30]
Basic oxygen furnace steelmaking	1.6–3.13	[3, 8, 18–32]
Finishing		
Casting	0.2–1.22	[3, 33]
Hot rolling	0.77	[3, 34–36]
Cold rolling	0.26	[3, 18, 34–36]
Passivation/coating/plating/blackening	0.15	[3, 18, 34–36]

Circular Steel Production: Pathways to Net-Zero Carbon Emissions, First Edition.
Sandra Kiessling, Hamidreza Gohari Darabkhani, and Abdel-Hamid Soliman.

5.2 BF/BOF route carbon capture

5.2.1 Steelmaking process improvements: CO_2 emissions avoiding, reducing and capturing technologies and processes

The steelmaking process has been thoroughly investigated [37], starting with the blast furnace (BF) – as it dominates the production of steel worldwide. Approximately 70% is produced by integrated BF/basic oxygen furnace (BOF) route plants and only 30% by the EAF route (71.5% and 28.0%, respectively, in 2017 [37]). With 'BF', a vertical shaft furnace is usually referred to, which produces liquid metals by the reaction of hot air being added under pressure into the bottom of the furnace to a mixture of metallic ore, coke and flux fed into the top section. Continued immediate combustion is maintained by the continuous addition of air under pressure. Furthermore, the BF relies on coke and injected coal as a source of energy and reducing agent. One tonne of coke production requires 1.2–1.6 tonnes of coal [1, 14, 38, 39]. Preheating coal before coke-making reduces coal consumption and increases coke oven productivity. The production of coke oven gas (COG) is about 300–360 m^3/tonne coke. COG has a low calorific value (6–8 GJ/tonne coke) and contains heavy hydrocarbons like tar, which are subsequently removed in such a way as to minimise downstream disturbances [1, 14, 38, 39]. The BF is a highly efficient thermal reactor. It operates with two countercurrent streams, namely descending solids and ascending gases, and behaves as a heat exchanger. The BF produces its own reducing gas, CO. Reduction reactions of iron oxide by CO produce:

$$CO_2 \,.\, C_{(s)} + CO_{2(g)} = 2CO_{(g)}Fe_3O_{4(s)} + 4CO_{(g)} = 3Fe_{(s)} + 4CO_{2(g)0-/*870}$$

The BF generates about 70% of the total amount of CO_2 in the integrated mill, and most of the CO will be burnt downstream into CO_2. The BF top gas contains ~25% CO_2. A medium-sized BF with a production rate of 4500 tonnes/day produces about 1600 Nm^3/tonne of off-gas. To reduce the consumption of chemical energy in the BF, one should decrease the amount of fuel. There are several approaches to decrease the consumption of coke, but there are two sustainable options: (i) recycling the off-gas and (ii) using biomass to replace coal injection. The consumption of coke in a modern BF is of the order of 300–350 kg/tonne of hot-metal, and including pulverised coal injection (PCI), the total is about 500 kg/tonne of hot-metal [1, 14, 38, 39].

In Table 5.2, a selection of decarbonisation projects and endeavours are detailed.

5.2.2 Discussion

The conventional linear steel manufacturing process, including all processes such as coal, iron ore and limestone extraction; crushing; pelletising; sintering; smelting; casting; rolling and finish machining, has been investigated and resulted in establishing that the BF/BOF route alone produces ~4.6 t of CO_2 emissions, which are one of the undesired by-products of every metric tonne of steel produced. The results to date of research into the CO_2 emissions at the various stages of the steelmaking process and related subprocesses have been established to be between 13 and 31 metric tonnes of CO_2 per tonne of crude steel produced, depending on the parameters within the calculations (such as including CO_2e oxygen, pressurising and heating). CO_2 emissions as a direct result of steel manufacture increased by 1.6% to 8.7 Gt CO_2 in 2020, although, in the net zero emissions by 2050 scenario, industry emissions are estimated to decrease by 2.3% annually to 6.9 Gt CO_2 by 2030 – despite expected industrial production growth [40–42]. Improved material and energy efficiency, the increased utilisation of renewable energy technologies and the development and deployment of low-carbon process routes (including CCS and hydrogen) are all considered to have an emission reduction effect of individually between 12% and 30% (Appendix 2). Fossil-fuel-derived energy generation and combustion of fossil fuels for transport – including aviation – are the two biggest sources of CO_2 emissions – worldwide [22, 23, 42–46]. Steel production is in close third place, followed by cement production and chemical and glass industries. One possible solution to combat CO_2 emissions in the steel industry is the BiSC as a model, based on circular tech economic principles. Steel scrap is considered a resource and, similar to recycling waste and by-products, is an integral part of the circular production process. Off-gases (CO_2 and other GHGs) are being captured and reutilised, and, alongside the implementation of steelmaking process improvements, furnace heat capture and utilisation, CAT, CCS and CCUS technologies and processes, and multidisciplinary external components, are closing the circle [22, 45, 47–53]. Technologies such as Reclamet (waste resource recovery, post-combustion) [54–57] and electrolysis projects, i.e. GrInHy and H2Future [58, 59] (direct water splitting: biomass > hydrogen, pre-combustion), all have an impact in the magnitude of between 12% and 25%, although further research is required to establish not only the most effective technology in terms of environmental impact but also which technology can be deployed the fastest and the most cost-efficient.

5.2.3 CO_2 capture mechanisms

The standard steel production (SSP) process in combination with the currently operational newly developed technologies can achieve a reduction of more than 50% [60] with successive implementation to less than eight metric tonnes per tonne of steel produced. By incorporating the BiSC components of CAT, CCS and CCUS into existing steel production sites, an almost 100% CO_2 emission reduction can be achieved immediately. The post-combustion capture of CO_2 (CCS) and other GHGs and the exploration of carbon scrubbing of flue gases have been explored. There are several possible technologies and processes to be considered for post-combustion carbon capture [8, 21, 22, 47, 61–64]:

- Mechanical capture
- Compression and dehydration

- Membrane installation
- Guiding off-gas through troughs of physical solvents/solid sorbents (such as Zeolite 13X) and chemical solvents
- As well as utilising metal–organic frameworks

Suitable literature was thoroughly investigated with regard to applied and innovative steelmaking procedures, CAT, CCS and CCUS processes (Chapter 4: Table 4.1: Process improvement technologies in development and installed), and improved management systems such as I4.0 [65, 66]. The key components within the BiSC are based on a circular production process and are functioning in an interactive manner. The basis for this system is the BF/BOF route and involves the aforementioned CAT, CCS and CCUS and process improvements where possible.

5.3 Renewable energy technologies

The individual technologies were already investigated in detail in previous chapters, with regard to the most widely used technologies being solar, wind, hydro and geothermal energy systems. At this point, it just needs to be emphasised that the implementation of the BiSC renewable energy technologies in (steel) production as isolated features of the BiSC has the potential to reduce CO_2 emissions by up to 65%. The results of process simulations in a range of software packages will illustrate the effects in more detail in the following text.

5.4 DAC woodlands

Direct air capture (DAC) of post-combustion CO_2 emissions is difficult to quantify as a percentage, as wind speed in Beaufort [57] has an impact on the density of the off-gas CO_2 concentrations (usually 400 ppm on average [61]), as well as wind direction, precipitation, air pressure and prevalent species. Bamboo [67], as an example, can absorb up to 17 tonnes of CO_2 emissions per hectare per year – native trees manage a maximum of >9 tonnes [58, 68–70].

5.5 CEPS and the Geomimetic© process

Carbon enrichment for plant stimulation (CEPS) uses filter elements, such as made of Zeolite 13X, as carriers to desorb CO_2, in a process costing between ca. £12 and £240 per tonne of CO_2 (without the cost for the filter elements [61]). The CEPS process has the potential to recycle >99% of CO_2 emissions and use the captured carbon to produce food. The same efficiency rating applies to the Geomimetic© [59] process, although the company behind this technology states that they are able to utilise direct post-combustion off-gas streams without the need to use filter elements and are able to absorb and recycle almost 100% of all post-combustion and other

CO_2 emissions flows. The captured carbon is then recycled into recycled aggregate concrete (RAC). In conclusion, the CEPS as well as the Geomimetic© process has the potential for almost 100% CO_2 emissions reduction, as the captured carbon is either used for food production or recycled into concrete, along with other input materials. The cost for the Geomimetic© process is site capacity and location dependent and can therefore not be reliably stated at this point.

5.6 Anaerobic digestion and sewage treatment

Another important component of the BiSC is the anaerobic digestion (AD) in conjunction with sewage treatment. The AD takes care of the debris from forestry activity, implemented to facilitate DAC, and waste products such as waste water and slurry. The by-product recycling CO_2 emissions were not part of this research but nonetheless are an important part of a holistic approach such as the circular economy and relevant processes. Ninety per cent of the fresh water used for cooling, pickling and cleaning can be recycled and play an important role in reducing the water footprint of steel production.

5.7 Biomass, biogas and hydrogen

As aforementioned, innovative technologies such as HIsarna [48], GrInHy [58, 59] and HDR have CO_2-saving potential in their own right and are therefore an integral part of the process improvement elements of the BiSC. Replacing energy derived from fossil fuels such as coal as the primary energy source and carbon reduction material in the BF with biomass can achieve an immediate 60% CO_2 emissions reduction [7, 13, 18, 23, 28, 38, 51, 54, 60]. Hydrogen direct reduction of iron ore eliminates the need for sintering and pelletising of iron ore and reduces the CO_2 emissions for iron ore preparation and pelletising to between 0.3 and 2.19 t CO_2 emissions and for the iron ore sintering process to 0.34–2.21 t of CO_2 emissions, per tonne of crude steel produced. Considering one cycle with all stages between mining and marketing, one tonne of crude steel generates up to 13 t of CO_2 emissions; these two steps alone can achieve a reduction in CO_2 emissions of >45%. According to Siemens [67, 68], a 50% carbon emissions reduction is immediately possible via the utilisation of green HDR.

5.8 CCUS

5.8.1 Introduction

The off-gas emissions within the steelmaking process can be captured, and the CO_2 is separated to form aggregate (pellets) and oxygen. Additionally, the AD unit produces biogas and hydrogen, which can be reused in the steelmaking process – hydrogen as a replacement for coal and biogas to operate microturbines. For this purpose, components had to be created displaying these processes in Aspen and INOSIM and corroborated in Simul8 processing. In the following, the sections displaying these processes are shown in Figures 5.1 and 5.2.

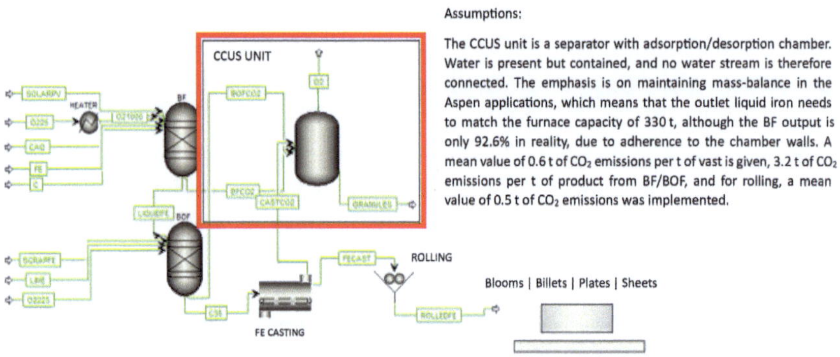

Figure 5.1 Aspen simplified layout of CCUS section (red).

The effects of switching to green energy providers, which causes a reduction in CO_2 emissions of 30%, and the installation of solar panels, which causes a further reduction of 30%, had to be incorporated into the simulations via CO_2 input streams, adhering to the achieved percentage reductions.

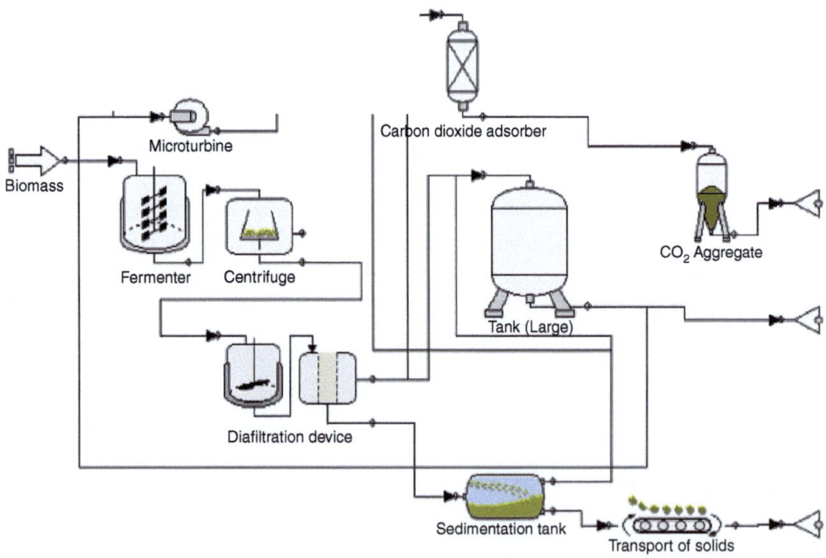

Figure 5.2 INOSIM simplified layout of CCUS section.

5.8.2 Results of process simulations

The simulation results confirmed the findings from the literature review, standard mathematical evaluation and analysis and MS Excel calculations. All steps within the BiSC are displayed in Figure 5.3.

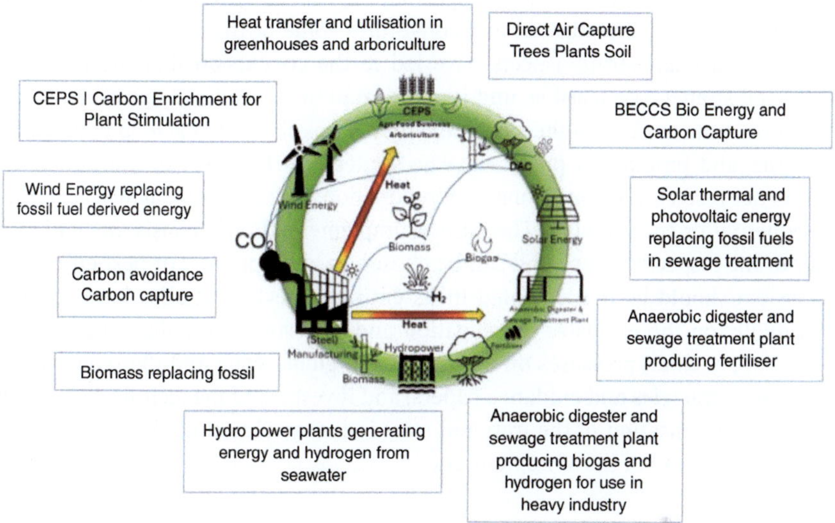

Figure 5.3 The BiSC components, all of which are either CAT, CCUS or CO_2 recycling.

5.8.3 Discussion

Establishing the CO_2 emissions at the various stages in steelmaking was the driving force for compiling these process simulations in the various software applications. Engineering aspects needed to be established, and parameters needed to be set, and for this purpose, the capabilities of four different software solutions (MS Excel, Aspen, Simul8 and INOSIM) were compared. This resulted in the realisation that most process simulation software packages are mostly working with heat, work and material flow and are functioning on the principle of mass balance: quantities in the input streams at the entry point need to be equal to quantities at the exit points. Therefore, the most feasible way to achieve mass balance was to work backwards from the exit points. A BF capacity and a BOF capacity of 330 t or 330,000 kg were assumed. The input stream resources might be transformed during the individual process stages, but essentially, the mass stream balance requirement means that the amount of input required is to be equal to the output stream.

5.8.4 CCUS conclusions

It was therefore considered essential that all quantities were accurate at every stage, working backwards from the exit and output points. The determining factor for quantities and volumes was the BF and BOF capacities, each standing at 330 t or 330,000 kg (kg being the unit of measure in the software applications). Having used all findings and parameters in the process simulations, it was established that arguably, the BiSC model and strategy [71] could be seen as the engineering solution to the pressing issue of requiring to reduce CO_2 emissions – as fast as possible. A range of decarbonisation technologies at TRLs 6–9 to avoid and reduce CO_2 emissions has been investigated and constitutes some of the components within. The analysed engineering aspects have led to the conclusion that the proposed model and strategy are feasible, and implementation is technically possible, in the short term, not confined to the steel industry but possibly applicable to most heavy industries and production processes. The strategy of the BiSC, detailed in the 7 steps to net-zero carbon emissions steel production, is easily implementable, either subsequently or individually. The ease of implementation is consistent with the order of the respective steps and can be chosen as appropriate. For best efficiency, the system should be implemented in its entirety. Previous studies have focused individually on the exclusive investigation and assessment of policy needs, skills needs, supply chain pressures on a regional and global scale and the requirement for models, strategies and guidance papers, and investigated the technical solutions for the decarbonisation of the iron and steel industry. This study is the first of its kind to combine all of the aforementioned.

5.9 Validating process flowcharts and simulations

Providing proof of concept and verification and proving the viability of the proposed BiSC model and 7-step strategy [71] are some of the crucial elements of any desktop research in the absence of a physical testing site or laboratory. There are sophisticated software tools available, which are helpful in demonstrating processes and establishing chemical and metallurgical reaction outcomes. Although the process simulations only entail a proportion of 20% of the pentagonal research approach, they are an integral part of providing a widely accepted method of research result verification. For this purpose, and as a base to start the software simulations from, an MS Excel flowchart was constructed. This contains the quantities of material flow, which most process simulation software packages are based on, and the reaction formulae, which are the conversion elements in the steelmaking process. It also provides a valuable tool to verify the results from the various simulations and subsequent reports on CO_2 emissions. The preliminary results of the literature review, mathematical standard calculations, MS Excel extrapolation and process simulations have been compared with contemporary research and verified and validated, such as by data from the German government. According to the German government and their environment agency, Germany has avoided CO_2 by using renewable energy technologies across the country by a staggering 237 million tonnes (Mte) in 2022 [72, 73]. Special attention was paid to the reduction in CO_2

emissions for electricity, as the steel manufacturing process at all stages uses a large quantity of electricity, derived from fossil fuel sources. The use of renewables has increased dramatically in Germany over the last three decades. In 2022 alone, 237 Mte was avoided by using renewable energy technologies, which constituted saving more than eight times as much compared to the CO_2e emissions savings in 1990. The biggest proportion of these savings was attributed to electricity generation from renewable sources, accounting for almost 78% of the avoided emissions in 2021, whereas 18% were attributed to the heating sector and 4% came from the transport sector – using biofuels and electricity from renewables [72, 73].

5.9.1 Process flowchart in MS excel

In order to construct process simulations in the various software solutions, a flowchart needs to be compiled with the materials and quantities involved, including the formulae and equations, which explain the chemical and metallurgical processes taking place at the different stages of the steelmaking process. For this purpose, MS Excel was deemed best suited to provide a framework for achieving this. Figure 5.4 demonstrates the flow of material, quantities of the materials involved and results per implementation of the respective steps within the BiSC.

The quantities and values in Figure 6.3 display the quantities of iron ore, limestone, lime, coke and 1000 °C oxygen required to be sent to the BF as the first furnace-based process. Following on, the product of the BF, the liquid iron, is sent to the BOF, where the hot-metal alongside lime, scrap steel and oxygen at 25 °C is being converted to liquid steel. The results on the right-hand side of Figure 5.4 summarise the quantities of CO_2 emissions during the steelmaking process, including resource extraction, calcination, coking, sintering, pressurising oxygen and through the BF/BOF route to casting, rolling and finally finish machining. The effects on the total emissions of all process stages leading to the production of steel have been summarised, as well as the effects of the BiSC components' implementation. The individual CO_2 emissions savings procedures have been applied with their percentage savings potential, and the resulting reduced totals are shown, respectively.

5.9.2 Process simulations in Aspen+V12.1, Simul8 and INOSIM

A similar approach was taken to corroborate the results from the MS Excel extrapolations, the mathematical formulation and calculations, and the Simul8 process by producing simulations in Aspen+V12.1. The calculations are set to work for a 330 t BF capacity, similar to the British Steel site in Scunthorpe [74]. The set-up parameters are shown in Figures 5.5–5.7.

Verifying the results of the preceding evaluations, computations and calculations, initially, 977,856 kg of CO_2e flow for the BF stage was reported by the Aspen system analysis, without any mitigating measures.

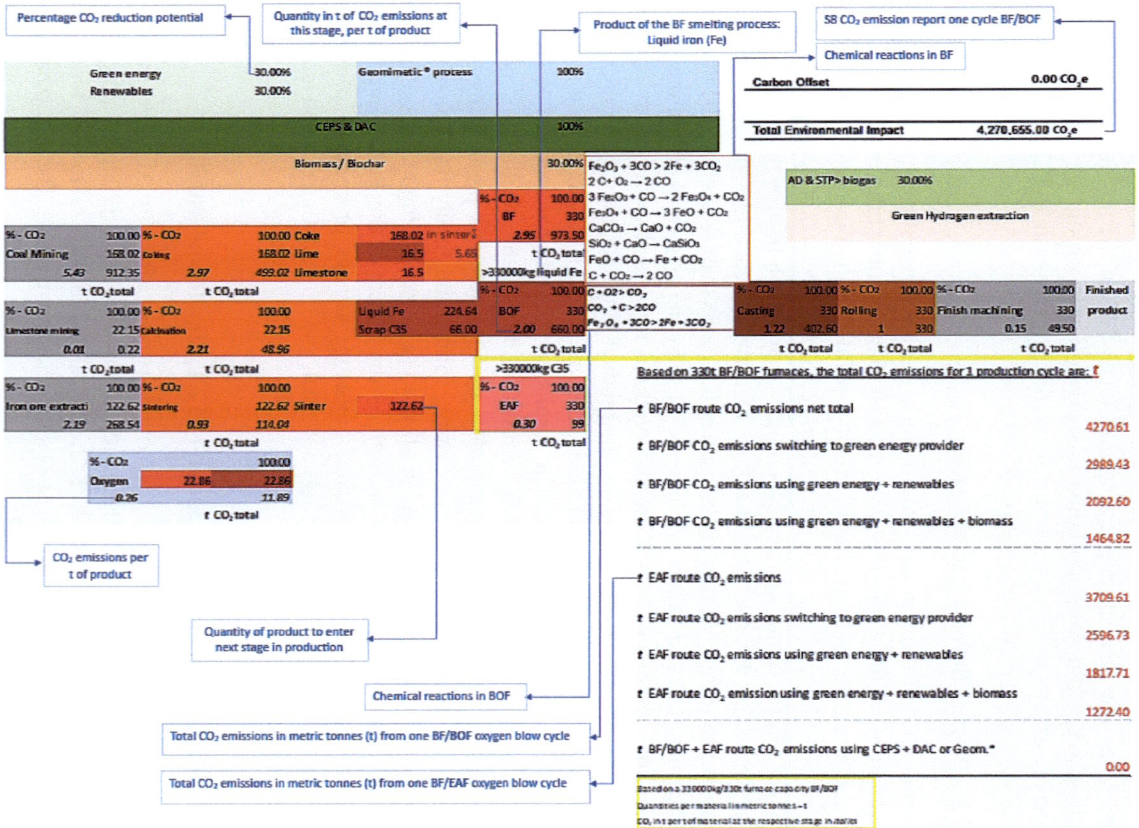

Figure 5.4 Flowchart steelmaking process with CO_2 emissions and explanatory comments.

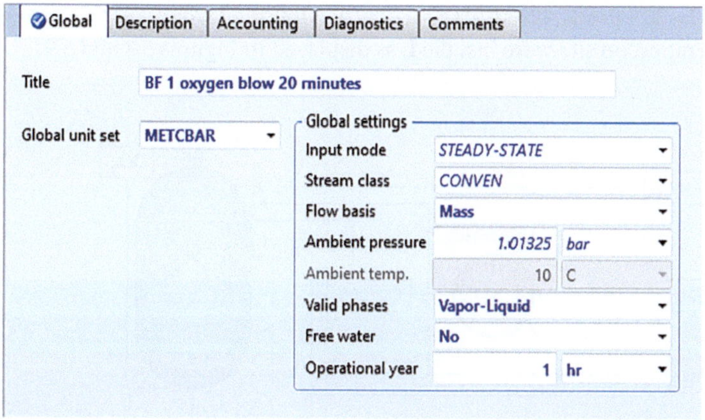

Figure 5.5 Set-up parameters in Aspen.

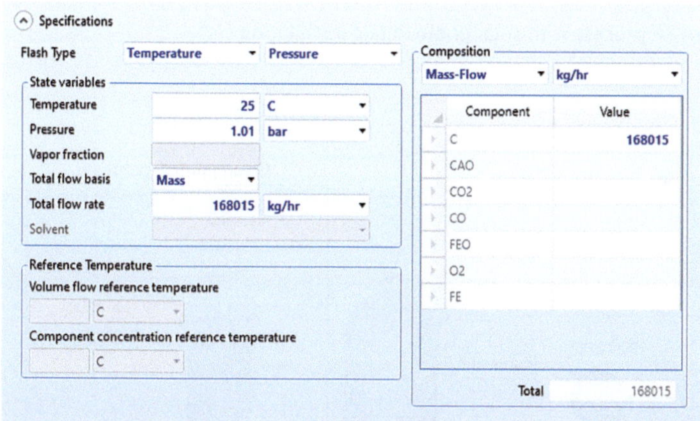

Figure 5.6 Material flow set-up parameters in Aspen with the example for coal/coke (chemical sign: C).

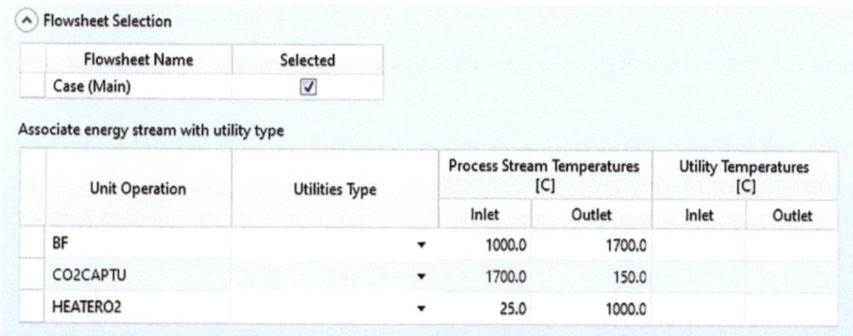

Figure 5.7 Input and outlet temperatures for the blast furnace configuration.

A reduction of ~30% to 684,499 kg of CO_2e flow was reported when solar PV panels for energy generation on site were installed, as displayed in Figures 5.8 and 5.9.

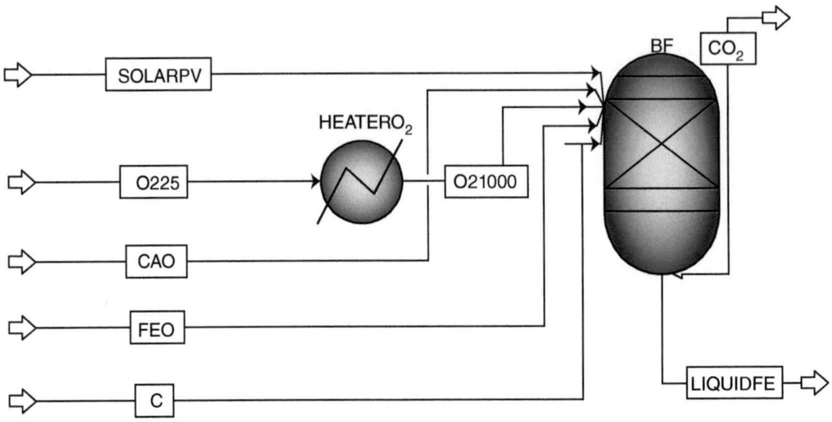

Figure 5.8 Aspen BF process with solar photovoltaic installation.

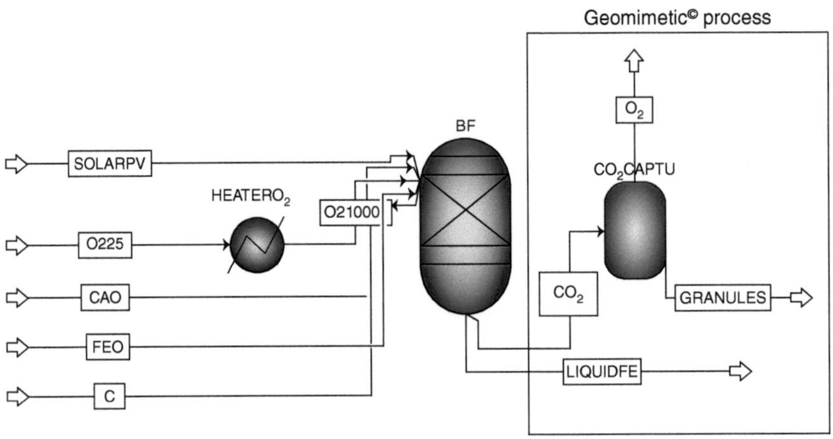

Figure 5.9 Aspen BF process with solar photovoltaic and Geomimetic© installation.

Encouraged by the verification of the preliminary research results from the literature review, mathematical investigations and MS Excel analysis by the initial process simulations in Aspen, the BF/BOF configuration was compiled, as displayed in Figure 5.10.

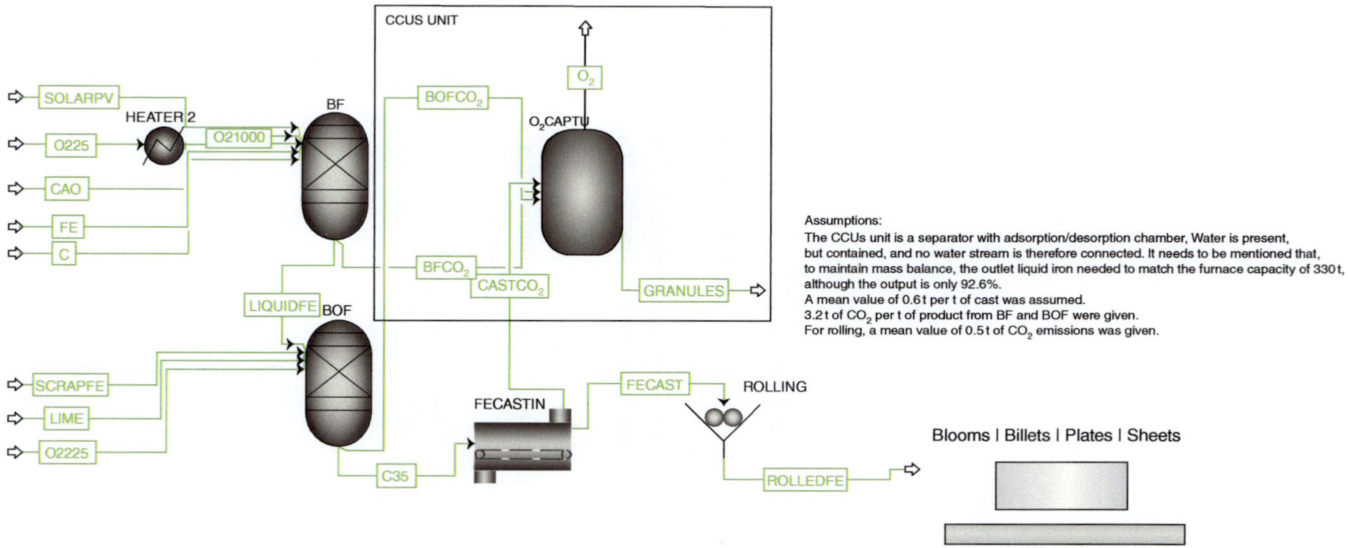

Figure 5.10 Aspen BF/BOF configuration with CCUS/BiSC.

Again, the findings were confirmed, as demonstrated in Table 5.2.

Table 5.2 Aspen stream flow report for BF.

	Units	C	CAO	FE	O21000	CO2	LIQUFE	GRANULES
Maximum Relative Error				1.1264e-16				
Cost Flow	$/hr							
− MIXED Substream								
Phase		Liquid Phase	Liquid Phase		Vapor Phase	Vapor Phase	Liquid Phase	Liquid Phase
Temperature	C	25	25	25	1000	1700	1700	150
Pressure	bar	1.01	1.01	1.01	1.01	1.01	1.01	1.01
Molar Vapor Fraction		0	0	0.927242	1	1	0	0
Molar Liquid Fraction		1	1	0.0727585	0	0	1	1
Molar Solid Fraction		0	0	0	0	0	0	0
Mass Vapor Fraction		0	0	0.909444	1	1	0	0
Mass Liquid Fraction		1	1	0.090556	0	0	1	1
Mass Solid Fraction		0	0	0	0	0	0	0
Molar Enthalpy	cal/mol	18030.2	-67175.3	-88612	7732.46	-72499.2	-1274.32	-18210.1
Mass Enthalpy	cal/gm	1123.89	-1197.9	-1974.81	241.648	-1647.34	-22.8181	-326.071
Molar Entropy	cal/mol-K	-37.3413	-13.5786	-0.0095936	11.2238	23.3392	13.6602	-3.76466
Mass Entropy	cal/gm-K	-2.32761	-0.24214	-0.000213804	0.350756	0.530317	0.2446	-0.0674103
Molar Density	mol/cc	0.280497	0.00115284	4.41842e-05	9.53974e-06	6.1555e-06	0.12173	0.129314
Mass Density	gm/cc	4.49994	0.0646482	0.00198259	0.00030526	0.000270903	6.79823	7.2218
Enthalpy Flow	cal/sec	5.24528e+07	-7.37164e+06	-5.34121e+07	1.53474e+06	-1.0822e+08	-468418	-260.959
Average MW		16.0428	56.0774	44.8711	31.9988	44.0099	55.847	55.847
+ Mole Flows	**kmol/hr**	**10473**	**395.054**	**2169.95**	**714.527**	**5373.76**	**1323.3**	**0.0515896**
+ Mole Fractions								
+ Mass Flows	**kg/hr**	**168015**	**22153.6**	**97368**	**22864**	**236499**	**73902.1**	**2.88112**

5.9.3 Process simulations in Simul8

Process simulations in Simul8 and Aspen+ were used to explain the individual production process implications.

The BF/BOF route is the most widespread method of steelmaking, representing c. 70% of current global crude steel production [75]. The layout is displayed in Figure 6.10.

In Simul8, the layout is more focused on the display of the framework and flow of streams, rather than specifying the individual components, as is possible in Aspen+, as demonstrated in Figure 5.11.

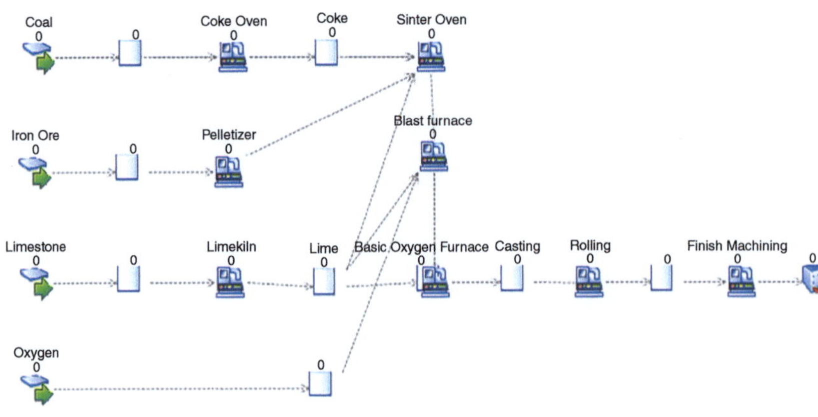

Figure 5.11 Simul8 process simulation of the BF/BOF steelmaking route.

It needs to be emphasised, though, that the following CO_2 emissions in metric tonnes represent the CO_2 emissions at their respective stage, per tonne of product, and not per tonne of steel, yet. The emissions burden on each of the input streams in the simulation software systems Simul8 and Aspen+ is already set within the system parameters and therefore calculated during the process simulations, providing a more detailed set of data per metric tonne of respective product produced. Additionally, in order to represent the overall mean CO_2 emissions burden in t per tonne of steel, the mean value of 4.95 t/t cs was allocated for entering the input stream in S8 as BF 2950 kg/2.9 t of CO_2 per metric tonne of liquid iron produced and BOF 2000 kg/2 t of CO_2 per metric tonne of steel produced. The resulting carbon report was created subsequently after operating this configuration and presented the following statement in Figure 5.12.

Carbon Emissions	21,000,00 CO_2e
Coal	5430.00 CO_2e
Iron Ore	2190.00 CO_2e
Limestone	35.00 CO_2e
Oxygen	260.00 CO_2e
Sinter Oven	930.00 CO_2e
Pelletizer	35.00 CO_2e
Limekiln	2210.00 CO_2e
Blast Furnace	2950.00 CO_2e
Basic Oxygen Furnace	2000.00 CO_2e
Casting	1220.00 CO_2e
Rolling	1000.00 CO_2e
Finish Machining	150.00 CO_2e
Coke Oven	2590.00 CO_2e
Carbon Offset	0.00 CO_2e
Total Environmental Impact	**21,000.00 CO_2e**

Figure 5.12 S8 carbon report for the BF/BOF route operation.

This was followed by the creation of a BF–EAF process route simulation, representing the second most common method of steelmaking, with a share of c. 20% of global production. Emphasis is again laid on the point that the CO_2 emissions declared are to be seen as per tonne of product at the respective stage. Figure 5.13 displays the configuration of the BF/EAF configuration of steelmaking, and the resulting CO_2 report after the configuration run is shown in Figure 5.14.

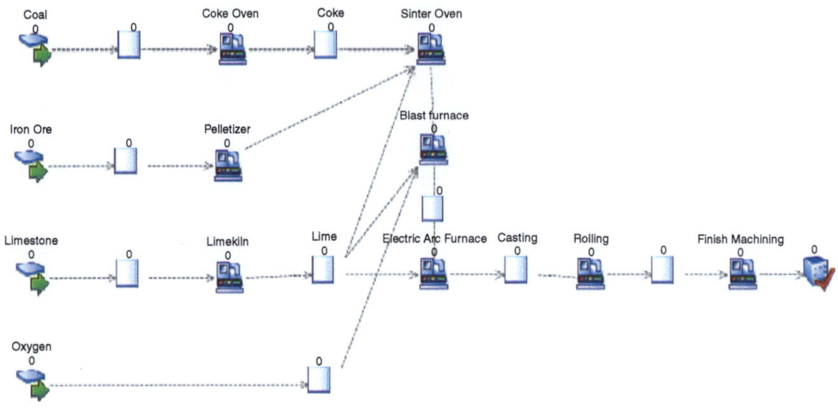

Figure 5.13 Simul8 BF/EAF configuration of steelmaking.

Carbon Emissions	19,300.00 CO_2e
Coal	5430.00 CO_2e
Iron Ore	2190.00 CO_2e
Limestone	35.00 CO_2e
Oxygen	260.00 CO_2e
Sinter Oven	930.00 CO_2e
Pelletizer	35.00 CO_2e
Limekiln	2210.00 CO_2e
Blast Furnace	2950.00 CO_2e
Electric Arc Furnace	300.00 CO_2e
Casting	1220.00 CO_2e
Rolling	1000.00 CO_2e
Finish Machining	150.00 CO_2e
Coke Oven	2590.00 CO_2e

Carbon Offset	0.00 CO_2e

Total Environmental Impact	**19,300.00 CO_2e**

Figure 5.14 BF/EAF route carbon emissions report S8.

Initially, as biomass has the potential to reduce CO_2 emissions by ~30%, the usage of biomass was implemented. The carbon reports displayed in Figures 5.14–5.16 provide the details in direct comparison with the MS Excel extrapolations and mathematical calculations (Figure 5.4).

The Simul8 process simulations and subsequent carbon reports demonstrate the effect of implementing biomass in the steelmaking process as one of the BiSC components. These simulations were followed by configurations including the implementation of the Geomimetic© process, as the implementation of this process, which includes off-gas flue stack filters, makes it possible to produce steel at almost '0' CO_2 emissions – for both the BF/BOF route and the BF/EAF process route. The carbon reports to this effect are shown in Figures 5.17 and 5.18.

Carbon Emissions	21,000,00 CO_2e
Coal	5430.00 CO_2e
Iron Ore	2190.00 CO_2e
Limestone	35.00 CO_2e
Oxygen	260.00 CO_2e
Sinter Oven	930.00 CO_2e
Pelletizer	35.00 CO_2e
Limekiln	2210.00 CO_2e
Blast Furnace	2950.00 CO_2e
Basic Oxygen Furnace	2000.00 CO_2e
Casting	1220.00 CO_2e
Rolling	1000.00 CO_2e
Finish Machining	150.00 CO_2e
Coke Oven	2590.00 CO_2e
Carbon Offset	7,000.00 CO_2e
Total Environmental Impact	**14,000.00 CO_2e**

Figure 5.15 S8 carbon report for the BF/BOF with the usage of biomass.

Carbon Emissions	19,300.00 CO_2e
Coal	5430.00 CO_2e
Iron Ore	2190.00 CO_2e
Limestone	35.00 CO_2e
Oxygen	260.00 CO_2e
Sinter Oven	930.00 CO_2e
Pelletizer	35.00 CO_2e
Limekiln	2210.00 CO_2e
Blast Furnace	2950.00 CO_2e
Electric Arc Furnace	300.00 CO_2e
Casting	1220.00 CO_2e
Rolling	1000.00 CO_2e
Finish Machining	150.00 CO_2e
Coke Oven	2590.00 CO_2e
Carbon Offset	6,400.00 CO_2e
Total Environmental Impact	**12,900.00 CO_2e**

Figure 5.16 S8 carbon report for the BF/EAF with the usage of biomass.

Carbon Emissions	21,000.00 CO_2e
Coal	5430.00 CO_2e
Iron Ore	2190.00 CO_2e
Limestone	35.00 CO_2e
Oxygen	260.00 CO_2e
Sinter Oven	930.00 CO_2e
Pelletizer	35.00 CO_2e
Limekiln	2210.00 CO_2e
Blast Furnace	2950.00 CO_2e
Basic Oxygen Furnace	2000.00 CO_2e
Casting	1220.00 CO_2e
Rolling	1000.00 CO_2e
Finish Machining	150.00 CO_2e
Coke Oven	2590.00 CO_2e
Carbon Offset	21,000.00 CO_2e
Geomimetic(c)	21,000.00 CO_2e
Total Environmental Impact	**0.00 CO_2e**

Figure 5.17 S8 carbon report for the BOF with CCUS.

Carbon Emissions	19,300.00 CO_2e
Coal	5430.00 CO_2e
Iron Ore	2190.00 CO_2e
Limestone	35.00 CO_2e
Oxygen	260.00 CO_2e
Sinter Oven	930.00 CO_2e
Pelletizer	35.00 CO_2e
Limekiln	2210.00 CO_2e
Blast Furnace	2950.00 CO_2e
Electric Arc Furnace	300.00 CO_2e
Casting	1220.00 CO_2e
Rolling	1000.00 CO_2e
Finish Machining	150.00 CO_2e
Coke Oven	2590.00 CO_2e
Carbon Offset	19,300.00 CO_2e
Geomimetic(c)	19,300.00 CO_2e
Total Environmental Impact	**0.00 CO_2e**

Figure 5.18 S8 carbon report for the EAF with CCUS.

For the process simulations in INOSIM, the same parameters as for MS Excel, Aspen and INOSIM have been applied. The result reports in INOSIM matched those produced by standard mathematical procedures, MS Excel, Aspen and Simul8. Figure 5.19 depicts an image of the INOSIM BF/BOF process simulation with CCUS, and Table 5.3 depicts the reports and the overview of the results.

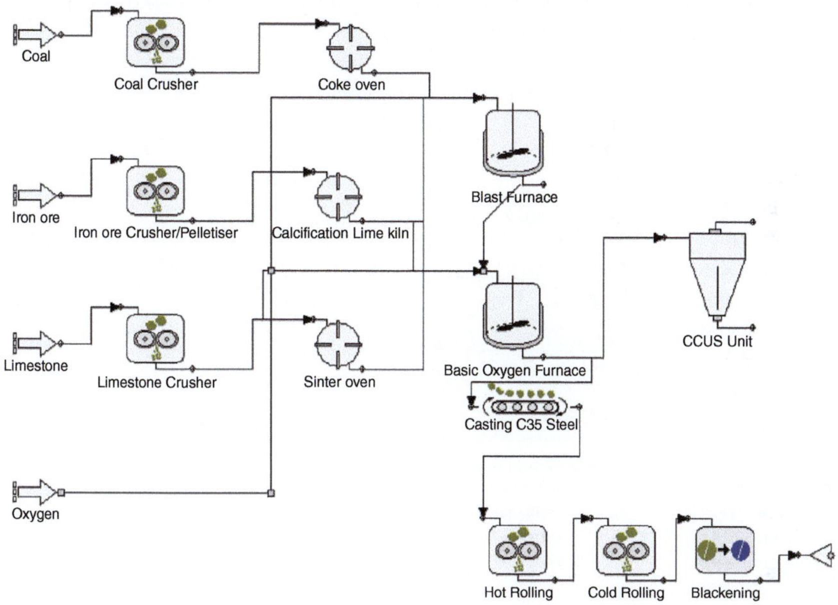

Figure 5.19 INOSIM BF/BOF standard process simulation with CCUS unit.

INOSIM provides a vast range of reactors and bioengineering components, which allows for easy integration of the innovative BiSC model. Components such as fermenters are provided, which have yet to be found to be included in any other process simulation software.

Table 5.3 BF/BOF process simulation reports' overview.

Reduction % change	Energy independence increase % change	Simul8 kg of CO_2 emissions reduction results	Aspen™ kg of CO_2 emissions successive change	Aspen™ kg of CO_2 emissions reduction result	INOSIM kg of CO_2 emissions reduction result
—	—	977,856.00	—	977,856.00	977,856.00
30	30.00	684,499.20	293,356.80	684,499.20	684,499.20
30	60.00	479,149.44	205,349.76	479,149.44	479,149.44
30	88.00	335,404.61	143,744.83	335,404.61	335,404.61
		0	143,744.83	0	0

5.9.3.1 Results

As demonstrated in the following overview Table 6.2, the results from the various process simulations are shown to be similar to those in standard mathematical calculations and MS Excel verification and extrapolations. As the results from the process simulations were almost identical, the different steps for achieving CO_2 emissions reduction were listed according to their savings potential.

Table 5.3 and Figure 5.20 provide a summary of the results reported by the process simulations in the three software applications.

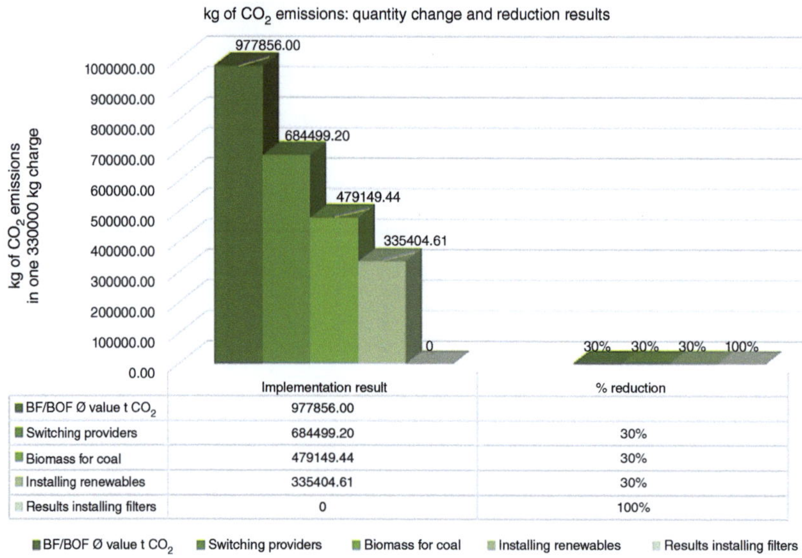

Figure 5.20 Process and flowchart simulations results' summary.

5.9.3.2 Conclusions and discussion

Conclusively, it can be said that the pentagonal approach to investigation, verification and validation of the research findings has proven to provide a range of viable results. The choice not to rely on just one process simulation software but to use three simultaneously has removed the pressure from one method to provide proof of concept. In addition, using mathematical standard procedure (algebra), statistical methods, MS Excel calculations and extrapolations has delivered proof of viability for hypothesis one: stating that it is possible to reduce CO_2 emissions in steel production by at least 30% and achieve a higher degree of energy independence at the same time.

5.10 Findings

5.10.1 Introduction

Compiling a report with the data and results of a three-year project has proven to be an interesting challenge, as the framework changed as a result of gaps in knowledge found during the literature review. As a pragmatist approach had been taken, driven by a contemporary issue, quantitative as well as qualitative data had to be organised in a meaningful way.

While reviewing the literature, gaps in the knowledge about the level of emissions in the iron and steel industry have been discovered. Influenced by events such as the Paris Agreement being ratified [76], the UN Climate Change Conferences (COP25: Madrid, Spain, from 2 to 13 December 2019 under the presidency of the Chilean government; COP26, 2–9 December 2019, Glasgow/UK) taking place and subsequent IPCC [44, 51, 52, 77, 78] reports, aims and objectives were set as a result of formalising the arising research questions:

- Which level of CO_2 emissions is being produced at which stage of the steel production process?
- How can CO_2 emissions be effectively reduced in steel production?
- Can the newly developed BiSC model and strategy support energy independence?

The methodology followed a pentagonal approach, and the research-derived results will be presented accordingly, as mentioned in Chapter 2: Methodology. Step 4: Analysis will use standard mathematical procedures and principles, further analysis and iterations via MS Excel. Step 5: Validation will be using standard mathematical procedures, principles and statistics (t-test), MS Excel functionality and process simulation results and reports.

The results of each section of the pentagonal pyramid are presented in graphs/figures, tables and calculations. The aim is to present all findings and results in a way which answers the three research questions adequately. Therefore, the questions are being set as the framework in which the findings are being presented.

5.10.2 Mathematical analysis

As already described in Chapter 2: Methodology, a set of formulae has been compiled for the purpose of calculating the total CO_2 emissions per metric tonne of crude steel produced and the emissions savings in the application of the various decarbonisation methods. For this purpose, the input data sheet for the process simulation software applications had to be compiled first to use the correct quantities of CO_2 emissions along the steelmaking process route, all levelled with CO_2 emissions per metric tonne of crude steel produced. Algebraic function calculation was deemed best suited to determine the overall CO_2 emissions of the steelmaking process, from mining to marketing, because, if the function is the solution of some polynomial (an expression of more than two algebraic terms, particularly the sum, expressed as $\sum$, of several terms), it is said to be algebraic. The decision to choose the algebraic function was made because they are functions which satisfy a polynomial equation whose coefficients are polynomials.

This means that the function $f(x)$, is an algebraic function if $y = f(x)$, and a solution of the polynomial [79].

$$p_n(x)y^n + \ldots + p_1(x)y + p_0(x) = 0.$$

The basis equation $ECO_2 = E_{\text{fos}} + E_{\text{flu}} - E_{\text{seq}}$ is expressed as:

$$CO_2E_{\text{total}} = \sum \times \Omega$$

$CO_2E_{total} = \sum \times \Omega$ ($\sum$ sigma as the sum of Ω omega products/terms) has been developed (formulae 1–4) [79–85] to simplify calculating the GHG emissions and the possible effect in real-term carbon savings and percentages of the identified technologies in steel production. For the BF/BOF, the value of 3.6–6.3 t (mean: 4.95 t) of CO_2 emissions and, for the BF/EAF route, 0.3–3.6 t (mean: 1.95 t) of CO_2 emissions per tonne of steel produced have been applied.

Theoretically, this approach will work, but there are the technical issues of different percentage-level CO_2 emissions of various reduction technologies, with different approaches to emission avoidance, mitigation and capture, at every stage along the linear steel-producing process (mining, preparation, coking, calcination, etc.). However, the research focus was to determine the levels of CO_2 emissions at all stages leading to and constituting steelmaking and following through up to the output point; Table 5.4 provides some insight into the different opinions on the matter of BF/BOF route steelmaking from a range of literary sources.

The ensuing calculations taking into consideration all CO_2 emissions from mining coal to finish machining and the BF/BOF route have therefore been used [79–85]. It needs to be emphasised that, at this point, the actual quantities determined by the literature review have been entered. The defined resource quantities (coal, lime, iron ore, oxygen, etc.) and resulting CO_2 emission will be detailed during the steelmaking process software simulations. Once this has been carried out, the values from the literature review may be replaced by the values determined as input data for the process simulations.

$$\boldsymbol{CO_2E_{total}} = \sum(\textbf{\textit{emissions per ton of steel}}) \times \boldsymbol{\Omega}\ (\textbf{\textit{output tonnes produced}}) \tag{5.1}$$

1 charge

$$\begin{aligned} CO_2E_{total,1c} = \sum(&CO_2RE_{coal} + CO_2RE_{ore} + CO_2RE_{oxy} + CO_2RE_{lime} \\ &+ CO_2PT_{coal} + CO_2PT_{lime} + CO_2ST_{sint} + CO_2BF + CO_2BOF \\ &+ CO_2C + CO_2M + CO_2FM) \times (CapBF \times t) \end{aligned}$$

1 day

$$\begin{aligned} CO_2E_{total,1c} = \sum(&CO_2RE_{coal} + CO_2RE_{ore} + CO_2RE_{oxy} + CO_2RE_{lime} \\ &+ CO_2PT_{coal} + CO_2PT_{lime} + CO_2ST_{sint} + CO_2BF \\ &+ CO_2BOF + CO_2C + CO_2M + CO_2FM) \times 330{,}000 \end{aligned}$$

30 days/1 month

$$\begin{aligned} CO_2E_{total,1c} = \sum(&CO_2RE_{coal} + CO_2RE_{ore} + CO_2RE_{oxy} + CO_2RE_{lime} \\ &+ CO_2PT_{coal} + CO_2PT_{lime} + CO_2ST_{sint} + CO_2BF \\ &+ CO_2BOF + CO_2C + CO_2M + CO_2FM) \times (330{,}000 \times 30) \end{aligned}$$

1 year (p.a.) (300,000 × 12)

$$CO_2E_{total,1c} = \sum(CO_2RE_{coal} + CO_2RE_{ore} + CO_2RE_{oxy} + CO_2RE_{lime}$$

Table 5.4 BF/BOF route emissions data sources.

Reference	a) BF t CO_2/t steel	BF/BOF as one unit	Notes	Quantity	No	CO_2 t	Class	Value
Toktarova (2020)	1.60				1	0.81		
Griffin Hammond (2019)	2.20				2	1.60		
OECD (1994)	5.85	8.35	70/30		3	1.75		
UNIDO (2014)	1.89				4	1.89		
OECD (2001)	4.65				5	2.20	Median	
Martelaro (2016), Li Zhu (2014)	7.52		SEC[c]		6	2.80	Median	2.5
Worrell et al. (2010), Prakash (2007)								
EERE (2021)	0.81	1.16	70/30		7	3.05		
Chen (2018)	1.75	2.5	70/30		8	4.65		
Pandit (2020)	3.05	4.35	70/30		9	5.85		
Hasanbeigi Springer (2019)	2.80				10	7.52		
Mandova (2019) (BF 70%/BOF 30% CO_2 emissions)	17.98	25.69	*Outlier*		Σ	32.11	3.21	
Kim Worrell (2002)	500.48	714.98	*Outlier*					
Totals Σ	32.11	757.03						
Mean	3.21		Rounded					
Median	2.50							
Mode	N/A			2.5–3.2t CO_2				
References	**b) BOF t CO_2/t steel**	**BF/BOF as one unit**	**Notes**	**Quantity**	**No**	**CO_2 t**	**Class**	**Value**
Toktarova et al. (2020)	2.20				1	0.35		
Khalid et al. (2021)	2.10				2	0.75		
OECD (1994) (BF 70%/BOF 30% CO_2 emissions)	2.51	8.35			3	1.31		
Griffin Hammond (2019)	2.76				4	1.49		
Martelaro (2016), Li Zhu (2014), Worrell et al. (2010), Prakash (2007)	1.60		SEC[c]		5	1.60		
Garvey Norman Barrett (2022)	2.20				6	2.10	Median	

(Continued)

Table 5.4 (Continued)

References	[b]BOF t CO_2/t steel	BF/BOF as one unit	Notes	Quantity	No	CO_2 t	Class	Value
UNIDO (2014)	2.27			200,000 m^3	7	2.20	Median	2.15
EERE (2021) (BF 70%/BOF 30% CO_2 emissions)	0.35	1.16			8	2.20	Mode	
Chen (2018) (BF 70%/BOF 30% CO_2 emissions)	0.75	2.50			9	2.27		
Pandit (2020)	1.31	4.35			10	2.51		
Hasanbeigi Springer (2019)	1.49	2.15			11	2.76		
Tian (2022)	3.13				12	3.13		
Mandova (2019) (BF 70%/BOF 30% CO_2 emissions)	7.71	25.69	*Outlier*		$\sum$	22.66	Mean	1.89
Kim Worrell (2002) (BF 70%/BOF 30% CO_2 emissions)	214.49	*714.98*	*Outlier*					
Totals $\sum$	22.66	759.18						
Mean	1.89		Rounded					
Median	2.15							
Mode	2.20			**1.89–2.2t CO_2**				
References	**[b]BOF t CO_2/t steel**		**By-products**	**Quantity**				
Greiner, T.H. (1997): 90% CO, 10% CO_2; Grammelis et al. (2016)	0.670		Carbon monoxide					
If CO is not recycled within the furnace, it turns to CO_2 in the air.	0.080		Carbon dioxide	180,000 m^3				
			Slag	20,000 m^3				
			Nitrous oxide					
			Magnesium sulphate					

a) Blast furnace.
b) Basic oxygen furnace.

$$+ CO_2PT_{coal} + CO_2PT_{lime} + CO_2ST_{sint} + CO_2BF + CO_2BOF + CO_2C + CO_2M + CO_2FM)) \times (330{,}000 \times 365)$$

To determine the effects of the carbon-saving or reduction methods, the following has been developed:

The base calculation for establishing the carbon reduction effects:

$$\boldsymbol{CO_2E_{total}} = \sum \boldsymbol{(emissions\ per\ ton\ of\ steel)} \times \boldsymbol{\Omega}\ \boldsymbol{(output\ tonnes\ produced)} \times \boldsymbol{Red\%} \tag{5.2}$$

In order to determine the effect of the chosen emissions reduction factors and components, the following has been calculated:

$$\boldsymbol{CO_2E_{total,1c}} = \sum \boldsymbol{(CO_2RE_{coal} + CO_2RE_{ore} + CO_2RE_{oxy} + CO_2RE_{lime} + CO_2PT_{coal} + CO_2PT_{lime} + CO_2ST_{sint} + CO_2BF + CO_2BOF + CO_2C + CO_2M + CO_2FM)} \times \boldsymbol{(CapBF \times t)} \times \boldsymbol{Red\%} \tag{5.3}$$

To support the identification of the equation components from within the text, a key to formulae and components has been compiled.

<u>Key to formula components and factors:</u>

CO_2E_{total} = CO_2 emissions total
CO_2RE_{coal} = CO_2 emissions resource extraction coal
CO_2RE_{ore} = CO_2 emissions resource extraction iron ore
CO_2RE_{oxy} = CO_2 emissions resource extraction oxygen
CO_2RE_{limes} = CO_2 emissions resource extraction limestone
CO_2PT_{coal} = CO_2 emissions primary resource transformation coal > coke
CO_2PT_{lime} = CO_2 emissions primary resource transformation limestone > lime
CO_2ST_{sint} = CO_2 emissions secondary resource transformation coke and iron > sinter
CO_2BF = CO_2 emissions blast furnace
CO_2BOF = CO_2 emissions basic oxygen furnace
CO_2EAF = CO_2 emissions electric arc furnace
CO_2C = CO_2 emissions casting
CO_2M = CO_2 emissions milling
CO_2FM = CO_2 emissions finish machining
CapBF = Capacity of blast furnace/basic oxygen furnace
t = time (charges per day) 400 tonnes per charge every 40 minutes, Ø of 10,000 tonnes/day
Red% = Reduction of CO_2 emissions in per cent (%)

The formulae are then adjusted to represent certain sections within the steelmaking process, such as the BF/BOF route:

1 full-day cycle

$$CO_2E_{total,1c} = \sum (CO_2BF + CO_2BOF) \times (CapBF \times t)$$

In order to answer research question 1, What are the current levels of CO_2 emissions of steelmaking?, the base formula has to be applied. The calculations are based on a 330-t capacity furnace, for one day's output.

For the BF/BOF route and the BF/EAF route, the mean values have been applied. The resulting operations for the BF/BOF route and the BF/EAF route are as follows:

BF/BOF route

$$CO_2E_{\text{total,1c}} = \sum(CO_2BF + CO_2BOF) \times (CapBF \times t)$$
$$CO_2E_{\text{total,1c}} = (4.95) \times (330 \times 1) =$$
$$CO_2E_{\text{total,1c}} = 4.95 \times 330 =$$
$$CO_2E_{\text{total,1c}} = 1633.50\,\text{t}$$

BF/EAF route

$$CO_2E_{\text{total,1c}} = \sum(CO_2BF + CO_2EAF) \times (CapBF \times t)$$
$$CO_2E_{\text{total,1c}} = (1.95) \times (330 \times 1) =$$
$$CO_2E_{\text{total,1c}} = 1.95 \times 330 =$$
$$CO_2E_{\text{total,1c}} = 643.50\,\text{t}$$

This means that, for the BF/BOF route, 1633.50 t of CO_2 emissions and, for the BF/EAF route, 643.50 t of CO_2 emissions per total capacity at 330 t of crude steel produced are being released.

In order to answer research question 2, How can a reduction in CO_2 emissions be achieved?, the formulae have to be adapted accordingly and will result in processing as follows. As a basis, the total charge capacity of 330 t has been assumed. Initially, the usage of biomass has been applied, resulting in a 30% reduction in CO_2 emissions.

BF/EAF route *1 charge*

$$CO_2E_{\text{total,1c}} = \sum(CO_2BF + CO_2EAF) \times (CapBF \times t) \times Red\%$$
$$CO_2E_{\text{total,1c}} = \sum(1.95) \times (330 \times 1) \times 0.7 =$$
$$CO_2E_{\text{total,1c}} = 1.95\ \text{x}\ 330 \times 0.7 =$$
$$CO_2E_{\text{total,1c}} = 450.45\,\text{t}$$

This means that, for the BF/BOF route, 1,143.45 t of CO_2 emissions and, for the BF/EAF route, 450.45 t of CO_2 emissions per charge of 330 t of crude steel produced are being released when biomass has been used instead of coal. In order to calculate the CO_2 emissions for the BF/BOF and the BF/EAF route process emissions reductions when (i) *biomass* has been used instead of coal, additionally, (ii) *renewable energy* is being installed, and subsequent to this, (iii) *biogas* is being circulated into the steelmaking process; the calculation process will be as follows:

BF/BOF route

$$CO_2E_{\text{total,1c}} = \sum(CO_2BF + CO_2BOF) \times (CapBF \times t) \times Red\%$$
$$CO_2E_{\text{total,1c}} = \sum(4.95) \times (330 \times 1) \times 0.7 =$$
$$CO_2E_{\text{total,1c}} = 4.95 \times 330 \times 0.7 =$$
$$CO_2E_{\text{total,1c}} = 1143.45\,t$$

BF/BOF route

$$CO_2E_{total,1c} = \sum (CO_2BF + CO_2BOF)\ x\ (CapBF\ x\ t)\ x\ Red\%$$
$$CO_2E_{total,1c} = (\sum (4.95) \times (330 \times 1) \times 0.7) \times 0.7 =$$
$$CO_2E_{total,1c} = ((4.95 \times 330 \times 0.7) \times 0.7 =$$
$$CO_2E_{total,1c} = 800.42\ t$$

BF/BOF route

$$CO_2E_{total,1c} = \sum (CO_2BF + CO_2BOF) \times (CapBF \times t) \times Red\%$$
$$CO_2E_{total,1c} = ((\sum (3.2 + 3.1) \times (330\ x\ 1) \times 0.7) \times 0.7) \times 0.7 =$$
$$CO_2E_{total,1c} = ((4.95 \times 330 \times 0.7) \times 0.7) \times 0.7 =$$
$$CO_2E_{total,1c} = 560.29\ t$$

This means that, for the BF/BOF route, the following t of CO_2 emissions per furnace capacity of 330 tonnes of crude steel produced has to be considered for scenarios:

(i) $CO_2E_{total,1c} = 1143.45\ t$
(ii) $CO_2E_{total,1c} = 800.42\ t$
(iii) $CO_2E_{total,1c} = 560.29\ t$

Implementation of most BiSC components and capturing CO_2 emissions via flue stack filters, such as with the Geomimetic© process, reduce the CO_2 emissions simply to almost '0'. For the BF/EAF route, the same applies. The calculations, based on the successive implementation of the BiSC components, are as follows:

BF/EAF route

$$CO_2E_{total,1c} = \sum (CO_2BF + CO_2BOF) \times (CapBF \times t) \times Red\%$$
$$CO_2E_{total,1c} = \sum (1.95) \times (330 \times 1) \times 0.7 =$$
$$CO_2E_{total,1c} = 1.95 \times 330 \times 0.7 =$$
$$CO_2E_{total,1c} = 450.45\ t$$

BF/EAF route

$$CO_2E_{total,1c} = \sum (CO_2BF + CO_2BOF) \times (CapBF \times t) \times Red\%$$
$$CO_2E_{total,1c} = (\sum (1.95) \times (330 \times 1) \times 0.7) \times 0.7 =$$
$$CO_2E_{total,1c} = ((1.95 \times 330 \times 0.7) \times 0.7 =$$
$$CO_2E_{total,1c} = 315.32\ t$$

BF/EAF route

$$CO_2E_{total,1c} = \sum (CO_2BF + CO_2BOF) \times (CapBF \times t) \times Red\%$$
$$CO_2E_{total,1c} = ((\sum (1.95) \times (330 \times 1) \times 0.7) \times 0.7) \times 0.7 =$$
$$CO_2E_{total,1c} = ((1.95 \times 330 \times 0.7) \times 0.7) \times 0.7 =$$
$$CO_2E_{total,1c} = 220.72\ t$$

This means that, for the BF/BOF route, the following t of CO_2 emissions per capacity of 330 tonnes of crude steel produced has to be considered for scenarios:

(i) Biomass (BG) $CO_2E_{total,1c} = 450.45$ t
(ii) Biomass + renewable energy (RE) $CO_2E_{total,1c} = 315.32$ t
(iii) Biomass, RE + biogas $CO_2E_{total,1c} = 220.72$ t

As with the BF/BOF route, implementation of most BiSC components within the BF/EAF route and capturing CO_2 emissions via flue stack filters, such as with the Geomimetic© process, reduce the CO_2 emissions simply to almost '0'. The previous step answers the third research question at the same time, How can a higher degree of greater energy independence be made a reality?, as the usage of biomass reduces the amount of coal and coke produced under the usage of energy derived from fossil fuels. The energy not required for coking does not have to be imported. Additionally, the installed renewable energy technologies (solar or wind, geothermal and hydropower) and the back-feeding of biogas from AD all provide the industrial processes with renewable energy from their own infrastructure. Thus, the energy produced on site will make most industries inevitably more energy independent.

5.10.3 Data modelling

As mentioned in Chapter 2: Methodology, the data gathered and insights gained during the literature review have been compiled in a comprehensive MS Excel database, which poses the foundation of the development of the BiSC model and strategy and the basis by which all presented data have been extracted.

- Figure 8.5 provides answers to question 1: What are the current levels of CO_2 emissions of steelmaking?
- Currently, the preliminary research findings are pointing towards a total of between 3.6 and 6.3 t (median: 4.95 t) of CO_2 emissions per metric tonne of crude (low-carbon) steel produced via the BF–BOF route only, which have been applied throughout the calculations and simulations.

There are some additional quantities to be considered, which will be discussed in Section 6.2.1. It seems that, within BF operations, the emissions for heating the oxygen from ambient temperatures to 1000 °C have not yet been included in any of the publications researched for answers during the literature review. One of the few exceptions was a report administered by the Australian Department for Resources, which confirmed that, indeed, not only process CO_2 emissions need to be considered but also the energy required for heating the oxygen (produced using fossil fuels) via the BF–BOF route [24]. Therefore, deductions with standard mathematical procedures have been applied to calculate the level of CO_2 emissions produced as a result of the heating process, with electricity used to increase the temperature of the injected oxygen, derived from fossil fuel sources. Additionally, the level of CO_2 emissions as a result of injecting the heated oxygen at 220 psi into the BF has also, so far, not been included in any of the sources of information. Therefore, at this point, mathematical principles have also been applied to determine the level of CO_2 emissions as a result of this exact procedure. Additionally, a question of logical deduction needs to be answered: the up to 30% scrap steel content in the basic oxygen would have to be considered as already having caused 66 t of CO_2 emissions (in a 330-t capacity furnace, similar to the British Steel BF in Scunthorpe [74]). One could argue that these would have to be added to the CO_2 emissions balance, too. Furthermore, in answering research question 2, How can a reduction in CO_2 emissions be achieved?, the literature review has provided a multitude of possible procedural and technical solutions (Chapter 4, Table 4.1), which have the potential to reduce the CO_2 emissions from ironmaking and steelmaking significantly. Particularly in Chapters 3 and 4: The Bio Steel Cycle engineering aspects, the 7-step strategy to net-zero CO_2 emissions steel production has been explored in more detail. Chapter 5 provides details of the CO_2 emissions reduction development in the application of emission reduction measures, whilst simultaneously increasing the levels of energy independence.

Figure 5.21 displays a section of the master data sheet; Table 5.1 answers the research question 1: What are the CO_2 emissions for all stages of the steelmaking process? Figure 5.22 points towards specific sections in literary sources of information for emissions along the steelmaking process. During the literature review, further insights were gained into the application of standard mathematical procedures. The purpose of this study was not to question and evaluate existing information but rather a mere compilation of established facts in a comprehensive manner to give an overview of the CO_2 emissions along the steelmaking process. Some answers in relation to research question 2, How can a reduction in the CO_2 emissions be achieved?, are given in Chapter 4, Table 4.1, and answers to research question 3 are given in Chapter 5: How can a higher degree of greater energy independence be made a reality?

CO_2 emissions from every stage of the steelmaking process have been considered. From mining to calcination of limestone, coking of coal, making sinter from lime and iron ore pellets and the BF, BOF, casting and finishing operations have all been investigated throughout – to the point of a product issue.

However, the issue with the data in the data sheet is that – at the time of compilation – the levels of CO_2 emissions represent the levels at every production stage, per metric tonne of the individual product, not per metric tonne of steel, as it was at a later stage and is presented in Table 5.1.

In order to determine the overall CO_2 emissions per resource extracted and produced, such as coal, oxygen, limestone, iron ore and their respective products along the transformation process, such as lime, iron ore pellets, concentrated oxygen and sinter, the requirement per metric tonne of steel, extrapolated to furnace capacity, will have to be determined, and the overall CO_2 emissions cannot be summarised from the data sheet as this is geared towards the emissions per tonne of product. Therefore, the input data sheet is required to enter the individual resources and materials into the software as quantified material streams and parameters as per production capacity, which enables the systems to determine the CO_2 emissions along the processes and finally per tonne of steel produced.

Figure 5.21 displays a screen copy of the MS Excel database with the parameters, data and findings for most processes associated with the steelmaking process, and Figure 5.22 comments on the timeline featuring screen copies from literary sources confirming the preliminary findings.

The conventional BF/BOF route CO_2 emissions net total seems to be varying between 3.6 and 6.3 t (mean: 4.95 t) of CO_2 emissions per tonne of steel produced. At this point, it is worth emphasising that there is a great range and difference in recorded levels of CO_2 emissions in steelmaking in existing literature due to (i) different approaches to collecting data, (ii) populations of any survey, (iii) different frameworks of valid data points and (iv) lack of guidance for collecting CO_2 emissions and how to measure the various levels. In Chapters 3 and 4: The Bio Steel Cycle, the mechanisms leading to a CO_2 emissions reduction to almost '0' CO_2 emissions are explored in greater detail. Some deductions on this basis are as follows.

Switching the energy provision to a commercial entity producing their energy entirely using renewable energy technologies leads to a reduction in fossil-fuel-derived energy consumption of 30%, with CO_2 savings at the same percentage rate – reducing the CO_2 emissions by 1.49 t from 4.95 t to 3.47 t. With the replacement of coal with biomass, a further 30% reduction from 3.47 t of CO_2 emissions leads to an emissions reduction of 1.39 t to 1.98 t of CO_2 emissions per metric tonne of crude steel, whilst simultaneously increasing the level of energy independence by 60%. At this point, it cannot be stressed enough that the replacement of coal with biomass will not only directly reduce the CO_2 emissions during the BF–BOF route but also eliminate the GHG emissions entirely for mining the coal, preparing and coking, with the elimination of processing emissions from energy usage at the same time. Biomass does have some carbon footprint but pales in comparison with coal-based steelmaking. Besides, biomass is a more sustainable carbon source, and the increased usage and production would have the positive side effects of potentially increased rewilding, habitat rehabilitation and providing an increased volume of CO_2 sinks. Therefore, it could be argued that the carbon capture capabilities of biomass should be taken into consideration when calculating the overall impact on CO_2 emissions, as the DAC of biomass can be between 9 and 17 t per hectare, per year, depending on the species. If renewable energy resources can be installed using the existing infrastructure, a further 30% of CO_2 emissions can be reduced, leading to a reduction of 0.18 t of CO_2 emissions per tonne of steel produced and '0' with the introduction of CO_2 flue stack CO_2 filters. At the same time, a higher degree of energy independence can be achieved, at 88%. The data processing in MS Excel provided meaningful insight into the development of the different scenarios of sustainable and renewable energy technology and process component applications, in ceteris paribus (Latin for 'all else being equal'), investigated in specific process stage emission component values. The MS Excel extrapolations demonstrated the impact of the different component implementations. The currently accepted value of between 3.6 and 6.3 t (mean: 4.95 t) of CO_2 emissions per tonne of steel produced (in answer to research question 1) was demonstrably reduced (in answer to research question

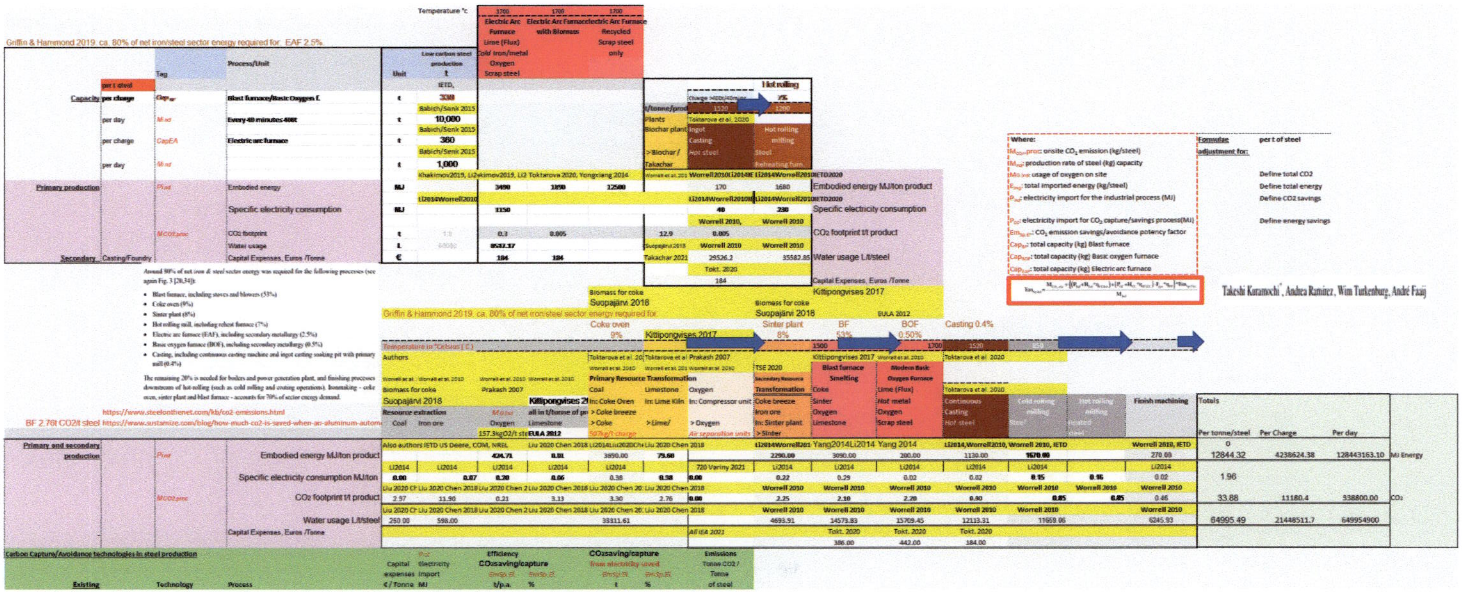

Figure 5.21 MS Excel compilation of gathered data.

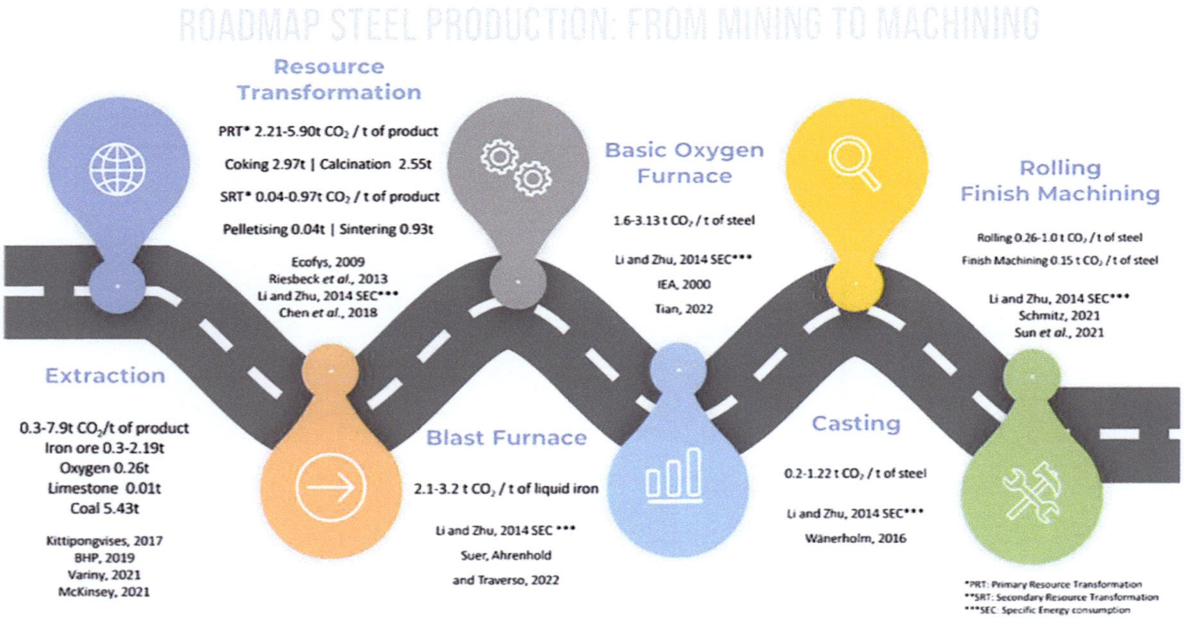

Figure 5.22 CO_2 emission evidence linked with the S8 carbon reports and Table 5.1.

2) by only switching to an energy provider who derives their energy entirely from renewable energy sources, replacing coal with biomass and installing renewable energy technologies for energy production on site. This will not only help to reduce CO_2 emissions but (in answer to research question 3) also supply the production site with renewable energy, where 30% of energy does not have to be imported from third parties and paid for, bringing the reduction in CO_2 emissions reduced to −60%. At the point in production where biomass has already replaced the use of coke and renewable energy technologies (solar, solar PV, wind and hydro) have been installed, the remaining energy requirement can be provided by using biogas produced in the link-connected anaerobic digester, which produces biogas from the connected biomass-producing agribusinesses. This might even provide an additional source of income, as surplus energy fed into the National Grid is financially rewarded, in the United Kingdom, with the current status of 'Smart Export Guarantees' (SEG).

5.10.4 Statistics: T-test

Statistical relevance was required to be added, and therefore, a t-test (Table 5.6) was performed in MS Excel. The data relevant for this test are a data extract from the workings reflected in the tables and data from all previous chapters.

Table 5.5 Data extract from BF/BOF and BF/EAF.

BOF	4.95	1633.50
BOF + B	4.95	1143.45
BOF + B/RE	4.95	800.42
BOF + B/RE/BG	4.95	560.29
EAF	1.95	643.50
EAF + B	1.95	450.45
EAF + B/RE	1.95	315.30
EAF + B/RE/BG	1.95	220.72

Legend:

B = Biomass

RE = Renewables

BG = Biogas

The directional hypothesis states that it is possible to reduce the CO_2 emissions per tonne of steel produced by at least 20%, possibly to '0', and achieve greater energy independence at the same time for the same percentage value.

The null hypothesis states that it is not possible to reduce the CO_2 emissions per tonne of steel produced by at least 20%, possibly to '0', and achieve greater energy independence at the same time, for the same percentage value.

Equations:

1 Hypothesis: H_1: >20%

0 Hypothesis: H_0: ≤20%

Interpretation of the two-sample t-test results is shown in Table 5.6.

Table 5.6 T-test.

T-test: paired two samples for means		
	Variable 1	*Variable 2*
Mean	3.45	720.9538
Variance	2.571428571	219618.8
Observations	8	8
Pearson correlation	0.71506427	
Hypothesised mean difference	0	
Df	7	
	–	
t Stat	4.341073061	
$P(T \leq t)$ one-tail	0.001695811	
t critical one-tail	1.894578605	
$P(T \leq t)$ two-tail	0.003391622	
t critical two-tail	2.364624252	

The output indicates that the mean CO_2 emissions quantity per tonne of crude steel for BF/BOF and BF/EAF is 3.45 t, and for the applied technologies and processes in BF/BOF and BF/EAF operation, it is 720.95 t per charge or production cycle. The *significance level* is 0.05. As the *p-value* is 0.000168 and, therefore, less than the significance level, the difference between means is statistically significant. Therefore, the 0 Hypothesis needs to be rejected, and the 1 Hypothesis needs to be accepted: it is possible to reduce the CO_2 emissions per tonne of steel produced by at least 20% and achieve greater energy independence at the same time, for the same percentage value.

5.10.5 Total CO_2 emissions from BF/BOF operating oxygen

As previously mentioned in Section 3.4, the energy required to inject the compressed 99% oxygen into the BF at 220 psi and 1000 °C and at 150 psi at ambient temperature does not seem to have been taken into consideration in any of the investigated sources of information. Therefore, the enthalpy, the energy required and the resulting CO_2 emissions have been calculated with the application of basic mathematical principles. In order to determine those values, additional questions needed answers, such as 'How much oxygen m^3 is being used in BOF per minute in a 20-minute blow?'. It was determined that 800 m^3 of oxygen is being used per minute, weighing in at 1143.20 kg [86]. Furthermore, 'How much CO_2 for use of energy is being emitted via the BF when heating 800 m^3 oxygen per minute for 20 minutes from 25 °C to 1000 °C?'. For this purpose, the required formula and equation is that of enthalpy

[87–89] using the heat capacity of oxygen at 918 J/kg°C [90]. Given that any 330 t furnace capacity will result in a 92.6% yield [91] of steel from liquid iron ladled into the BOF from the BF, resulting in 370.4 t of liquid steel.

1000 J = 0.0002777778 kWh; 0.46 kg of CO_2 per kWh produced [86]. 1 minute of 800 m^3 oxygen = 1143.20 kg. Therefore, 20 minutes of 800 m^3 oxygen per minute = 22,864.0 kg in weight of oxygen per oxygen blow.

Enthalpy = a thermodynamic quantity equivalent to the total heat content of a system. It is equal to the internal energy of the system plus the product of pressure and volume [92, 93].

Heat energy calculation

$$
\begin{aligned}
\boldsymbol{Q = mcs\Delta T}\\
&= mcs(T_{final} - T_{initial})\\
&= (918\ \mathrm{J/kg^\circ C}) \times (2.2864 \times 10^4\ \mathrm{kg}) \times (1000 - 25)^\circ\mathrm{C}\\
&= (918\ \mathrm{J/kg^\circ C}) \times (2.2864 \times 10^4\ \mathrm{kg}) \times (925)^\circ\mathrm{C}\\
&= 918 \times 2.2864 \times 10^4 \times 925\\
&= 1.94145 \times 10^{10}\ \mathrm{J}\ \text{(per 20-minute blow)}
\end{aligned}
$$

Conversion to kWh $1.94145 \times 10^{10} \times 2.7778 \times 10^{-4} = 5.393 \times 10^6$ kWh

CO_2 emissions calculation $(5.393 \times 10^6\ \mathrm{kWh}) \times (0.46\ \mathrm{kg\,CO_2/kWh}) = 2.481 \times 10^6$ kg of CO_2

CO_2 emissions per metric tonne of steel

$$\frac{\boldsymbol{2.481 \times 10^6\ kg\ CO_2}}{\boldsymbol{3.3 \times 10^5\ kg\ steel}} = \boldsymbol{7.52\ t\ CO_2/metric\ tonne\ of\ steel}$$

Now, to determine how much CO_2 emissions are being produced for the injection of the oxygen into the BF at 220 psi and into the BOF at 150 psi, it is required to calculate the pressure in relation to oxygen value in weight and heat value (J):

$$\text{pressure} \times \text{volume} = \text{energy}$$

Blast furnace (BF) energy calculation

$$(220\ \mathrm{psi}) \times (2.2864 \times 10^4\ \mathrm{kg})\ \mathrm{kg} = 5.03008 \times 10^6\ \mathrm{J}$$

$$(5.03008 \times 10^6\ \mathrm{J}) \times (2.777778 \times 10^{-4}) = 1.397 \times 10^0\ \mathrm{kWh}$$

Basic oxygen furnace (BOF) energy calculation

$$(150\ \mathrm{psi}) \times (2.2864 \times 10^4\ \mathrm{kg}) = 3.4296 \times 10^6\ \mathrm{J}$$

$$(3.4296\ \mathrm{x}\ 10^6\ \mathrm{J}) \times (2.7778 \times 10^{-4}) = 9.527 \times 10^{-1}\ \mathrm{kWh}$$

CO_2 emissions calculation [94]

$$0.46\ \mathrm{kg\,CO_2/kWh}$$

Therefore, 7.52 t/CO_2/metric tonnes of steel and an additional 0.64 kg CO_2 for the 220 psi exerted pressure of 220 psi for the injection of oxygen at 1000 °C into the BF will have to be added to the CO_2 balance, per metric tonne of crude steel produced. As the oxygen injection is operated at 150 psi in the BOF and unheated, an additional 0.44 kg of CO_2 needs to be considered. As the oxygen does not need to be heated for injection into the BOF, no energy requirement or enthalpy needs to be determined. The established values have been added to the calculations for the determination of the mean values for BF/BOF and BF/EAF route operations.

5.10.6 Validation and verification of the preliminary research results

For the purpose of providing proof of replicability, verifiability and therefore viability of the reported results, the same input dataset was used for the purpose of compiling the process simulation in different sets of simulation software packages. Calculations and process simulations were based on a sample plant with 330 metric tonne capacity, similar to British Steel's Scunthorpe plant. Although the same data have been used for each application, interestingly, every simulation software resulted in different, but similar, reported results. The same quantities and values were applied; specifically, the CO_2 emissions, as this was tested for, were mostly quite close. The fluctuations can largely be attributed to the different functionalities of the process simulation software packages and to what different process parameters these are operating with. Table 5.7 displays the individual values entered into the three simulation software solutions.

Blast furnace reactions:

Step 1: Bottom blown 1000 °C/O_2 @ 220 psi;
$2C + O_2 \rightarrow 2CO$ and 1700–2300 °C

Step 2: Furnace top ~200 °C – 700 °C;
$3\,Fe_2O_3 + CO \rightarrow 2\,Fe_3O_4 + CO_2$

Step 3: Furnace middle 850 °C; $Fe_3O_4 + CO \rightarrow 3\,FeO + CO_2$

Step 4: Air + N_2 (up) flux (down); $CaCO_3 \rightarrow CaO + CO_2$

Step 5: Slag forming $SiO_2 + CaO \rightarrow CaSiO_3$

Step 6: Iron oxide (down) to bottom 1200 °C;
$FeO + CO \rightarrow Fe + CO_2$

Step 7: The added coke causes $C + CO_2 \rightarrow 2CO$
Boudouard reaction controlling the gas atmosphere;
$2CO \rightleftharpoons CO_2 + C$

Basic oxygen furnace reactions:

BOF oxidation at the jet impact zone
$C + O_2 \rightarrow CO_2$
$Si + O_2 \rightarrow SiO_2$
$2Fe + O_2 \rightarrow O_2 \rightarrow 2MnO$

Table 5.7 Input data for entry I process simulation software.

Stage	Process	Contents of	Resource	Chemical formula	State	Reaction	Result	Chemical formula	Quantity	Result weight (kg)	Temperature (°C)	Pressure atm/psi/bar	State	Electricity kWh/t of product/charge	CO_2 emissions energy/combustion per t product/kiln
Preparation	Coking	Coal	Coal, bituminous	$C_{137}H_{97}O_9NS$	Solid		Coke	$C_{135}H_{96}O_9NS$	Kiln capacity	36,000 kg/ 36t	1100–2000	1/14.69	Solid	937.82 kWh/t coke	**2.97 t/ t coke**
Preparation	Calcification		Limestone	$CaCO_3$	Solid	$CaCO_3 \rightarrow CaO + CO_2$	Lime	CaO	Kiln capacity	50,000 kg/ 50 t		1/14.69	Solid	1180.57 kWh/t lime	
			Water	H_2O	Liquid							1/14.69			
											900				**2.21 t/ t lime**
Preparation	Sintering		Oxygen	O_2	Gaseous				Ambient			1/14.69	Gaseous	2100 kWh/t of sinter	
			Iron ore	FeO	Solid				62.85%	207,405		1/14.69	Solid		
			Lime	CaO	Solid	$\Delta Fa^{a)} = \Delta E - T\Delta S$	Sinter	FeO	24.15%	79,695		1/14.69	Solid		
			Coke	$C_{135}H_{96}O_9NS$	Solid				5.00%	16,500		1/14.69	Solid		
			Limestone	$CaCO_3$	Solid				8.00%	26,400	1480	1/14.69	Solid		**0.93 t/ t sinter**
Blast furnace	Smelting reduction		Coke	$C_{135}H_{96}O_9NS$	Solid				50.9% capacity	168015.25		1–2.5 bar	Solid		
			Limestone	$CaCO_3$	Solid	$CaCO_3 \rightarrow CaO$	Quick lime			16500			Solid	22153.62	
			Oxygen	O_2	Gaseous				800 m^3/ min = kg	22864		220 psi	Gaseous		**7.52 t CO_2/t cs**

(Continued)

Table 5.7 (Continued)

Stage	Process	Contents of	Resource	Chemical formula	State	Reaction	Result	Chemical formula	Quantity	Result weight (kg)	Tempera-ture (°C)	Pressure atm/psi/bar	State	Electricity kWh/t of product/charge	CO_2 emissions energy/combustion per t product/kiln
		Sinter	Lime	CaO	Solid				4.5–19.5% of [a]	5653.62			Solid		
			Iron ore	Fe	Solid	$FeO > Fe_2O_3$ partly			<22.5% of [a]	28268.11			Solid		
			Iron oxide	Fe_2O_2	Solid				>55.0% of [a]	69099.8			Solid		
			Silicon dioxide	SiO_2	Solid				2.5–10.5% of [a]	3140.9			Solid		
			Magnesium oxide	MgO	Solid				<4.5% of [a]	5653.62			Solid		
			Aluminium oxide	Al_2O_3	Solid				<3% of [a]	3769.08			Solid		
			Zinc	Zn	Solid					1256.36			Solid		
			Titanium	Ti	Solid				<5% of [a]	1256.36			Solid		
			Potassium	K_2O	Solid					1256.36			Solid		
			Chromium	Cr	Solid					1256.36			Solid		
			Manganese	Mn	Solid					1256.36			Solid		
			Water	H_2O	Liquid				0–0.6%	753.82			Liquid		
											1650				**3.2 t/t steel**
Basic oxygen	Oxygen		Liquid Fe	FEO	Liquid					224,636		1/14.69	Liquid		
furnace	converter		Lime		Solid				50 kg/t liquid steel	16,500		1/14.69	Solid		
	process		Oxygen 99%		Gaseous					22,864	25	150 psi	Gaseous		
			Scrap metal		Solid					66,000	1700		Solid		**66 t × 4.95 tCO_2**
										330,000					**3.1t/t steel**

a) Percentages and quantities calculations based on a 330,000 kg/330 t furnace capacity – similar to the British Steel plant in Scunthorpe. Source: Ref. [74].

BOF desulphurisation
$(FeS) + CaO \rightarrow CaS + FeO$
$(FeS) + MaO \rightarrow MaS + FeO$
$3Fe_2O_3 + CaS \rightarrow CaO + 6FeO + SO_2$
$2CaS + 3O_2 \rightarrow 2CaO + 2SO_2$

BOF steelmaking:
$C + O2 \rightarrow CO_2$
$CO_2 + C \rightarrow 2CO$
$Fe_2O_8 + 3CO \rightarrow 2Fe + 3CO_2$

BOF desilication
$CaCO_8 \rightarrow CaO + CO_2$
$CaO + SiO_2 \rightarrow CaSiO_3$

BOF dephosphorisation
$4P + 5O_2 \rightarrow 2P_2O_5$
$P_2O_3 + 3CaO \rightarrow Ca_3(PO_4)_2$

In Figure 5.22, the most prominent sources stating the CO_2 emissions at every stage of the steelmaking process are shown, down to detailed passages and images shown in the text. Great efforts have been made to ensure that the input data reflect current common practices in the iron and steel industry. In order to provide comparable evidence, the basic input materials were used, which were available across the software packages. All process simulation software packages have one issue in common: they only operate on the material, work or heat flow principle. Components such as solar panels work on the conversion of UV rays into electric energy, which are not yet available as building blocks. There are ways to work around this lack of provision, as will be demonstrated later in the images of the various configurations. The latest environmental dataset available on Aspen+V12.1 used for CO_2 reporting stems from the 2009 USEPA report, although the 2023 report is available [95]; so, the data and results presented by Aspen+ need to be viewed cautiously, bearing this in mind.

5.10.6.1 Simul8

Staffordshire University provides students with a range of process simulation software packages, one of which is Simul8 (Simulate). It is a quite intuitively laid-out simulation tool, where even beginners can become proficient fairly quickly. As one of the first simulations, the BiSC layout has been created using Simul8. The different components for simulation had to be identified, for them to determine the parameters of the input data, as previously described and displayed. To understand the process of working with Simul8, initially, a BF–BOF process route simulation has been created, as the BF/BOF route is the most widespread method of steelmaking, representing c. 70% of current global crude steel production [3, 75]. It needs to be emphasised, though, that the following CO_2 emissions in metric tonnes represent the CO_2 emissions at their respective stage, per tonne of product, and *not* per tonne of steel. The emissions burden on each of the input streams in the simulation software systems Simul8, Aspen+ and INOSIM is already set within the system parameters and therefore calculated during the process simulations, providing a more detailed set of data, per metric tonne of steel produced. Additionally, in order to represent the overall mean CO_2 emissions burden in t per tonne of steel, the mean value of 4.95 t/t cs was allocated for entering the input stream in S8 as BF 2950 kg and BOF 2000 kg. The shown process simulation layouts will display the sections relevant to the CO_2 S8 carbon report, at every stage in the steelmaking process, in t per tonne of product.

The carbon reports for the BF/BOF route were created subsequently after operating the configurations, which was followed by the creation of a BF–EAF process route simulation, representing the second most common method of steelmaking, with a share of c. 20% of global production. The emissions are per tonne of product at the respective stage. Initially, as biomass has the potential to reduce CO_2 emissions by ~30%, the usage of biomass was implemented. The carbon reports displayed provide the details in direct comparison with the mathematical calculations and the data modelled from the MS Excel main spreadsheets and input data. The switching to energy suppliers who derive 100% of their energy from renewable energy sources is external to these configurations. The overall effect of switching to a green energy supplier, implementing renewable energy technology and using biogas from own production in on-site turbines, on CO_2 emissions and the achieved degree of energy independence are displayed in Chapter 6. The Simul8 process simulations and subsequent carbon reports demonstrate the effect of implementing biomass in the steelmaking process as one of the BiSC components. These simulations were followed by configurations including the implementation of the CCUS process and also the implementation of the Geomimetic© process, including off-gas flue stack filters, which makes it possible to produce steel at almost '0' CO_2 emissions – for both the BF/BOF route and the BF/EAF process route.

The simulations and resulting reports provided proof of concept of the BiSC; insofar as with the implementation of only these two steps out of seven (biomass and the carbon capture with filters and the Geomimetic process), the steel production CO_2 emissions could be reduced by 30% and to almost '0', therefore proving the hypothesis that it is possible to produce steel without CO_2 emissions if the novel BiSC concept and strategy was being implemented.

5.10.6.2 Aspen+V12.1

The work with the Aspen+ software suites has been a steep learning curve, as this particular software was originally designed for the chemical industry. Similar to other process simulation software packages, Aspen+ works on the principle of material, heat and workflows, whereas components, such as solar panels – due to their UV rays to electricity conversion principle – are not available just yet. There are, however, ways to construct a working configuration, nonetheless. To demonstrate a comparable configuration in either system, the basic components of steelmaking were used every time. For proof of concept, the basic input stream components were chosen to be used to simulate the one-hour BF process in a single blow of 20 minutes, based on a 330-t capacity BF. The Aspen+ GHG emissions calculations are also based on combustion events, which in the case of ironmaking are not the reality: the CO_2 emissions are the result of the O_2 being injected into the BF charge at 220 psi and 1000 °C, also causing the exothermic reaction and creating a 1700 °C environment. The BF operation is simulated in Aspen+ with an RGibbs reactor. Some general assumptions had to be made to simulate the BF/BOF process, particularly focusing on the BF operation. As Aspen+ works with continuous flows and the calculations are in a steady state, the amount of gas emitted per tonne of hot-metal produced in the BF has been converted to a flow. 1.01325 bar BF vessel pressure was set. In the simulation, it has been considered that the production of one tonne of hot-metal lasts one hour, as the time to inject the 1000 °C oxygen via tuyeres into the charge, called a blow, usually lasts ~20 minutes. The overall process of producing liquid iron usually lasts ~40 minutes. Therefore, the gas emitted per tonne of hot-metal produced has been introduced in Aspen+ as a flow per hour. This assumption is making scaling of the simulation to the size of a real steel mill, such as the 330-t capacity British Steel furnace in Scunthorpe, significantly easier. It will not have any impact on the final results, but it will be helpful in understanding the analysis of the final results in that the energy produced in one hour – or the moles emitted per hour – is equal to the energy or mass per tonne of hot-metal. The use of process simulation software to simulate situations in the iron and steel industry has been a useful tool for recreating otherwise costly pilot plants. Scientists, academics and professionals have used this technology to their advantage. As a relevant example, attempts have been made to incorporate biomass in steelmaking [29, 30]. Encouragingly, also the integration of technical advances in the iron and steel industry has been recreated in a range of process simulation software packages. As documented [29, 30], the usage of biomass and the processes involved have already been given some attention. The aim was to reduce the amount of coal and coke used by introducing hydrogen and syngas. It seems apparent that a higher amount of coke (in weight) could be saved if the injection rate of syngas was increased from 10 kg·tHM-1 to 40 kg·tHM-1. The replacement ratio seems to vary after 40 kg·tHM1. Potentially, this could be associated with the higher percentage of hydrogen, alongside the increased injection of syngas. As is the nature of these elements, in the case of injecting 100% hydrogen into the BF, a smaller amount of coke can be replaced per injected amount of hydrogen, considering an injection rate of >20 kg·tHM-1 [29, 30]. In conclusion, it was found that lower hydrogen injection rates are a better mechanism to save on coke usage. Publications like these were the inspiration for the creation of a working model and 7 steps to net-zero steel strategy, and attempts have been made to use the different already tried and tested systems and combine them into one circular iron and steel production process: the BiSC [71]. In order to demonstrate proof of concept, process simulation configurations were built in Aspen+V12.1. To establish whether Aspen+ was suitable for trialling selected stages of CO_2 emissions reduction measures implementation, the calcination and sintering processes were taken as initial trials. RGibbs reactors in steady state were chosen, with an inside vessel pressure of 1.01325 bar. 50,000 kg capacity is assumed. The section is displayed in Figure 5.23.

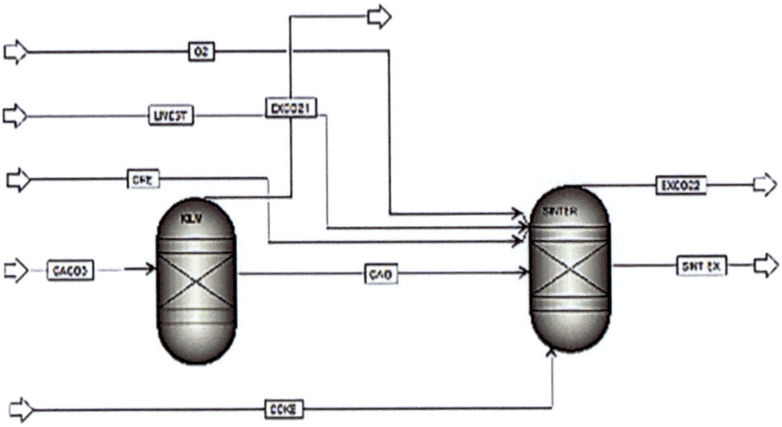

Figure 5.23 The lime-kiln and sinter oven section of the steelmaking process.

Preliminary research findings pointed towards a CO_2 emissions burden of 2.760 t and 2.25 t of CO_2/t of product, which was reported by the Aspen+ calculations, as displayed in Figure 5.24. Encouraged by these results, the basic layout of the BF operation was constructed as the starting configuration for the simulations, as displayed in Figure 5.25.

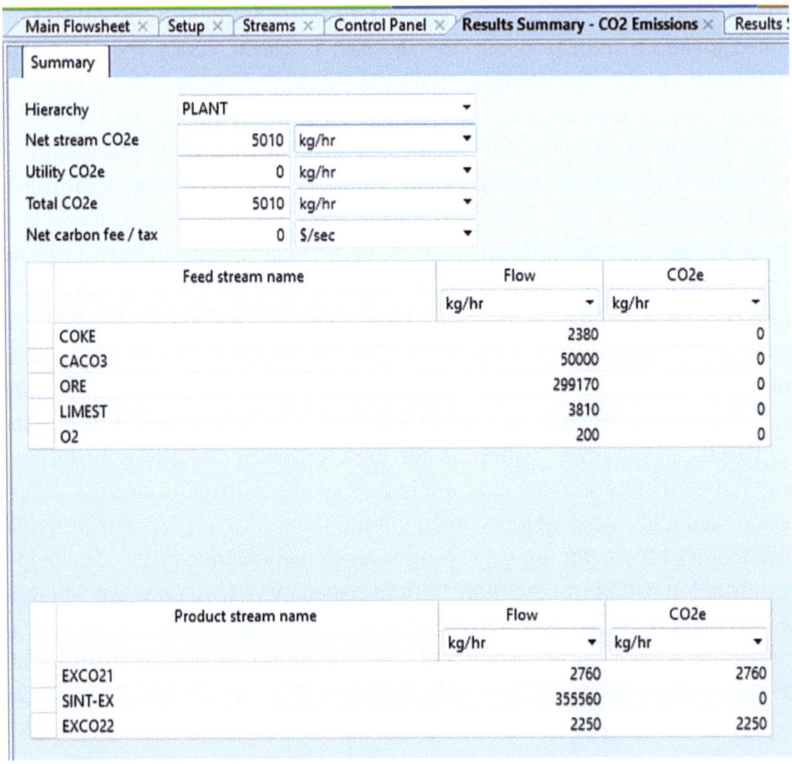

Figure 5.24 Aspen+ overview of material flow and CO_2 emissions.

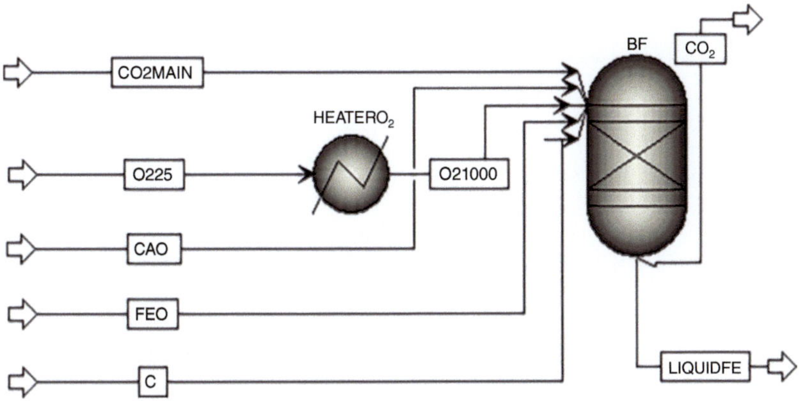

Figure 5.25 Blast furnace basic configuration for the production of liquid iron.

Similar to the established range between 2.1 and 3.2 t as a result of preliminary research findings, the Aspen BF process simulation effectuated 977,856 t of CO_2 per daily output, meaning 2.96 t of CO_2 emissions per tonne of liquid iron (later steel), tapped from the bottom of the BF, thus confirming the findings of the literature review and mathematical corroborations of between 2.1 and 3.2 t of CO_2e /t cs. The calculations were set to work for a 330,000 kg/330 t BF, similar to the British Steel site in Scunthorpe. The set-up parameters are shown in Figures 5.24, 5.26 and 5.27.

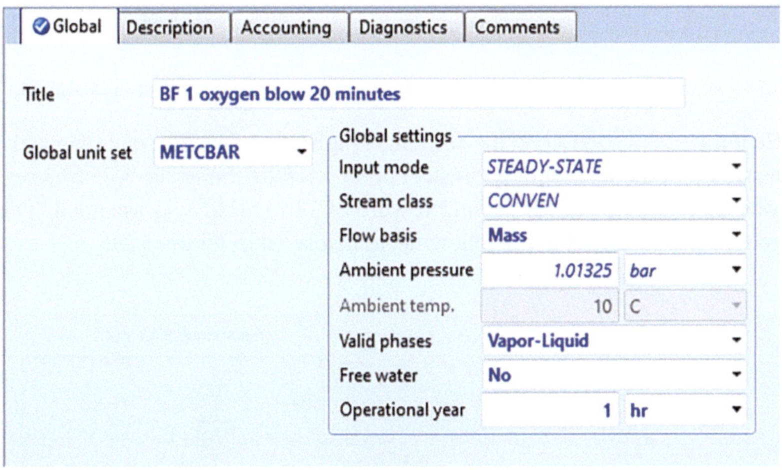

Figure 5.26 Material flow set-up parameters in Aspen.

The next step was to establish the reduction in CO_2 emissions when solar PV systems were introduced, resulting in this configuration, as displayed in Figure 5.28.

Based on research findings from the literature review, evaluation and extrapolations in MS Excel and mathematical validation, a 30% reduction in CO_2 emissions was the expected result to be reported. With 684,499 t of CO_2e reported, the 30% potential CO_2 emissions reduction was confirmed.

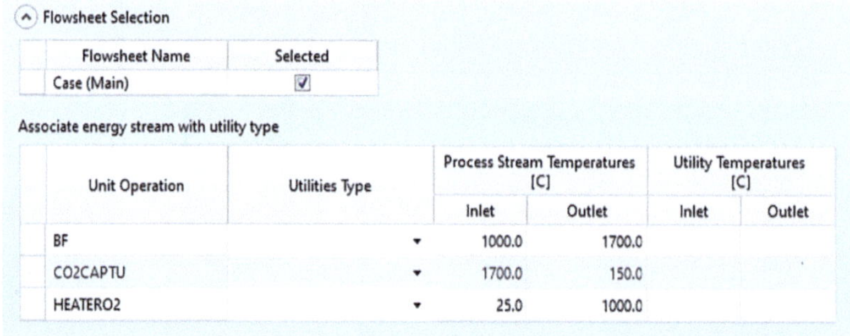

Flowsheet Selection

Flowsheet Name	Selected
Case (Main)	☑

Associate energy stream with utility type

Unit Operation	Utilities Type	Process Stream Temperatures [C]		Utility Temperatures [C]	
		Inlet	Outlet	Inlet	Outlet
BF		1000.0	1700.0		
CO2CAPTU		1700.0	150.0		
HEATERO2		25.0	1000.0		

Figure 5.27 Input and outlet temperatures of the blast furnace configuration.

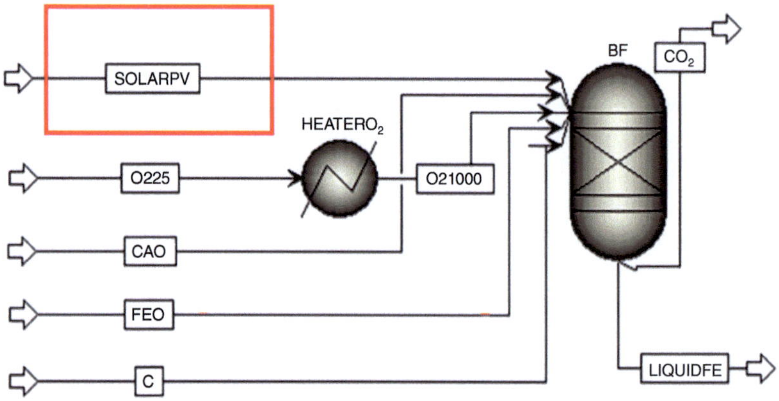

Figure 5.28 BF basic configuration with solar PV panels for energy production installed.

The Geomimetic© process [59] has the single highest impact, as it can reduce the CO_2 emissions (in theory, not only) in steel production to almost '0'. A Flash2 separator was used, and an inside vessel pressure of 1.01325 bar was assumed. This configuration was extended to the following, as displayed in Figure 5.29.

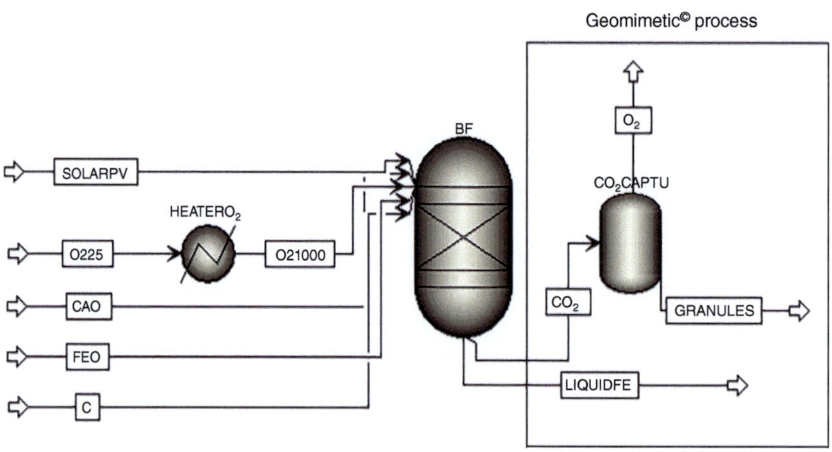

Figure 5.29 BF configuration with solar PV and the Geomimetic© process installed.

As initially mentioned, all simulation software packages work in a similar, yet different, way. In Aspen Plus, the confirmed 100% CO_2 capture with the installed Geomimetic© system is displayed in Figure 5.29, where the captured off-gas was split into 72.71% O_2 and 27.29% CO_2 granules for further usage in a different environment. Interestingly, though, the CO_2e result has increased again by the implementation of the Geomimetic© process from 684,499 kg to 758,019 kg of CO_2 emissions, as displayed in Figure 5.30.

Product stream name	Flow	CO_2e
	kg/hr	kg/hr
LIQUIDFE	236872	0
O2	758019	758019

Figure 5.30 Increased CO_2 emissions after splitting the BF CO_2 gas into O_2 gas and CO_2 granules.

Encouraged by these results, the configuration for the BOF was constructed, also containing the section of the Geomimetic© process, and additionally, twice the 30% reduction in CO_2 emissions was considered due to switching to a 'green' energy provider and installing solar PV panels for energy generation. Additionally, the Geomimetic© process was also implemented here, splitting the off-gas CO_2 into O_2 gas and CO_2 granules, resulting in a CO_2 GHG emission of effectively '0', as the CO_2 from the BOF was captured at 100% in the Geomimetic© unit, as displayed in Figure 7.22. The resulting CO_2 emissions report calculated 469,243 kg of CO_2 emissions per 20-minute oxygen blow at 25 °C of 22,864 kg of oxygen at 150 psi in the BOF, as shown in Figure 5.31.

Product stream name	Flow	CO_2e
	kg/hr	kg/hr
C35STEEL	146633	0
O2	469243	469243
GRANULES	5.71657	0

Figure 5.31 BOF configuration with 'green energy' provider, solar PV panels installed and the Geomimetic© process in place.

The integration of BiSC components in Aspen followed subsequently, as displayed in Figure 5.32.

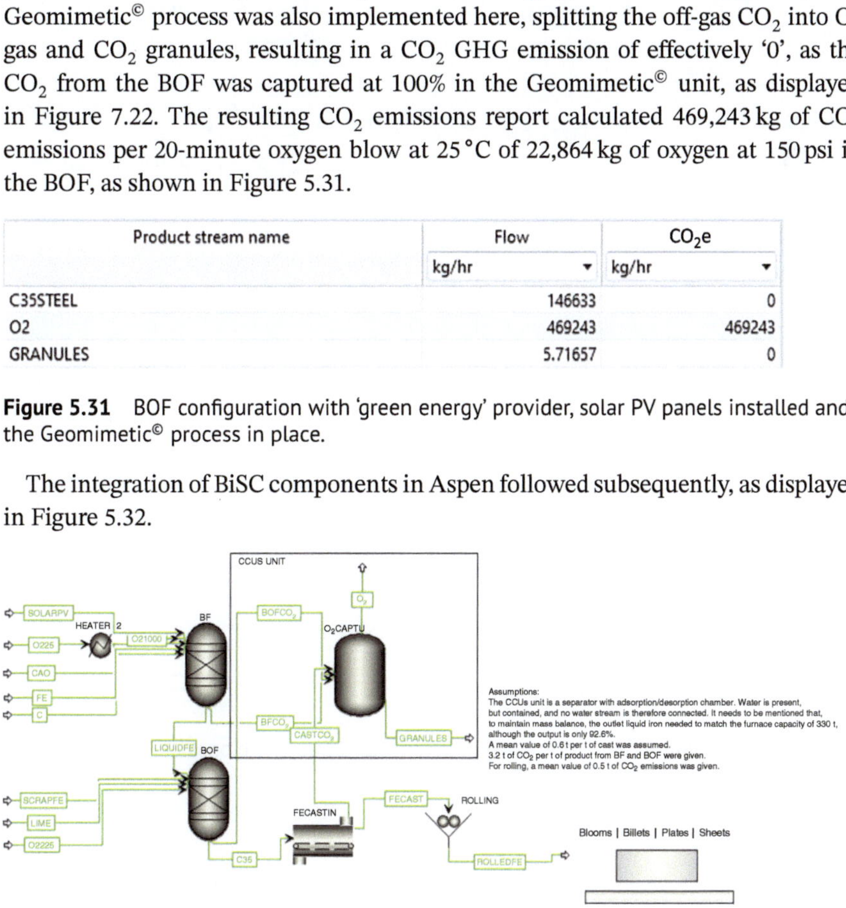

Figure 5.32 Aspen Bio Steel Cycle component integration layout.

5.10.6.3 INOSIM

As previously mentioned, most simulation software solutions work in a similar, yet different, way. Most operate with an MS Excel-like structure underneath all the surface applications and input screens, and hence, the results are similar. INOSIM has no exemption yet has some features, which are not available in other applications, such as Simul8 or Aspen: there are preinstalled functions for bioprocess engineering features, such as autoclave, fermenter, heat steriliser and UV steriliser. Most elements of the BiSC could be demonstrated using INOSIM (Figure 5.33).

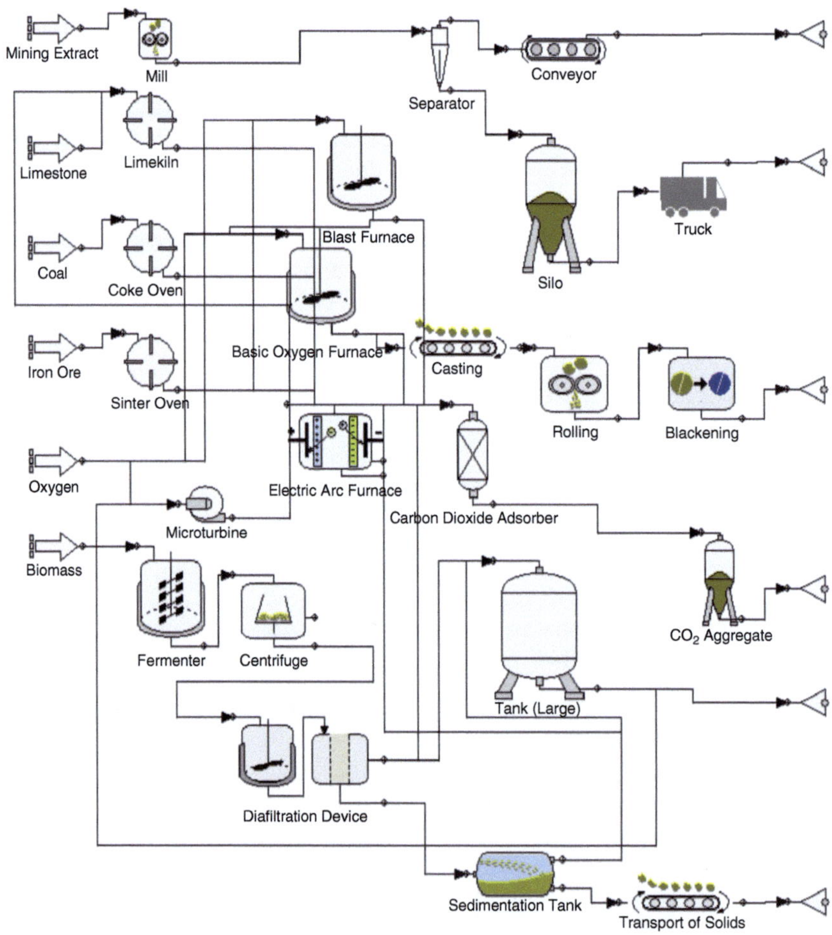

Figure 5.33 INOSIM BF/BOF process simulation with CCUS and biogas utilisation.

Confirmed 100% CO_2 capture with the installed CCUS system, insofar as the captured off-gas was split into 72.71% O_2 and 27.29% CO_2 granules for further usage in a different environment. Interestingly, though, the CO_2e result has increased again with the implementation of the CCUS process from 684,499 kg to 758,019 kg of CO_2 emissions, which, upon investigation, was identified as the energy required to separate the O_2 and CO_2. INOSIM provides a vast range of reactors and bioengineering components, which allows for easy integration of the innovative BiSC model. Components such as fermenters are provided, which have yet to be found to be available from any other process simulation software.

5.11 Process simulations verification and validation

The CO_2 data accumulated as a result of the literature review, mathematical analysis of the aforementioned, MS Excel data compilation, evaluation, extrapolations and analysis, statistical validation and the process software simulations in Aspen, Simul8 and INOSIM have provided congruent data and results for verification and validation purposes.

Table 5.8 and Figure 5.20 provide a summary of the results reported by the process simulations in the three software applications, whereas Table 5.9 gives insight into the verification and validation protocol.

Table 5.8 BF/BOF process simulation reports' overview.

Reduction % change	Energy independence increase % change	Aspen kg of CO_2 emissions successive change	Aspen kg of CO_2 emissions implementation result
—	—	—	977856.00
30	30.00	293356.80	684499.20
30	60.00	205349.76	479149.44
30	88.00	143744.83	335404.61
		143744.83	0

For verification and validation purposes, the model process simulations and available experimental data in Aspen Plus were compared with peer-reviewed publications of process simulations in other industries [96–101, 104–106], where carbon avoidance or reduction technologies were applied, similarly to what has been achieved during this research. The available experimental data as a result of the process simulations in Aspen Plus V12.1 displayed in this work align with the aforementioned Aspen Plus process simulations flowcharts, diagrams and research results.

Table 5.9 Verification and validation protocol.

Verification and validation techniques	Description	Applied, verified and deemed valid Simul8, INOSIM and Aspen + V12.1
Animation	The model's operational behaviour during a test-run through the steelmaking process is depicted graphically. The results are declared as the simulation moves through time during the production process.	✓
Comparison to other models	The output (reports) of the models as they move through time is being compared to other valid models. The simple models of this research were compared to known results of analytic modes within the three used software packages and compared with similar, previously verified and validated simulation models [96–101].	✓
Degenerate test	Degeneracy of the model behaviour was tested against a selection of values (input/output parameters: Tables 7.1 and 7.4 and Section 6.7; data as a basis for Figures 2.6 and 3.3), i.e. does the average emission quantity increase over time with increased throughput (material)?	✓
Event validity	The stages of the BiSC steelmaking process within the model were compared to existing steelmaking models in Aspen (various versions) to compare if they were similar [96–101].	✓
Extreme condition test	The model output (report; Chapters 6 and 7) should plausibly show significant differences under exposure to extreme conditions, such as input (i.e. iron ore, lime, coal, oxygen and electricity); inventory set at '0' should display carbon emissions at '0'.	✓
Face validity	Discussion with and receiving expert confirmation and advice who are proficient in the application of the software packages, whether they think the model and its behaviour are reasonable and within known parameters and determining whether the logic in the conceptual model with respect to input–output relationships is reasonable.	✓
Fixed values	Varying set values (constants) are utilised in a variety of model inputs and internal parameters. These were mathematically checked against calculations provided by the individual system.	✓

Historical methods	It is accepted that the three historical methods are described as (1) Rationalism (2) Empiricism (3) Positive economics With applied rationalism, it is accepted that everyone knows that the underlying assumptions of a model are true. The author displayed all entries made with a description of the values and components entered into all software packages, so all audiences would be in a position to know the underlying model assumptions. With the application of empiricism, it is assumed that every parameter has been empirically validated. Every value and figure used have been referenced and researched for evidence as to their correctness and applicability. Positive economics tests against the model's ability to predict future outcomes, testing causal relationships and applied mechanisms.	✓
Internal validity	A range of repetitions (simulation model runs) of a stochastic (technical) process simulation model was carried out to establish the variability of the model under investigation. The consistency in the levels of carbon emissions, reported from all three systems, proved the models' validity.	✓
Multistage validation	Model validation method which combines historical methods (rationalism, empiricism and positive economics) with three steps: (i) Developing the model parameters based on theory, general knowledge, observations and operational (stochastic/technical) function. (ii) Testing the model's assumptions against empirical data. (iii) Comparing the model's input–output results with existing, real-world systems. This has been successfully carried out and in comparison with systems such as British Steel in Scunthorpe (Page 3 of the full accounts as of 2023, for 2021, [102] has provided a positive validation result for all three simulation systems).	✓

(Continued)

Table 5.9 (Continued)

Verification and validation techniques	Description	Applied, verified and deemed valid Simul8, INOSIM and Aspen + V12.1
Operational graphics	Observation of performance measures as the model moves through time, i.e. the amount of material in the queue point and different components (such as coal coking, blast furnace and basic oxygen furnace) at the different production stages. The dynamic (time-related) system behaviours of performance indicators were shown as the simulation models moved through time in the three software solutions.	✓
Parameter variability–sensitivity analysis	Similar to the extreme condition test, the model parameters are observed as to the changes of the input–output parameters and their effect on the model's behaviour.	✓
Predictive validation	The created initial flowchart models as a basis for the simulations in the three software applications are utilised to forecast the system's behaviour, and the model's forecasts are compared for similarity. Flowcharts were created in MS Word and then filled with data in MS Excel, and the modelled calculations in MS Excel were used for input–output parameter setting in the process software simulation models.	✓
Traces	The behaviour of different parameters (set quantities of iron ore, limestone, oxygen, coal, liquid iron and steel) was traced during their journey through the process simulation models in the three applications, and it was found that the model's logic was correct. The required accuracy was ascertained.	✓
Turing tests	In adaptation of the Turing test method [103], people sufficiently knowledgeable about the three simulation software solutions were asked if they were able to discriminate between their system and the newly created model behaviours and outputs and responded in the affirmative. They commented on the fact that there is a positive response to the emissions levels, and the models were deemed correctly displayed in relation to input–output tables. Correct display of the carbon emissions reports was confirmed, therefore validating the models.	✓

5.11.1 Validation and verification of the preliminary research results under the application of process simulation software models

5.11.1.1 Introduction

First of all, it needs to be determined what process simulation model verification and validation actually mean:

- *Verification*: the process of determining that a model implementation in a simulation software solution accurately represents the author's conceptual description of the model and the solution to the model [107–112].
- *Validation*: the process of determining if a model is the accurate representation of a real-world application with regard to the intended uses of the model [107–112].

There are three basic approaches for determining whether a verified simulation model is valid or invalid.

(1) Firstly, the model development author will conduct verification and validation during the model development process. It needs to be emphasised that this is a subjective decision based on the results of the various reports as a result of running this research project's process simulation models, statistical tests and mathematical evaluations conducted as a phase of the model development process.
(2) Secondly, during IV and V (independent verification and validation), an independent third party decides whether the model can be considered valid. This procedure is usually used for large-cost projects, where credibility is of paramount importance in relation to return on investment (ROI).
(3) The third approach for determining the validity of a model is using a range of scoring systems. Scores or weights are determined subjectively when carrying out the different aspects of the validation process and then combined to establish first a category score and then an overall score for the simulation model. The validity of a simulation model is considered established if the overall and category scores are greater than a random passing score. It needs to be emphasised that this approach is infrequently used in practice, as determining the weights and scores is highly subjective.

This work as part of the PhD research project was entirely self-funded, and the first approach was chosen for the verification and validation of the developed model, strategy and simulation models. The IV and V methods and scoring were not deemed necessary, as partly sections of previously verified and validated process simulations have been utilised and modified [96–101]. The first and second methods were chosen for the verification and validation of the developed model and process simulations using the three different software packages.

5.11.1.2 Verification of process simulations

Model verification via static analysis and model verification via dynamic analysis are the two main methods used in process simulation model testing [109–112]. During verification via static testing, the computer program used to create the computerised model is analysed for correctness, which (in this case: the settings) were read against existing industry standards, such as those from the International Organization for Standardization (ISO), and investigated whether the model structure and output represent the author's process simulation model layout, also, using methods such as correctness proofs (in this case: runtime reports output observed to be without error or fault messages), structure walkthrough (in this case: checking against gathered data and literature against the input parameters) and examining the structure properties of the program (in this case: checking the existing program settings and finding that Aspen+V12.1 in 2023 still works with the 2009 US CO_2 emission data at the time of thesis submission, whereas INOSIM and Simul8 are operating with current data) [107–112]. Dynamic testing of the process simulation software model involves execution of the model under different stresses (i.e. extreme condition test, Table 7.3). The subsequent output is then used to determine if the computer program, the process simulation program and its implementations in line with the model structure are correct [109–112]. This involves both the values observed during the program execution and the final report output. In dynamic testing, three different strategies are commonly used:

- Bottom-up testing (initially testing the subcomponents [sub-models] and then the complete model)
- Top-down testing (testing programming stubs [sets of data] for each of the subcomponents and then testing the sub-models)
- Mixed testing (a combination of bottom-up and top-down testing [113])

The methods commonly used in dynamic testing are traces, investigations of input–output relations using different validation techniques, internal consistency checks and reprogramming critical components to determine if the same results are obtained; in this case, the integrated RGibbs reactors were set at different temperatures, material quantities and CO_2 per tonne of throughput. If there are a large number of variables, one might aggregate some of the variables to reduce the number of tests needed or use certain types of design of experiments [114], e.g. use factor screening experiments to identify the key variables in order to reduce the number of experimental conditions that need to be tested. It is necessary to be aware whilst checking the correctness of the computer program and its implementation (Table 5.7).

5.11.1.3 Validation of simulation models

The validation of software-based process simulation models follows a range of validation techniques and tests [107, 108, 115]. These can be used objectively by using either statistical test methods, such as the t-test, or standard mathematical procedures, such as the hypothesis tests, and facilitating confidence intervals. Usually, a combination of these techniques is being used, choosing the most suitable for the individual project. The models in the different software solutions have been verified and validated against statistical tests, mathematical calculations and existing verified and validated models, as well as compared against each other. For the purpose of providing proof of replicability, verifiability and therefore validity of the reported results, the same input dataset was used for the purpose of compiling the process simulation in different sets of simulation software packages. Calculations and process simulations were based on a sample plant with 330 metric tonnes capacity, similar to British Steel's Scunthorpe plant. Although the same data have been used for each application, interestingly, every simulation software resulted in different, but similar, reported results. The same quantities and values were applied; specifically, the CO_2 emissions, as this was tested for, were mostly quite close. The fluctuations can largely be attributed to the different functionalities of the process simulation software packages and to what different process parameters these are operating with. Table 5.7 displays the individual values entered into the three simulation software solutions.

5.11.1.4 Validation of simulation models

Moreover, the validation of all three software-based process simulation models and extrapolations in MS Excel follows a range of validation techniques and tests [107, 108]. These were used objectively by using statistical test methods, such as the t-test, or standard mathematical procedures, such as the hypothesis tests, and facilitating confidence intervals. As a combination of these techniques was being used, it was paramount to choose the most suitable for this project. The models in the different software solutions and the experimental data have been verified and validated using statistical tests, mathematical calculations and existing verified and validated models, as well as having been compared against each other. For the purpose of providing proof of replicability, verifiability and therefore validity of the reported results [107, 115, 116], the same input data set was used for the purpose of compiling the process simulation in three different sets of simulation software packages. Calculations and process simulations were based on a sample plant with 330 metric tonnes capacity, similar to British Steel's Scunthorpe plant.

Several publications have been identified as scientific benchmark publications for the steel industry, using the three process simulation software applications, as shown in Table 5.10.

Table 5.10 Benchmark publications.

Software	Title benchmark publication	Reference
Aspen	Eco-Techno-Economic Analysis of Steel Manufacturing Off-gas Valorisation	Deng [117]
	Benchmarking of developed steelmaking chain models	Matino et al. [118]
	Carbon dioxide capture in the iron and steel industry: Thermo-dynamic analysis.	Mio et al. [119]
Simul8	Discrete Event Simulation – Production Model In SIMUL8	Fousek et al. [120]
	Simulation of a real life problem in flexible manufacturing system without breakdown and having alternate machines using SIMUL8	Debta et al. [121]
INOSIM	EmKus: Process simulation for steel and iron industries	Inosim [122]
	The Bio Steel Cycle: 7 Steps to Net-Zero Emissions Steel Production	Kiessling et al. [71]

Although the same data have been used for each application, interestingly, every simulation software resulted in different, but similar, reported results, with the same parameters and material flow quantities applied. The same quantities and values were applied, specifically the CO_2 emissions, as this was tested for, which were quite close. The fluctuations can largely be attributed to the different functionalities of the process simulation software packages and to what different process parameters these are operating with. Table 5.7 displays the individual values entered into the three simulation software solutions, and the table below displays the technique, description of the process and the respective software used. It needs to be emphasised, though, that mostly the process simulation layout and some simulation parameters could be validated against, as most publications do not state their input for establishing material balance and mass flow. Table 5.11 describes the verification and validation protocol and validation of the various process simulations, resulting experimental data and the BiSC principle of carbon emissions reduction using various standard mathematical methods, software and machine algorithms. Table 5.12 provides details of comparable evidence and benchmark simulations to further demonstrate the validity of the current work.

For detailed validation, the software input data from the current work were additionally validated against work stating the actual parameters, mass balance date, information on operational pressure and temperature and data on CO_2 emissions.

Deng [117], Mio et al. [119] and Matino et al. [118] compiled process simulation models with which the current work is very closely aligned to. Here is an example from Matino et al. [118], who created a simulation model very similar to the current work but continued with the flowsheet development by adding BF gas dedusting and cleaning operation simulations, as displayed below:

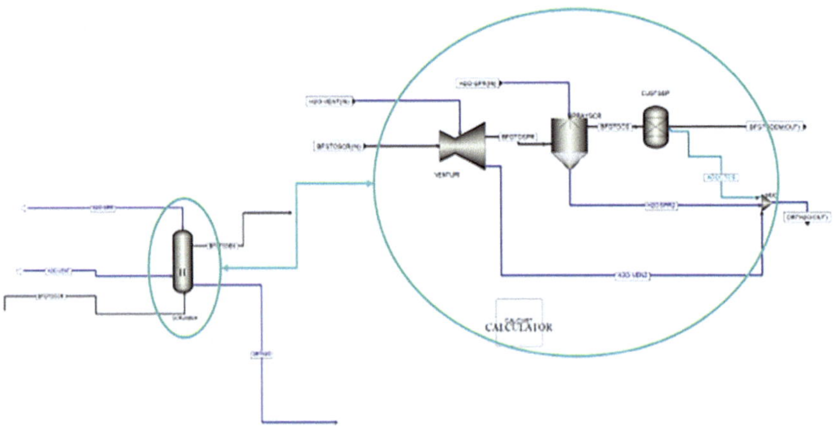

Figure 5.34 Internal flowsheet BFG dedusting/cleaning system model [118].

5.11.1.5 Graphical display of simulation models validation results

The following tables and graphical displays demonstrate the validation of experimental data and results for the models and process simulations of the different stages during the steelmaking process:

Table 5.11 Verification and validation protocol.

Verification and validation techniques	Description	Applied, verified and deemed valid Simul8, INOSIM and Aspen+V12.1
Animation	The model's operational behaviour during a test-run through the steelmaking process is depicted graphically. The results are declared as the simulation moves through time during the production process. The results of the different animated software solutions were compared and found to corroborate the research findings.	✓
Comparison to other models	The output (reports) of the models as they move through time is being compared to other valid models (Section 6). The simple models of this research were compared to known results of analytic modes within the three used software packages and were compared with similar, previously verified and validated simulation models [94, 98, 99, 105, 106, 117–119, 123–127] The results in the different software solutions were compared to other models and were found to corroborate the research findings.	✓
Degenerate test	Degeneracy of the model behaviour was tested against a selection of values. I.e. Does the average emission quantity increase over time with increased throughput (material)? The degenerate test established a congruent behaviour upon increase and decrease in material flow.	✓
Event validity	The stages of the BiSC steelmaking process within the models were compared to existing steelmaking models in Aspen (multiple versions) and others to compare if they were similar [94, 98, 99, 105, 106, 117–119, 123–127] The similarity between the compared models was deemed sufficient to arrive at event validity.	✓
Extreme condition test	The model output should plausibly show significant differences under exposure to extreme conditions, such as input (i.e. iron ore, lime, coal, oxygen and electricity) inventory set at '0' should display carbon emissions at '0'. Extreme condition testing was carried out in all three software solutions, in the application of all possible scenarios, reducing the material flow by 30% step changes, in cascade, to '0'. Biomass, renewables, biogas and carbon capture with flue stack filters. The differences in material flow showed significant congruent differences.	✓

(Continued)

Table 5.11 (Continued)

Verification and validation techniques	Description	Applied, verified and deemed valid Simul8, INOSIM and Aspen+V12.1
Face validity	Discussion with and receiving expert confirmation and advice who are proficient in the application of the software packages, whether they think the model and its behaviour are reasonable and within known parameters and determining whether the logic in the conceptual model with respect to input–output relationships is reasonable. The process simulation experts at Simul8, Aspen and INOSIM confirmed the face validity of the simulations.	✓
Fixed values	Varying set values (constants) are utilised in a variety of model inputs and internal parameters. These were mathematically checked against calculations provided by the individual system. For validity, replicability and verification purposes, the same material flow values were applied throughout all applications.	✓
Historical methods	It is accepted that the three historical methods are described as 1) Rationalism 2) Empiricism 3) Positive economics With applied rationalism, it is accepted that everyone knows that the underlying assumptions of a model are true. The author displayed all entries made with a description of the values and components entered into all software packages, so all audiences would be in a position to know the underlying model assumptions. With the application of empiricism, it is assumed that every parameter has been empirically validated. Every value and figure used have been referenced and researched for evidence as to their correctness, applicability and validity. Positive economics tests were carried out against the models' ability to predict future outcomes, testing causal relationships and applied mechanisms. The results confirmed the models' validity.	✓
Internal validity	A range of repetitions (simulation model runs) of stochastic (technical) process simulation models was carried out to establish the variability of the models under investigation. The consistency in the levels of carbon emissions, reported from all three systems, proved the models' validity.	✓

Multistage validation	Model validation method which combines historical methods (rationalism, empiricism and positive economics) with three steps: (i) Developing the model parameters based on theory, general knowledge, observations and operational (stochastic/technical) function. (ii) Testing the model's assumptions against empirical data. (iii) Comparing the model's input–output results with existing, real-world systems. This has been successfully carried out and in comparison with systems such as British Steel in Scunthorpe (Page 3 of the full accounts as of 2023, for 2021 [102], has provided a positive validation result for all three simulation systems).	✓
Operational graphics	Observation of performance measures as the model moves through time, i.e. the amount of material in the queue point and different components (such as coal coking, blast furnace and basic oxygen furnace) at the different production stages. Evidence of the models' validity was provided by the observation of dynamic (time-related) system behaviours of performance indicators shown as the simulation models moved through time in the three software solutions.	✓
Parameter variability–sensitivity analysis	Similar to the extreme condition test, the model parameters were observed as to the changes of the input–output parameters and their effect upon the models' behaviour.	✓
Predictive validation	The created initial flowchart models as a basis for the simulations in the three software applications were utilised to forecast the systems' behaviour, and the models' forecasts were compared for similarity. Flowcharts were created in MS Word and then filled with data in MS Excel, and the extrapolated calculations in MS Excel were used for input–output parameter setting in the process software simulation models.	✓
Traces	The behaviour of different parameters (set quantities of iron ore, limestone, oxygen, coal, liquid iron and steel) was traced during their journey through the process simulation models in the three applications, and it was found that the models' logic was correct. The required accuracy was ascertained.	✓
Turing tests	In adaptation of the Turing test method [103], people sufficiently knowledgeable about the three simulation software solutions were asked if they were able to discriminate between their system and the newly created model behaviours and outputs and responded in the affirmative. They commented on the fact that there is a positive response to the emissions levels, and the models were deemed correctly displayed in relation to input–output. Correct display of the carbon emissions reports was confirmed, therefore validating the models accordingly.	✓

Table 5.12 Validation of the various process simulations, resulting experimental data and the BiSC principle of carbon emissions reduction using various standard mathematical methods, software and machine algorithms.

No.	Author	Title	Description	Validation and similarity	Text, figures and tables
1	He et al. [94]	Vinyl Chloride Distillation Process Simulation Optimisation Evaluation	This research used Aspen Plus to optimise the vinyl chloride distillation process, aiming to reduce energy consumption and thus CO_2 emissions. The study employed the non-dominated sorting genetic algorithm III (NSGA-III) to optimise key parameters and improve process safety.	A precedent providing validation of the author's work via comparison of the CO_2 capture process simulations using Aspen Plus. All three studies and the author's research aim to reduce CO_2 emissions in their respective fields, emphasising the importance of sustainable practices. Each study highlights the need for improving energy efficiency to achieve their goals, whether through process optimisation, renewable energy integration or innovative (production process) technologies. The overarching theme of sustainability is present in all three studies, promoting practices that reduce environmental impact and enhance resource efficiency. Advanced technologies and methodologies are employed in all studies to achieve their objectives, from carbon capture and hydrogen utilisation to optimisation algorithms and risk analysis. Optimisation of parameters to improve CO_2 capture efficiency is present, similar to the author's conducted experiments and validation protocol (Table 6.3). The validation is also based on the similarity of principles, process simulations and experimental data similarity, which indicates a common commitment to reducing carbon emissions and promoting sustainable practices.	'...Based on the actual operation of the vinyl chloride distillation process in enterprises, this research employs the Non-dominated Sorting Genetic Algorithm III (NSGA-III) to optimise key parameters of the distillation operation, aiming to achieve multiple objectives such as improving product quality, reducing energy consumption, decreasing CO_2 emissions, and enhancing process safety....' 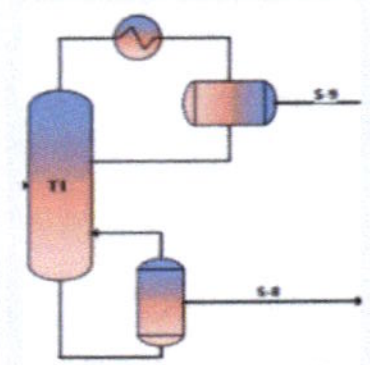 '...achieve multiple objectives such as improving product quality, reducing energy consumption, decreasing CO_2 emissions, and enhancing process safety...' '...The results highlight that the optimised schemes significantly decreased CO_2 emissions compared to the original design...'. 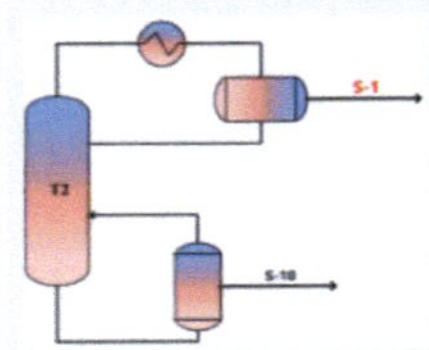

2	Hu, Wang and Fang [123]	Simulation and analysis of CO_2 capturing from converter gas using Monoethanol-amine	This study used Aspen Plus to simulate a post-combustion capture process for CO_2 from converter gas in the steel industry. The results showed that increasing the depletion flow rate, temperature and MEA concentration can improve the CO_2 capture rate.	A precedent providing validation of the author's work via comparison of the CO_2 capture process simulation using Aspen Plus and the ELECNRTL thermodynamic model, similar to the author's various process simulations. This study carried out an analysis of the absorption and desorption performance of MEA on converter gas, which is in line with the author's BiSC component CEPS. Optimisation of parameters to improve CO_2 capture efficiency is present, similar to the author's conducted experiments and validation protocol (Table 6.3). The validation is also based on the similarity of principles, process simulations and experimental data similarity, which indicate a common commitment to reducing carbon emissions and promoting sustainable practices.	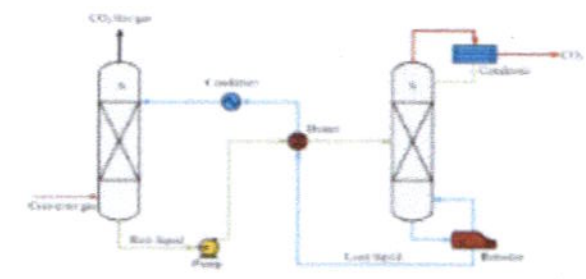 '…Increasing the depletion flow rate, temperature, and MEA concentration can help to improve the CO_2 capture rate; the feed position and desorption pressure affect the CO_2 desorption effect, and the best desorption effect is achieved when the feed position is the second tray, and the desorption tower pressure is 1.9 bar. The results of the study can provide a new idea for CO_2 capture at the end of converter steelmaking, …'.
3	Xin et al. [124]	Process Simulation and Economic Analysis of Pre-combustion CO_2 Capture With Deep Eutectic Solvents	This paper developed an Aspen Plus rate-based absorber model to simulate CO_2 capture from a coal gasification combined cycle power plant. The study compared different mass transfer correlations and evaluated the economic feasibility of using deep eutectic solvents for CO_2 separation.	A precedent providing validation of the author's work via comparison of the CO_2 capture process simulation using Aspen Plus. All studies either directly use or imply the use of advanced simulation tools like Aspen Plus for modelling complex processes. Both the BiSC and the deep eutectic solvents study focus on carbon capture technologies, and a variety of simulation models are used for designing and optimising these systems with the aim to maximise CO_2 capture efficiency. Each study takes a slightly different approach but emphasises that improving energy efficiency and process simulations were used to identify opportunities for energy savings and to optimise operating conditions. Simulation tools	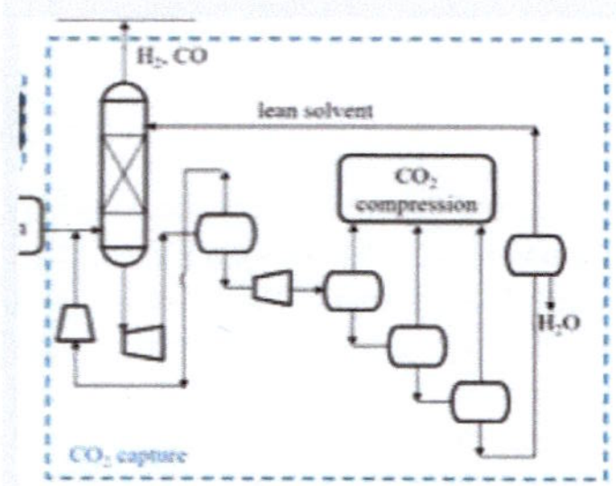

(Continued)

Table 5.12 (Continued)

No.	Author	Title	Description	Validation and similarity	Text, figures and tables
				were also used not only to optimise processes but also to assess their economic and environmental impacts. This helps in making informed decisions that balance cost and sustainability. Optimisation of parameters to improve CO_2 capture efficiency is present, similar to the author's conducted experiments and validation protocol (Table 6.3). The validation is also based on the similarity of principles, process simulations and experimental data similarity, which indicates a common commitment to reducing carbon emissions and promoting sustainable practices.	'…To incorporate CO_2 capture, the IGCC plant design entails two additional subsystems. … The first additional system is a two-stage WGS reaction unit (including a high-temperature reactor and a low temperature reactor) located downstream of the gasifier quench chamber. The second sub-system is the CO_2 capture unit, which is a commercial physical absorption process, similar to the Selexol process used for sulphur (H_2S) removal in plants without CCS…'.
4	Özge and Zeng [99]	Challenges and opportunities of modelling biomass gasification in Aspen Plus: A review	This review focuses on recent developments and studies on modelling biomass gasification in Aspen Plus, including key aspects such as tar formation and model validation. Accordingly, challenges in modelling due to specific assumptions and limitations will be highlighted to provide a useful basis for researchers and end users for further process modelling of biomass gasification in Aspen Plus.	A precedent providing validation of the author's work via comparison of the CO_2 capture process simulation using Aspen Plus. This review focuses on recent developments and studies modelling biomass gasification in Aspen Plus, including key aspects such as tar formation and model validation. Whilst the review does not provide specific carbon emissions percentage reductions, it highlights the potential of biomass gasification to significantly reduce CO_2 emissions by converting biomass into cleaner energy sources, which is also a component of the BiSC. Optimisation of parameters to improve CO_2 capture efficiency is present, similar to the author's conducted experiments and validation protocol (Table 6.3). The validation is also based on the similarity of principles, process simulations and experimental data similarity, which indicates a common commitment to reducing carbon emissions and promoting sustainable practices.	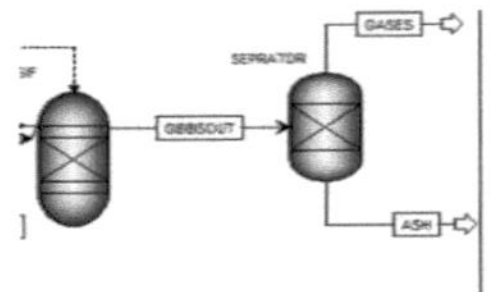

5	Abdul Azeez et al. [106]	Aspen Plus simulation of biomass gasification: a comprehensive model incorporating reaction kinetics, hydrodynamics and tar production	Process simulation and analysis of fluidised bed gasification using Aspen Plus software are considered in this study. Among the three Aspen Plus models developed in the present study, model 1 and model 2 are developed based on equilibrium and kinetic approach. Reaction kinetics, gasifier geometry and bed hydrodynamics are considered in model 3.	A precedent providing validation of the author's work via comparison of the CO_2 capture process simulation using Aspen Plus. This study focuses on biomass gasification as a method to reduce CO_2 emissions and emphasises the role of biomass as a sustainable energy source to reduce greenhouse gas emissions, which is a key component of the BiSC model and strategy. Optimisation of parameters to improve CO_2 capture efficiency is present, similar to the author's conducted experiments and validation protocol (Table 6.3). The validation is also based on the similarity of principles, process simulations and experimental data similarity, which indicates a common commitment to reducing carbon emissions and promoting sustainable practices.	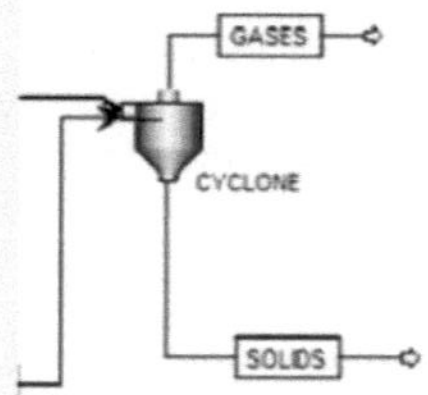
6	Madeddu et al. [98]	The CO_2 reduction potential for the European industry via direct electrification of heat supply (power-to-heat)	This work presents the results from a comprehensive bottom-up analysis of the energy use in 11 industrial sectors (accounting for 92% of Europe's industry CO_2 emissions) and estimates the technological potential for industry electrification in three stages.	A precedent providing validation of the author's work via using a variety of calculations of the CO_2 capture process. The study estimates that direct electrification of heat supply could cut CO_2 emissions by 78%, almost entirely abating energy-related CO_2 emissions. The work also covers 11 industrial sectors, accounting for 92% of Europe's industry CO_2 emissions. It was found that at least 78% of the energy demand is electrifiable with technologies that are already established, and 99% electrification can be achieved with the addition of technologies currently under development and at TRL 6-7. With an	'…CO_2 reductions for diverse industrial processes are unknown'….'78% of the energy demand is electrifiable with technologies that are already established, while 99% electrification can be achieved with the addition of technologies currently under development. Such a deep electrification reduces CO_2 emissions already based on the carbon intensity of today's electricity (~300 g CO_2/kWhel). With an increasing decarbonisation of the power sector (IEA: 12 gCO_2/kWhel in 2050), electrification could cut CO_2 emissions by 78%, and almost entirely abate the energy-related CO_2 emissions…'.

(Continued)

Table 5.12 (Continued)

No.	Author	Title	Description	Validation and similarity	Text, figures and tables
				increasing decarbonisation of the power sector (projected to be 12 g CO_2/kWhel in 2050), electrification could significantly reduce CO_2 emissions: deep electrification would reduce the industry bottleneck to only residual process emissions. The study underscores the significant potential of direct electrification (power-to-heat) in reducing CO_2 emissions across European industries, and by leveraging established and developing technologies, industries can achieve substantial emissions reductions, which could be contributing to achieving the EU's climate neutrality goals. Optimisation of parameters to improve CO_2 capture efficiency is present, similar to the author's conducted experiments and validation protocol (Table 6.3). The validation is also based on the similarity of principles, process simulations and experimental data similarity, which indicates a common commitment to reducing carbon emissions and promoting sustainable practices.	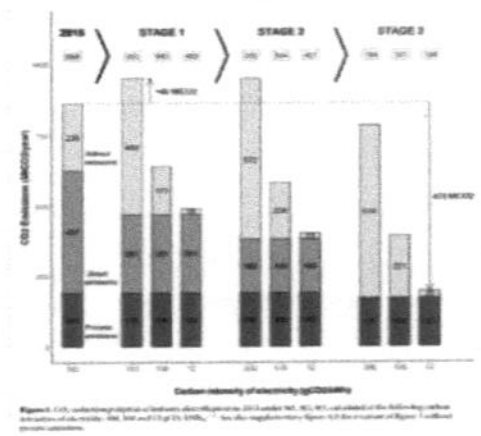
7	Ahmed [105]	Power-to-X: Modelling of Fischer-Tropsch synthesis in Aspen Plus	In this work, first, a summary of some PtX technologies is given, where their contribution to defossilising the energy sector is investigated. Then,	A precedent providing validation of the author's work via comparison of experimental data results and simulation of the CO_2 capture process using Aspen Plus. All three studies focus on significant CO_2 emissions reductions through advanced technological and process improvements. The author's work and both 'The Bio Steel Cycle' and 'Power-to-X' studies emphasise the use of carbon capture technologies to achieve emissions reductions. The importance of renewable energy sources is highlighted in 'The Bio Steel Cycle' and	'…The second approach is direct hydrogenation of carbon dioxide to produce C5+ liquid fuels, where RWGS reaction and FT synthesis are activated by the same catalyst in the same reactor simultaneously [61, 62]. The direct hydrogenation of CO_2 approach is less mature than the two-step approach. Many research attempts are being currently done to develop catalysts that can improve the selectivity of C5+liquid fuels in the case of direct hydrogenation of carbon dioxide…'.

			a techno-economic analysis of carbon-neutral FT synthesis is conducted, where Aspen Plus is used to model the process with having hydrogen and carbon dioxide as feedstock.	'Greater Energy Independence with Sustainable Steel Production', while 'Power-to-X' utilises renewable electricity for hydrogen production. Each study employs advanced process optimisation techniques, whether it is through process simulations, integrating renewable technologies or optimising chemical synthesis pathways. Both 'The Bio Steel Cycle' and 'Greater Energy Independence with Sustainable Steel Production' have long-term objectives of achieving near- or net-zero CO_2 emissions, while 'Power-to-X' aims for significant reductions through innovative fuel synthesis. Optimisation of parameters to improve CO_2 capture efficiency is present, similar to the author's conducted experiments and validation protocol (Table 6.3). The validation is also based on the similarity of principles, process simulations and experimental data similarity, which indicates a common commitment to reducing carbon emissions and promoting sustainable practices.	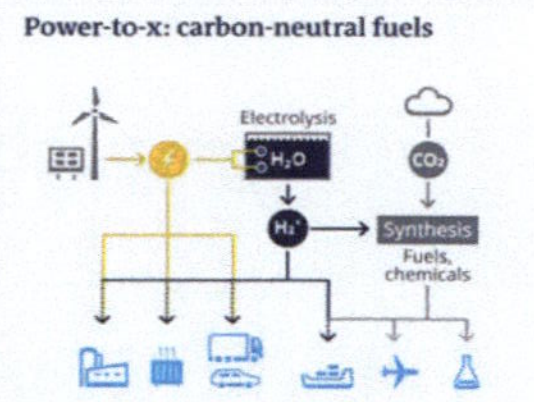 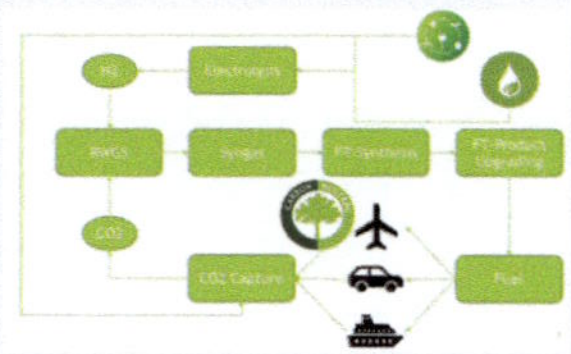
8	Zhao et al. [125]	Cutting parameter optimisation for reducing carbon emissions using digital twin	This paper presents a method of cutting parameter optimisation based on the construction of a digital twin of a CNC machine tool. The method dynamically optimises machining process parameters to reduce carbon emissions.	A precedent providing validation of the author's work via creating a digital twin with a focus on CO_2 emissions reduction. Both studies emphasise the importance of reducing CO_2 emissions, though they apply to different industrial processes (steel production versus machining). Both studies highlight the use of advanced technologies (carbon capture and renewable energy in steel production and digital twin optimisation in machining) to achieve emissions reductions. Optimisation of parameters to improve CO_2 capture efficiency is present, similar to the author's conducted experiments and validation protocol (Table 6.3). The validation is also based on the similarity of principles, process simulations and experimental data similarity, which indicates a common commitment to reducing carbon emissions and promoting sustainable practices.	Similar to the BiSC model, this paper is demonstrating a method of process optimisation. This is conducted by cutting parameter optimisation on basis of constructing a digital twin of a CNC machine tool. A dynamic optimisation method on cutting parameters is presented according to the simulation and optimisation of the virtual twin with the dynamic perception of the machining conditions of the physical machine. This work at last presents a case study, which validates this method for effectively optimising the cutting parameters and decreasing carbon emissions.

(Continued)

Table 5.12 (Continued)

No.	Author	Title	Description	Validation and similarity	Text, figures and tables
9	Liu et al. [126]	Modeling and assessment of carbon emissions in additive-subtractive hybrid manufacturing	This research assesses the carbon emission characteristics of additive-subtractive integrated hybrid manufacturing (ASIHM) and compares it with conventional subtractive manufacturing. The results show that ASIHM has a significantly lower carbon footprint.	A precedent providing validation of the author's work via calculations and flowcharts demonstrating the CO_2 capture process. Objective: Assess carbon emissions in additive-subtractive integrated hybrid manufacturing (ASIHM) and compare it with conventional subtractive manufacturing (CSM). Results: The study found that ASIHM has the highest proportion of carbon emissions during the additive-forming stage, reaching over 54%. However, compared to conventional milling, ASIHM has an 80% lower carbon footprint. The work of all authors focuses on CO_2 emissions reduction, and all studies emphasise the importance of reducing CO_2 emissions, though they apply to different industrial processes (steel production versus hybrid manufacturing). All studies highlight the use of advanced technologies to achieve emissions reductions. All documents suggest process improvements and technological advancements to reduce carbon emissions. Optimisation of parameters to improve CO_2 capture efficiency is present, similar to the author's conducted experiments and validation protocol (Table 6.3). The validation is also based on the similarity of principles, process simulations and experimental data similarity, which indicates a common commitment to reducing carbon emissions and promoting sustainable practices.	'… the carbon emissions of the ASIHM process were analysed, and comparative research on the carbon emissions in material and energy consumption was also carried out with the conventional subtractive manufacturing (CSM) process. The results have revealed that ASIHM has the highest proportion of carbon emissions during the additive forming stage, reaching over 54%. Compared to conventional milling, ASIHM has an 80% lower carbon footprint…'.

10	Michailos and Gibbins [127]	A Modelling Study of Post-Combustion Capture Plant Process Conditions to Facilitate 95–99% CO_2 Capture Levels From Gas Turbine Flue Gases	The principal purpose of this study is to examine the changes in process conditions that might be needed to achieve up to 99% capture levels in amine post-combustion capture (PCC) plants for combined cycle gas turbine (CCGT) flue gases.	A precedent providing validation of the author's work via calculations and flowcharts focusing on the CO_2 capture process. All authors' work emphasises the importance of reducing CO_2 emissions, though they apply to different industrial processes (steel production versus gas turbine flue gases). All studies highlight the use of advanced technologies to achieve emissions reductions and suggest process improvements and technological advancements to reduce carbon emissions, using mathematical extrapolations as well as Aspen Plus process simulations to demonstrate post-combustion carbon capture. Optimisation of parameters to improve CO_2 capture efficiency is present, similar to the author's conducted experiments and validation protocol (Table 6.3). The validation is also based on the similarity of principles, process simulations and experimental data similarity, which indicates a common commitment to reducing carbon emissions and promoting sustainable practices.	'…For a given amount of CO_2 captured and constant rich loading, lower lean loadings also increase specific energy consumption, … for a capture level of 95% using the modified Sherman case (i.e., a fixed CO_2 capture rate of 39.985 kg/s), reports calculated specific energy consumption for three rich loadings, 0.422, 0.446, and 0.463 $molCO_2/molMEA$. …'.

Table 5.13 Coking process experimental input and results data.

Stage	Process	Resource	Quantity	Result Weight kg	Quantity confirmed by	Quantity kg cited	Temperature °C	Temperature °C confirmed or range stated	Pressure	Pressure confirmed by	CO_2 emissions energy/combustion per tonne product/kiln	CO_2 emissions confirmed by	CO_2 emissions stated in t per tonne of product
Preparation	Coking	Coal, bituminous	Kiln capacity	36000 kg/ 36 t	Neuwirth and Redemann [128]	45,000			1/14.69/ 1.01	Desai and Camobreco [129]		Chen et al. [7]	2.59
					Neuwirth, [130]	55,000		1,275		Pandey, 2025		Firsbach, Senk and Babich, 2022	0.27–3.93
										Kertcher and Linsky, 1974 15,000	1100–1400		0.27–3.93 t/t coke
										Hao et al., 2024 28,000–36,000		1050–1300	
										Ergul and Selimli, 2024 20,000		1200–1300	

Coking The input data for the coal coking process are displayed in Table 5.13 and the data are presented in graphs in Figures 5.35–5.37:

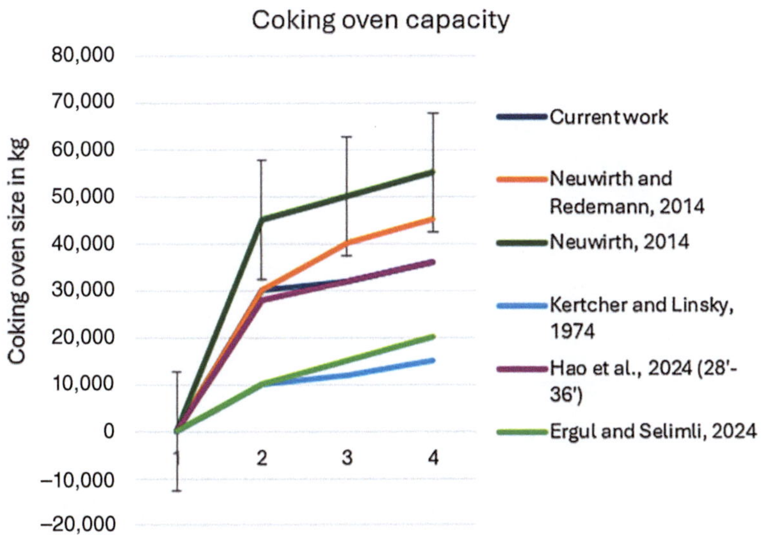

Figure 5.35 Coal coking oven size.

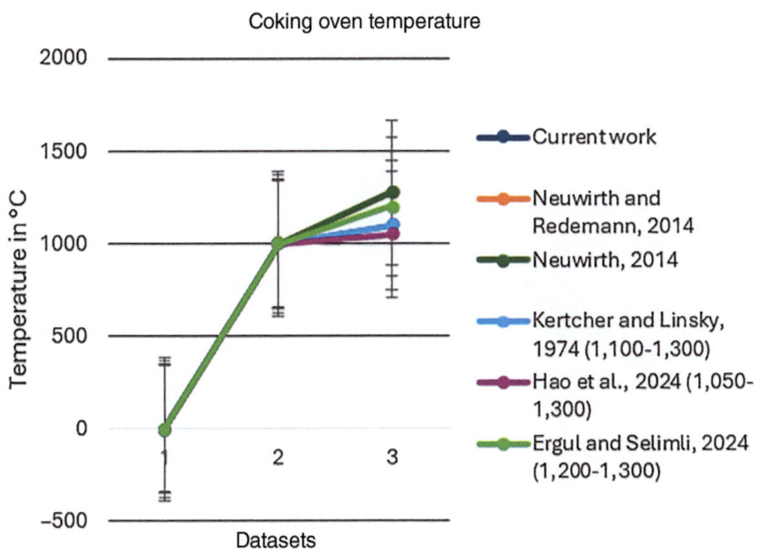

Figure 5.36 Operational temperature in the coking oven.

The current works' input and experimental data for coking coal, represented by a dark blue line, are partly obscured by the lines representing the authors confirming the current work's data, as they represent the same values.

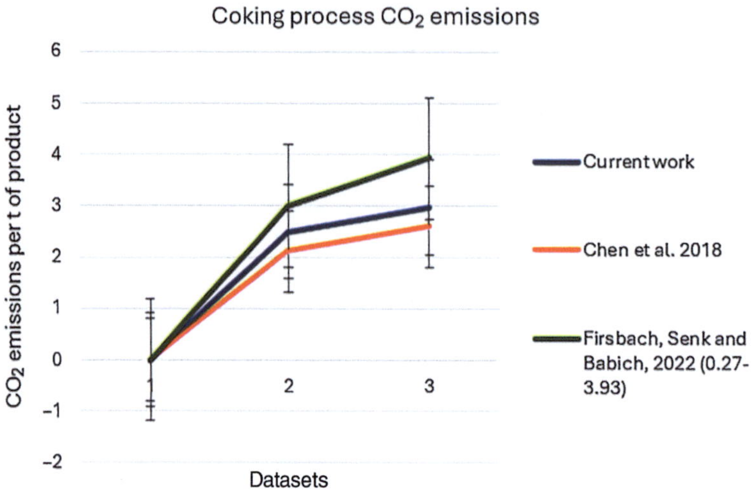

Figure 5.37 Variations of CO_2 emissions from coking coal.

Calcification The input data for the calcification process are displayed in Table 5.14 and the data are presented in graphs in Figures 5.38–5.40:

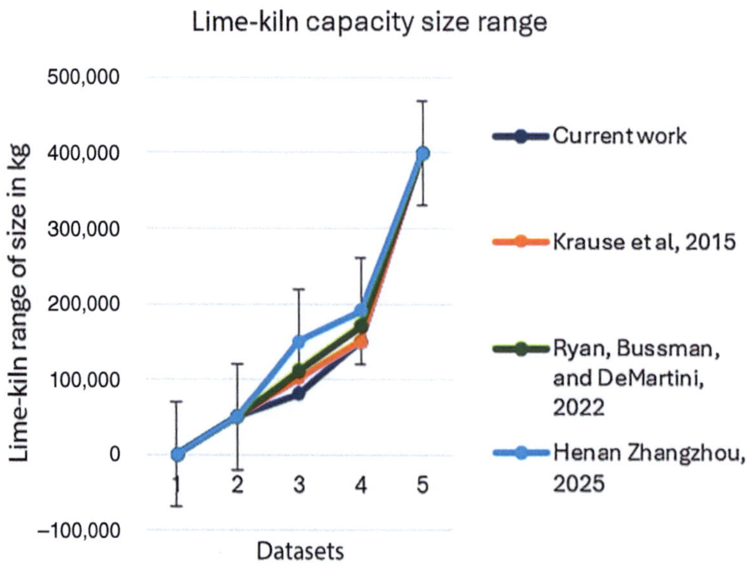

Figure 5.38 Lime-kiln size variations.

The current works' input and experimental data for calcification, represented by a dark blue line, are partly obscured by the lines representing the authors confirming the current work's data, as they represent the same values.

Table 5.14 Calcification process experimental input and results data.

Stage	Process	Resource	Quantity	Result Weight kg	Quantity confirmed by	Quantity kg cited	Temperature °C	Temperature °C confirmed or range stated	Pressure	Pressure confirmed by	CO_2 emissions energy/ combustion per tonne product/kiln	CO_2 emissions confirmed by	CO_2 emissions stated in t per tonne of product
Preparation	Calcification	Limestone	Kiln capacity	50,000 kg/50 t	Krause et al. [136]	50'–400,000	900	900	1/14.69/1.01	Liu [137]	1.26–2.21 t/t lime	Chen et al. [7]	2.21
		Water			Ryan, Bussman, and DeMartini [138]	50'–400,000		900–2000				Wu, Wang, Ke and Zang [139]	1.10–1.26
					Henan Zhengzhou, [140]	50'–400,000		1050–1150					

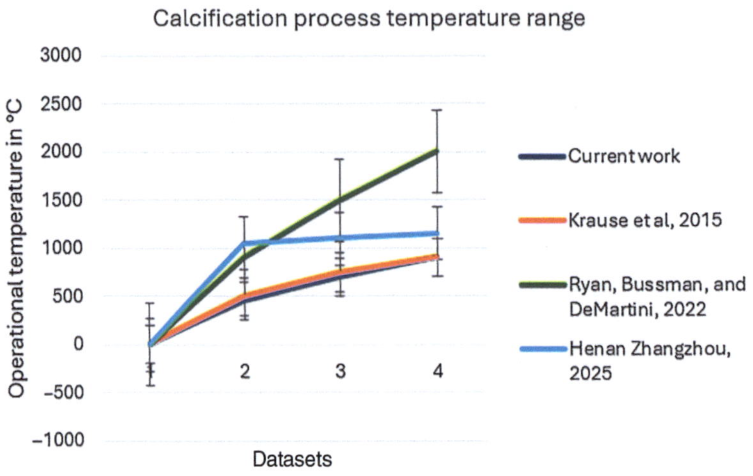

Figure 5.39 Lime-kiln temperature variations.

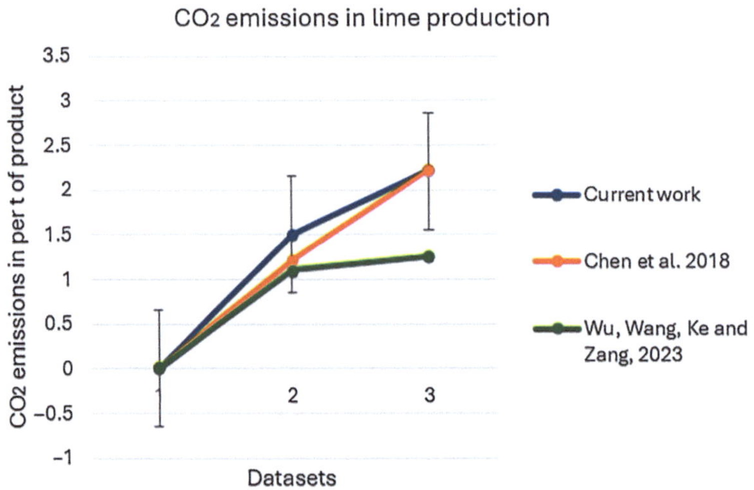

Figure 5.40 CO_2 emission of the calcification process.

Sintering The input data for the sintering process are displayed in Table 5.15 and the data are presented in graphs in Figures 5.41–5.43:

The current works' input and experimental data for the sintering process, represented by a dark blue line, are partly obscured by the lines representing the authors confirming the current work's data, as they represent the same values.

Table 5.15 Sintering process experimental input and results data.

Stage	Process	Resource	Quantity	Result Weight kg	Temperature 1480 °C confirmed by	Temperature °C confirmed or range stated	Pressure 1/14.69/1.01 confirmed by	CO_2 emissions energy/combustion per tonne product/kiln	CO_2 emissions confirmed by	CO_2 emissions stated in t per tonne of product
Preparation	Sintering	Oxygen	Ambient		Liang et al. [141]	1020–1080	Singh et al. [64]		Riesbeck et al. [16]	0.02
		Iron ore	62.85%	207405	Cores et al. [142]	1250–1350	Liang et al. [141]		Ecofys [7]	0.28
		Lime	24.15%	79695	Zhou et al. [143]	1000–1500		0.02–0.97 t/t sinter	Mohammad, Patra	
		Coke	5.00%	16500	Wang et al. [144]	1000–1250			and Harichandan [145]	0.37
		Limestone	8.00%	26400	Fernández-González et al. [146]	1300–1480			Wang et al. [29]	0.97
					Worrell et al. [18]	1300–1480				

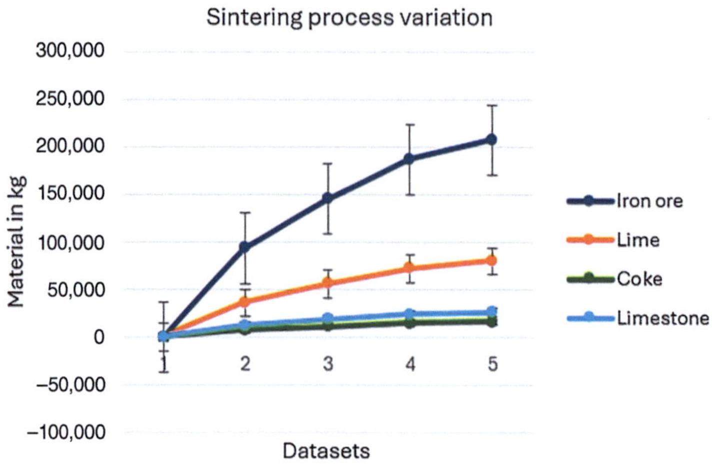

Figure 5.41 Sintering material quantity variation.

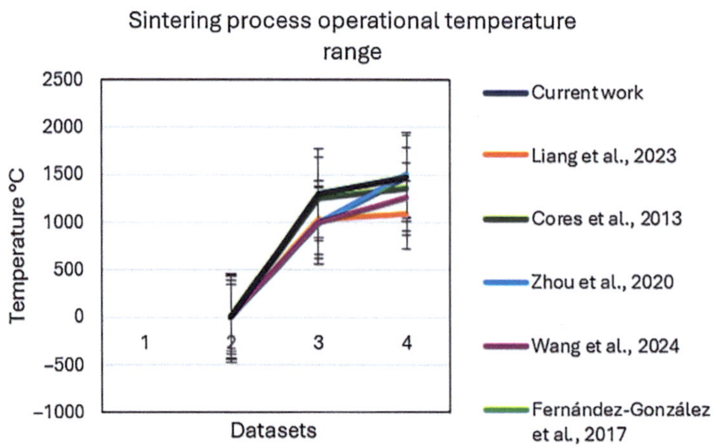

Figure 5.42 Sintering operational temperature variation.

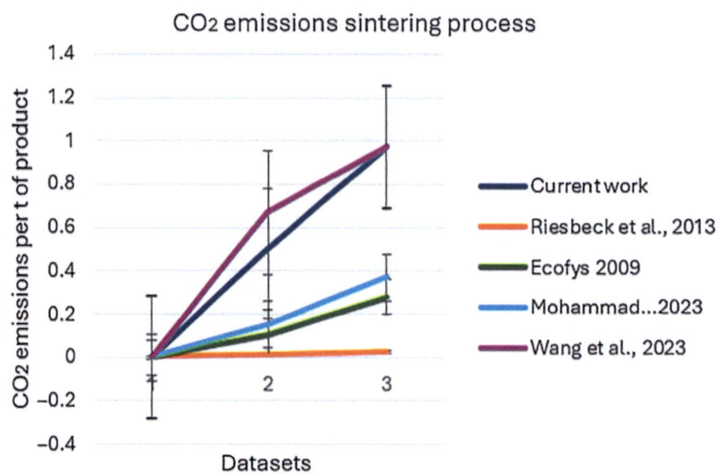

Figure 5.43 CO_2 emission of the sintering process.

Blast furnace operation The input data for the blast furnace operation are displayed in Table 5.16 and the data are presented in graphs in Figures 5.44–5.46:

The current works' input and experimental data for the blast furnace operation are partly obscured by the lines representing the authors confirming the current work's data, as they represent the same values.

Basic oxygen furnace operations The input data for the basic oxygen furnace operation are displayed in Table 5.17 and the data are presented in graphs in Figures 5.47–5.49.

The current works' input and experimental data for the basic oxygen furnace operation are partly obscured by the lines representing the authors confirming the current work's data, as they represent the same values.

The current Aspen Plus, Simul8 and INOSIM simulations largely align with the simulation models, as far as process design, flowcharts, component selection and layout are concerned. However, as the identified benchmark simulations did not provide detailed input data for comparison of mass balance, temperature or furnace pressure, the validation against scientific literature, confirming the parameters and input data used, was carried out.

The differences between the CO_2 emissions stated in the current work and the literature confirming the levels of emissions and material quantities used can be explained by (i) a difference in approach to gathering emissions data, (ii) a difference in the furnaces or oven size in coking, calcification, sintering and BF/BOF vessels and (iii) inclusion of the emissions as a result of the specific energy requirements at the different stages.

It is my understanding from undertaking this research that the findings are pointing to a research gap: the energy (derived from fossil fuels, primarily) required to heat the oxygen for the BOF. To date, there seems to be a lack of scientific publications, which combine the elements of providing data on CO_2 emissions at all stages of the steelmaking process from mining to finish machining, and process simulations using either Aspen Plus, Simul8 or INOSIM, and also providing input data for mass balancing and validation purposes. Further research seems to be necessary in this regard, to identify mechanisms to further reduce carbon emissions in steel production, possibly in connection with TGRBF, EAF and oxy-fuel combustion systems.

5.11.1.6 Conclusions and discussion verification and validation

Conclusively, it can be said that the pentagonal approach to investigation, verification and validation of the research findings has proven to provide a range of viable results. The choice not to rely on just one process simulation software but to use three simultaneously has removed the pressure from one method to provide proof of concept. In addition, using mathematical standard procedure (algebra), statistical methods, MS Excel calculations and extrapolations has delivered proof of viability for hypothesis one: stating that it is possible to reduce CO_2 emissions in steel production by at least 20% and achieve a higher degree of energy independence at the same time. Table 5.18 displays the values and sources in literature, as a result of mathematical extrapolations and extrapolations in MS Excel, leading to establishing the values

Table 5.16 Blast furnace process experimental input and results data.

Stage	Process	Resource	Quantity	Result Weight kg	Confirmed current quantities/ % range	System Temperature 1650–2300 °C confirmed by	Temperature °C confirmed or range stated	Pressure 1–2.5 atm	Pressure confirmed by	CO_2 emissions energy/ combustion per tonne product/kiln	CO_2 emissions confirmed by	CO_2 emissions stated in t per tonne of product
Blast furnace	Smelting	Coke	28–50.9% capacity	168,015.25	Li et al. [75]	Kawashiri et al. [147]	2000–2400	~1–2.5 bar	Kawashiri et al. [147]		Toktarova [23]	1.60–2.20
	Reduction	Limestone		16,500	Yang et al. [148]	Satyendra [149]	1650		Babich and Senk [150]		Griffin Hammond [151]	0.90–3.20
		Oxygen	800 m^3/min = kg	22,864	Chen et al. [7]	Bailera et al. [152]	1550–2300	220 psi		7.52 t CO_2/t cs	Mandova [22]	17.98
		Lime	4.5–19.5%	5,653.62	Wang et al. [153]						OECD [154]	5.85
		Iron ore	20–22.5%	28,268.11	Kawashiri et al. [147]						EERE [155]	1.16
		Iron oxide	50–55.0%	69,099.8	Satyendra [149]						Chen [7]	2.50
		Silicon dioxide	2.5–10.5%	3,140.9	Babich and Senk [150]						Hasanbeigi Springer [156]	1.49–2.80
		Magnesium oxide	3.5–4.5%	5,653.62							Wang et al. [29]	1.90
		Aluminium oxide	1–3%	3,769.08							Suer [25]	2.10–3.00
		Zinc		1,256.36						2.1–3.2 t/t steel		
		Titanium	2–5%	1,256.36								
		Potassium		1,256.36								
		Chromium		1,256.36								
		Manganese		1,256.36								
		Water	0–0.6%	753.82								

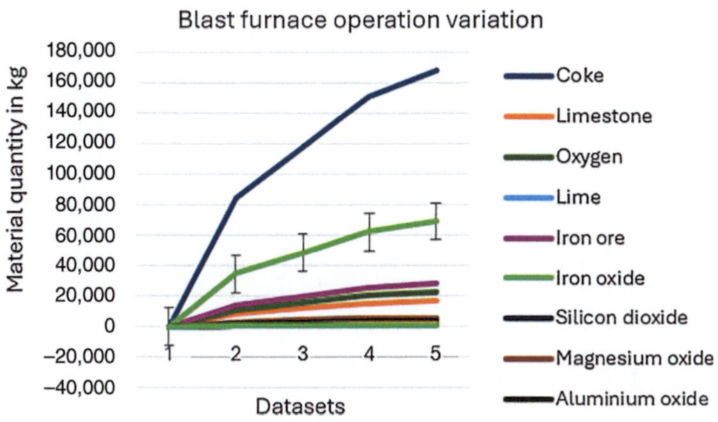

Figure 5.44 Blast furnace material quantity variation.

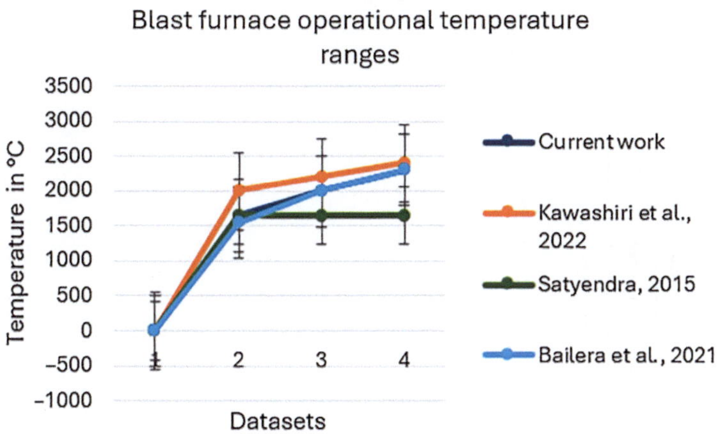

Figure 5.45 BF temperatures.

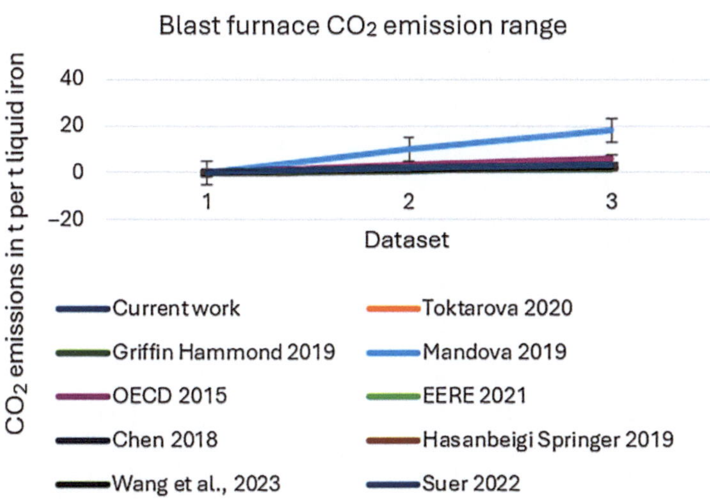

Figure 5.46 BF CO_2 emissions.

Table 5.17 Basic oxygen furnace process experimental input and results data.

Stage	Process	Resource	Quantity	Result Weight kg	Confirmed current quantities/ % range	System Temperature 1650–2300 °C confirmed by	Temperature °C confirmed or range stated	Pressure 1–2.5 atm	Pressure confirmed by	CO_2 emissions energy/ combustion per tonne product/kiln	CO_2 emissions confirmed by	CO_2 emissions stated in t per tonne of product
Basic oxygen	Oxygen	Liquid Fe		224,636	Hasanbeigi Springer [156]		1600–1650	1 atm/ 14.69 psi			Toktarova et al. [23]	2.20
furnace	Converter	Lime	50 kg/t liquid steel	16,500	Griffin Hammond [151]		1600		Griffin Hammond [151]		Mandova [22]	7.71
	Process	Oxygen 99%		22,864	Garvey Norman and Barrett [31]	25	~1600	150	Garvey Norman and Barrett [31]		Khalid et al. [157]	2.10
		Scrap metal		66,000	Hasanbeigi Springer [156]	1700	1600–1650			66t × 4.95 tCO_2	Griffin Hammond [151]	2.76
				330,000						1.6–3.13 t/t steel	Martelaro [158]	1.60
											Garvey Norman and Barrett [31]	1.85–2.20
											Hasanbeigi Springer [156]	1.49–2.80
											Napp et al. [159]	2.27

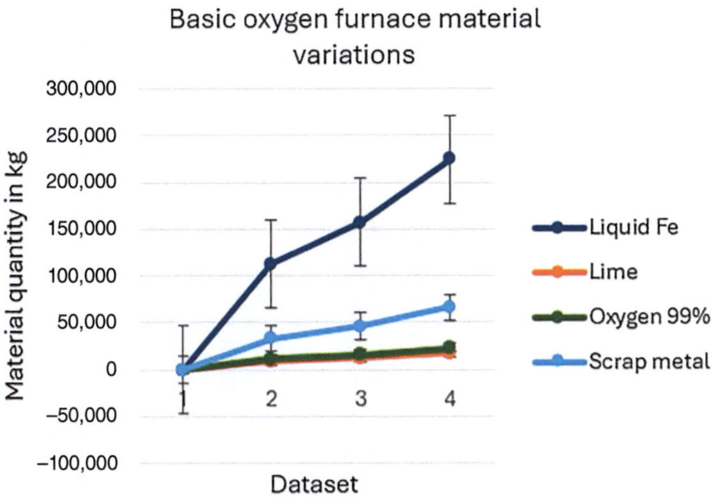

Figure 5.47 BOF material quantity variation.

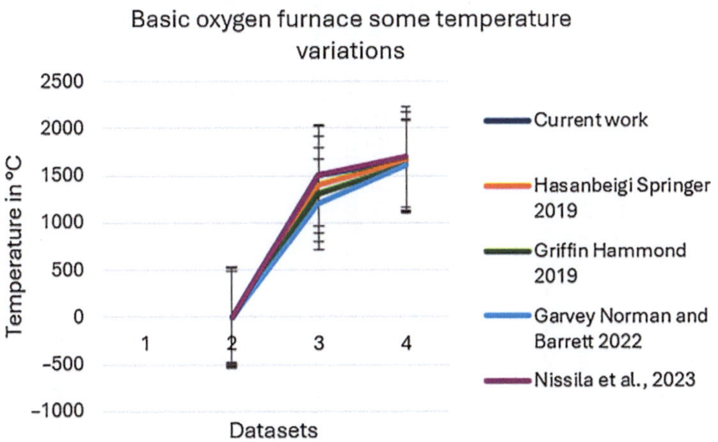

Figure 5.48 BOF temperature ranges.

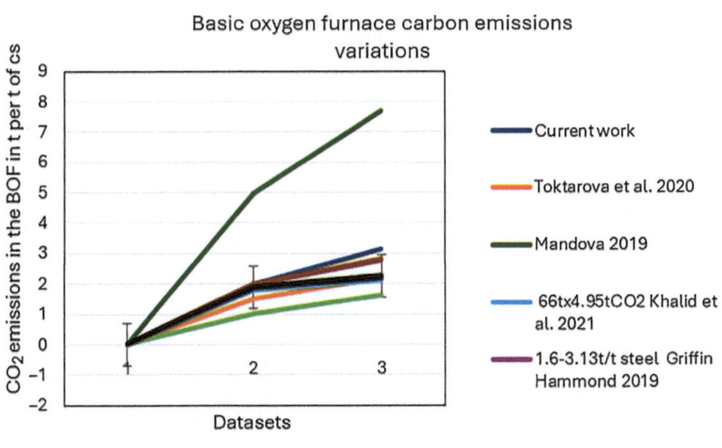

Figure 5.49 BOF CO_2 emissions.

Table 5.18 Table overview of CO_2 emissions values applied and literature sources/URLs.

Steel production stage	LR data CO_2/t Product[a)]	Values applied In simulations	URL to source	References
Oxygen	0.26 t	0.26 t	https://www.ncbi.nlm.nih.gov/pmc/articles/PMC8508159/	[10]
Limestone	0.035–2.31 t	0.035 t	rtuect-2017-0011 (sciendo.com)	[13]
Iron ore	2.19–6.4 t	2.19 t	untitled (bhp.com) Report page 9 and 292	[96]
Coal	5.43–12.31 t	5.43 t	untitled (bhp.com) Report page 9 and 292	[96]
Coking	2.97–3.14 t	2.59 t	https://www.sciencedirect.com/science/article/pii/S0306261918303726	[3, 7]
Calcination	2.21–2.32 t	2.21 t	https://media.rff.org/documents/WP_22-16_M10.pdf	[3, 7]
Iron ore pelletising	0.035–0.45 t	0.035 t	http://ebookcentral.proquest.com/lib/staffordshire/detail.action?docID=6821497	[104]
Sintering	0.04–0.97 t	0.93 t	http://www.divaportal.org/smash/get/diva2:1004013/FULLTEXT01.pdfpage 5	[16, 97]
Blast furnace	2.1–3.2 t	2.95 t	Carbon Footprint and Energy Transformation Analysis of Steel Produced via a Direct Reduction Plant with an Integrated Electric Melting Unit | SpringerLink	[25]
Basic oxygen furnace	1.6–3.13 t	2.0 t	https://onlinelibrary.wiley.com/doi/abs/10.1002/srin.202100852	[32]
Casting	0.2–1.22 t	1.22 t	https://www.divaportal.org/smash/get/diva2:1140576/FULLTEXT01.pdfPage 3	[33]
Rolling	0.26–1.0 t	1.0 t	https://www-sciencedirectcom.ezproxy.staffs.ac.uk/science/article/pii/S0360544221016777 Page 7	[34]
Finish machining	0.15	0.15t	https://www-sciencedirectcom.ezproxy.staffs.ac.uk/science/article/pii/S0360544221016777 Page 7	[35, 98]

a) Specific energy consumption (SEC) was added where appropriate [3].

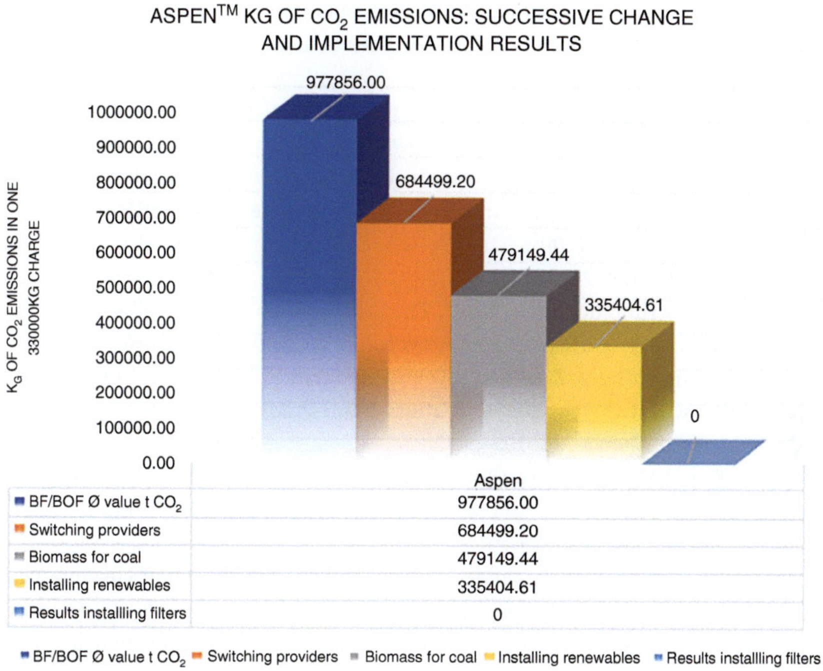

Figure 5.50 CO_2 emissions reduction results were exported from Aspen to MS Excel.

later applied in the process simulations and confirmed as reported results from the process simulation runs.

The BiSC components were applied, and the total CO_2 emissions reduction results were exported from Aspen and sent to MS Excel. The values were then modelled and expressed in a graph, as displayed in Figure 5.50.

Conclusively, it is safe to say that the different simulations provided proof of concept and rendered the 1 hypothesis valid: it is possible to reduce CO_2 emissions in steel production to almost '0' when the BiSC model and strategy is being implemented, in sections or entirely.

References

1 US Energy Information Administration (EIA) (1994). Carbon dioxide emission factors for coal [Internet]. https://www.eia.gov/coal/production/quarterly/co2_article/co2.html (accessed 6 July 2023).

2 Bogunovic, D., Kecojevic, V., Lund, V. et al. (2009). Analysis of energy consumption in surface coal mining. *Transactions of the Society for Mining, Metallurgy, and Exploration, Inc* 326: 79–87.

3 Li, Y. and Zhu, L. (2014). Cost of energy saving and CO_2 emissions reduction in China's iron and steel sector. *Applied Energy* 130: 603–616.

4 Liu, Z., Ciais, P., Deng, Z. et al. (2020). Near-real-time monitoring of global CO_2 emissions reveals the effects of the COVID-19 pandemic. *Nature Communication* 11 (1): 5172.

5 McKinsey & Company (2021). Metals & mining practice: the future of the European steel industry. A road map toward economic and environmental sustainability [Internet]. https://www.mckinsey.com/~/media/mckinsey/industries/metals%20and%20mining/our%20insights/the%20future%20of%20the%20european%20steel%20industry/the-future-of-the-european-steel-industry_vf.pdf (accessed 7 July 2023).

6 Griffin, P. and Hammond, G. (2021). The prospects for 'green steel' making in a net-zero economy: a UK perspective. *Global Transitions* 1 (3): 72–86.

7 Chen, Q., Gu, Y., Tang, Z. et al. (2018). Assessment of low-carbon iron and steel production with CO_2 recycling and utilization technologies: a case study in China. *Applied Energy* 220: 192–207.

8 Griffin, P. and Hammond, G. (2019). Industrial energy use and carbon emissions reduction in the iron and steel sector: a UK perspective. *Journal of Applied Energy* 249: 109–125.

9 Hegemann, K.R., Guder, R. Linde-Fränkl-Verfahren der Sauerstoff-Gewinnung. In: Hegemann, K.R., Guder, R., editors. *Stahlerzeugung: Integrierte Hüttenwerks- und Gasreinigungsanlagen [Internet].* Wiesbaden: Springer Fachmedien; 2020 p. 47–9. https://doi.org/10.1007/978-3-658-29,091-7_5 (accessed 8 July).

10 Variny, M., Jediná, D., Rimár, M. et al. (2021). Cutting oxygen production-related greenhouse gas emissions by improved compression heat management in a cryogenic air separation unit. *IJERPH* 18 (19): 10370.

11 Intergovernmental Panel for Climate Change (IPCC) *Revised 1996 IPCC Guidelines for National Greenhouse Gas Inventories: Workbook [Internet].* Geneva: IPCC; 1996. https://www.ipcc-nggip.iges.or.jp/public/gl/guidelin/ch2wb1.pdf (accessed 7 July 2023).

12 Lin, S., Kiga, T., Wang, Y., and Nakayama, K. (2011). Energy analysis of $CaCO_3$ calcination with CO_2 capture. *Energy Procedia* 4: 356–361.

13 Kittipongvises, S. (2017). Assessment of environmental impacts of limestone quarrying operations in Thailand. *Environmental and Climate Technologies* 20 (1): 67–83.

14 Yang, W., Pudasainee, D., Gupta, R. et al. (2021). An overview of inorganic particulate matter emission from coal/biomass/MSW combustion: sampling and measurement, formation, distribution, inorganic composition and influencing factors. *Fuel Processing Technology* 2013: 106657.

15 Shan, Y., Liu, Z., and Guan, D. (2016). CO_2 emissions from China's lime industry. *Applied Energy* 166: 245–252.

16 Riesbeck, J., Hooey, L., Kinnunen, K., et al. (2013). Global effects of closing down sinter plant at Ruukki Raahe integrated steel works [Internet]. http://ltu.diva-portal.org/smash/record.jsf?pid=diva2%3A1004013&dswid=-4624 (accessed 7 July 2023).

17 Fraunhofer Institute (2009). Methodology for the free allocation of emission allowances in the EU ETS post 2012 Sector report for the iron ore industry

By order of the European Commission Study Contract: 07.0307/2008/515770/ ETU/C2 Ecofys project Number: PECSNL082164 [Internet]. p. 1–15. https://climate.ec.europa.eu/system/files/2016-11/bm_study-iron_ore_en.pdf (accessed 7 June 2023).

18 Worrell, E., Blinde, P., Neelis, M, et al. (2010). Energy efficiency improvement and cost saving opportunities for the U.S. iron and steel industry an ENERGY STAR(R) guide for energy and plant managers [Internet]. p. LBNL-4779E, 1026806. Report No.: LBNL-4779E, 1026806. http://www.osti.gov/servlets/purl/1026806/ (accessed 2 November 2022).

19 Zakkour, P. and Cook, G. (2010). *UNIDO CCS Roadmap for Industry: High-Purity CO_2 Sources. Sectoral Assessment and Final Draft Report*. Vienna, Austria: United Nations industrial development Organization.

20 Office of Energy Efficiency & Renewable Energy (EERE) Energy.gov. (2018) Manufacturing energy and carbon footprints (2018 MECS). https://www.energy.gov/eere/iedo/manufacturing-energy-and-carbon-footprints-2018-mecs (accessed 6 July 2023).

21 Mandova, H., Leduc, S., Wang, C. et al. (2018). Possibilities for CO_2 emission reduction using biomass in European integrated steel plants. *Biomass and Bioenergy* 1 (115): 231–243.

22 Mandova, H., Patrizio, P., Leduc, S. et al. (2019). Achieving carbon-neutral iron and steelmaking in Europe through the deployment of bioenergy with carbon capture and storage. *Journal of Cleaner Production* 218: 118–129.

23 Toktarova, A., Karlsson, I., Rootzén, J. et al. (2020). Pathways for low-carbon transition of the steel industry – a Swedish case study. *Energies* 13 (15): 3840.

24 Pandit, J.K., Qader, A., and Lim S. (2021). *Cross-Technology Scheme Options to Reduce Greenhouse Gas Emissions in a Steel Industry [Internet]*. Rochester, NY. https://papers.ssrn.com/abstract=3821438 (accessed 7 July 2023).

25 Suer, J., Ahrenhold, F., and Traverso, M. (2022). Carbon footprint and energy transformation analysis of steel produced via a direct reduction plant with an integrated electric melting unit. *Journal of Sustainable Metallurgy* 8 (4): 1532–1545.

26 Prakash, R. and Muller M. (2007). Industrial oxygen: its use and generation [Internet]. https://www.eceee.org/library/conference_proceedings/ACEEE_industry/2007/Panel_6/P6_13/ (accessed 7 July 2023).

27 Office of Energy Efficiency & Renewable Energy (EERE) (2021). Hydrogen production: natural gas reforming [Internet]. (Hydrogen Production: Natural Gas Reforming). https://www.energy.gov/eere/fuelcells/hydrogen-production-natural-gas-reforming (accessed 8 November 2022).

28 Toktarova, A. (2021). *The Low-Carbon Steel Industry-Interactions Between the Hydrogen Direct Reduction of Steel and the Electricity System – Proquest [Internet]*. Gothenburg, Sweden: Department of Space, Earth and Environment CHALMERS UNIVERSITY OF TECHNOLOGY. https://www.proquest.com/openview/c2a4d77d71403fdbffd33f250b13251d/1?pq-origsite=gscholar&cbl=2026366&diss=y (accessed 7 July 2023).

29 Wang, G., Li, R., Dan, J. et al. (2023). Preparation of biomass hydrochar and application analysis of blast furnace injection. *Energies* 16 (3): 1216.

30 Wang, C., Walsh, S., Haynes, M.W. et al. (2022). From Australian iron ore to green steel: the opportunity for technology-driven decarbonisation [Internet]. https://ecat.ga.gov.au/geonetwork/srv/api/records/5377741d-ece0-4e01-b87e-6bf49da11c03 (accessed 7 July 2023).

31 Garvey, A., Norman, J.B., and Barrett, J. (2022). Technology and material efficiency scenarios for net zero emissions in the UK steel sector. *Journal of Cleaner Production* 333: 130216.

32 Tian, B., Zhu, R., Zhou, Y. et al. (2022). A static balance model and analysis of 430 stainless steel produced by basic oxygen furnace–argon oxygen decarburization furnace process. *Wiley Online Library: Steel Research International [Internet].*; 93(7). https://onlinelibrary.wiley.com/doi/abs/10.1002/srin.202100852 (accessed 7 July 2023).

33 Wänerholm, M. (2016). Rapport nr 2016–008 – Climate impact of metal-casting [Internet]. p. 1–26. https://www.diva-portal.org/smash/get/diva2:1140576/FULLTEXT01.pdf (accessed 7 August 2023).

34 Schmitz, N., Sankowski, L., Kaiser, F. et al. (2021). Towards CO_2-neutral process heat generation for continuous reheating furnaces in steel hot rolling mills – a case study. *Energy* 120: 155.

35 Sun, Y., Tian, S., Ciais, P. et al. (2022). Decarbonising the iron and steel sector for a 2 °C target using inherent waste streams. *Nature Communication* 13 (1): 297.

36 SSAB (2022). Sustainability at SSAB. https://www.ssab.com/en/company/sustainability/first-in-fossil-free-steel/environmental-benefits (accessed 7 July 2023).

37 Conejo, A.N., Birat, J.P., and Dutta, A. (2020). A review of the current environmental challenges of the steel industry and its value chain. *Journal of Environmental Management* 1 (259): 109782.

38 Benetto, E., Popovici, E.C., Rousseaux, P., and Blondin, J. (2004). Life cycle assessment of fossil CO_2 emissions reduction scenarios in coal-biomass based electricity production. *Energy Conversion and Management* 45 (18): 3053–3074.

39 Yi, Q., Zhao, Y., Huang, Y. et al. (2018). Life cycle energy-economic-CO_2 emissions evaluation of biomass/coal, with and without CO_2 capture and storage, in a pulverized fuel combustion power plant in the United Kingdom. *Applied Energy* 1 (225): 258–272.

40 International Energy Agency (IEA) (2023). IEA data and statistics. https://www.iea.org/data-and-statistics (accessed 7 July 2023).

41 International Energy Agency (IEA) (2018). Global energy demand grew by 2.1% in 2017, and carbon emissions rose for the first time since 2014 – News. https://www.iea.org/news/global-energy-demand-grew-by-21-in-2017-and-carbon-emissions-rose-for-the-first-time-since-2014 (accessed 7 July 2023).

42 International Energy Agency (IEA) (2021). Global CO_2 emissions rebounded to their highest level in history in 2021 – news. https://www.iea.org/news/

global-co2-emissions-rebounded-to-their-highest-level-in-history-in-2021 (accessed 7 July 2023).

43 U.S. Energy Information Administration (EIA) (2021). Independent statistics and analysis – international energy outlook 2021 [Internet]. https://www.eia.gov/outlooks/aeo/data/browser/#/?id=1-IEO2021&sourcekey=0 (accessed 11 July 2023).

44 Intergovernmental Panel for Climate Change (IPCC) (2021). *Climate Change 2021 – The Physical Science Basis Summary for Policymakers [Internet]*. Cambridge: Cambridge University Press. https://www.ipcc.ch/report/ar6/wg1/downloads/report/IPCC_AR6_WGI_SPM.pdf (accessed 7 July 2023).

45 International Renewable Energy Agency (IRENA) (2021). World energy transitions outlook: 1.5 °C pathway [Internet]. https://www.irena.org/publications/2021/Jun/World-Energy-Transitions-Outlook

46 International Renewable Energy Agency (IRENA) (2022). Renewable power remains cost-competitive amid fossil fuel crisis [Internet]. https://www.irena.org/news/pressreleases/2022/Jul/Renewable-Power-Remains-Cost-Competitive-amid-Fossil-Fuel-Crisis (accessed 13 July 2023).

47 Bataille, C., Åhman, M., Neuhoff, K. et al. (2018). A review of technology and policy deep decarbonization pathway options for making energy-intensive industry production consistent with the Paris Agreement. *Journal of Cleaner Production* 187: 960–973.

48 Bataille, C., Nilsson, L.J., and Jotzo, F. (2021). Industry in a net-zero emissions world: new mitigation pathways, new supply chains, modelling needs and policy implications. *Energy and Climate Change* 2: 100059.

49 National Renewable Energy Laboratory (NREL) (2019). 2019 year in review – NREL's top 20 stories [Internet]. https://www.nrel.gov/news/program/2019/2019-year-in-review-nrel-top-20-stories.html (accessed 11 July 2023).

50 Waisman, H., Bataille, C., Winkler, H. et al. (2019). A pathway design framework for national low greenhouse gas emission development strategies. *Nature Climate Change* 1 (9): 261.

51 Intergovernmental Panel for Climate Change (IPCC) (2022). Climate change 2022: impacts, adaptation and vulnerability. [Internet]. https://www.ipcc.ch/report/ar6/wg2/ (accessed 3 January 2023).

52 Intergovernmental Panel for Climate Change (IPCC) (2022a). Climate change 2022: mitigation of climate change [Internet]. https://www.un.org/en/climatechange/reports?gclid=EAIaIQobChMIxsyF3ZSl-AIVoo1oCR19Dw3IEAAYAiAAEgIty fD_BwE (accessed 30 April 2022).

53 Santos, M.P.S. and Hanak, D.P. (2022). Carbon capture for decarbonisation of energy-intensive industries: a comparative review of techno-economic feasibility of solid looping cycles. *Frontiers of Chemical Science and Engineering* 16 (9): 1291–1317.

54 Tata Steel (2024). Tata Steel in Europe. Zeremis. https://www.tatasteeleurope.com/sustainability/green-steel-solutions/zeremis (accessed 9 February 2024).

55 Tata Steel Europe (2024). Zeremis. https://www.tatasteeleurope.com/sustainability/green-steel-solutions/zeremis (accessed 7 July 2023).

56 Tata Steel Europe (2020). Tata Steel in Europe sustainability report 2019/2020 [Internet]. https://www.tatasteeleurope.com/sites/default/files/TSE%20Sustainability%20report%202,019-20%20(EN).pdf (accessed 7 July 2023).

57 Meteorological Office UK (2023). Beaufort wind force scale. https://www.metoffice.gov.uk/weather/guides/coast-and-sea/beaufort-scale (accessed 12 February 2024).

58 Nix, J. (2023). *John Nix Pocketbook – For Farm Management*, 53rde. Leicestershire: The Anderson Centre.

59 Blue Planet Systems (2023) Technology. https://www.blueplanetsystems.com/technology (accessed 6 July 2023).

60 Wilkerson, J. (2018). HARVARD UNIVERSITY BLOG biomass over coal: burning different carbon to mitigate climate change [Internet]. *Science in the News*. https://sitn.hms.harvard.edu/flash/2018/biomass-over-coal-burning-different-carbon-to-mitigate-climate-change/ (accessed 7 July 2023).

61 Bao, J., Lu, W.H., Zhao, J., and Bi, X.T. (2018). Greenhouses for CO_2 sequestration from atmosphere. *Carbon Resources Conversion* 1 (2): 183–190.

62 Williams, R., Jack, C., Gamboa, D., and Shackley, S. (2021). Decarbonising steel production using CO_2 Capture and Storage (CCS): results of focus group discussions in a Welsh steel-making community. *International Journal of Greenhouse Gas Control* 104: 103218.

63 Lu, X., Tian, W., Li, H. et al. (2022). Decarbonization options of the iron and steelmaking industry based on a three-dimensional analysis [Internet]. http://ijmmm.ustb.edu.cn/article/doi/10.1007/s12613-022-2475-7 (accessed 7 July 2023).

64 Singh, V., Buelens, L.C., Poelman, H. et al. (2022). Decarbonisation of steel mill gases in an energy-neutral chemical looping process. *Energy Conversion and Management* 254: 115248.

65 Franciosi, C., Voisin, A., Miranda, S., and Iung, B. (2020). Integration of I4.0 technologies with maintenance processes: what are the effects on sustainable manufacturing? *IFAC-PapersOnLine* 53 (3): 1–6.

66 Jahani, N., Sepehri, A., Vandchali, H.R., and Tirkolaee, E.B. (2021). Application of industry 4.0 in the procurement processes of supply chains: a systematic literature review. *Sustainability* 13 (14): 7520.

67 Seethalakshmi, K.K., Jijeesh, C.M., and Balagopalan, M. (2016). *Bamboo Plantations: An Approach to Carbon Sequestration [Internet]*. Kerala/Thrissur: Kerala Forest Research Institute https://www.researchgate.net/publication/215475397_Bamboo_plantations_An_approach_to_Carbon_sequestration.

68 Forestry Commission (2003). *Forests, Carbon and Climate Change: The UK Contribution*. Edinburgh.

69 Forest Research (2023). Tools and resources woodland statistics [Internet]. https://www.forestresearch.gov.uk/tools-and-resources/statistics/statistics-by-topic/woodland-statistics/ (accessed 7 December 2023).

70 Nix, J. (2020). *John Nix Pocketbook – For Farm Management*, 50the. Leicestershire: The Anderson Centre.

71 Kiessling, S., Darabkhani, H.G., and Soliman AH. (2022). Energies | Free Full-Text | The bio steel cycle: 7 steps to net-zero CO_2 emissions steel production [internet]. https://www.mdpi.com/1996–1073/15/23/8880 (accessed 11 July 2023).

72 Umweltbundesamt, W.S. (2017). Indicator: Greenhouse gas emissions avoided by renewable energies. https://www.umweltbundesamt.de/en/data/environmental-indicators/indicator-ghg-emissions-avoided-through-the-use-of (accessed 6 July 2023).

73 Umweltbundesamt, W.S. (2023). Indicator: Greenhouse gas emissions avoided by renewable energies. https://www.umweltbundesamt.de/en/data/environmental-indicators/indicator-ghg-emissions-avoided-through-the-use-of (accessed 12 July 2023).

74 British Steel (2023) How we make steel. https://britishsteel.co.uk/what-we-do/how-we-make-steel/ (accessed 6 July 2023).

75 Li, Z., Li, J., Spooner, S., and Seetharaman, S. (2022). Basic oxygen steelmaking slag: formation, reaction, and energy and material recovery. *Steel Research International* 93 (3): 2100167.

76 United Nations (UN) (2016). Paris Agreement 2016 [Internet]. https://www.un.org/en/climatechange/paris-agreement#:~:text=To%20tackle%20climate%20change%20and,2015%3A%20the%20historic%20Paris%20Agreement (accessed 2 March 2020).

77 Intergovernmental Panel for Climate Change (IPCC) (2023a). AR6 synthesis report climate change 2023 [Internet]. Report No.: AR6. https://www.ipcc.ch/report/ar6/syr/downloads/report/IPCC_AR6_SYR_SPM.pdf (accessed 5 January 2023).

78 Intergovernmental Panel for Climate Change (IPCC) (2023b). Summary for policymakers. In: *Climate Change 2021 – The Physical Science Basis: Working Group I Contribution to the Sixth Assessment Report of the Intergovernmental Panel on Climate Change [Internet]*, 3–32. Cambridge University Press. https://www.cambridge.org/core/books/climate-change-2021-the-physical-science-basis/summary-for-policymakers/8E7A4E3AE6C364220F3B76A189CC4D4C (accessed 12 July 2023).

79 Kench R.; study.com (2023). Algebraic function | definition, types & examples – video & lesson transcript. https://study.com/learn/lesson/algebraic-function-examples-types.html (accessed 16 November 2023).

80 Zhao, Q., Chu, X., Mei, X. et al. (2020). Co-treatment of waste from steelmaking processes: steel slag-based carbon capture and storage by mineralization. *Frontiers in Chemistry [Internet]*. 8. https://www.frontiersin.org/articles/10.3389/fchem.2020.571504/full (accessed 5 March 2021).

81 Kuramochi, T., Ramírez, A., Turkenburg, W., and Faaij, A. (2011). Techno-economic assessment and comparison of CO_2 capture technologies for industrial processes: preliminary results for the iron and steel sector. *Energy Procedia* 4: 1981–1988.

82 Kuramochi, T., Höhne, N., Schaeffer, M. et al. (2018). Ten key short-term sectoral benchmarks to limit warming to 1.5 °C. *Climate Policy* 18 (3): 287–305.

83 Khakimov, H.T., Shayumova, Z.M., Kurbanbaeva, Z.K.H., and Khusanov, B.M. (2019). Development of optimal modes and mathematical models of energy performance of electric steelmaking production. Voropai, N., Senderov, S., Michalevich, A., and Guliev, H. editors. *E3S Web Conference* 139:01076.

84 Julia, G. (1918). Sur les equations fonctionelles. *Journal Mathématiques – Pure Application* 47–245.

85 Arnold, V.I. (1980). *Mathematical Methods of Classical Mechanics (2nd Corrected Printing)*. New York: Springer International Publishing.

86 AVCalc LLC (2023). Liquid oxygen volume to weight conversion [Internet]. https://www.aqua-calc.com/calculate/volume-to-weight/substance/liquid-blank-oxygen (accessed 14 July 2023).

87 Treptow, R.S. (1999). How thermodynamic data and equilibrium constants changed when the standard-state pressure became 1 bar. 76: 212.

88 Canagaratna, S.G. (2000). A visual aid in enthalpy calculations. *Journal of Chemical Education* 77 (9): 1178.

89 McClary, W.D., Sumida, J.P., Scian, M. et al. (2016). Membrane fluidity modulates thermal stability and ligand binding of cytochrome p4503a4 in lipid nanodiscs. *Biochemistry* 55 (45): 6258–6268.

90 Giauque WF, Johnston HL. (2002). The heat capacity of oxygen from 12 k. To its boiling point and its heat of vaporization. The entropy from spectroscopic data. *ACS Publications. American Chemical Society* 51(8):2300–2321. https://pubs.acs.org/doi/pdf/10.1021/ja01383a003 (accessed 14 July 2023).

91 Encyclopedia Britannica (2023). Steel – basic oxygen, refining, alloying | britannica [Internet]. https://www.britannica.com/technology/steel/Basic-oxygen-steelmaking (accessed 14 July 2023).

92 Meador, W.E. and Smart, M.K. (2005). Reference enthalpy method developed from solutions of the boundary-layer equations. *AIAA Journal* 43 (1): 135–139.

93 A-Level Chemistry (2023). Enthalpy and entropy | a-level chemistry revision notes [Internet]. https://alevelchemistry.co.uk/notes/enthalpy-and-entropy/ (accessed 14 July 2023).

94 He, M., Chen, F., Wen, P. et al. (2024). Vinyl chloride distillation process simulation optimization evaluation: optimization based on NSGA-III Algorithm and Quantitative Risk Analysis. *Processes* 12 (11): 2413.

95 United States Environmental Protection Agency (USEPA) (2023). Greenhouse Gas (GHG) emissions and removals [Internet]. https://www.epa.gov/ghgemissions (accessed 2 January 2023).

96 Madeddu, C., Baratti, R., and Errico, M. (2016). *Modeling of a CO_2-MEA absorption system: a new view in the steady-state analysis*. Anacapri, Italy. https://www.researchgate.net/publication/308970987_MODELING_OF_A_CO2-MEA_ABSORPTION_SYSTEM_A_new_view_in_the_steady-state_analysis (accessed 23 May 2020).

97 Madeddu, C., Errico, M., and Baratti, R. (2019). Process modeling in Aspen Plus®: modeling, analysis and design. In: *CO_2 Capture by Reactive Absorption-Stripping*, 13–30. Springer.

98 Madeddu, S., Ueckerdt, F., Pehl, M. et al. (2020). The CO_2 reduction potential for the European industry via direct electrification of heat supply (power-to-heat). *Environmental Research Letter [Internet]* 15. https://iopscience.iop.org/article/10.1088/1748-9326/abbd02.

99 Özge, C.M.and Zeng, T. (2020). Challenges and opportunities of modeling biomass gasification in Aspen Plus: A review [Internet]. https://scholar.google.com/citations?viewop=viewcitation&hl=tr&user=PoLzhQgAAAAJ&citationforview=PoLzhQgAAAAJ:LkGwnXOMwfcC (accessed 7 July 2023).

100 Fermeglia, M. (2021). Physical properties estimation using Aspen Plus. https://moodle2.units.it/pluginfile.php/431350/mod_resource/content/1/08-PhysicalProperties-Aspen%2B.pdf

101 Luderer, G., Madeddu, S., Merfort, L., et al. (2021). Impact of declining renewable energy costs on electrification in low-emission scenarios Nat [Internet]. https://www.nature.com/articles/s41560–021-00937-z.pdf

102 British Steel (2021). British Steel limited annual report and financial statements 2021 [Internet]. https://find-and-update.company-information.service.gov.uk/company/12303256/filing-history (accessed 19 August 2022).

103 Oppy, G. and Dowe, D. (Winter 2021). The Turing test. In: Zalta, E.N. editor. *The Stanford Encyclopedia of Philosophy [Internet]*. Metaphysics Research Lab, Stanford University. https://plato.stanford.edu/archives/win2021/entriesuring-test/ (accessed 19 August 2024).

104 Arachchige, U., Rasenthiran, K., and Liyanage, M. (2019). Modelling of CO_2 capture using aspen plus for coal fired power plant. *International Journal of Scientific & Technology Research* 19 (8): 316–318.

105 Ahmed, H.S. (2021). Power-to-X: modelling of Fischer-Tropsch synthesis in Aspen Plus. https://www.semanticscholar.org/paper/Power-to-X%3A-Modelling-of-Fischer-Tropsch-synthesis-Ahmed/5e7a161fb334a5e74866403eb56e585285748ccd (accessed 25 October 2024).

106 Abdul Azeez, K.T., Suraj, P., Muraleedharan, C., and Arun, P. (2023). Aspen plus simulation of biomass gasification: a comprehensive model incorporating reaction kinetics, hydrodynamics and tar production. *Process Integration and Optimization for Sustainability* 7 (1): 255–268.

107 Sargent, R.G. (1998). Verification and validation of simulation Models. Syracuse University [Internet]. College of Engineering and Computer Science. https://surface.syr.edu/cgi/viewcontent.cgi?referer=&httpsredir=1&article=1006&context=eecs

108 Thacker, B.H., Doebling, S.W., Hemez, F.M. et al. (2004). *Concepts of Model Verification and Validation [Internet]*. Los Alamos National Lab (LANL), Los Alamos, NM (United States); Report No.: LA-14167. https://www.osti.gov/biblio/835920 (accessed 6A).

109 Aggarwal, A. and Jalote P. (2006). Integrating static and dynamic analysis for detecting vulnerabilities. *30th Annual International Computer Software and Applications Conference (COMPSAC'06) [Internet]*. p. 343–350. https://ieeexplore.ieee.org/abstract/document/4020095 (accessed 19 August 2024).

110 Zhang, S., Zhao, C., and Huang, B. (2019). Simultaneous static and dynamic analysis for fine-scale identification of process operation statuses. *IEEE Transactions on Industrial Informatics* 15 (9): 5320–5329.

111 de Paula, F.W., Armellini, F., and De Santa-Eulalia, L.A. (2020). Simulation in industry 4.0: a state-of-the-art review. *Computers & Industrial Engineering* 1 (149): 106868.

112 Zou, X. and Zhao, C. (2020). Concurrent assessment of process operating performance with joint static and dynamic analysis. *IEEE Transactions on Industrial Informatics* 16 (4): 2776–2786.

113 Caspari, A., Tsay, C., Mhamdi, A. et al. (2020). The integration of scheduling and control: top-down vs. bottom-up. *Journal of Process Control* 1 (91): 50–62.

114 Kleijnen, J.P.C. (1997). *Experimental Design for Sensitivity Analysis, Optimization and Validation of Simulation Models*, 52. New York: Wiley.

115 Sargent RG. (2020). Verification and validation of simulation models: an advanced tutorial. Bae, K.H., Feng, B., Kim, S. et al. *Proceedings of the 2020 Winter Simulation Conference*; 14.

116 Lavery, M.R., Bostic, J.D., Kruse, L. et al. (2020). Argumentation surrounding argument-based validation: a systematic review of validation methodology in peer-reviewed articles. *Educational Measurement: Issues and Practice* 39 (4): 116–130. https://doi.org/10.1111/emip.12378 (accessed 22 July 2021).

117 Deng, L. (2023). Eco-technoeconomic-analysis of steel manufacturing off-gas valorization. A thesis submitted to the School of Graduate Studies in partial fulfilment of the requirements for the Degree Doctor of Philosophy. McMaster University. Deng_Lingyan_202004_PhD.pdf (accessed 5 February 2025).

118 Matino, I., Vignali, A., Mocci, C., Colla, V., Haikarainen, C., Saxen, H., Olie, N., Breukelman, N., Sun, J., and Yapar, G. (2023). Benchmarking of Developed Steelmaking Chain Models - Ref. Ares(2023)8199099 - 30/11/2023. MaxH2DR_D10_Benchmarking-of-developed-steelmaking-chain-models.pdf (accessed 5 December 2023).

119 Mio, A., Petrescu, L., Luca, A.-V. et al. (2023). Carbon dioxide capture in the iron and steel industry: thermodynamic analysis, process simulation, and life cycle assessment. *Chemical and Biochemical Engineering Quarterly* 36 (4): (PDF) Carbon Dioxide Capture in the Iron and Steel Industry: Thermodynamic Analysis, Process Simulation, and Life Cycle Assessment (accessed 5 February 2025).

120 Fousek, J., Kuncova, M., Fabry, J. (2017). *Discrete Event Simulation – Production Model In SIMUL8*, 229–234. (PDF) Discrete Event Simulation – Production Model In SIMUL8 (accessed 7 February 2025).

121 Debta, M.K., Mishra, S.S., and Mishra, D. (2017). Simulation of a real life problem in flexible manufacturing system without breakdown and having alternate machines using SIMUL8. *International Research Journal of Engineering and Technology (IRJET)* 04 (05). Simulation of a Real Life Problem in Flexible Manufacturing System without Breakdown and Having Alternate Machines Using SIMUL8 | Request PDF (accessed 8 February 2025).

122 INOSIM (2025). *EmKus: Process Simulation for Steel and Iron Industries.* EmKus: Process Simulation for Steel and Iron Industries - INOSIM (accessed 16 February 2025).

123 Hu, Q., Wang, S., and Fang, H. (2023). Simulation and analysis of CO_2 capturing from converter gas using monoethanolamine. *Theoretical Foundations of Chemical Engineering* 57 (6): 1524–1533.

124 Xin, K., Hashish, M., Roghair, I., and Annaland, M. (2020). Process simulation and economic analysis of pre-combustion CO_2 capture with deep eutectic solvents. *Frontiers in Energy Research.* 8.

125 Zhao, L., Fang, Y., Lou, P. et al. (2021). Cutting parameter optimization for reducing carbon emissions using digital twin. *The International Journal of Precision Engineering and Manufacturing* 22 (5): 933–949.

126 Liu, Z., Zhao, Y., Wang, Q. et al. (2024). Modeling and assessment of carbon emissions in additive-subtractive integrated hybrid manufacturing based on energy and material analysis. *The International Journal of Precision Engineering and Manufacturing-Green Technology.* 11 (3): 799–813.

127 Michailos, S. and Gibbins, J. A modelling study of post-combustion capture plant process conditions to facilitate 95–99% CO_2 capture levels from gas turbine flue gases. *Frontiers in Energy Research [Internet]* 10: https://www.frontiersin.org/journals/energy-research/articles/10.3389/fenrg.2022.866838/full (accessed 24 May 2022).

128 Neuwirth, R. and Redemann, F. (2014). 'High-capacity coke ovens - cost efficient and environmentally friendly production of large quantities of coke'. ThyssenKrupp Industrial Solutions AG, EuroCoke Summit 2014 in Edinburgh. https://www.researchgate.net/publication/297937873_High_capacity_coke_ovens (accessed 8 May 2020).

129 Desai, M. and Camobreco, V. (2020). *Inventory of US Greenhouse Gas Emissions and Sinks 1990 - 2018*, 733. https://www.epa.gov/sites/default/files/2020-04/documents/us-ghg-inventory-2020-main-text.pdf (accessed 7 June 2021).

130 Neuwirth, R. (2014). High capacity coke ovens. *Steel Times International* 38: 7. High capacity coke ovens - ProQuest (accessed 8 May 2020).

131 Pandey, T. (2025). *Coal Ecology and Environmental Impacts.* Google Books.

132 Firsbach, F., Senk, D., and Babich, A. (2022). Multi-step recycling of BF slag heat via biomass for CO_2 mitigation. *Minerals* 12 (2): 136. https://www.mdpi.com/2075-163X/12/2/136 (accessed 7 June 2023).

133 Kertcher, L.F. and Linksy, B. (1974). Economics of coke oven charging controls. *Journal of the Air Pollution Control Association* 24 (8): 765–771. http://www.tandfonline.com/doi/abs/10.1080/00022470.1974.10469967 (accessed 20 June 2020).

134 Hao, L., Zhang, J., Cheng, H. et al. (2014). Study of the influence of industrial coke oven size on the quality of metallurgical coke. *Processes* 12 (8): 1637. https://www.mdpi.com/2227-9717/12/8/1637 (accessed 27 March 2025).

135 Ergul, M. and Selimli, S. (2024). An applied study on energy analysis of a coke oven. *Science and Technology for Energy Transition* 79: https://www.stet-review.org/articles/stet/abs/2024/01/stet20230198/stet20230198.html (accessed 24 March 2025).

136 Krause, B., Liedmann, B., Wiese, J. et al. (2015). Coupled three dimensional DEM-CFD simulation of a lime shaft kiln - calcination, particle movement and gas phase flow field. *Chemical Engineering Science* 134: 834–849. https://www-sciencedirect-com.ezproxy.staffs.ac.uk/science/article/pii/S000925091500408X?via%3Dihub (accessed 18 June 2020).

137 Liu, Z., Ciais, P., Deng, Z. et al. (2020). Near-real-time monitoring of global CO_2 emissions reveals the effects of the COVID-19 pandemic. *Nature Communications* 11 (1): 5172. https://doi.org/10.1038/s41467-020-18922-7 (accessed 7 January 2021).

138 Ryan, J., Bussmann, M., and DeMartini, N. (2022). *CFD Modelling of Calcination in a Rotary Lime Kiln* 10 (8): 1516. Available at: https://www.mdpi.com/2227-9717/10/8/1516 (accessed 14 November 2023).

139 Wu, E., Wang, Q., Ke, L., and Zhang, G. (2023). Study on carbon emission characteristics and emission reduction measures of lime production - a case of enterprise in the Yangtze River Basin. *Sustainability* 15 (13): 10185. https://www.mdpi.com/2071-1050/15/13/10185 (accessed 26 March 2025).

140 Zhengzhou, H. (2025). *Lime Kiln* https://www.zkcomp.com/product/lime-kiln.html (accessed 28 March 2025).

141 Liang, Z., Chen, J., Huang, Z., and Huang, B. (2023). Characteristics and sintering mechanisms of iron ores with a high proportion of High-AI203 Limonite'. *ACS Omega* 8 (18): 15951–15959. https://doi.org/10.1021/acsomega.2c07659 (accessed 28 March 2025).

142 Cores, A., Verdeja, L., Fereirra, S. et al. (2013). Iron ore sintering, part 1. Theory and practice of the sintering process. *Dyna (Medellin Colombia)* 80: 152.

143 Zhou, M., Yu, Z., Wang, P. et al. (2020). Thermodynamic analysis of iron ore sintering process based on biomass carbon. *Energies* 13 (922): 5988.

144 Wang, L., Tian, H., Lei, W. et al. (2024). Development of high-strength ceramsite via sintering of iron ore tailings: process optimization and properties. *Chemical Engineering Journal* 457: 139440. https://linkinghub.elsevier.com/retrieve/pii/S0950061824045823 (accessed 28 March 2025).

145 Mohammad, P., Patra, S., and Harichandan, B. (2023). Reductants in iron ore sintering: a critical review. *Fuel* 332: 126194.

146 Fernández-González, D., Ruiz-Bustinza, I., Mochon, J. et al. (2017). Iron ore sintering: process. *Mineral Processing and Extractive Metallurgy Review* 38 (4): 215–227.

147 Kawashiri, Y., Ozawa, S., Takahashi, K., Nouchi, T. and Morita, Y. (2022) *Blast furnace operation method and auxiliary equipment of blast furnace*. European Union. PN EP4083234A1. https://patents.google.com/patent/EP4083234A1/en (accessed 14 March 2023).

148 Yang, Y., Holappa, L., Saxen, H., and van der Stel, J. (2024). Chapter 1.2 – Ironmaking. In: *Treatise on Process Metallurgy*, 2e (ed. S. Seetharaman, R. Guthrie, A. McLean, et al.). London: Elsevier.

149 Satyendra, K.S. (2015). 'Blast furnace process automation, measurement, and control system'. https://www.ispatguru.com/blast-furnace-process-automation-measurement-and-control-system/ (accessed 9 July 2020).

150 Babich, A. and Senk, D. (2015). Recent developments in blast furnace iron-making technology. In: *Iron Ore*, 505–547. Elsevier. https://doi.org/10.1016/B978-1-78242-156-6.00017-4 (accessed 22 September 2020).

151 Griffin Hammond (2019).

152 Bailera, M., Nakagaki, T. and Kataoka, R. (2021). 'Revisiting Rist diagram for predicting operating conditions in blast furnaces with multiple injections'. Draft Article, Open *Research Europe*. https://zaguan.unizar.es/record/118047/files/texto_completo.pdf (accessed 25 March 2024).

153 Wang, R., Liang, L., Wang, Y. and Roskilly, A.P. (2019). 'Process simulation of blast furnace operation with biomass syngas injection for clean production', *International Conference on Applied Energy 2019*, Västerås, Sweden (12-15 August 2019), Paper ID: 9. Proceedings of (accessed 12 July 2023).

154 Organisation for Economic Co-operation and Development (OECD) (2015). *OECD 12 May 2015 Steel Committee meeting: ETP (Energy Technology Perspective) 2015: Iron & Steel Findings*. Energy Technology Perspectives 2015 | OECD (accessed 7 July 2023).

155 United States Office of Energy Efficiency & Renewable Energy (EERE). (2021). *Hydrogen Production: Natural Gas Reforming*. https://www.energy.gov/eere/fuelcells/hydrogen-production-natural-gas-reforming (accessed 5 July 2022).

156 Hasanbeigi, A. and Springer, C. (2019). How clean is the U.S. Steel Industry? An International Benchmarking of Energy and CO_2 intensities. *Global Efficiency Intelligence* How Clean is the U.S. Steel Industry?—Global Efficiency Intelligence. (accessed 3 April 2020).

157 Khalid, Y., Wu, M., Silaen, A. et al. (2021). Oxygen enrichment combustion to reduce fossil energy consumption and emissions in hot rolling steel production. *Journal of Cleaner Production* 320: 128714. https://www.sciencedirect.com/science/article/pii/S0959652621029139 (accessed 12 April 2022).

158 Matelaro, N. (2016). *Energy Use in US Steel Manufacturing. Submitted as coursework for PH240*. Stanford University. http://large.stanford.edu/courses/2016/ph240/martelaro1/?t (accessed 16 May 2020).

159 Napp, T.A., Gambhir, A., Hills, T.P. et al. (2014). A review of the technologies, economics and policy instruments for decarbonising energy-intensive manufacturing industries. *Renewable and Sustainable Energy Reviews* 30: 616–640.

6

Greater energy independence with sustainable steel production

6.1 Introduction

Global energy market price volatility and an upward trajectory of prices per unit of electricity have sent all industrial sectors and many economies to the brink of recession. Alongside the urgent need for decarbonisation of all industries, achieving a globally higher level of energy independence across all sectors seems imperative. A multidisciplinary approach with a proposed system of CO_2 emissions reduction and capture technologies has the potential for short-term emission reduction to near zero in the steel industry – although some of the mechanisms can be implemented across most heavy industries. The findings of this research show a CO_2 emission reduction of ~30% from 977 t of CO_2 to 684 t in one single blast furnace production cycle (based on 330 tonnes of liquid iron production capacity, with the mean of 2.1–3.2 tonnes CO_2/t of steel and chemical reactions emissions applied), by switching the electricity provider for operating the electric heaters to providers generating energy exclusively from renewable sources. Replacing coal with biomass and adding post-combustion capture units to the blast furnace operation will add carbon neutrality into the process – resulting in CO_2 emission reduction to near-zero. Bio energy carbon capture and storage (BECCS) from biomass utilisation will add the benefit of carbon-negative emissions to the cycle. Simultaneously, energy-saving and process improvement measure implementation (up to 60% efficiency increase), excess heat recovery <30% of energy savings and retrofitting renewable energy technology resulted in an energy independence of 88%. Engineering solutions, partly subsidised in the United Kingdom, are readily available for implementation in the iron and steel manufacturing industry. The volatility of the global energy market and recent price-hikes by energy producers have caused never-before-seen levels of profit for energy companies, and untold pressures for businesses and the population in the most developed economies. Numerous countries are on the brink of recession across geographical Europe at the time of writing, and energy price increases have made a strong case for the urgent need to achieve greater energy independence – the demonstration of the feasibility being the purpose of this paper. This could be considered one of the foremost important contemporary endeavours. The energy sector and the iron and steel industry, along with heavy industry and petroleum refineries [1], are by far the largest emitters of CO_2 emissions, due to their high fuel and

Circular Steel Production: Pathways to Net-Zero Carbon Emissions, First Edition.
Sandra Kiessling, Hamidreza Gohari Darabkhani, and Abdel-Hamid Soliman.

energy demand. The steel industry accounts for between 7% and 11% [1–7] of global CO_2 emissions as a result of steelmaking, and China is responsible for 50% of these GHGs [7], due to their heavy reliance on coal. The increased use of coal in energy generation, due to imposed oil and gas shortages, was found to be the main factor [1, 8] driving up global energy-related anthropogenic CO_2 emissions by over 2 billion tonnes, their largest ever rise in absolute terms. This research focused on the technical solutions currently available for the decarbonisation of the steel industry and how decarbonisation could lead to a higher degree of energy independence, based on the currently known energy consumption trends. The consumption trend evaluation or influence of renewables implementation on consumption trends on a global scale was not the key focus of this project, as these are two completely independent, separate issues altogether but pose an interesting challenge for future research. This research focused on the steel industry and connected subsidiary industries only.

6.2 Materials and methods

Throughout this project, global datasets in connection with renewable energy technology implementation in different settings were utilised [9–14]. Information on factual CO_2 emissions in steel production [15] and manufacturing [16–18] has been considered, as well as data from other industrial sectors [4, 9, 19–21]. In order to visualise the opportunities of a circular steel production process, implementation of sustainable elements, and opportunities for achieving great energy independence, a comprehensive steel manufacturing overview has been compiled, as displayed in Figure 6.1. The data were accumulated, analysed and used for modelling using MS Excel and simultaneously analysed by applying standard mathematical principles and followed for proof of concept with steel production simulations in Simul8, INOSIM and Aspen. Although there are currently several projects in progress concerned with filtering CO_2 emissions from post-combustion off-gas, a solution has yet to be found that could achieve a 100% filtering of CO_2 emissions.

There are two major hurdles to the implementation of sustainable steel and other production techniques: (i) lack of willingness across all industries to incorporate the required changes in their financial and production planning and managing the installation of suitable technology and (ii) ensuring a viable financing of all efforts using appropriate MCDA and costing and incentives provided by the respective governments. The first hurdle relies on industry leaders to act and drive the decarbonisation of the steel industry to avert climate disaster, as politicians worldwide seem to lack the willingness to implement and enforce adequate policies – the signs are all around us. There are continued research and campaigning efforts taking place in this direction. The second hurdle has been partly addressed by governments across the globe and the UK government with grants, loans and subsidies.

The preceding publication "The 7 Steps to Net-Zero CO_2 Emissions Steel Production" [15] strategy can be seen as a strategic guidance paper for the decarbonisation of steelmaking, in tandem with this publication as a guideline for achieving higher energy independence. The system implementation, as described in [15], will

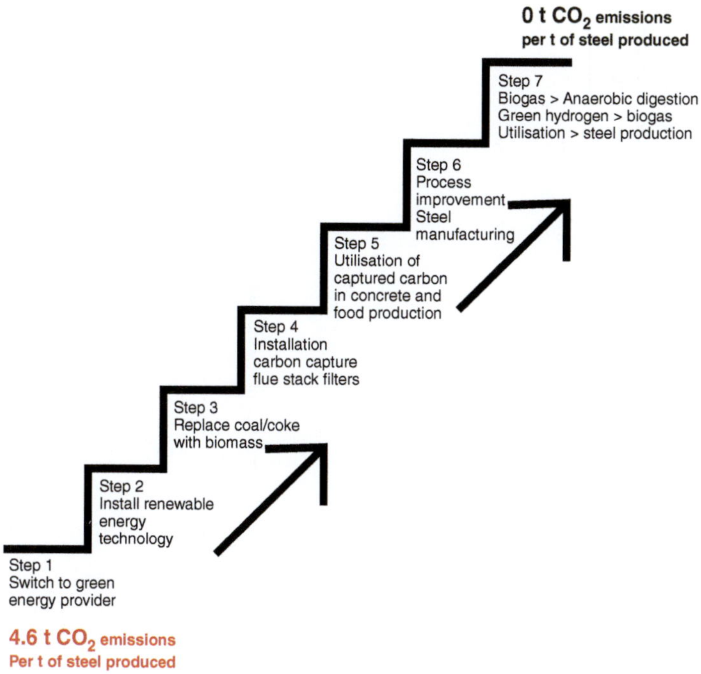

Figure 6.1 The Bio Steel Cycle – 7 Steps to Net-Zero Carbon Emissions Steel Production.

likely achieve a higher degree of energy independence, in the short term. It could be achieved in seven easy-to-follow steps, even if only some sections of the following are being applied during the BF/BOF route, as shown in Figure 6.1.

Step 1: Switching to a 100% green energy provider
Step 2: Installing renewable energy technologies
Step 3: Replacing coal and coke with biomass (biochar)
Step 4: Installation of post-combustion carbon capture technology
Step 5: Utilisation of CO_2 in food and building projects
Step 6: Further process improvement in steel manufacturing
Step 7: Implementation of AD > biogas > green hydrogen

Conventional energy use and renewable energy component implementation points have been incorporated to highlight the simplicity of achieving a higher degree of energy independence, whilst simultaneously decarbonising the steelmaking process.

Table 6.1 gives an overview of the established green energy providers who evidently are generating their electricity and gas exclusively from renewable energy sources.

The flowchart in Figure 6.2 illustrates the journey of steel, from mining to marketing, in a circular economic system – where everything is reused or recycled, and nothing is wasted or requires disposal at landfill sites.

Table 6.1 UK green energy providers.

Energy provider	UK headquarters address	Renewable sources	Green electricity	Green gas	Carbon offsetting
Ecotricity, 2023	Lion House, Rowcroft, Stroud, Gloucestershire, GL5 3BY https://www.ecotricity.co.uk/our-green-energy/green-electricity	Wind (98%), solar (0.12%) and hydro (0.7%)	100%	Yes	Yes
Good Energy UK, 2023	Monkton Park Offices, Monkton Park Chippenham SN15 1GH https://www.goodenergy.co.uk/	49.41% = Wind 32.71% = Bio generation 13.60% = Solar 4.28% = Hydro	100%	No	Yes
Green Energy UK, 2023	Green Energy (UK) plc Black Swan House, 23 Baldock Street Ware, Herts, SG12 9DH https://www.greenenergyuk.com	Hydro, solar and wind	100%	100%	No
Octopus Energy, 2023	UK House, 5th floor, 164–182 Oxford Street, London, W1D 1NN https://octopus.energy/	Anaerobic digestion, solar, wind and hydro	100%	0%	Yes

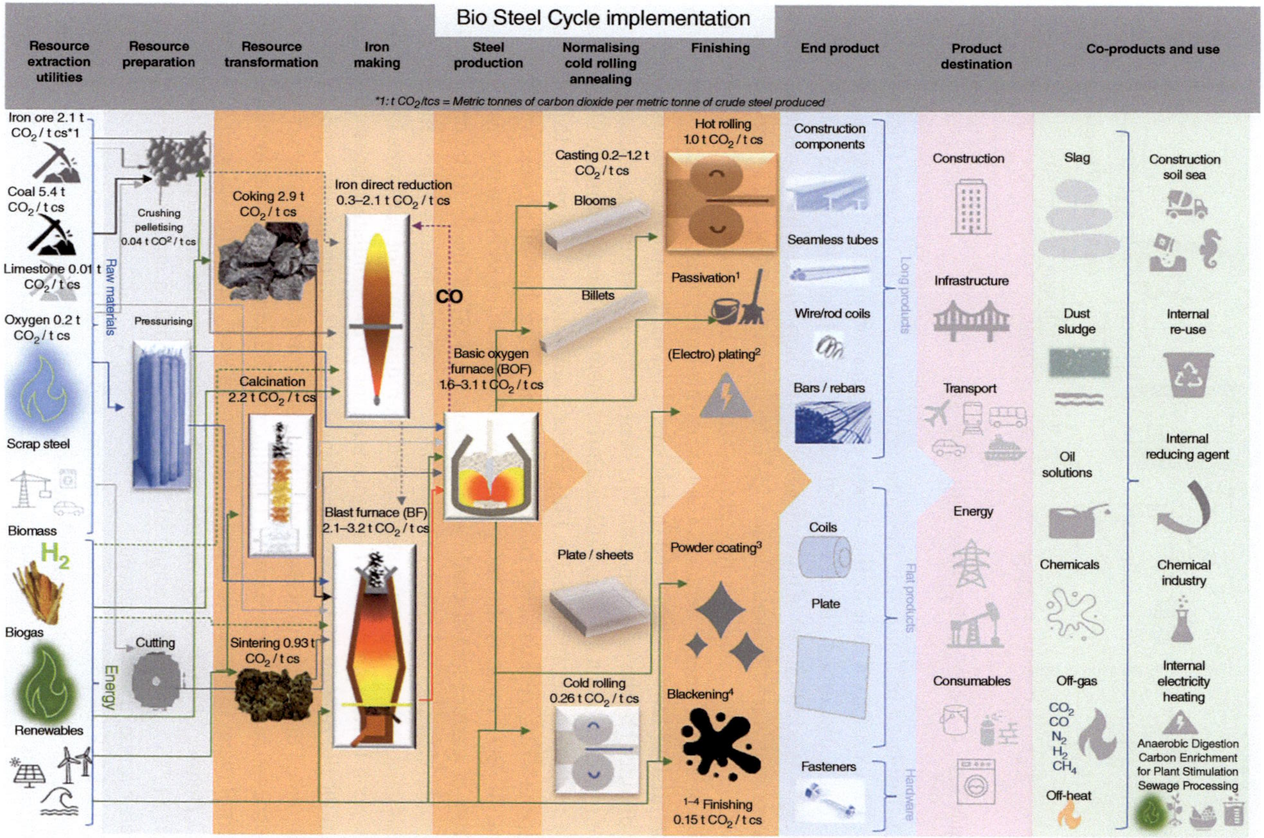

Figure 6.2 Steelmaking table flowchart with BiSC.

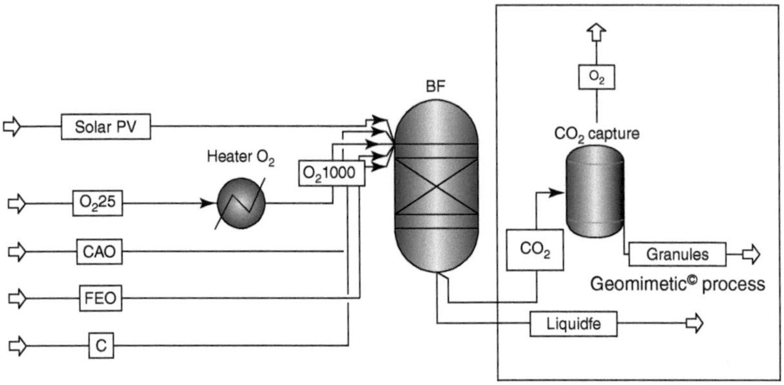

Figure 6.3 Aspen BF configuration with CCUS unit.

As displayed in Figure 6.1, the same principle applies to the steel (Figure 6.3), in the BF with CCUS unit in the Aspen configuration production process, as far as off-heat is concerned. It was established that a total potential of 425 PJ (1 PJ [peta-joule] = 31.6 million m^3 of natural gas or 278 million kilowatt hours of electricity) of excess heat is readily available at a 95 °C temperature, and 960 PJ at approximately 25 °C [22]. This amount is thought to represent between 4% and 9% of the total industrial final energy demand. Capturing this excess heat means utilising energy potential we have already used in industry, thus, reducing the amount of energy to be produced by the same amount. The benefits for agri-businesses utilising off-heat from production and CO_2 in Carbon Enrichment for Plant Stimulation (CEPS) are the subject of ongoing research [22]. Process simulations in Simul8, INOSIM, and Aspen+V12.1 were used to explain the individual production process implications. The BF/BOF route is the most widespread method of steelmaking, representing ~70% of current global crude steel production [23]. It needs to be emphasised, though, that the displayed CO_2 emissions in metric tonnes represent the CO_2 emissions at their respective stage, per t of product, and not per tonne of steel. The emission burden on each of the input streams in the simulation software systems Simul8 and AspenV.12.1 is already set within the system parameters and, therefore, calculated during the process simulations, providing a more detailed set of data per metric tonne of respective product produced. Additionally, in order to represent the overall mean CO_2 emission burden in t per t of steel, the mean value of 4.95 tonnes CO_2/tonnes of crude steel was allocated for entering the input stream in S8 as BF 2.95 t of CO_2 per metric tonne of liquid iron produced and BOF 2 t/t of CO_2 per metric tonne of steel produced.

The BF/EAF process route simulation, representing the second most common method of steelmaking, has a share of c. 20% of global steel production. Initially, as biomass [21, 24] has the potential to reduce CO_2 emission by ~30%, the usage of biomass was implemented [21, 24] and filters as well as CCUS units were installed (Figures 6.3 and 6.4), in direct comparison with the MS Excel modelling and mathematical calculations. The calculations are set to work for a 330-t blast furnace, similar to the British Steel site in Scunthorpe, and for the operational year of one hour, to

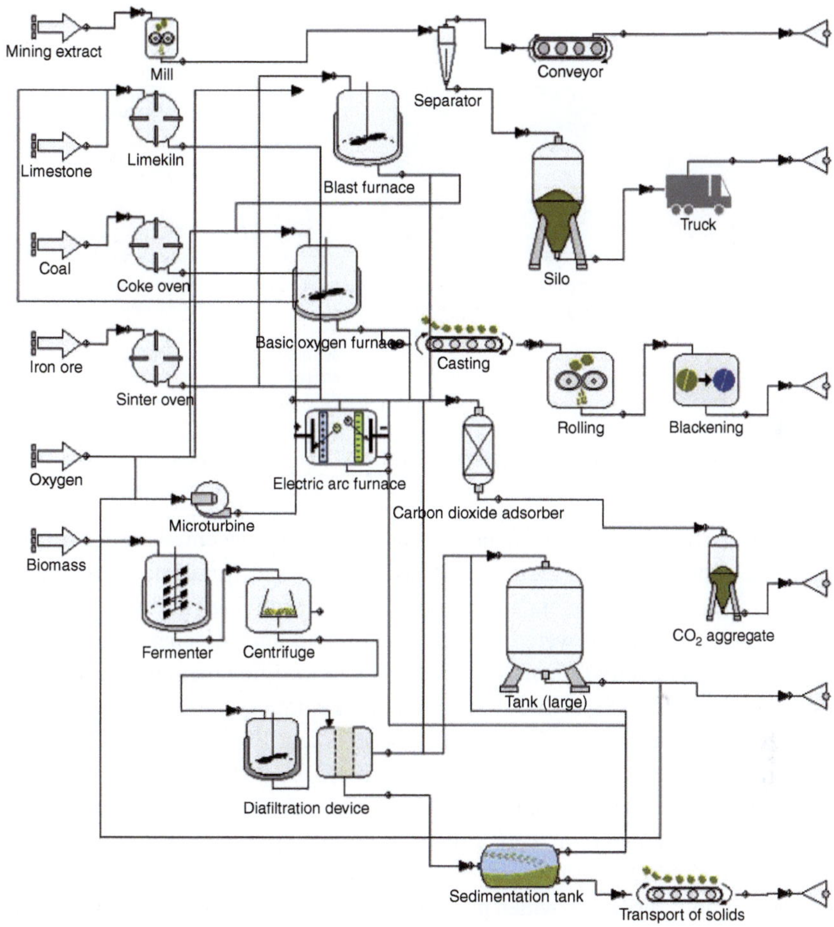

Figure 6.4 INOSIM configuration mining to marketing with CCUS.

reflect the 40-minute average charge processing and discharge time (Figure 6.5). Verifying the results of the preceding evaluations, computations, and calculations initially, 977 t of CO_2 flow was reported by the Aspen system analysis, without any mitigating measures. The CO_2 reductions concerning a 330-t blast furnace vessel have been modelled in three separate process simulation software packages and have been carried out considering the production parameters of the British Steel Scunthorpe site. Having implemented steps of the Bio Steel Cycle model and strategy, the findings of this research show a CO_2 emission reduction of ~30% from 977 t of CO_2 to 684 t in one single blast furnace production cycle (based on 330 tonnes of liquid iron production capacity that means 2.1–3.2 tonnes CO_2/t of steel), by switching the electricity provider for operating the electric heaters to providers generating energy exclusively from renewable sources and replacing coal with biomass added carbon neutrality into the process. Adding post-combustion capture units to the blast furnace resulted in CO_2 emission reduction to near-zero. Carbon capture from

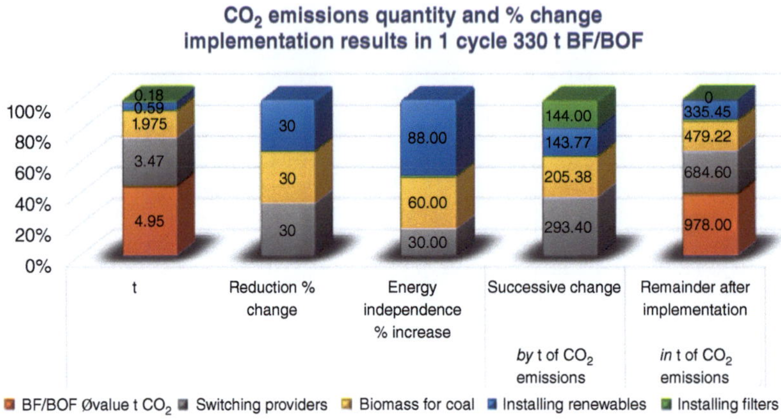

Figure 6.5 CO_2 emission change in tonnes (t) with CAT and CCUS implementation.

biomass utilisation also added the benefit of carbon-negative emission to the cycle (Figure 6.2). The process simulations and carbon reports in Simul8 demonstrate the effect of implementing biomass and carbon capture technology in the steelmaking process. These simulations were followed by configurations in Aspen and INOSIM, including the implementation of CCUS processes, as the implementation of this process, which includes off-gas flue stack filters, makes it possible to produce steel at almost zero CO_2 emissions – for both, the BF/BOF and BF/EAF process route. To corroborate the results from the MS Excel (Microsoft 365, 2020 version, on Windows 10) modelling, the mathematical formulation and calculations, and Simul8 (version 2020) process simulation were utilised by producing configurations in Aspen+ (V12.1) and INOSIM (Expert Edition 13.0), as displayed in Figure 6.4.

The simulations reports provided proof for the concept of the Bio Steel Cycle, insofar as with the implementation of both biomass and the Geomimetic process, and the steel production CO_2 emissions could be reduced by 30% and to almost zero, as detailed in Figure 6.4, therefore proving the hypothesis that it is possible to produce steel without CO_2 emissions if the novel concept and strategy were being implemented. At the same time, up to 88% higher degree of energy was utilised to produce steel without CO_2 emissions if the novel concept and strategy were being implemented. At the same time, up to 88% higher degree of energy independence can be achieved when other components of the Bio Steel Cycle, including renewable energy sources, are being installed, which is further elaborated in the following text; the summary results are shown in Figure 6.5.

The simulation results are detailed in metric tonnes, occurring during one production cycle, based on 330 metric tonnes of BF and BOF furnaces. The first column demonstrates the emission levels in metric tonnes (t) as per legend underneath from BF/BOF average value in CO_2 emissions, the reduced level of CO_2 emissions after switching to an energy provider deriving their energy at 100% from renewable sources, the reduced level of CO_2 emissions after replacing coal with biomass in the BF, the reduced level of CO_2 emissions after installation of renewable energy technologies (solar, wind and hydro as appropriate), and the reduced level of CO_2 emissions after the installation of post-combustion CO_2 emissions filters. The second column shows the individual percentage proportion in reduction as a result of implementation. The third column shows the successive percentage increase in energy independence. The fourth column shows the successive change in CO_2 emissions in t, based on one cycle of a 330 t BF/BOF process. The fifth column shows the remainder of the CO_2 emissions after successive implementation of the suggested technologies – resulting in 0.18 t of CO_2 emissions but displaying as 0 due to set software parameters, after successful installation of post-combustion CO_2 filters at the individual sites of off-gas outlets.

6.3 Heat loss recovery – energy- and CO_2-saving protocols

Already since the 1990s, scientists have been convinced [25] that 30% of the heat energy entering any production process is: (i) lost and (ii) could be recovered. The required technology has experienced a steep learning curve and is now commercially viable and available. Excess heat from any production or manufacturing process can be reused to supply any production site with heat and warm water, partly due to the simplicity of the technology required. The energy basis and flow have been investigated thoroughly, and an energy industry defining and telling report was produced by Moran and Sciubba [25]. The theory of exergy analysis is based on the fact that if 100% of energy is being inserted into any energy-requiring unit, the amount of 70% will be effectively used for the intended purpose, whilst 30% is lost due to inefficiencies and deficits within the production and processing infrastructure. The strategy for making existing infrastructure more efficient is to (i) upgrade the existing energy infrastructure, (ii) reduce energy/heat loss by improving energy efficiency with focused energy utilisation and (iii) capture and re-utilise lost energy/heat. The technologies utilised for the strategy points (i)–(iii) are as follows: (i) repair energy/heat escape points and insulate as a preventive measure; (ii) using the most focused technology: instead of the BF/BOF-route, using the BF/EAF route for steel production and (iii) installation of baffles, plate heat exchangers and other energy/heat-capturing technology where technically feasible. On a practical level, this means insulation of the heat-bearing infrastructure (furnaces and pipework). Via a connected network of pipes and lines, the energy or heat captured can be exported to nearby agri-businesses or transferred to neighbouring homes and industries through a district energy system. Excess heat is a hidden resource of energy, and it is all around us. Utilising excess heat means enabling >95% energy efficiency. According to the International Energy Agency [1, 5], it is apparent that energy demand is set to grow dramatically in the near future, due to population growth and rising lifestyle energy demands. Without urgent action to tackle the demand side of our lifestyle choices, and decarbonisation requirements of the climate crisis, by using energy more efficiently, we will not get on track to meet global climate goals. Global emissions [26] of CO_2 – including land use and fossil CO_2 – will remain relatively high at 40.5 Gt CO_2 in 2022, but still below their 2019 peak of 40.9 Gt CO_2. A global push for more efficient use of energy can reduce CO_2 emissions by an additional 5 gt per year by 2030 [1, 8, 27–29], based on current energy demand. The global electricity consumption is displayed in Figure 6.5. However, energy efficiency increases constitute merely 30% of the required CO_2 reduction needed to meet the Net-Zero by 2050 Scenario [1, 8, 27–29]. As far as energy security and greater energy independence are concerned, these energy savings are set to avoid having to produce almost 30 million barrels of oil per day (three times Russia's average annual production, based on the 2021 data), and 650 billion m^3 of natural gas per year – around four times of EU imports from Russia in 2021. Although there has been a steady decline in overall electricity consumption in the iron and steel manufacturing sector (Figure 6.6), due to already implemented structural changes towards more sustainable production processes in the industry, along with the installation of energy efficiency measures, and improved energy efficiency measures [30], there is still a substantial amount of electricity that could be saved (Figure 6.7). Additionally, this would render the existing infrastructure utilisation more efficient. One of the most important factors, the required short-term solutions, is helped by using suitable existing infrastructure, as it can be easily retrofitted with technology to prevent heat loss and decarbonise production process technology at the same time.

Simultaneously, it could be utilised to (i) capture >98% [31] of CO_2 emissions and use the captured CO_2 in ancillary industries, and (ii) install renewable energy resources (solar, wind and hydro) in suitable locations to increase the level of energy independence. What is being produced on-site does not have to be imported from somewhere else, at fluctuating prices [1, 8, 27]. As shown in Figure 5.7, the development of electricity usage in steelmaking has been on a downward trajectory since the year 2000, partly due to process and efficiency improvements across most industries. The 2023 figures are provisional, as the reported figures are yet to be

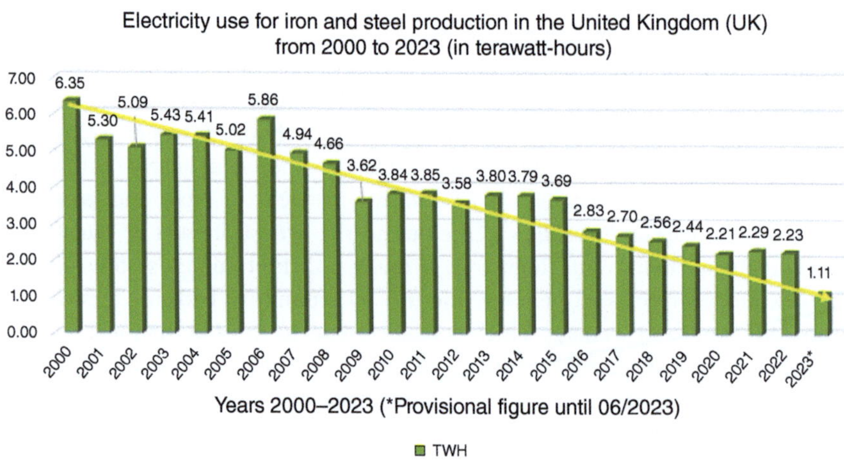

Figure 6.6 Electricity use for iron/steel industry from 2000 to 2023 [29, 30].

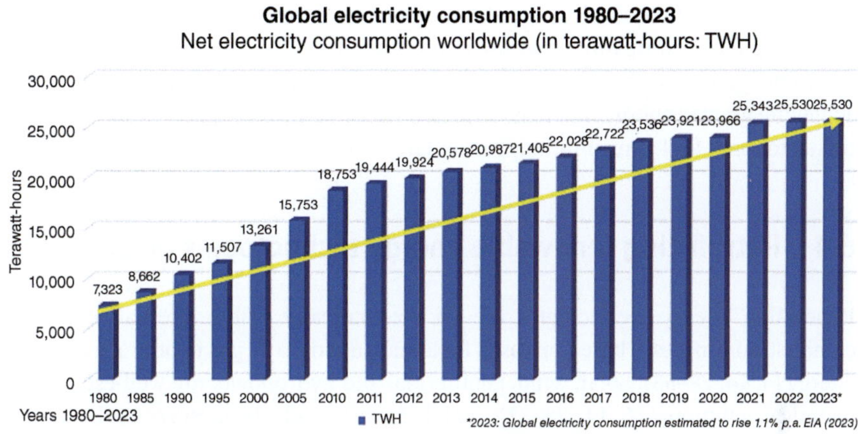

Figure 6.7 Global electricity consumption 1980–2023 [1, 5, 8, 27–29].

confirmed – hence the *. In stark contrast, quite the opposite observation was made for global electricity consumption [1, 5, 8, 27–29] as previously demonstrated in Figure 5.6. Global electricity consumption has continuously increased during the last 50 years, arriving at an estimated 25,530 terawatt-hours in 2023 [28, 32]. Since 1980 and up to 2021, global electricity consumption has increased threefold and the global population increased by approximately 75% simultaneously. In line with extended industrialisation and infrastructural improvement, these factors caused a threefold increased electricity demand, with an upward trajectory prognosis, as of the end of 2021. Since the year 2000, China's gross development product (GDP) was recorded as developing a 16-fold increase [7, 28, 29] – therefore, establishing China as the second largest economy, after the United States. The development of its billion-strong population and manufacturing industries has caused China to require increased levels of energy more than any other country. Thus, it has become the largest consumer of electricity worldwide. China and other BRIC (Brazil, Russia, India and China) countries are still vastly outpaced by developed economies with smaller population sizes, in terms of per capita electricity consumption. To place this in context, Iceland, with a population of less than half a million inhabitants, consumes the most electricity per capita (per person) in the world, followed by Norway, Qatar, Canada and the United States [7, 28, 29]. Contributing factors such as the existence of power-intensive industries, household sizes, living situations, appliance and efficiency standards, and access to alternative heating fuels have been identified as the determinants of the amount of electricity the average person requires in the cited countries. Therefore, given these developments and the looming climate catastrophe, greater energy efficiency, and exergy loss prevention, is a technically simple and effective short-term solution. There is vast potential to simultaneously save energy by making existing infrastructure more efficient by reducing the energy/heat loss and, therefore, saving energy at the same percentage (30%), meaning that this is 30% of the energy that the industry does not have to import and pay for from external sources. At the same time, 30% of CO_2 emissions for the energy not required, as saved, would not have to be produced. Consequently, improving the existing infrastructure to prevent energy and heat loss would mean a more energy-secure and more sustainable production cycle, in any industry, whilst achieving greater energy independence and reducing the greenhouse gas emissions linked to fossil fuel consumption, particularly of energy derived from fossil fuels.

6.4 Retrofitting renewable energy technologies on site

Renewable energy technologies are an economically viable alternative [12] to combustion processes based on fossil fuels such as coal and gas to produce heat and energy [4, 14, 28–33]. Solar, wind, geothermal and hydropower are well-established technical solutions [4, 14, 28–33], which have already been successfully implemented in a range of countries and settings [34–39]. In the United Kingdom, there are a range of schemes accessible to businesses, such as finance and support from the Department for Business, Energy and Industrial Strategy (2023) and others, as displayed in Table 6.2. Some countries and industries are supplying their entire energy needs via renewable energy solutions [4, 30, 32] – hence why this component is one of the cornerstones of the BiSC, as producing electricity and heating energy accounts for 36% of the UK's CO_2 emissions [14, 27, 28, 30, 40–42]. Besides the emission savings, using renewable energy technologies exclusively could provide greater independence to businesses across all sectors and increase the UK's energy self-sufficiency. The first step to greener production and greater energy independence on a fossil fuel base is the switch to an energy provider that derives its energy 100% from renewable sources. Utilising existing buildings on industrial and production sites, suitable locations can easily be retrofitted with photovoltaic (PV) solar panels, producing energy from daylight and sunshine. These have the additional benefit of monetary grants (non-repayable), provided by the UK government, and government-backed loans and subsidies [43, 44]. Additionally, the same applies to wind turbines, technology generating electricity using either biomass or hydro-turbines (water-based) and anaerobic digestion (AD) systems. Table 6.2 shows the relevant schemes and their descriptions, mostly detailing how the different technologies could potentially qualify for the individual schemes. The choice and implementation of any of these technologies are entirely dependent on the individual site parameters and need to be thoroughly assessed with regard to their suitability for the identified location and viability with an outlook over the next 30 years. The accurate potential return on investment (ROI) can only be determined for every individual site, as the current contracts with energy providers are one of the determining factors and the authors have not been privy to this kind of information.

Renewable energy technology is market-ready—now the implementation is key.

Some technical solutions for achieving a higher degree of energy independence have been intensely researched [23, 45–48] and implemented in a range of industrial sectors. Projects containing components, such as direct air capture (DAC),

Table 6.2 UK support schemes for renewable energy technologies.

Scheme	Description
England Woodland Creation Offer	Landowners, land managers, and public bodies can apply to the England Woodland Creation Offer (EWCO) for support to create new woodland. Over GBP 10,000 per hectare.
Greening Eden	The CBEN Partnership will complete the calculations using data provided by each company and site visits to provide practical and cost-effective advice on how to reduce emissions. A GBP 400,000 grant fund has been established to help capital investment projects deliver emission reductions.
Green Heat Network Fund	Commercialisation and construction of new low- and zero-carbon (LZC) heat networks (including the supply of cooling). Retrofitting and expansion of existing heat networks. Funding will support the uptake of low-carbon technologies like heat pumps, solar and geothermal energy as a central heating source. The GHNF is open to organisations in the public, private and third sectors in England (no individuals, households or sole traders).
Green Gas Support Scheme	Funding support for biomethane injection to the national grid.
Smart Export Guarantee (SEG)	The SEG funds for the low-carbon electricity exporters, feeding back to the National Grid. Anyone with an installation of one of the following technology types is eligible to apply: solar photovoltaic (solar PV), wind, micro- combined heat and power (micro-CHP), hydro, anaerobic digestion (AD) support and grants for SMEs to help them to reduce carbon emissions.
SMEES	SME Energy Efficiency Scheme (SMEES): guidance and funding for businesses looking to improve their energy efficiency.
Energy for Business	Support and grant funding for SMEs with projects to reduce carbon emissions or save energy.
HNIP	Heat Networks Investment Project (HNIP) government-backed funding.
Low Carbon Dorset	Free support to help businesses in Dorset reduce their carbon emissions, improve energy efficiency and aid the development of new low-carbon products.
Business Energy Efficiency Programme	Energy reviews and grants to help businesses in the West Midlands manage and reduce energy costs.
Low Carbon Workspaces	Offers grants to implement energy efficiency measures, save money and cut waste.
Horizon Europe funding	Funding for research or innovation that is groundbreaking, improves European research standards or responds to challenges like climate change or food security.
Coventry and Warwickshire Green Business Programme	Grants, free energy audits and low-carbon product development support for businesses.

The current UK decarbonisation grants schemes are being monitored by Laurea Tech Consulting Ltd. and courses to equip you with the knowledge and skills to successfully apply for up to more than 6-figure finance for your projects are available here: https://laurea-tech.com/decarbonisation-grants-course

re-directing heat and utilising anaerobic digesters to produce biogas (methane and hydrogen), and producing energy on site with suitably sized turbines [49], can be directly linked to one of the incentives shown in Table 5.1, such as the 'Farming Investment Fund', and include funding for those in agricultural businesses such as farmers, foresters, growers and agri-contractors with grants for investing in new technologies, equipment and infrastructure [44]. The second largest dominant technology after solar in the United Kingdom is wind energy. Implementation of sustainable and renewable energy components into any production cycle, such as direct air capture (DAC) to (i) capture off-gas carbon and (ii) produce biomass for the production of biogas in anaerobic digestion, CEPS (promoting growth in greenhouses for food production), anaerobic digestion (AD) to produce biogas and hydrogen, and capturing and utilising excess heat can individually make significant contributions to a higher degree of energy independence for the individual commercial entity. The MCDA in Table 5.2 includes service and maintenance time and cost, the likelihood of repairs and the availability of suitable service providers to carry out said repairs, servicing and maintenance. Projections for policy-making, ROI or how this could lead to the development of new incentives in the realm of renewable energy and sustainable industrial practices were not the substance of this research. Some examples are given in Table 6.3.

The MS Excel extrapolations established how the different components of the '7 Steps to Net-Zero Carbon Emissions Steelmaking' strategy can be implemented and installed. Installing renewable energy technologies can not only help to reduce CO_2 emissions but also supply the production site with renewable energy, where 30% of energy does not have to be imported from third parties and paid for, bringing the reduction in CO_2 emissions down to −49%. At this point in production, where biomass has already replaced the use of coke, and renewable energy technologies (solar, solar PV, wind and hydro) have been installed, an additional 30% of energy can be saved and, thus, does not have to be bought in, by using biogas, bringing the reduction in CO_2 emissions to −65.7% [50]. This will have been produced in the link-connected anaerobic digester, which produces biogas from connected agri-businesses. The negative percentage reduction in emissions means, in reverse, that at these points in the (steel) production process, a greater degree of energy independence can be achieved at 30%, 49%, 65.7% and 88% [15, 50]. This implies that energy at the same percentage point levels is not required to be imported from external sources, as it is produced either on site or link connected. Besides energy and heat saving, generating their own energy will inevitably lead to achieving a higher level of independence, at least by 30% and ideally, at 88%. Additionally, there are savings to be had by not being forced to import and pay for energy from conventional suppliers, energy derived from renewable sources or not. We have been made painfully aware that private and business users of energy and fuel are at the mercy of corporate stakeholder interests and, thus, vulnerable to high price velocity. Retrofitting the existing industrial building infrastructure with renewable energy technology, with components of 'The 7 Steps to Net-Zero Carbon Emissions

Table 6.3 MCDA excerpt implementation details and cost.

Step 1–7		**Project or**		
BiSC	Technique	Process	Company	System/Performance
1	Switching > green energy	Energy provider	See Appendix A	100% renewable energy
2	Installing renewables: solar	Solar PV panels	Internal contracts	1 kWh/4 panels = 25,667 panels
2	Installing renewables: wind	Horizontal axis wind turbine	Norvento nED100	100 kW/£317,655.27 × 65
2	Installing renewables: wind	Horizontal axis wind turbine	Enercon E53	800 kW/£807,581.80 × 8
2	Installing renewables: wind	Horizontal axis wind turbine	EWT DW61	1 MW/£981,368.75 × 6
2	Installing renewables: wind	Horizontal axis wind turbine	Enercon E82	3 MW/£1,829,271.35 × 3
2	Installing renewables: wind	Horizontal axis wind turbine	Enercon E126 EP3	3.5 MW/£2,458,302.00 × 2
2	Installing renewables: wind	Vertical axis wind turbine	Patriot Modular	70 kW/£188,196.00/ × 92
2	Installing renewables: hydro	Small closed-loop system	Helios Atlas	6.5 MW
3	Using biomass/green H_2	H_2ermes: H_2 from seawater	HyCC/Tata Steel	15,000 t H_2/p.a.
4	CO_2 filters installation	$CaCO_3$-based CO_2 absorber	Giammarco Vetrocoke	Hot potassium carbonate (HPC) solution-based filter
5	Utilisation of captured CO_2	Geomimetic: CCUS in aggregate	Blue Planet	100% CCUS
6	Process improvement	HIsarna; ironmaking by simultaneous iron ore reduction and scrap melting combined with biomass and limestone instead of lime	Tata Steel, Horizon 2020, Horizon Europe	3300 t hot metal per day
7	Anaerobic digestion > biogas utilisation in steelmaking	Biogas and H_2 from anaerobic digestion	Biogen	Biogas production

Steelmaking' strategy, can support (i) achieve greater energy efficiency and independence, (ii) turbo-charge the decarbonisation of energy production, (iii) decarbonise steel and industrial production and (iv) provide savings opportunities via excess heat recovery. The MCDA analysis of different renewable energy systems, taking into consideration the incentives in Table 6.1, leaves much hope for the establishment of the United Kingdom as a green energy hub, as the provided initiatives can not only the stakeholders willing to invest in green energy to make the green industrial revolution in the United Kingdom happen but also could enable the investors to do so at minimal cost. Table 6.2 shows an extract of the extensive BiSC implementation MCDA analysis. Observing economic principles and baselines, it can be assumed that, overall, there is a third of the cost involved, with two-thirds of saving on energy cost over a 30-year investment period. Additional significant positive milestones can be reached, such as investment into a workforce with 'green' skills, future-proofing the business against energy price hikes, besides the positive effects of greater energy independence, decarbonisation of production and rehabilitation of the natural world, which has been disrupted beyond recognition by the Anthropocene. The political and legal landscape and countries' governmental guidance will have to change dramatically [51, 52] in order to meet the targets set by the Paris Agreement [53] and COP15 [52] and the dire warnings issued by the recent IPCC reports (2023) [54]. By improving their carbon capture and off-heat utilisation capabilities, and investing in renewable energy technology, businesses are:

- Supporting the decarbonisation of production
- Reaching a higher degree of energy independence
- Achieving a higher level of asset efficiency
- Training workforce in required 'green' skills
- Reducing their energy costs
- Creating a viable additional income stream

Additionally, besides achieving a higher degree of energy independence, economic advantages in monetary terms are almost inevitable when renewable energy systems are being installed. Doing so will also support limiting the global temperature rise to below 2 °C and thus avoiding climate disaster. The CO_2 emissions by industry sector were analysed [34, 40, 55–63], and besides the iron and steel industry, there are other heavy industries that are CO_2 emitters and generate copious amounts of off-heat, co-products and therewith resources, which could be harvested and used to power energy-dependent devices: transport, chemical industry, energy supply, residential/commercial buildings sector, agriculture, and waste management. There are currently globally 89 projects in process with the aim to achieve climate neutrality at the latest by 2050 [64].

6.5 Conclusions

Achieving a higher degree of energy independence is, within reach of all sectors of society, made possible by technological progress and incentives and grants provided by the respective governments and countries, including the United Kingdom. The opportunities are manifold, and they merely require political willingness and implementation across all industries. The process simulations compiled within the systems S8, AspenV.12.1 and INOSIM aligned with the mathematical analysis and modelling in MS Excel and provided almost matching reports, stating a reduction of 30% from 977 t of CO_2 emissions to 684 t of CO_2 flow in one single blast furnace production cycle (based on 330 tonnes liquid iron production capacity and a mean of 2.1–3.2 tonnes CO_2/t of steel and emissions from chemical reactions considered), by switching the electricity provider for operating the electric heaters to providers generating energy exclusively from renewable sources. Replacing coal with biomass added carbon neutrality into the process. Adding the post-combustion capture unit to the blast furnace resulted in a CO_2 emission reduction to near zero. Carbon capture from biomass utilisation added the benefit of carbon-negative emissions to the cycle. Simultaneously, energy-saving and process improvement measure implementation (up to 60% efficiency increase), excess heat recovery <30% of energy savings and retrofitting renewable energy technology resulted in an energy independence of 88%. Engineering solutions that are partly subsidised in the United Kingdom – as described in Table 5.1 – are readily available for implementation in the iron and steel manufacturing industry. Furthermore, the application points of renewable energy technology within the steel production process were established where achieving a 30%, 49% and up to 88% higher energy independence is achievable, in the short term. The cost risk of installing renewable energy and decarbonisation technology was minimised to less than a third of the initial outlay in individual cases, as the UK government is providing grants, funding and loan schemes to encourage all industrial sectors to work towards Net-Zero. The research preceding the compilation of the current paper has provided answers that are reaching over into multiple other heavy industries and additionally resulting in a further opportunity for a range of research directions, which are listed as follows:

- Utilisation of CO_2 in the building industry (CCUS), agriculture (CEPS), chemical industry, food and drinks industry and pharmaceuticals
- Utilisation of waste products from steelmaking in infrastructural projects
- Other GHG (CO and H_2) captured in TGRB, can also be reused in the chemical industry, thus eliminating the need for waste management

The upgrading and retrofitting of existing and new steel plants and other industrial sites provide an immense opportunity for further research, as there is a vast range of active and abandoned production plants available in the United Kingdom, not limited to:

- Adding solar foil, panels, tiles and shingles to buildings and car parks
- Adding biogas, replacing fossil fuel and a hydrogen network from anaerobic digestion for hydrogen direct reduction of iron ore (DRI) in steel production, in addition to reformers and electrolysers
- Installing wind turbines at brownfield sites not suitable for human habitation

- Developing filters and capture units to be retrofitted to existing production plants
- Emitting GHGs

Further work is currently underway and will provide more details on the more salient points of this research upon publication.

References

1 International Energy Agency (2023). Data and Statistics. https://www.iea.org/data-and-statistics (accessed 7 July 2023).

2 Baena-Moreno, F., Cid-Castello, N., Arellano-García, H., and Reina, T. (2021). Towards emissions free steel manufacturing—Exploring the advantages of a CO_2 methanation unit to minimize CO_2 emissions. *Sci. Total Environ.* 781: 146776.

3 Hasanbeigi, A. (2022). Steel Climate Impact—An International Benchmarking of Energy and CO_2 Intensities. https://static1.squarespace.com/static/5877e86f9de4bb8bce72105c/t/624ebc5e1f5e2f3078c53a07/1649327229553/Steel+climate+impact-benchmarking+report+7April2022.pdf (accessed 7 July 2023).

4 COM (European Commission); CORDIS (2023). Steel Industry Boost Research into Cleaner Technologies. https://cordis.europa.eu/article/id/29184-steel-industry-boost-research-into-cleaner-technologies (accessed 6 July 2023).

5 International Energy Agency (2021). Data and Statistics. https://www.iea.org/data-and-statistics/data-browser?country=WORLD&fuel=Energy%20transition%20indicators&indicator=ETISharesInPowerGen (accessed 4 December 2023).

6 Swalec, C. (2021). Guest Post: These 553 Steel Plants Are Responsible for 9% of Global CO_2 emissions. Carbon Brief. https://www.carbonbrief.org/guest-post-these-553-steel-plants-are-responsible-for-9-of-global-co2-emissions/ (accessed 7 July 2023).

7 Ren, L., Zhou, S., Peng, T., and Ou, X. (2021). A review of CO_2 emissions reduction technologies and low-carbon development in the iron and steel industry focusing on China. *Renew. Sustain. Energy Rev.* 143: 110846.

8 International Energy Agency (2018). Global Energy Demand Grew by 2.1% in 2017, and Carbon emissions Rose for the First Time Since 2014—News. https://www.iea.org/news/global-energy-demand-grew-by-21-in-2017-and-carbon-emissions-rose-for-the-first-time-since-2014 (accessed 7 July 2023).

9 The Anderson Centre (2020). *John Nix Pocketbook—For Farm Management*, 50e. The Anderson Centre: Leicestershire, UK.

10 The Anderson Centre (2023). *John Nix Pocketbook—For Farm Management*, 53e. The Anderson Centre: Leicestershire, UK.

11 Dastoor, P. (2021). Solar Foil. Centre for Organic Electronics Department of Physics, University of Newcastle Callaghan, Callaghan, NSW, Australia. 18 October 2021.

12 Caminiti, G., Anderson, P., Pulido, T. et al. (2021). Implementation of Solar Energy. ArcGIS StoryMaps. https://storymaps.arcgis.com/stories/70d4da4a613a4829b4f258ad3e724923 (accessed 6 July 2023).

13 El-Khozondar, H. and El-Batta, F. (2022). Solar energy implementation at the household level: Gaza Strip case study. *Energy Sustain. Soc.* 12: 17.

14 IRENA (2022). Renewable Power Remains Cost-Competitive amid Fossil Fuel Crisis. https://www.irena.org/news/pressreleases/2022/Jul/Renewable-Power-Remains-Cost-Competitive-amid-Fossil-Fuel-Crisis (accessed 13 July 2023).

15 Kiessling, S., Darabkhani, H.G., and Soliman, A.H. (2022). Energies Free Full-Text The Bio Steel Cycle: 7 Steps to Net Zero CO_2 emissions Steel Production. https://www.mdpi.com/1996.1073/15/23/8880 (accessed 11 July 2023).

16 World Steel Association (2019). Worldsteel 2019. https://worldsteel.org/wp-content/uploads/2019-World-Steel-in-Figures.pdf (accessed 7 July 2023).

17 World Steel Association (2021). World steel in Figures. https://worldsteel.org/world-steel-in-figures-2021/ (accessed 7 July 2023).

18 World Steel Association (2022). World Steel in Figures 2022—A Healthy Economy Needs a Healthy Steel Industry. https://worldsteel.org/steel-topics/statistics/world-steel-in-figures-2022/#:~:text=However%2C%20we%20have%20forecasted%20that,Safety%20remains%20our%20top%20priority (accessed 7 July 2023).

19 European Cement Research Academy (2009). TR-ECRA 106.2009 ECRA CCS Project and Report about Phase II. Düsseldorf, Germany: ECRA. https://ecra-online.org/fileadmin/redaktion/files/pdf/ECRA__Technical_Report_CCS_Phase_II.pdf (accessed 4 March 2020).

20 European Lime Association (2012). Summary of the technical report A Competitive and Efficient Lime Industry—Cornerstone for a Sustainable Europe. https://www.eula.eu/wp-content/uploads/2019/02/A-Competitive-and-Efficient-Lime-Industry-Summary_0.pdf (accessed 4 December 2023).

21 COM (European Commission) (2019). BIONICO: A Pilot Plant for Turning Biomass Directly into Hydrogen BIONICO Project Results in Brief H2020 CORDIS European Commission. https://cordis.europa.eu/article/id/394984-bionico-a-pilot-plant-for-turning-biomass-directly-into-hydrogen (accessed 6 July 2023).

22 Fleiter, T., Manz, P., Neuwirth, M., et al. (2020). Documentation on Excess Heat Potentials of Industrial Sites Including Open Data File with Selected Potentials: D5.1. https://urn.kb.se/resolve?urn=urn:nbn:se:hh:diva-42463 (accessed 6 July 2023).

23 Li, J., Li, C., Zhang, W. et al. (2022). Material, energy and exergy flows of the oxygen blast furnace process with sintering flue gas injection. *J. Clean. Prod.* 371: 133294.

24 Mandova, H., Patrizio, P., Leduc, S. et al. (2019). Achieving carbon-neutral iron and steelmaking in Europe through the deployment of bioenergy with carbon capture and storage. *J. Clean. Prod.* 218: 118–129.

25 Moran, M.J. and Sciubba, E. (1994). Exergy Analysis: Principles and Practice. *J. Eng. Gas Turbines Power* 116: 285–290.

26 World Economic Forum (2019). The $86 Trillion World Economy—In One Chart. https://www.weforum.org/agenda/2019/09/fifteen-countries-represent-three-quarters-total-gdp/ (accessed 14 July 2023).

27 International Energy Agency (2021). Global CO_2 emissions Rebounded To their Highest Level in History in 2021—News. https://www.iea.org/news/global-co2-emissions-rebounded-to-their-highest-level-in-history-in-2021 (accessed 7 July 2023).

28 STATISTA (2023). Global Net electricity consumption worldwide in select years from 1980 to 2022 (in terawatt-hours). https://www.statista.com/statistics/280704/world-power-consumption/ (accessed 28 November 2023).

29 STATISTA (2023). Electricity use for iron and steel production in the United Kingdom (UK) from 2000 to 2023 (in terawatt-hours). https://www.statista.com/statistics/323391/electricity-use-for-iron-and-steel-production-in-the-united-kingdom-uk/ (accessed 29 November 2023).

30 Langley, K.F. (1986). Energy efficiency in the UK iron and steel industry. *Appl. Energy* 23: 73–107.

31 Yurcent Environmental Technology Co., Ltd. (2023). Zeolite Rotor-Product. https://en.yurcent.com/Products_3/20.html (accessed 7 July 2023).

32 International Energy Agency (2023). World Energy Outlook 2023—Analysis and key findings. A report by the International Energy Agency. Paris, France: IEA. https://www.iea.org/reports/world-energy-outlook-2023 (accessed 25 January 2024).

33 Chen, Q., Gu, Y., Tang, Z. et al. (2018). Assessment of low-carbon iron and steel production with CO_2 recycling and utilization technologies: A case study in China. *Appl. Energy* 220: 192–207.

34 Gozens, J., Turnbull, A., and Jotzo, F. (2021). An Installation-Level Model of China's Coal Sector Shows How Its Decarbonization and Energy Security Plans Will Reduce Overseas Coal Imports. Canberra, Australia: CCEP Working Paper; Crawford School of Public Policy, Centre for Climate and Energy Policy, Australia National University. https://ccep.crawford.anu.edu.au/sites/default/files/publication/ccep_crawford_anu_edu_au/2021-12/ccep2109_-_gosens_turnbull_jotzo_-_china_coal_model.pdf (accessed 3 August 2022).

35 COM (European Commission) (2023). Circular Economy: Faster Progress Needed to Meet EU Resource-Efficiency Targets, Ensure Sustainable Use of Materials and Enhance Strategic Autonomy. https://environment.ec.europa.eu/news/circular-economy-faster-progress-needed-meet-eu-resource-efficiency-targets-ensure-sustainable-use-2023-05-15_en#:~:text=In%20recent%20years%2C%20circular%20economy,0.8%25%20of%20the%20EU's%20GDP (accessed 3 January 2023).

36 COM (European Commission); CORDIS European Commission (2023). Hydrogen Meeting Future Needs of Low Carbon Manufacturing Value Chains H2Future Project Fact Sheet H2020. https://cordis.europa.eu/project/id/735503 (accessed 6 July 2023).

37 COM (European Commission) (2023). EU Climate Targets: How to Decarbonise the Steel Industry. https://joint-research-centre.ec.europa.eu/jrc-news-and-updates/eu-climate-targets-how-decarbonise-steel-industry-2022-06.15_en (accessed 6 July 2023).

38 US Department of Energy Efficiency & Renewable Energy (2021). Hydrogen Production: Natural Gas Reforming. https://www.energy.gov/eere/fuelcells/hydrogen-production-natural-gas-reforming (accessed 4 December 2023).

39 Tata Steel Europe (2020). Tata Steel in Europe Sustainability Report 2019/2020. https://www.tatasteeleurope.com/sites/default/files/TSE%20Sustainability%20report%202019-20%20(EN).pdf (accessed 7 July 2023).

40 Tata Steel (2024). Tata Steel in Europe. Sustainability Report 2021–2023 Tata Steel's UK business. https://www.tatasteeleurope.com/sites/default/files/tata-steel-uk-sustainability-report-2021-2023.pdf (accessed 25 January 2024).

41 Toktarova, A. (2021). The Low-Carbon Steel Industry-Interactions Between the Hydrogen Direct Reduction of Steel and the Electricity System—ProQuest. Gothenburg, Sweden: Department of Space, Earth and Environment, Chalmers University of Technology. https://www.proquest.com/openview/c2a4d77d71403fdbffd33f250b13251d/1?pq-origsite=gscholar&cbl=2026366&diss=y (accessed 7 July 2023).

42 Kittipongvises, S. (2017). Assessment of Environmental Impacts of Limestone Quarrying Operations in Thailand. *Environ. Clim. Technol.* 20: 67–83.

43 Seethalakshmi, K.K., Jijeesh, C.M., and Balagopalan, M. (2016). Bamboo Plantations: An Approach to Carbon Sequestration. Trichur, India: Kerala Forest Research Institute. https://www.researchgate.net/publication/215475397_Bamboo_plantations_An_approach_to_Carbon_sequestration (accessed 4 December 2023).

44 Kuramochi, T., Ramírez, A., Turkenburg, W., and Faaij, A. (2011). Techno-economic assessment and comparison of CO_2 capture technologies for industrial processes: Preliminary results for the iron and steel sector. *Energy Procedia* 4: 1981–1988.

45 Kuramochi, T., Höhne, N., Schaeffer, M. et al. (2018). Ten key short-term sectoral benchmarks to limit warming to 1.5°C. *Clim. Policy* 18: 287–305.

46 Office for National Statistics (2021). Population Estimates. https://www.ons.gov.uk/peoplepopulationandcommunity/populationandmigration/populationestimates (accessed 14 July 2023).

47 International Renewable Energy Agency (2021). World Energy Transitions Outlook: 1.5°C Pathway. https://www.irena.org/publications/2021/Jun/World-Energy-Transitions-Outlook (accessed 4 December 2023).

48 Intergovernmental Panel for Climate Change (2021). Climate Change 2021—The Physical Science Basis Summary for Policymakers. Cambridge, UK: Cambridge University Press. https://www.ipcc.ch/report/ar6/wg1/downloads/report/IPCC_AR6_WGI_SPM.pdf (accessed 7 July 2023).

49 OFGEM (2023). Smart Export Guarantee (SEG). https://www.ofgem.gov.uk/environmental-and-social-schemes/smart-export-guarantee-seg (accessed 7 July 2023).

50 OFGEM (2023). Find Business Energy Efficiency Grants and Schemes. https://www.ofgem.gov.uk/information-consumers/energy-advice-businesses/find-business-energy-efficiency-grants-and-schemes (accessed 7 July 2023).

51 Wang, R. (2022). Energy Saving Technologies and Optimisation of Energy Use for Decarbonised Iron and Steel Industry. PhD thesis, Durham University, Durham, UK. http://etheses.dur.ac.uk/14289/ (accessed 7 July 2023).

52 Wang, C., Walsh, S.D., Haynes, M.W. et al. (2022). From Australian Iron Ore to Green Steel: The Opportunity for Technology-Driven Decarbonisation. https://ecat.ga.gov.au/geonetwork/srv/api/records/5377741d-ece0-4e01-b87e-6bf49da11c03 (accessed 7 July 2023).

53 Wang, R., Liang, L., Wang, Y., Roskilly, A.P. (2019). ID: 9. Process Simulation of Blast Furnace Operation with Biomass Syngas Injection for Clean Production. Newcastle, UK: Sir Joseph Swan Centre for Energy Research, Newcastle University. https://www.energy-proceedings.org/wp-content/uploads/2020/02/9_Paper_0705055709.pdf (accessed 10 January 2020).

54 Kempken, T.; Hauck, T.; De Santis, M. et al. (2021). Collection of Possible Decarbonisation Barriers (Deliverable D1.5). Brussels, Belgium: Centre for European Policy Studies (CEPS); Green Steel for Europe Consortium. https://www.estep.eu/assets/Uploads/D1.5-Collection-of-possible-decarbonisation-barriers.pdf (accessed 2 March 2022).

55 Bazooyar, B. and Darabkhani, H.G. (2020). Design, manufacture and test of a micro-turbine renewable energy combustor. *Energy Convers. Manag.* 213: 112782.

56 Devlin, A. and Yang, A. (2022). Regional supply chains for decarbonising steel: energy efficiency and green premium mitigation. *Energy Convers. Manag.* 254: 115268.

57 United Nations Environment Program (2020). The Six-Sector Solution to the Climate Crisis. https://www.unep.org/interactive/six-sector-solution-climate-change/ (accessed 13 July 2023).

58 UNEP United Nations Environment Program (2022). UN Biodiversity Conference (COP 15). http://www.unep.org/un-biodiversity-conference-cop-15 (accessed 13 July 2023).

59 United Nations (2016). Paris Agreement 2016. New York: United Nations. https://unfccc.int/process-and-meetings/the-paris-agreement (accessed 25 January 2024).

60 Intergovernmental Panel for Climate Change (2023). CLIMATE CHANGE 2023 Synthesis Report—Summary for Policymakers. Report No.: AR6. https://www.ipcc.ch/report/ar6/syr/downloads/report/IPCC_AR6_SYR_SPM.pdf (accessed 5 January 2023).

61 Jahani, N., Sepehri, A., Vandchali, H.R., and Tirkolaee, E.B. (2021). Application of Industry 4.0 in the Procurement Processes of Supply Chains: A Systematic Literature Review. *Sustainability* 13: 7520.

62 Office for National Statistics (2021). Population Estimates for the UK, England and Wales, Scotland and Northern Ireland. National and subnational mid-year population estimates for the UK and its constituent countries by administrative area, age, sex and components of population change. https://www.ons.gov.uk/peoplepopulationandcommunity/populationandmigration/populationestimates/bulletins/annualmidyearpopulationestimates/mid2018 (accessed 14 July 2023).

63 Office for National Statistics (2023). Foreign direct investment, experimental UK subnational estimates: 2021. https://www.ons.gov.uk/releases/foreigndirectinvestmentexperimentaluksubnationalestimatesmarch2023 (accessed 14 July 2023).

64 Leadership Group for Industrial Transition (2023). Green Steel Tracker. https://www.industrytransition.org/green-steel-tracker/ (accessed 25 January 2024).

7

Technological challenges to and opportunities of the BiSC concept implementation

7.1 Introduction

In this chapter, the technological challenges to and opportunities of the Bio Steel Cycle (BiSC) concept implementation and the technological challenges to the integration of the individual improved processes, as well as barriers and hurdles, are being discussed. With details on retrofitting existing steel plants and how to conceptualise newly conceived projects, a particular focus is here on renewable energy technologies, carbon capture, biomass and green hydrogen, and hydrogen direct reduction (HDR).

7.2 Challenges

7.2.1 Steelmaking process improvements: implementation of TGR in BF, BOF and EAF

Top gas recycling (TGR) in blast furnace (BF) or basic oxygen furnace (BOF) is a promising technology to significantly reduce the CO_2 emissions from BF/BOF operations [1]. It includes the injection of carbon content reducing top gas components CO and H_2 into the shaft via tuyeres, lowering coke rates and using 99% of pure O_2 instead of hot blast air through the tuyeres. Additionally, the capture of CO_2 at 400 ppm [2] and carbon enrichment for plant stimulation (CEPS) from the off-gas stream are useful technologies for both the BF and BOF operations. The traditional BF is, for these purposes, retrofitted with a TGR unit, where the CO_2 is separated from the post-combustion off-gas and recycled – ideally in a CEPS component or as base material to produce recycled aggregate concrete (RAC). Furthermore, some remaining CO_2-stripped gas is diverted back into the BF, after being reheated using off-heat from the BF operation – as displayed in Figure 7.1. Reusing the CO_2-stripped off-gas in the BF operation reduces direct CO_2 emissions from the BF by 26% and an additional 52% can be avoided (CAT) by combining the TGR with carbon capture, utilisation and storage (CCUS). The CO_2 reduction potential of the TGRBF is therefore set at 78% [3].

The TGRBF operation works in several stages, beginning with the capture of the post-combustion off-gas. The BF gas (BFG) is cleaned and fed into a mixing unit,

Circular Steel Production: Pathways to Net-Zero Carbon Emissions, First Edition.
Sandra Kiessling, Hamidreza Gohari Darabkhani, and Abdel-Hamid Soliman.

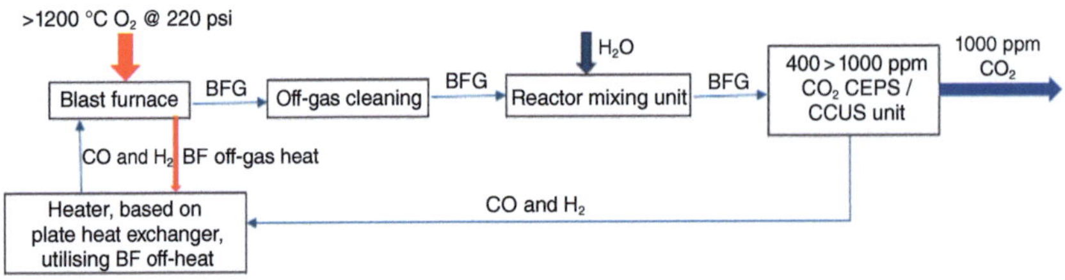

Figure 7.1 TGRBF operation schematic diagram.

where water is added to the mixture. In the CEPS/CCUS unit, the BFG is stripped of its CO_2 contents and fed into an off-heat utilising heater element, from where the BFG (stripped of the CO_2 contents), is being fed back into the BF. The CO_2 capture and utilisation costs were estimated to be between £28 and £65/tCO_2 [1], depending on the end-use being either reusage at the plant or further recycling in RAC or agri-food production. The ULCOS TGRBF with CCS is included in the ultra-low CO_2 steel (ULCOS) [3–6] European Union (EU) programme, and in an early phase of the ULCOS experimental BF in Luleå, the heat and mass balance as well as the tuyeres parameters were mathematically modelled, process simulations were compiled, and resulting modifications influenced the subsequent two test campaigns. Interestingly, the campaign results obtained were similar to the simulated and calculated test configurations [7–12]. Commercial deployment is ongoing [13]. As the cost has reduced significantly since the conception of the technology, the financial barriers to the implementation of TGR have been largely removed. The hurdles to overcome are now the willingness of industry to accept their responsibilities for the environmental devastation caused and to decarbonise all heavy industries to remedy the resulting damages.

7.2.2 Utilising excess heat from furnaces

7.2.2.1 Challenges of excess heat harvesting implementation

Excess heat recovery is still in its infancy, although the energy and carbon emissions savings potential are significant, as discussed in Chapter 6. Currently, off-gas waste heat recovery in electric arc furnace (EAF) operation of steelmaking is not a widely used technology; thus, almost a third of energy input is lost via off-gas waste heat, in most EAFs [14–16]. Novel technologies for excess heat harvesting are currently being developed as we speak [17], as discussed in Chapter 6. One of the proposed systems is the trigeneration for waste heat recovery from slag in steelmaking plants are in-cascade-installed turbines and input streams [16]. This process involves biogas produced from the anaerobic digestion process of wastewater and sewage of residential areas. Additionally, the combined heating, cooling and power cycle aimed to recover the thermal energy of waste steel slag. Interestingly, the energy, exergy and economic (3E) analyses and performances of this method under different objectives resulted in the following findings: the exergy unit cost of output power, cooling and heating was determined to be ca. £24/h, ca. £22/h and ca. £22/h, respectively. [16]. Additionally, the power generation capacity of this process was established to be 700 kW, the cooling effects were 40 tonnes of refrigeration, and hot water production was produced at 29,400 kg/h. For the organic Rankine cycle [18], R123 was investigated and found to have the highest efficiencies (16–28%) in the turbine inlet temperature range of 200–420 °C [16]. The highest output power was found to be ca. 50 kW. The highest congruent increase in exergy efficiency and power generation at the 200–420 °C temperature range was observed when HFE7000 was utilised as the operating fluid [14–16]. Significant obstacles hindering the implementation of waste heat recovery processes are the instable temperatures and mass flow streams, as well as the high dust content (>25 g/m^3) of the EAF off-gas and high capital expenditure (CapEx) amortisation rates [14–16]. A technical hurdle is the requirement for process-related heat storage devices, to allow for balancing the system during tapping and charging. In most established facilities, waste heat collection chambers, boilers and steam accumulators are used for waste heat extraction and storage rates [14–16]. Issues identified during the operation were corrosion and erosion of the waste plate heat exchangers, the low detected temperature of the heat extracted and storage mass density. It was concluded, therefore, that a combined heat and power generation system or direct use of redirected steam or heat must be made technically and economically feasible [1, 14–16, 19]. This is partly being realised now in some BF–BOF route operations, where the CO from the BOF is directed to be used in the BF [19–22]. A promising waste heat recovery concept as an option for retrofitting at existing steel plants is currently developed at the 'Institute of Combustion and Power Plant Technology' (IFK) [15] in Stuttgart, Germany. The aims of the process are a higher heat storage density rate and an increase in heat extraction temperature for efficient power generation. A selection of the following given parameters was utilised to arrive at the resulting findings.

The European steel industry as a world leader in innovation and environmental sustainability has an annual turnover (per 2023, using 2022 data) of ~€130 billion and is a direct employer of ~306,000 highly skilled people in the EU [23, 24]. They are producing ca. 152 million tonnes of steel, every year, at 500+ steel production sites across the 22 EU member states [25]. Steel billets of an average size (240 × 240 mm) were used as emitter bodies for calculation. The resulting data were used for the calculation of heat emissions from surface area in m^2, per linear metre, and for the radiant power measured in Joules as heat energy. The radiant power as per surface unit was calculated in line with the Stefan–Boltzmann law [26] for grey bodies, where emissivity is being used as a calculation factor. According to the findings as a result of simulations and calculations as per data presented in Table 7.1, the radiant heat potential in Europe is estimated to be ca. 10,549 MWh/p.a. for excess heat recovered from steelmaking, measured at the section between continuous casting outlet and oxy-cutting procedure [16]. The argument could be made that CO_2 punishments would have to be factored into the efficiency and emissions reduction calculations, as the infrastructure for heat recovery needs to be built specifically first for this purpose. However, these can be considered a one-time expenditure, as the environment would only have to be built once and would reap the benefits of excess heat recovery continuously over the average estimated asset lifetime of 30 years.

Table 7.1 Parameters for determination of heat values in EAF continuous cast-steel.

Description	Values of continuous cast-steel
Surface temperature in °C	1000
Emissivity value in Joules	0.82
Radiant surface emissivity in Joules per m^2 of linear m	0.72
Radiant energy in kW/linear m	88
EAF production rate in metric tonnes of steel per hour	150
Total line-length linear meters per hour	24
Available radiant heat in kWh per metric tonne	14
Annual heat radiation in MWh/p.a.	10,549

7.2.3 Biomass to replace coal and coke

Biochar derived from agricultural waste and its application as an energy source in BF can be compared in its efficiency with powdered coal injection (PCI) coal – but the influence of pyrolysis temperature on the combustion indexes and physicochemical properties of the different types of biochars must be considered. Different organic matters that undergone pyrolysis have varying fixed yields of carbon content binding properties, but the higher the temperature during pyrolysis, the lower the mass yield (carbon binding properties) – which means that, at the same time, the carbon penalty increases. Coconut shell and corn-cob biochars, transformed in pyrolysis at lower temperatures, exhibited improved combustion and carbon binding properties and could serve as an equivalent substitute to PCI coal in a BF. Organic matters with higher ash content are deemed unsuitable, such as tree bark biochar [27]. Therefore, the challenge of replacing fossil fuels with biomass is partly to identify which feedstock has the highest yield and generates the lowest carbon penalty.

Pyrolysis is one process to transform waste or by-products into biochar; there are others, too, such as hydrothermal carbonisation (HTC) technology. This methodology upgrades combustible waste (CW) to high-quality fuels called hydrochar [28]. The HTC process can effectively remove harmful elements (K, Na, Cl and S) from feedstocks and transform these into their liquid and gas phases without adding other reagents and will additionally reduce harmful effects in the BF. Interestingly, the removal rates for alkali metals were more than 80% and >73.9% for Cl. It could reach >98.18% for polyvinyl chloride hydrochar. The CO_2 emissions reduction of replacing bituminous coal or PCI with 40% HTC-treated maize straw can reach 94.7 kg per metric tonne of hot metal – bearing in mind that the BF capacities are ranging from 100 t to more than 330 tonnes charge capacity in up to 33 cycles per day, with one charge finished every 40 minutes. Even when the CO_2 penalties for the HTC process are taken into consideration, hydrochar can be seen as a clean energy source and direct clean alternative compared with fossil fuels and is suitable for BF injection in place of PCI [28]. However, the replacement of coal-based fuels by renewable fuels such as charcoal is still considered the most widely used alternative and an opportunity to reduce net greenhouse gas emissions from the integrated steelmaking route [29] by between 32% and 58%. When the combustibility of four types of charcoal with PCI in a BF was compared, it was found that burn-outs were superior to the high volatile powdered coal injected. Hardwood charcoals had burn-outs ~40% higher than those of equivalent coals, and the softwood charcoal studied was ~50% higher. Incremental increases in oxygen enrichment proved effective in increasing burn-out. The fragmentation of the charcoals during combustion was found to be boosting the already high surface areas and combustibility.

Biomass to replace fossil fuels in industrial operations can come from a variety of resources, with charcoal being only one of many [30]. Energy crops, such as Miscanthus (Miscanthus X Giganteus) – a tall perennial woody grass of East Asian origin, which has a high biomass yield potential [31] – play a significant role due to their higher conversion of solar radiation into biomass (photosynthesis) and higher water and nitrogen (N) conversion efficiency [30, 31]. The conversion of grass species for biochar and energy generation and research into the processes and effects involved are being constantly monitored to improve yield, as they are considered a potential negative emissions technology [30].

When the injection of biofuels was modelled, using real BF data as reference [32], it was found that for charcoal and torrefied and pelletised biomass as a replacement of fossil fuels, replacement limits were established, comparing all parameters when the conditions were equal. CO_2 emissions and energy savings were substantial and in the region of >30%, when any biomass was used. Charcoal from pyrolysis was established to be able to fully replace PCI, while torrefied material could replace 22.8% and pelletised wood could replace 20.0%, respectively, measured in weight. During the creation of an energy and mass balance model (MASMOD), some interesting additional findings could be established, which were quite unexpected: (1) the more the BFG is being generated, the higher is the reduction in fuel consumption in an integrated plant, (2) there is a lower requirement for flux (limestone and lime) than was expected, and (3) the resulting amount of slag is lower, too. Additionally, (4) a reduced fuel consumption for heating the hot BF charge was observed – hence the high reduction rate of >30% of carbon emissions reduction and reduced need for energy [32].

Now, when the economic assessment of options for biomass pretreatment and their use in the BF operation was carried out [33], it was found that it was economically advantageous to use the hot stove flue gas (as discussed in TGRBF operation) for drying the biomass used for iron smelting. The use of recycled top gas has the potential to reduce the annual operating costs of preheating the biomass to replace PCI or coke, which is considered to be ca.0.5 M€ for a steel plant with an annual production capacity of 1.4 Mt. of crude steel [33]. Based on the findings in experiments carried out at Finnish steel works [34], bio-reducers, such as charcoal, torrefied wood and wood-based synthetic natural gas (Bio-SNG) (generated as a result of logging residues, small-diameter wood and stumps, as a vital component within the BiSC, presented valuable findings. Based on the Finnish example, the global CO_2 reduction potential could be up to 0.99–1.01 Mt./p.a. without by-product credits and 1.29–1.39 Mt./p.a. *with* by-product credits. The CO_2 mitigation costs for charcoal, torrefied wood and Bio-SNG without by-product credits are 26–55 £/t CO_2, 17–46 £/t CO_2 and 77–120 £/t CO_2 and with by-product credits are 21–43 £/t CO_2, 17–42 £/t CO_2 and 85–121 £/t CO_2, respectively. Although the CO_2 mitigation cost of bio-reducers established in the Finnish trials is relatively high, the costs are reduced considerably when TGRBF excess heat is being used for pretreatment of biomass [33].

7.2.4 CO_2 capture technology and filtering systems

Scientific analysis has resulted in identifying eight major methodologies proposed for carbon capture. Those are as follows [35]:

- Membrane
- Oxy-fuel combustion
- Absorption
- Adsorption
- Multiphase absorption
- Chemical looping combustion
- Calcium looping
- Cryogenic

Currently, there are nine processes associated with the above-mentioned methodologies. In the following, the carbon capture technologies will be explained in more detail.

7.2.4.1 Direct air capture (DAC)

Direct air capture (DAC) technology is based on processes naturally occurring in nature, where the CO_2 is being absorbed by plants in photosynthesis to produce plant matter on land and at sea, and by soil and mineral structures in a chemical reaction. Translated into anthropologically conceived technology, air filters are installed to capture carbon dioxide from the surrounding air. Currently, there are companies such as Yurcent [36] that are manufacturing Zeolite 13X-based rotator carbon filters in China, whereas other companies, such as Heirloom [37] in California (USA), are using tall towers of trays containing limestone powder.

7.2.4.2 Enhanced rock weathering (ERW)

Enhanced rock weathering (ERW) [35] is the process where crushed silicate minerals and rocks (basalt and glauconite) are spread on arable soil, thus accelerating a natural process which usually takes over millions of years to only two or more decades, whilst simultaneously increasing the amount of CO_2 captured in soil. This technology is not only the most cost-efficient technology but also improves soil fertility.

7.2.4.3 Aqueous amine-based CO_2 capture

This technology [38] involves streaming industrial off-gases through an amine-based fluid bed, which absorb the CO_2 before it is released into the atmosphere. The CO_2

is subsequently separated from the amine and reused. Although this technology is mature, it is still very costly and energy-intensive.

7.2.4.4 Membrane gas separation

Gas mixtures are efficiently separated by being directed through synthetic membranes consisting of polymers (polyamide) or cellulose acetate [39], or from ceramic-based materials. It operates at relatively low temperatures and pressures (ambient/20 °C, 1 ATM), and it is energy-efficient and cost-effective. This technology holds a lot of promise in the capture, separation and storage of CO_2.

7.2.4.5 Carbon capture and conversion

This technology involves carbon capturing and effectively recycling the CO_2 into usable products such as RACs [40, 41], fuels, industrial chemicals and polymers. The carbon capture and conversion technology is a sustainable approach to decarbonisation of production whilst reutilising undesired by-products.

7.2.4.6 Bioenergy with carbon capture and storage (BECCS)

Bioenergy with carbon capture and storage (BECCS) [42] is a carbon-negative process, where CO_2 from biomass energy production (biogas) is captured and stored or circled back into the system. This circular process not only captures CO_2 but also generates electricity and can therefore be considered as a carbon-negative process. Bioenergy with carbon capture and storage is a low-carbon energy source that can make a significant contribution to decarbonising the economy.

7.2.4.7 Chemical looping

Using calcium looping (CaL) to capture CO_2, whilst simultaneously dry reforming methane (DRM) [43], has been developed and can now be seen as front-runner methods for carbon capture because of its low cost, easy availability, high adsorption capacity, fast adsorption rate in CaL processes and value-added products in the DRM process. Carbon dioxide (CO_2) and methane (CH_4) decrease simultaneously. The by-products of this process, namely syngas in the form of carbon monoxide (CO) and hydrogen and H_2 in the DRM process, can directly be used in the production of liquid fuels and chemicals. The challenges of this technology are still manifold; however, as CaO-based adsorbents for CaL processes are very energy-intensive, the occurrence of pore collapse is difficult to overcome, and sintering problems ensue after some cycles. Additionally, methane decomposition and reverse water gas shift (RWGS) are still an issue, as the initial activity is being deactivated during the DRM reaction, as coke forms on the active Ni surface. Fortunately, coupling CaL–DRM processes with an increase in operating temperature can alleviate the aforementioned challenges [38, 43].

7.2.4.8 Cryogenic carbon capture (CCC)

When cryogenic carbon capture (CCC) is mentioned [44], a method is referred to which involves cryogenic distillation in conventional natural gas (NG) liquefaction plants and constitutes another method of capturing carbon currently in development. The results of experimental plants show that due to the azeotropic properties of ethane and CO_2 mixtures, the CO_2 component in NG can be reduced to below 50 ppm by cryogenic distillation. The findings of thermodynamic analysis suggest that with a dual mixed refrigerant (DMR) liquefaction process – in tandem with ethane recovery – this method of carbon capture can be a solution for NG if it is rich in ethane. How it works: First, the cryogenic distillation singles out the CO_2 from NG, as well as the ethane, and then, the ethane is further separated by extractive distillation from the CO_2–C_2H_6 mixture. The energy analysis reports identified that NG contains 2–20 mol% ethane, and the specific energy consumption (SEG) for reaching 99.5% of CO_2 and ethane is ca. 0.40 kWh/Nm^3 (NG). So far, this method has shown a better performance than chemical absorption in both energy consumption and cleaner production aspects. CCC is a relatively new carbon capture technology that relies on cryogenic cooling to capture and remove CO_2 from gas streams. It can extract CO_2 at a higher rate than conventional systems and can store it at a reduced volume. The application of CCC is still limited, but as the technology matures, it shows great potential in mitigating carbon emissions from various sources [44].

7.2.4.9 Carbon capture using nanotechnology

When carbon capture using nanotechnology is mentioned, we are talking about thin-film composite (TFC) membranes, which feature excellent separation properties for H_2/CO_2, CO_2/N_2 and CO_2/CH_4 [45]. The thickness of the chosen layers becomes gradually thinner (~100 nm or less) to increase the efficiency of gas permeation and can be surface-engineered to increase the gas separation properties. Identified scalable nanotechnologies and their efficiency at TRL6-9 depend on the following parameters and processes: modified membrane surfaces, atomic layer deposition preparedness, chemical vapour deposition, direct fluorination, plasma treatment, ion/electron beam treatment, ozone treatment and surface-initiated polymerisation. This technology involves using nanomaterials, such as carbon nanotubes, to capture and store CO_2 at much lower pressures than other technologies. This method uses fewer resources and energy and produces less waste than other carbon capture technologies. Although carbon capture using nanotechnology is still at its experimental stage, it holds a lot of promise for its scalability and efficiency [45].

7.2.4.10 The current status quo

Every single one of these technologies has its challenges, benefits and limitations, which render them more or less effective in the varying applications and scenarios. It can be said, though, that they feature one unifying factor: they are all in their own right a suitable solution for reducing the CO_2 emissions which have brought the climate as we know it close to irreversible tipping points – once they have been reached, this would be the point of no return. Therefore, research and development of suitable technologies must continue as a top priority and must be moved along with supportive legislation across the globe and incentives to achieve limiting the temperature rise to 1.5 °C of the temperatures before industrial activity, according to the Paris Agreement [46].

There is currently considerable activity occurring in developing technologies for carbon capture and storage [47] across the United States and worldwide [17], and the state of the art as far as barriers to more widespread deployment are concerned – and the hurdles to overcome – has been investigated, resulting in the findings, as follows.

Carbon capture with sequestration and storage (CCS) is the process of capturing CO_2 generated by the combustion of fossil fuels during power generation and industrial processes and sequestering it permanently, to prevent it from ever reaching the atmosphere. There is a range of CCS technologies already widely used in industry and some in power generation (as discussed in Chapter 4), which have the potential to reduce CO_2 emissions in industrial processes and energy generation between 90% and some at 100% (TGRBF as an example). The Global CCS Institute stated in their 2021 report that projects currently underway in the United States have the capacity to capture 40 Mt. of CO_2 p.a. Across the United States , in 2019, there were ten commercial operational facilities [47].

In 2021, only two years later, 102 CCS facilities were either in advanced or early stages of development. Combined with facilities already under construction or in operation, these facilities could capture 149.3 million metric tonnes of CO_2 per year. In the context of the CO_2 emissions reported in 2019 of 5 billion t of CO_2 emissions, this carbon capture capacity of around 150 Mt. looks terribly inadequate, and it is obvious to the informed reader that a lot more needs to be done [47].

7.2.4.11 Barriers to deployment

One of the single most impactful barriers to increased application of CCS technologies is the CapEx. All deployment sites vary widely, but the highest costs are typically associated with the equipment and energy needed for the capture, separation, absorption/desorption and compression phases. Capturing CO_2 is known to decrease energy and resource efficiencies and increase a plant's water use, and the increased costs added by these, and other factors can render CCS projects financially uneconomical and nonviable. Additionally, the increased requirement for water in operation causes further issues in areas which are already facing water scarcity. Furthermore, ROI on CCS projects are riskier, as more unpredictable, due to their innovative nature. Consequently, investors are forced to request higher risk premiums, which further increases the absolute cost of the investment. Hence, the mitigation of risk for investors is crucial to motivate investors and increase involvement in CCS. Subsequently, research, development and deployment (RDD) policies that incentivise and support investment with grants, low-interest loans and research and development tax credits or allowances which have the potential to de-risk such investments are thus highly desirable, and policies that stimulate innovation are exactly what requires to be created, as soon as possible, so the necessary technology to reduce CO_2 emissions significantly can be widely deployed [47].

7.2.4.12 Transportation challenges

Besides the high CapEx, there are other challenges, such as those associated with transporting CO_2 after capture. A high amount of energy is required to compress and chill CO_2 and it is necessary to maintain high pressure and low temperatures throughout the network of pipelines, with the pipelines themselves being costly to build and maintain, as existing oil and gas pipelines are unsuitable to be used. Any impurities in the CO_2 stream, even water, can potentially cause damage to pipelines. This can lead to dangerous leaks and even explosions, as the compressed fluid, exposed to ambient temperatures, rapidly expands into a gas. Furthermore, the exceedingly cold temperatures cause metal pipes and ancillary equipment to become brittle [47].

7.2.4.13 Storage considerations

All sources of captured CO_2 have to be connected to an appropriate storage facility via a pipeline, which is another cost factor making CCS more difficult and expensive to implement when the site is located in areas at a distance from geological formations that are appropriate to use for long-term storage (sequestration). It is estimated that there is probably plenty of storage for CO_2, worldwide, for at least the next century. Another concern is the potential for leakage in long-term storage sites and the risk of causing seismic activity as a result of the underground injection of CO_2. Researchers are therefore busy finding solutions which can reduce these risk factors and are looking at ways how above-ground carbon dioxide mineralisation can be deployed as an alternative to underground storage [47].

7.2.4.14 Uncertain public support

Another, not-to-be-underestimated, hurdle is public opinion. In Denmark, the majority of the population has consistently favoured government decarbonisation efforts and methods for reducing air pollution from fossil-fuel-based power plants [48]. Across the globe, siting pipelines either for fossil fuels or for CO_2 transport is highly contentious – due to their environmental impact and the infringement on landowners' rights. There are a range of elements which seem to play a role in forming a public opinion about CCS: the tangible benefits to the human population and the environment of reducing CO_2 emissions; the suggestion that the rollout of CCS plants actually extends and not reduces the use of fossil fuels for energy generation and other applications; the impact pipelines have on the natural environment and, in particular, the fragmentation of ecologically sensitive areas; and the assumed and actual safety of transportation and storage of CO_2 [49, 50].

7.2.4.15 Leading carbon capture companies

What seemed like a fantasy some 30 years ago has turned into what is called now the sustainability industry. Companies concerned with decarbonising industry aim to support the energy transition towards green energy and the dependence on the fossil fuel industries and are worth a closer look [51]. Large investments have been poured into carbon capture projects and have resulted in cheaper and easier-to-install methods, which are making the future climate outlook better every day [47–51]. There are immense opportunities to keep the temperature rise below 1.5 °C before industrial activity and reduce greenhouse gas emissions, crucial to keep in line with the Paris Agreement [46]. It is worth having a look at some of the leading carbon capture companies, in order of capture capacity:

1) Heirloom
   Carbon capture per year (in tonnes of CO_2): 1000
   Carbon capturing and sequestration was still a far-fetched idea 30 years ago. Now, in 2024, scientists and in particular political leaders see it as an absolute necessity to avert climate disaster [52] and as an essential method to avert climate breakdown. The current US Biden administration has committed to spending at least $3.7 billion to power DAC development and other CCUS projects in the United States.
   Heirloom's facility [37] uses limestone, one of the most abundant minerals on Earth – to capture carbon from the air. However, making a noticeable difference could take some time, as their facility at the moment stands to absorb a maximum of 1000 metric tonnes of CO_2, per year [53, 54]. Considering the reported US 5 billion t of CO_2 emissions in 2019 [47], this is considered undersized to make a significant contribution to cutting CO_2 emissions. The company is set to increase their capacity and is hoping to remove 1 billion tonnes of carbon dioxide by 2035. This means they would have to increase their capacity by three times, each year.
2) Global thermostat
   Carbon capture per year (in tonnes of CO_2): 4000
   Global Thermostat's DAC technology uses a patented method called the 'Carbon dioxide Removal Assembly' (CDRA) [55, 56]. This method is based on using a proprietary adsorbent material regenerated by a low-grade (renewable) heat source to capture CO_2. The next step is the compression of the captured CO_2, to be used for a variety of purposes:
   - Enhanced oil recovery
   - Cement and concrete production
   - Resource for the production of renewable fuels

   The signs of a shifting public opinion and a shift in companies' strategies are certainly clearly visible, as 'old guard' oil companies, such as ExxonMobil, are invested in Global Thermostat, as a way of diversifying their portfolio and moving away from fossil fuels to ensure future-proofing of shareholder profits.
3) Climeworks
   Carbon capture per year (in tonnes of CO_2): 4000
   Climeworks, established in 2010, is located in Switzerland and is specialised in using a DAC process, which is based on extracting carbon dioxide from the surrounding atmosphere [57, 58]. The company has several DAC facilities dotted around the world, besides the headquarters in Switzerland, there are sites in Italy and Iceland, and in addition, partnerships were formed with organisations and companies such as Audi, Microsoft and Swiss Re, to ensure further research and development, increase efficiency and advance the development and deployment of DAC.

4) CO_2 solutions by SAIPEM
Carbon capture per year (in tonnes of CO_2): 11,000
SAIPEM, an Italian energy services company [59], has been using 'CO_2 Solutions' since 1997. The technology is based on using an enzyme-based fluid bed system to capture carbon dioxide from post-combustion industrial greenhouse gas emissions, such as those from power plants, cement factories and steel mills. The most distinctive element of the CO_2 Solutions technology is the lower cost in comparison with other, quite costly (i.e. amine-based solvents fluid beds), carbon capture techniques. Furthermore, using enzymes is considered a more sustainable and environmentally friendly process, as it does not need to use harmful substances and does not produce toxic by-products.
5) LanzaTech
Carbon capture per year (in tonnes of CO_2): 150,000
LanzaTech, established in 2005 in New Zealand [60], is a biotechnology-based company which specialises in researching, developing and commercialising carbon capture and utilisation (CCU) methods. The company has since expanded to locations around the world: the United States, India, China and Europe. LanzaTech's CCU technology can be compared with the TGRBF method, as it is capturing post-combustion waste gases from industrial processes, and subsequently converting these via microorganisms into chemicals and fuels. Additionally, LanzaTech is also researching and developing other technologies which are helpful to tackle the issues generated by a changing climate: they are converting the captured carbon dioxide into polymers and plastics as a resource for further manufacturing.
6) Carbon clean
Carbon capture in 2022 (in tonnes of CO_2): 335,745
Carbon Clean Solutions Limited, a company established in 2009 in London, UK [61], is also known as Carbon Clean. They have developed a low-cost carbon capture technology which can help reduce carbon emissions. Since it was founded in 2009 at the London headquarters, additional offices have been opened in India and the United States. Carbon Clean's technology utilises a solvent for CCUS to capture carbon dioxide emissions from power plants and cement factories, to name just a few. The captured CO_2 is then separated from the solvent and compressed into a pure form, which can be transported and stored for reuse or permanently sequestered underground.
7) Aker carbon capture
Carbon capture per year (in tonnes of CO_2): 400,000
Aker Carbon Capture, a subsidiary of Aker Solutions [62], is a leading player in the carbon sequestration industry. The company went public in 2020, with a listing on the Oslo Stock Exchange. They use their own proprietary carbon capture technology to capture CO_2 from post-combustion waste flue gases generated by various industries. The company has developed onshore and offshore modular and integrated solutions which have been designed with easy transport and installation in mind.

8) Carbon engineering

Carbon capture per year (in tonnes of CO_2): 1 million

Carbon Engineering applies a unique liquid DAC technique [63], which uses a potassium hydroxide fluid bed to trap CO_2. They have licensed their DAC technology to 1PointFive with the objective of constructing a megaton-scale plant in 2024 at the Permian Basin in the United States. It needs to be emphasised that this project is still under construction and the actual performance and carbon capture efficiency and capability have not been physically proven yet. All performance simulations suggest a powerful, 1 million tonnes of CO_2 capturing capacity, but this remains to be seen.

9) CarbonFree

Carbon capture per year (in tonnes of CO_2): 800 million

CarbonFree [64] uses patented technologies to capture CO_2 emissions directly from the emitters and converts these into (claimed) carbon-negative chemicals. Their innovative products, namely SkyCycle™ and SkyMine®, are based on utilising the captured CO_2 to produce resource materials and chemicals, such as:

- Sodium bicarbonate ($NaHCO_3$) (baking powder)
- Precipitated calcium carbonate, or PCC ($CaCO_3$), a synthetic calcium carbonate product which features the same chemical formula as natural calcium carbonate (limestone) although it is chemically purer than the natural limestone
- Hydrochloric acid (HCl)

CarbonFree's technology is modular and scalable and the technology can be used in situ, without the requirement for building costly infrastructures or transporting and storing the captured CO_2. The company's technology is applicable in most industrial manufacturing plants.

10) CarbFix

Carbon capture per year (in tonnes of CO_2): 1 billion/lifetime

Founded and established in 2006, CarbFix [65] is an Icelandic company specialising in carbon capture and storage (CCS) technology. Their approach is based on the acceleration of a natural process, during which in water dissolved CO_2 is injected into the basaltic rock formations which interact with the reactive rock formations to create stable carbonate minerals and are thought not to be prone to leakage of any kind. They have therefore created a safe and permanent storage and sequestration solution for $CO_{2,}$ deep underneath the Earth's surface, where it is sequestered away for the foreseeable future.

11) NeoCarbon GmbH

Carbon capture per year (in tonnes of CO_2): 1.3 Gt

NeoCarbon [66] is a startup based in Germany and was only established in 2022. Their methods capture carbon at scale in the form of waste heat and cooling towers. Their process leverages those industrial machines in combination with a (patented) reactor design, which decreases the costs involved dramatically, as no new infrastructure requires to be built to drastically cut down costs and deployment speed for DAC.

12) Meloon
    Carbon capture per year (in tonnes of CO_2): 35 Gigatonnes
    Meloon [67] is a leading climate tech startup dedicated to revolutionising carbon capture solutions. Our pioneering technology focuses on the development of advanced membrane-based systems that efficiently capture and store CO_2 emissions. We are driven by the mission to combat climate change and provide sustainable solutions to industries seeking to reduce their carbon footprint. By leveraging cutting-edge 2D materials and nanoporous structures, our membrane technology offers high performance, scalability and eco-friendliness. At Meloon, we cater to diverse industries, including cement production, manufacturing and power generation, offering customised carbon capture solutions tailored to each client's specific needs. Together with our clients, we aim to make a lasting positive impact on the environment, paving the way for a greener and more sustainable future.
13) ecoLocked
    Carbon capture through avoidance per year: almost limitless
    The basis for ecoLocked's technology is biochar/biocarbon from biowaste [68], and their carbon capture process has been acknowledged by the IPCC as one of the few technologies actually delivering on the promise of negative emissions. Carbon capture and storage, ecoLocked found, require a more scalable and permanent storage solution than what has been practically implemented in other companies, to date. Steel, cement and concrete are the most used materials worldwide and are responsible for between 8% and 11% of global CO_2 emissions. ecoLocked creates biocarbon-based concrete admix materials which are reducing the CO_2 emissions by the amount of embedded CO_2, effectively transforming the building sector into a verified carbon sink.

7.2.4.16 CO_2 filter systems

CO_2 filter systems are a crucial part of the BiSC system and are currently being developed by a range of companies, taking a wide range of approaches. A CO_2 filter system suitable for retrofitting into existing infrastructure at industrial manufacturing and production sites seems currently only being provided by Yurcent [36] at the time of writing. Besides membrane inserts and mineral traps or fluid beds in industrial applications situated CCUS technologies, DAC is the single most efficient method with the capacity to deliver CO_2 capture [69], at the required scale. The high cost and energy requirements have been identified as the hurdles to the rollout of DAC technology at the scale required. Hence, why the global research community continues in their efforts to identify more efficient methods of minimising the cost and required energy per tonne of CO_2 captured using DAC technology. It makes economic sense to look into utilising existing infrastructures and using traditional heating, ventilation and air conditioning (HVAC) systems and modifying these to capture CO_2 directly from ambient air, as it would have the effect that the energy required for DAC would be generated by the utilised HVAC system. In air conditioning, air purification devices utilise air filtration systems for maintaining optimum interior air quality [70]. As the pollutants contained in our surrounding air are varied and unstable, air-filtering materials made of glass fibres or non-biodegradable polymers have been found to be entirely unsuitable in removing particulate pollutants of finer sizes, such as CO_2 particles. Additionally, these air filtration materials are non-biodegradable and they will inevitably be disposed of in landfills. It is therefore imperative that sustainable, biodegradable or renewable filter inserts are being researched and developed, such as cellulose. It is a natural biopolymer, and with its porous structure and malleable quality, it has come to be used in high-performance air filtration systems worldwide. Cellulose can be seen as an essential polymeric material in the air purifying industries [69, 70]. Already in 1998 [71], a three-stage filtration system was developed which was able to capture hydrogen sulphide from an off-gas stream of CO_2 supply during a long-term tree growth experiment. This system features the ability to filter a mix of 75% CO_2 in ambient air at a flow rate of 30 m^3/h, resulting from biogas production at a waste-water treatment plant. The content of H_2S in the raw gas is 2500 μl/l, and this is reduced to less than 1 μl/l by this filtration method. The process contains three essential elements, such as a recirculating liquid system, a self-regenerating solution of iron chelate and a dry iron oxide box, alongside drums containing activated carbon (charcoal). Following the filtration process, ambient air is added to large open-top vaults, built around the surrounding trees. The CO_2 concentration is ca. 650 μl/l, and H_2S is less than 0.001 μl/l, deemed suitable and beneficial during the tree growth studies. This constitutes a low-cost option for supplying experimental field sites in the long-term with CO_2 to enhance plant growth – similar to CEPS systems [2].

7.2.5 Challenges of renewable energy implementation

In line with the Paris Agreement [46] and subsequently set sustainable development goals (SDGs) [72], the principal threats to human well-being have been identified and the target SDG reflects suitable measures to combat these. The expansion of using renewable energy sources to generate energy is hampered by a lack of accessible finance, ecological impact and lobbying from nonrenewable energy producers. Despite all this, environmental innovation and renewable energy deployment have been observed to have a positive cumulative effect on green growth and expansion of a green economy, and in some countries, like Germany and the United Kingdom, green energy is increasing in market share. The positive results seem to have shifted public opinion and are providing an increasing amount of vacancies in the renewable energy sector. Energy price volatility and a weakening supply structure have caused politicians and industry leaders to bring energy independence into the foreground of political and economic debates [73–75]. Analyses carried out for the G7 countries for the time between 1980 and 2018 have revealed a unidirectional connection between renewable energy to energy security and generating energy from renewable sources. It was found that the increased deployment of renewable energy technology exponentially reduces the risks of energy insecurity for the foreseeable future [73].

Various techno-economic analyses of photovoltaic (PV) systems, residential PV (RPV) and farm-based PV systems (FPV) in Indonesia [76], Sweden [8, 9] and China [77, 78] have led to the conclusion that with government-provided incentives, loans and grants, the uptake of these technologies grows exponentially. This translates into the conclusion that the CapEx is the main hurdle to the widespread deployment of renewable energy technology installations.

The role of the digital economy in the deployment of renewable energy technologies was found to remain insignificant [77], although a positive association between digital economy and renewable energy development was identified in Indonesia [74]. Most peculiarly, the digital economy seems to exert a stronger positive impact on hydroenergy system installation in comparison with solar or wind energy. Again, political influence with guidelines and incentives was found to have a positive effect if deployed wisely. Political willingness to support the green economy seems to be the single biggest factor in the widespread deployment of renewable energy technologies, as the systems in themselves have reached maturity and generated a wide range of diversity, readily serving any given site with the best- suited renewable energy source.

Companies in countries with a younger population and therefore a high level of human resources, such as India, and particularly Bangladesh, have achieved remarkable accomplishments in the solar energy sector – namely solar PV power unit production [79]. The importance of densely populated countries, such as India and China, in the move towards a greener energy market cannot be emphasised enough [77, 78]. They are making significant progress toward renewable-energy-derived electricity deployment, which will inevitably guarantee a more sustainable and environmentally beneficial energy structure, lower energy costs per unit and improve energy security in countries with rising energy consumption.

7.3 Opportunities

7.3.1 Utilising captured carbon

7.3.1.1 CO_2 in agri-food production

Not only can CCUS technologies be seen as having the potential to play a crucial role in mitigating climate change by capturing CO_2 emissions from industrial processes and power generation but also play an important role in increasing yields in industrial agriculture [2]. It is common knowledge that light, water and CO_2 are the most impactful resources required to increase crop yields in protected agriculture [80, 81]. Increasing the CO_2 content of ambient air and monitoring the photosynthesis rate were used as the yield index in CEPS application trials [78, 79]. The results were astonishing in that a more than 180% yield increase could be detected, at a balanced regulation cost of only about 51%. Although in the past, the ecological effects of dramatically increased atmospheric CO_2 levels as a result of anthropological and industrial activity were little understood, various studies have confirmed that abruptly elevated levels of CO_2 supplied to plants [82] have shown to increase yields in a variety of crops. Increased CO_2 concentration in the atmosphere as a result of industrial processes is thought to be modulating the response of plants to water and nitrogen, but unfortunately, the existing knowledge is not providing sufficient information for the fine management of water and nitrogen under elevated CO_2 levels in the atmosphere [83] and more research needs to be carried out in this area. A range of trials provided clear evidence [77–80, 84] that widespread implementation of the BiSC components and deployment of renewable energy technologies, and CCUS in combination with industrial agriculture is a way to ensure food security, besides the other positive effects of reaching net-zero CO_2 emissions and achieving greater energy independence in the short term.

7.3.1.2 CO_2 utilisation in the chemical industry

CO_2 utilisation pathways There are currently two categories of CO_2 valorisation: conversion of CO_2 into fuels and chemicals and direct utilisation of CO_2 [85]. The earliest utilisation of CO_2 and most widely known application is using CO_2 in the beverage industry to produce carbonated drinks. Additionally, CO_2 is used to manufacture dry ice, fire extinguishers, solvents, refrigerants and welding accessory (MIG) [82], or in industrial agriculture for enhancing plant growth. CO_2 is also used indirectly to modify the enhanced oil recovery (EOR) and enhanced gas recovery (EGR), and it is also used in enhanced geothermal systems (EGS) for HVAC application. In indirect chemical utilisation, CO_2 enhances the synthesis process and adds value. It needs to be emphasised that CO_2 utilisation in a range of applications in the chemical industry is classed as direct and indirect utilisation [82]. The sheer scale of anthropogenic CO_2 emissions generated is a critical issue and the world population is close to reaching tipping points, after which the effects on the natural environment are irreversible [86]. The sources of CO_2 are well known, but the capture, recycling and utilisation pathways are less defined or established. This is an industry in its infancy, and more research and development needs to be established to make sure that all industries adhere to the stipulations in the Paris Agreement [46] and create utilisation opportunities for waste and by-products. But the probably single most important factor is potentially the design stage, where every product to be developed needs to be designed with reusage and recyclability already built in. Products which do not fulfil this criterion should simply not be brought into circulation. At present, 40 national and 25 subnational jurisdictions have set carbon tax policies [87]. It could be argued that this does not go far enough, as only eight gigatonnes of CO_2e are affected by these policy implementations, which is equal to only slightly more than a quarter of global GHG emissions, whereas currently 37.15 billion tonnes of CO_2 are emitted annually. As CCU technologies are being researched and developed at a considerable pace, the opportunities for the development of CO_2-related policies and utilisation approaches in all industries provide ample opportunities for creating jobs and opportunities for investment at a breathtaking scale.

7.3.1.3 CO_2 in concrete production

Producing concrete releases a high level of greenhouse gas emissions, particularly during the manufacturing of the different materials required to produce concrete, such as cement. One metric tonne of cement releases ~0.8 t of CO_2 [79, 88–92].

As of 2020, the cement industry stood responsible for ca. 7% of global annual carbon emissions [86, 87, 93]. As discussed, there are already many companies dotted across the globe that use captured, high-purity CO_2 to generate RAC , and the use of which – at the same time – has the added positive effect of increasing the compressive strength and durability of concrete. Currently, three methods can be seen as the main approaches to the stages of using CO_2 in concrete production, as displayed in Table 7.2.

The three methods of sequestration mentioned before are injecting CO_2 in three different ways into cement-based materials, and they all have a different effect on the performance of cement-based materials. As an example: during the preparation of cement pastes, the $CaCO_3$ suspension over a wide range of concentrations was applied, as it directly incorporates nano-level $CaCO_3$ powders into the mix. $CaCO_3$ suspension has been found to improve the compressive strength of cement pastes, and the strength gains were increased congruently with the rise in $CaCO_3$ concentrations when the carbon injection (CI) method was used. In comparison with plain cement paste, CO_2-treated cement paste showed a 28-day compressive strength increase of 16% by the CI method, when a CO_2 concentration of 2% was used, and by 8% when the CP (pressurised carbon) method was used, at a CO_2 concentration of 4%. Additionally, the hydration heat kinetics and thermogravimetric analysis (TGA) both resulted in reports demonstrating that the hydration of cement paste was accelerated when it was prepared with $CaCO_3$ suspension. The CI method was found to generate a 5–10% higher hydration heat and 2–5% more hydration in the products, in comparison with the products manufactured using the CP method. It was also established that the microstructures of cement paste were densified when prepared with $CaCO_3$ suspension. Furthermore, the CI method also resulted in between 8% and >20% less porosity than the CP method. Overall, the CI method was found to be more efficient and economical, as it is less costly than the CP method, and the cement paste produced using the CI method also showed a more uniform particle size distribution, producing a higher-quality cement paste [79, 88–92].

7.3.1.4 CO_2 in glass manufacturing

Due to the high thermal energy use and the sheer volume of raw material required to produce glass [94], this industry and the manufacturing processes involved can be seen as difficult to abate in terms of GHG emissions. However, efforts have been made to achieve net zero in line with the Paris Agreement [46], and there are a variety of methods to reduce GHG emissions, such as CO_2 mineralisation, as this process captures CO_2 and converts it into a desired product. For this purpose, caustic NaOH (sodium hydroxide) is used as the absorbent, and as a result of a chemical reaction, a fraction of the resulting Na_2CO_3 (sodium carbonate) can be added back to the glass mixture effectively replacing soda ash. This is a very valuable aspect, as soda ash is a major component of glass manufacturing and a large source of direct and indirect GHG emissions. Therefore, combining glass manufacturing with CO_2 mineralisation seems like a feasible method to drastically reduce CO_2 emissions: CO_2 is captured and immediately converted into $NaHCO_3$ and Na_2CO_3 via absorption and electrolysis of sodium chloride into a NaOH

Table 7.2 Three methods of CO_2 sequestration in concrete production.

Method	Material	Contain	Generate	Positive effects	Negative effects
Sequestration of CO_2	Solid wastes containing CO_2: recycled concrete and steel slag	Oxides (CaO and MgO)	Carbonates; sulphate	Permanently locks CO_2	Reduces mechanical strength and chemical stability of some waste products; leaching of toxic chemicals into the environment.
Sequestration of CO_2	High concentration and pressurised CO_2		Precast fresh concrete; Dense carbonates on a thin surface of concrete		Costly; low mineralisation rate; the presence of formwork hinders diffusion of CO_2 into concrete
Sequestration of CO_2	Injects gaseous or liquid CO_2 into fresh concrete during mixing	Injected CO_2 dissolves in the mixing water and reacts with the cement to produce $CaCO_3$			Sequestration of CO_2 is low: <0.3% of the mass of cement; low reaction rate between CO_2 and alkali

solution. This process has the added benefit of generating the desirable by-products chlorine and hydrogen. The trials which based their work on mineralisation showed to generate net-negative CO_2 emissions, at between −152 and −528 kg CO_2/t of glass, in comparison with the unabated glass process. The amortisation rate for the CapEx required for the mineralisation ranges from 3.8 to 6.0 years.

When aluminium-combined window panels were chosen as a case trial [91], aspects such as the energy and CO_2 requirements for quantifying the impact of innovative solid-state recycling strategies for aluminium alloys were taken into consideration. It was found that glass-fibre-reinforced polymers have provided the best CapEx returns and reduction in environmental impact.

7.4 Conclusions

The technological challenges of the BiSC component implementation seem to be manifold, but the research required to compile this book has shown the following positive outcomes:

1) The technological challenges are mostly overcome through considerable efforts made in research and development
2) Many technologies are now at TRL9 or already in industrial application
3) In most industries, there seems to be a willingness to implement BiSC elements and any measures which aim to reduce GHG emission, to achieve a net-zero socio-economic situation
4) In many countries, a range of incentives, grants and subsidies is making it easier for entities willing to 'go green' and the ones who promote renewable energy technologies and carbon capturing or reducing greenhouse gas emissions are making it easier for all industries to overcome the individual technological challenges

It is therefore safe to say that there are no technological challenges which industry leaders cannot overcome and with the right support, implementation of the BiSC and carbon-free iron and steel production sounds not so far-fetched as it did in previous decades.

References

1 Elf, J.S., Espinosa, K.W., Martin, A., and Nemanova, V. (2017). Carbon capture and utilisation in the steel industry (Koldioxidåtervinning inom stålindustrin). https://www.diva-portal.org/smash/get/diva2:1201583/FULLTEXT01.pdf (accessed 15 February 2024).

2 Bao, J., Lu, W.H., Zhao, J., and Bi, X.T. (2018). Greenhouses for CO_2 sequestration from atmosphere. *Carbon Resources Conversion* 1 (2): 183–190.

3 Directorate-General for Research and Innovation (European Commission), Feiterna, A., Zagaria, A.M., et al. (2014). *ULCOS Top Gas Recycling Blast Furnace Process (ULCOS TGRBF): Final Report [Internet]*. Publications Office of the European Union. https://data.europa.eu/doi/10.2777/59481 (accessed 12 February 2024).

4 Birat, J.P. (2010). *ULCOS Program: Status & Progress*. In Brussels: European Commission; 1–30. https://www.eesc.europa.eu/sites/default/files/resources/docs/estep_ulcos_nov_2010.pdf (accessed 4 April 2020).

5 COM (European Commission). CORDIS | European Commission (2010). Ultra-low CO_2 steelmaking | ULCOS Project | Fact Sheet | FP6. https://cordis.europa.eu/project/id/515960 (accessed 10 July 2023).

6 Quader, A.M., Ahmed, S., Dawal, S.Z., and Nukman, Y. (2016). Present needs, recent progress and future trends of energy-efficient Ultra-Low Carbon Dioxide (CO_2) Steelmaking (ULCOS) program. *Renewable & Sustainable Energy Reviews* 1 (55): 537–549.

7 Johansson, M.T. and Söderström, M. (2011). Options for the Swedish steel industry: energy efficiency measures and fuel conversion. *Energy* 36 (1): 191–198.

8 Toktarova, A., Karlsson, I., Rootzén, J. et al. (2020). Pathways for low-carbon transition of the steel industry: a Swedish Case Study. *Energies* 13 (15): 3840.

9 Toktarova, A. (2021). *The Low-Carbon Steel Industry-Interactions Between the Hydrogen Direct Reduction of Steel and the Electricity System – ProQuest [Internet]*. Gothenburg, Sweden: Department of Space, Earth and Environment Chalmers University of Technology. https://www.proquest.com/openview/c2a4d77d71403fdbffd33f250b13251d/1?pq-origsite=gscholar&cbl=2026366&diss=y (accessed 7 July 2023).

10 Perpiñán, J., Peña, B., Bailera, M. et al. (2023). Integration of carbon capture technologies in blast furnace based steel making: a comprehensive and systematic review. *Fuel* 15 (336): 127074.

11 Shahabuddin, M., Brooks, G., and Rhamdhani, M.A. (2023). Decarbonisation and hydrogen integration of steel industries: recent development, challenges and technoeconomic analysis. *Journal of Cleaner Production* 395: 136391.

12 Ren, L., Zhou, S., and Ou, X. (2023). The carbon reduction potential of hydrogen in the low carbon transition of the iron and steel industry: the case of China. *Renewable & Sustainable Energy Reviews* 171: 113026.

13 WSA World Steel Association (2023). *Carbon Capture and Storage (CCS) [Internet]*, 2. Brussels: WSA https://worldsteel.org/wp-content/uploads/Carbon-Capture-Storage_2023.pdf.

14 Haghanimanesh, M., Baniasadi, E., Karimi Kerdabadi, J., and Yu, X. (2022). Exergoeconomic analysis of a novel trigeneration cycle based on steel slag heat recovery and biogas production in steelmaking plants. *Energy Conversion and Management* 263: 115688.

15 Hartfuß, G., Schmid, M., and Scheffknecht, G. (2020). Off-gas waste heat recovery for electric arc furnace steelmaking using Calcium hydroxide ($Ca(oH)_2$) dehydration. *Steel Research International* 91 (11): 2000048.

16 Iturralde, J., Gómez de Arteche, M., Aguirre, P. et al. (2019). *Radiant Waste Heat Recovery from Steelmaking and Glass Industry*, 29. Les Ulis: EDP Sciences.

17 Leadership Group for Industry Transition (LEADIT) (2023). Green steel tracker. https://www.industrytransition.org/green-steel-tracker/ (accessed 18 September 2023).

18 Tian, H. and Shu, G.Q. (2017). Organic Rankine Cycle – an overview | ScienceDirect Topics [Internet]. https://www.sciencedirect.com/topics/engineering/organic-rankine-cycle (accessed 20 February 2023).

19 Kildahl, H., Wang, L., Tong, L., and Ding, Y. (2023). Cost effective decarbonisation of blast furnace: basic oxygen furnace steel production through thermochemical sector coupling. *Journal of Cleaner Production* 20 (389): 135963.

20 Zhao, J., Li, B., Wei, X. et al. (2024). New green and low-carbon technology for all-sensible heat recovery of converter gas. *Journal of Cleaner Production* 20 (438): 140699.

21 Zhao, Q., Chu, X., Mei, X., et al. (2020). Co-treatment of waste from steelmaking processes: steel slag-based carbon capture and storage by mineralization. *Frontiers in chemistry [Internet]* 8. https://www.frontiersin.org/articles/10.3389/fchem.2020.571504/full (accessed 5 March 2021).

22 Pragya, S.A.A., Gupta, M.M., Shukla, A., and Atul. (2023). Waste heat recovery from industry through thermal energy storage mediums. In: *Thermal Energy*. CRC Press.

23 European Commission Access2Markets Carbon Border Adjustment Mechanism (CBAM) (2023). Carbon border adjustment mechanism (CBAM) | Access2Markets. https://trade.ec.europa.eu/access-to-markets/en/news/carbon-border-adjustment-mechanism-cbam (accessed 20 February 2024).

24 COM (European Commission) (2023). Circular economy: faster progress needed to meet EU resource-efficiency targets, ensure sustainable use of materials and enhance strategic autonomy [Internet]. https://environment.ec.europa.eu/news/circular-economy-faster-progress-needed-meet-eu-resource-efficiency-targets-ensure-sustainable-use-2023-05-15_en#:~:text=In%20recent%20years%2C%20circular%20economy,0.8%25%20of%20the%20EU's%20GDP (accessed 3 January 2023).

25 European Steel Association (EUROFER) (2023). Members of the European Steel Association [Internet]. https://www.eurofer.eu/about-steel/members/ (accessed 4 July 2023).

26 Tomka, S. (2023). *The Single Unified Field Theory, Wien's Law, Stefan Boltzmann's Law and Irradiance by Dr SKK Tomka [Internet]*. doi: 10.13140/RG.2.2.34191.66727/1 (accessed 20 February 2024).

27 Løvenskiold, A.C., Hu, X., Zhao, W., and Cherubini, F. (2022). Comparing the climate change mitigation potentials of alternative land uses: crops for biofuels or biochar vs. natural regrowth. *Geography and Sustainability* 3 (4): 347–357.

28 Ye, L., Zhang, J., Wang, G. et al. (2023). Feasibility analysis of plastic and biomass hydrochar for blast furnace injection. *Energy* 15 (263): 125903.

29 Mathieson, J.G., Rogers, H., Somerville, M.A., and Jahanshahi, S. (2012). Reducing net CO_2 emissions using charcoal as a blast furnace Tuyere injectant. *ISIJ International* 52 (8): 1489–1496.

30 Heinrich, T., Park, H., Orozco, R. et al. (2023). Biochar production from late-harvest grass: challenges and potential for farm-scale implementation. *Sustainable Production and Consumption* 1 (37): 256–267.

31 Teagasc (2017). Miscanthus energy crop – Teagasc | Agriculture and Food Development Authority [Internet]. https://www.teagasc.ie/rural-economy/rural-development/diversification/miscanthus-energy-crop/ (accessed 21 February 2024).

32 Wang, G., Li, R., Dan, J. et al. (2023). Preparation of biomass hydrochar and application analysis of blast furnace injection. *Energies* 16 (3): 1216.

33 Wiklund, C.M., Helle, M., and Saxén, H. (2016). Economic assessment of options for biomass pretreatment and use in the blast furnace. *Biomass & Bioenergy* 1 (91): 259–270.

34 Suopajärvi, H., Pongrácz, E., and Fabritius, T. (2014). Bioreducer use in Finnish blast furnace ironmaking: analysis of CO_2 emission reduction potential and mitigation cost. *Applied Energy* 1 (124): 82–93.

35 Verde Agritech (2023). The top 10 carbon capture technologies explained – Verde AgriTech – Blog (english) [Internet]. https://blog.verde.ag/en/top-10-carbon-capture-tech/ (accessed 20 February 2024).

36 Yurcent Environmental Technology Co., Ltd (2023). Zeolite Rotor-Product-Xi'an Yurcent Environmental Technology Co., Ltd. [Internet]. https://en.yurcent.com/Products_3/20.html (accessed 7 July 2023).

37 Gammon, K. (2023). Suck carbon from the air? US facility launches novel climate solution. *The Guardian [Internet].* https://www.theguardian.com/environment/2023/nov/19/carbon-dioxide-direct-air-capture (accessed 21 February 2024).

38 Yamada, H. (2021). Amine-based capture of CO_2 for utilization and storage. *Polymer Journal* 53 (1): 93–102.

39 Gugliuzza, A. and Basile A. (2014). 3 – Membrane processes for biofuel separation: an introduction. In: Gugliuzza, A. and Basile A, editors. *Membranes for Clean and Renewable Power Applications [Internet].* Woodhead Publishing; p. 65–103. https://www.sciencedirect.com/science/article/pii/B9780857095459500033 (accessed 22 February 2024).

40 Blue Planet (2024). Blue Planet systems: technology. https://www.blueplanetsystems.com/technology (accessed 22 February 2024).

41 Chakrabortty, S., Kumar, R., Nayak, J. et al. (2023). Green synthesis of MeOH derivatives through in situ catalytic transformations of captured CO_2 in a membrane integrated photo-microreactor system: a state-of-art review for carbon capture and utilization. *Renewable & Sustainable Energy Reviews* 1 (182): 113417.

42 Lefvert, A. and Grönkvist, S. (2024). Lost in the scenarios of negative emissions: the role of bioenergy with carbon capture and storage (BECCS). *Energy Policy* 1 (184): 113882.

43 Wang, E., Zhu, Z., Li, R. et al. (2024). Ni/CaO-based dual-functional materials for calcium-looping CO_2 capture and dry reforming of methane: progress and challenges. *Chemical Engineering Journal* 15 (482): 148476.

44 He, T., Si, B., Gundersen, T. et al. (2024). Integrated ethane recovery and cryogenic carbon capture in a dual mixed refrigerant natural gas liquefaction process. *Energy* 1 (290): 130125.

45 Hu, L., Bui, V.T., Esmaeili, N., and Lin, H. (2024). Nanoengineering membrane surfaces: a new paradigm for efficient CO_2 capture. *Carbon Capture Science & Technology* 1 (10): 100150.

46 UN United Nations (2016). Paris Agreement 2016 [Internet]. https://www.un.org/en/climatechange/paris-agreement#:~:text=To%20tackle%20climate%20change%20and,2015%3A%20the%20historic%20Paris%20Agreement (accessed 2 March 2020).

47 Gonzales, V., Krupnick, A., Dunlap, L., and Resources for the Future (2024). Carbon Capture and Storage 101. https://www.rff.org/publications/explainers/carbon-capture-and-storage-101/ (accessed 22 February 2024).

48 Ladenburg, J., Kim, J., Zuch, M., and Soytas, U. (2024). Taking the carbon capture and storage, wind power, PV or other renewable technology path to fight climate change? Exploring the acceptance of climate change mitigation technologies: a Danish national representative study. *Renewable Energy* 1 (220): 119582.

49 Otto, D. (2023). Public perception of CO_2 capture and storage | Helmholtz Climate Initiative [Internet]. https://www.helmholtz-klima.de/en/aktuelles/public-perception-co2-capture-and-storage (accessed 23 February 2024).

50 Pörtner, H.O., Roberts, D.C., Tignor, M.M.B. et al. (ed.) (2022). *Climate Change 2022: Impacts, Adaptation and Vulnerability*. Contribution of Working Group II to the Sixth Assessment Report of the Intergovernmental Panel on Climate Change.

51 Ahmad, M. (2023). Top 10: leading carbon capture companies [Internet]. https://energydigital.com/top10/top-10-leading-carbon-capture-companies (accessed 20 February 2024).

52 UK Science Museum (2024). Can carbon capture technology help address the climate crisis? | Science Museum [Internet]. https://www.sciencemuseum.org.uk/objects-and-stories/can-carbon-capture-technology-help-address-climate-crisis (accessed 23 February 2024).

53 Heirloom (2023). Heirloom [Internet]. https://www.heirloomcarbon.com/ (accessed 5 December 2023).

54 Reuters Climate tech company Heirloom opens US commercial carbon capture plant | Reuters [Internet] https://www.reuters.com/sustainability/climate-energy/climate-tech-company-heirloom-opens-us-commercial-carbon-capture-plant-2023-11-09/ (accessed 23 February 2024).

55 Lee, Y.Y. (2023). *Composite Materials of Reactive Ionic Liquids for Selective Separation of CO_2 at Low Concentration [Internet]*. Case Western Reserve University. https://etd.ohiolink.edu/acprod/odb_etd/etd/r/1501/10?clear=10&p10_accession_num=case1664981643517872 (accessed 24 February 2024).

56 Fasihi, M. and Efimova, O. (2019). Techno-economic assessment of CO_2 direct air capture plants. *Journal of Cleaner Production* 1 (224): 957–980.

57 Rezaei, S., Liu, A., and Hovington, P. (2023). Emerging technologies in post-combustion carbon dioxide capture & removal. *Catalysis Today* 1 (423): 114286.

58 Chowdhury, S., Kumar, Y., Shrivastava, S. et al. (2023). A review on the recent scientific and commercial progress on the direct air capture technology to manage atmospheric CO_2 concentrations and future perspectives. *Energy & Fuels* 37 (15): 10733–10757.

59 SAIPEM (2023). Our commitment to being sustainable energy company | Saipem [Internet]. https://www.saipem.com/en/about-us (accessed 27 February 2024).

60 LanzaTech (2023). LanzaTech – recycling carbon with biology [Internet]. https://lanzatech.com/ (accessed 27 February 2024).

61 Carbon Clean (2024). About us | Carbon Clean [Internet]. https://www.carbonclean.com/about-us (accessed 27 February 2024).

62 Slb Capturi (2024). Aker carbon capture. https://akercarboncapture.com/ (accessed 27 February 2024).

63 Carbon Engineering (2024). Direct air capture technology. https://carbonengineering.com/our-technology/ (accessed 27 February 2024).

64 Carbon Free (2023). CarbonFree. https://carbonfree.cc/ (accessed 27 February 2024).

65 Carbfix (2021). Our story [Internet]. https://www.carbfix.com/our-story (accessed 27 February 2024).

66 NeoCarbon (2024). NeoCarbon [Internet]. https://www.neocarbon.tech/ (accessed 27 February 2024).

67 Meloon(2024). Mission – Meloon [Internet]. https://meloon.de/mission/ (accessed 26 February 2024).

68 EcoLocked (2024). Don't waste the waste: sustainable biomass sourcing for biochar carbon removal [Internet]. https://www.ecolocked.com/blog-posts/waste-biomass (accessed 27 February 2024).

69 Sodiq, A., Abdullatif, Y., Aissa, B. et al. (2023). A review on progress made in direct air capture of CO_2. *Environmental Technology and Innovation* 1 (29): 102991.

70 Rana, A.K., Mostafavi, E., Alsanie, W.F. et al. (2023). Cellulose-based materials for air purification: a review. *Industrial Crops and Products* 1 (194): 116331.

71 Rogers, G.N.D. and Whitehead, D. (1998). Design of a three-stage filtration system to remove contaminant sulphur compounds from a CO_2-enriched gas supply for use in long-term tree growth experiments. *Environmental Technology* 19 (1): 103–107.

72 Saqib, N., Usman, M., Ozturk, I., and Sharif, A. (2024). Harnessing the synergistic impacts of environmental innovations, financial development, green growth, and ecological footprint through the lens of SDGs policies for countries exhibiting high ecological footprints. *Energy Policy* 1 (184): 113863.

73 Tansel Tugcu, C. and Menegaki, A.N. (2024). The impact of renewable energy generation on energy security: evidence from the G7 countries. *Gondwana Research* 1 (125): 253–265.

74 Kiessling, S., Darabkhani, H.G., and Soliman, A.H. (2022). Energies | free full-text | the bio steel cycle: 7 steps to net-zero CO_2 emissions steel production [Internet]. https://www.mdpi.com/1996-1073/15/23/8880 (accessed 11 July 2023).

75 Kiessling, S., Gohari Darabkhani, H., and Soliman, A.H. (2024). Greater energy independence with sustainable steel production. *Sustainability.* 16 (3): 1174.

76 Huda, A., Kurniawan, I., Purba, K.F. et al. (2024). Techno-economic assessment of residential and farm-based photovoltaic systems. *Renewable Energy* 1 (222): 119886.

77 Zheng, M. and Wong, C.Y. (2024). The impact of digital economy on renewable energy development in China. *Innovation and Green Development* 3 (1): 100094.

78 Chen, Q., Gu, Y., Tang, Z. et al. (2018). Assessment of low-carbon iron and steel production with CO_2 recycling and utilization technologies: a case study in China. *Applied Energy* 220: 192–207.

79 Hussain, M.N., Zaman, M.R., Halim, M.A. et al. (2024). A comprehensive review of renewable and sustainable energy sources with solar photovoltaic electricity advancement in Bangladesh. *Control Systems and Optimization Letters* 2 (1): 1–7.

80 Gao, P., Lu, M., Yang, Y. et al. (2024). Greenhouse light and CO_2 regulation considering cost and photosynthesis rate using i-nsGA II. *Expert Systems with Applications.* 1 (237): 121680.

81 Gao, X., Zhang, R., You, Z. et al. (2022). Use of hydrogen–rich gas in blast furnace ironmaking of V–bearing titanomagnetite: mass and energy balance calculations. *Materials (Basel)* 15 (17): 6078.

82 Yang, K., Huang, Y., Yang, J. et al. (2024). Do rice growth and yield respond similarly to abrupt and gradual increase in atmospheric CO_2? *Science of the Total Environment* 1 (906): 167658.

83 Cai, Z., Bai, J., Li, R. et al. (2023). Water and nitrogen management scheme of melon based on yield–quality–efficiency matching perspective under CO_2 enrichment. *Agricultural Water Management* 1 (285): 108379.

84 Taghizadeh-Hesary, F., Vandercamme, L., and Phoumin, H. (2024). Enhancing the Economic Feasibility of Carbon Capture, Utilisation, and Storage (CCUS) Projects. *Journal of Environmental Assessment Policy and Management* 6: 2350024.

85 Rafiee, A., Rajab Khalilpour, K., Milani, D., and Panahi, M. (2018). Trends in CO_2 conversion and utilization: a review from process systems perspective. *Journal of Environmental Chemical Engineering* 6 (5): 5771–5794.

86 Armstrong McKay, D., Lade, S., and Rockström, J. (2023). New report: tipping point threats and opportunities accelerate [Internet]. https://www.stockholmresilience.org/research/research-news/2023-12-06-new-report-tipping-point-threats-and-opportunities-accelerate.html (accessed 28 February 2024).

87 UNFCCC (2024). About carbon pricing | UNFCCC [Internet]. https://unfccc.int/about-us/regional-collaboration-centres/the-ciaca/about-carbon-pricing (accessed 28 February 2024).

88 Liu, Z., Du, J., and Meng, W. (2022). Achieving low-carbon cementitious materials with high mechanical properties using $CaCO_3$ suspension produced by CO_2 sequestration. *Journal of Cleaner Production* 1 (373): 133546.

89 European Cement Research Academy (ECRA) (2009). *ECRA CCS Project and Report About Phase II [Internet].* Düsseldorf: ECRA. https://ecra-online.org/fileadmin/redaktion/files/pdf/ECRA__Technical_Report_CCS_Phase_II.pdf (accessed 4 March 2020).

90 European Cement Research Academy (ECRA) (2009). Development of state of the art-techniques in cement manufacturing: trying to look ahead [Internet]. https://ecra-online.org/fileadmin/redaktion/files/pdf/CSI_ECRA_Technology_Papers_2009.pdf (accessed 2 January 2021).

91 Global Cement and Concrete Association (GCCA) (2021). Global cement and concrete industry announces roadmap to achieve groundbreaking 'net zero' CO_2 emissions by 2050 – includes milestone commitment to cut CO_2 emissions by a quarter by 2030 [Internet] https://gccassociation.org/news/global-cement-and-concrete-industry-announces-roadmap-to-achieve-groundbreaking-net-zero-co2-emissions-by-2050/#:~:text=%28London%3A%2012%20October%202021%29%20Forty%20of%20the%20world%E2%80%99s,the%20race%20to%20%E2%80%98Net%20Zero%E2%80%99%20concrete%20by%202050 (accessed 1 March 2022).

92 Driver, J.G., Paul, T.H.P.H.M.S., and Fennel, S. (2022). Simulation of direct separation technology for carbon capture and storage in the cement industry. *Chemical Engineering Journal* (449): 1385–8947.

93 Gale, W.F. and Totemeier, T.C. (2004). Smithells metals reference book | ScienceDirect [Internet]. https://www.sciencedirect.com/book/9780750675093/smithells-metals-reference-book (accessed 28 February 2024).

94 Caudle, B., Taniguchi, S., Nguyen, T.T.H., and Kataoka, S. (2023). Integrating carbon capture and utilization into the glass industry: economic analysis of emissions reduction through CO_2 mineralization. *Journal of Cleaner Production* 1 (416): 137846.

8

Macroeconomic and microeconomic challenges to implementation of the BiSC concept

8.1 Introduction

This chapter includes outlines of the significance of implementation hurdles, such as the political will of the major steel-producing countries, a UK focus with attention directed towards the issues that require addressing to achieve successful implementation of the innovative concept. The political and legal framework, adherence to the Paris Agreement [1], sustainability goals and amortisation budgets as well as other industry-specific financial implications are being considered. Together, these factors shape the broader context highlighting both the challenges and the potential pathways to successful implementation of the Bio Steel Cycle (BiSC).

8.2 Part I policy and BiSC implementation

Politics in any country in recent years seem only to serve the purpose of either (i) keeping those currently in government in place or (ii) creating a legal and political situation where those on the far right of the political spectrum lobby and influence those in power to achieve their own objectives. Democracy is at risk in many developed countries [2, 3]. Therefore, the need for sustainability and greenhouse gas (GHG) emissions reduction is insofar only recognised and acted upon as far as it is serving the political parties' agendas. However, the outlook is not generally bleak: the general populace and free industry have recognised the urgent need for immediate action, and there are a lot of efforts on the way to globally reduce GHG emissions [1, 4–10] and ensure the world economy achieves remaining within the 1.5 °C temperature rise limit in comparison with pre-industrial temperatures. To that effect, sustainability certifications have been developed, which provide the framework for the industry to act within and responsibly respond to the challenges at hand.

8.2.1 CO_2 mitigation and sustainability adaptation – US' evolving policy framework

The Intergovernmental Panel on Climate Change (IPCC) 2022 reports clearly stated the dire state of our climate and what we as a species are able to do to avert

Circular Steel Production: Pathways to Net-Zero Carbon Emissions, First Edition.
Sandra Kiessling, Hamidreza Gohari Darabkhani, and Abdel-Hamid Soliman.

climate disaster, in particular the acceleration of reducing carbon emissions [11]. In the United States, several policies and guidelines support and encourage the use of carbon capture and storage (CCS) technology. As an example, national tax credits have been instated for carbon sequestration in section 45Q of the Internal Revenue Code (IRC). Further incentives and crediting mechanisms have been created for California, Texas, Louisiana, Montana and North Dakota. The Bipartisan Infrastructure Law from 2021 (Infrastructure Investment and Jobs Act) ring-fenced $5.7 billion of funding for infrastructure projects. The Build Back Better Act [12], introduced in 2021 by the congressional Democrats, increased the value of the tax credit associated with 45Q by increasing the subsidy rate from $50/tonne (American measurement = 0.91 metric tonnes) of captured carbon dioxide (CO_2) to ca. $85/tonne. The proposal set out clearly the minimum carbon capture requirement for existing production sites and any plant capturing less than 75% of its emissions would be ineligible for the tax credits [12], which makes this proposal interesting to key decision makers who are still opposed to the phase-out of fossil fuels, due to vested interests.

8.2.2 CO_2 mitigation and sustainability adaptation – UK's evolving policy framework

8.2.2.1 The UK Government's Green Industrial Revolution

Steel is one of the most abundantly used materials worldwide [13]. Carbon dioxides are the so-called GHGs, which trap heat by building a gaseous barrier between the Earth's atmosphere and space and the subsequently trapped heat makes the planet warmer [14, 15]. This very principle was investigated, and it was established that anthropological activities and, specifically, the combustion of fossil fuels for electricity, heat and transportation can be considered solely responsible for the increased GHGs in the atmosphere for the time span of the last 170 years [15–18].

The six sectors identified in the UK government's Ten-Point-Plan [19] for being the most polluting, ordered by pollution intensity from highest to lowest, are as follows:

- Energy/Electricity
- Transportation
- Industry
- Commercial/Residential
- Agriculture
- Land Use/Forestry

The UK government seems to be proactively trying to reduce GHG emissions by issuing suitable guidance papers and policies, such as the 'Ten-Point-Plan for a Green Industrial Revolution' [19]. The ten points in question are dealing with the following issues:

8.2.2.2 Offshore wind

Quadrupling offshore production to achieve 40 GW of electric power by 2030 and powering every residential building whilst additionally supporting up to 60,000 jobs.

Planned investment of £160 million into modern ports and manufacturing infrastructure, with the delivery of 60% UK content in offshore wind projects. 5% reduction in carbon emissions.

8.2.2.3 Hydrogen

Aiming to generate 5 GW of low-carbon hydrogen-produced electricity capacity by 2030 for industrial and residential applications and having the first town heated entirely by hydrogen power by the end of the decade. This is being supported by a £240 million Net Zero Hydrogen Fund, whilst encouraging private sector investment and endeavour to create 8000 jobs across all industries. 9% reduction in carbon emissions.

8.2.2.4 Nuclear

Developing the next generation of small and advanced clean energy reactors, potentially supporting up to 10,000 jobs. £385 million already set aside in the Advanced Nuclear Fund, with £215 million investment already earmarked for the development of small modular reactors, whilst relying on unlocking up to £300 million private sector match funding. Increasing carbon emissions by building nuclear infrastructure.

8.2.2.5 Electric vehicles

Backing car manufacturers in the United Kingdom to accelerate the transition to electric vehicles and transforming the national infrastructure to support electric vehicles. Up to £1 billion committed to support the electrification of UK vehicle supply chains, with the first £500 million of investment being announced within the lifetime of this Parliament to drive the electrification of the UK automotive sector. Provision of £582 million to extend the Plug-in Car, Van, Taxi and Motorcycle grants to 2022–2023. Investment of £20 million next year in freight trials to pioneer zero-emission lorries. Endeavour to save 5 $MtCO_2$ emissions by 2032. 1.1% reduction in carbon emissions.

8.2.2.6 Public transport

Investment in zero-emission public transport and promoting cycling and walking. £4.2 billion have been set aside for city public transport and £5 billion for buses and to encourage cycling, whilst also electrifying more railway lines. Investment of £120 million in 2021 to begin the introduction of at least 4000 more British-built zero-emission buses. 0.44% reduction in carbon savings.

8.2.2.7 Jet zero

Supporting difficult-to-decarbonise industries through *research projects* for zero-emission planes and ships. £15 million was set aside for FlyZero, a 12-month *study*, into the strategic, technical and commercial issues in designing future zero-emission aircraft, and currently running a £15 million *competition* to support the production of sustainable aviation fuels (SAF) in the United Kingdom. 0.22% reduction in CO_2 emissions.

8.2.2.8 Homes and public buildings

Providing £1 billion to install 600,000 heat pumps every year by 2028, creating 50,000 jobs by 2030 and making residential buildings, schools and hospitals more energy efficient. The Green Homes Grant will be extended into 2022, homeowners can apply for the Homes Upgrade Grant, and funding for the social housing decarbonisation and Public Sector Decarbonisation Scheme will be provided. 16% reduction in carbon emissions.

8.2.2.9 Carbon capture

Becoming a world leader in CCS, with a target to remove 10 MT of CO_2 by 2030. The £1 billion Carbon Capture, Utilisation and Storage(CCUS) Infrastructure Fund will provide the industry with the certainty required to deploy CCUS at pace and at scale. 9% reduction in carbon emissions.

8.2.2.10 Nature

Protecting and restoring the natural environment and planting 30,000 hectares of trees every year, whilst creating and retaining thousands of jobs. The Green Recovery Challenge Fund has been increased to £80 million over 100 nature projects to be delivered over the next two years, with an additional 1.5% of AONB (area of outstanding natural beauty) natural land in England, contributing to achieve the target of protecting 30% of UK land by 2030.

8.2.2.11 Innovation and finance

Developing the technologies to make the City of London the global centre of green finance. The first phase of a £100 million investment has already been launched, applying state-of-the-art GHG removal technology, such as direct air capture. Additionally, £100 million will be provided for energy storage and flexibility innovation challenges. Additionally, the commercialisation of fusion energy technology is being supported, by providing £222 million for the visionary Spherical Tokamak for Energy Production (STEP) programme, on the grid in the United Kingdom by 2040, and £184 million for new fusion facilities, infrastructure and apprenticeships to lay the foundations of a global hub for fusion-related innovation in the United Kingdom.

In addition to these efforts, a multitude of projects and policies were designed to provide the interested reader with a road map for a sustainable future ahead.

8.2.2.12 Findings

The 10-point plan and all the efforts within amount to a total of only 40.76% in carbon emission savings projected to have been achieved by 2030, notwithstanding the unfortunate issue that nuclear power plant creation will actually be *adding* an incalculable amount of GHG and CO_2 emissions, through construction, building and operation of required infrastructure – besides the known risks of operating nuclear power plants, as demonstrated by the Chernobyl disaster and the potential for illegal weaponising of by-products, such as plutonium.

At first glance, it so seems as if the three most polluting sectors in the United Kingdom are not sufficiently focused on within the UK Green Industrial Revolution agenda, with most CO_2-saving opportunities in industrial processes, transport/

aviation, residential dwellings almost and (animal) agriculture completely being left out. It was established [15–17, 20–22] that animal agriculture, such as beef production, accounts for >25% of all three sectors such as global land use, land-use change and forestry emissions – and these issues do not appear to have been factored into the 10-point agenda. Especially for residential dwellings, there are various concepts available, which could contribute to significant CO_2 savings: with the German 'Passivhaus' [23], CO_2 from residential dwellings could possibly be reduced by more than 85%.

As far as agriculture is concerned, the implementation of robust guidance and financial incentives for sustainable farming could revolutionise UK farming within the next five years:

- Making it compulsory for every meat and dairy producer to be assigned to the 'Good farming protocol' [24] and to have an anaerobic digester (AD) unit on site for recycling organic waste. Support farmers' efforts with grants to build AD's and support them with mentoring, i.e. technical support, to build the infrastructure on site to utilise the biogas as a by-product of anaerobic digester operation.

Farm effluents such as slurry and organic waste would then not just be disposed of on existing farmland, endangering the freshwater table and increasing the nitrogen levels in natural water courses, but could be utilised as biomass and a valuable resource for energy production.

The required technical support and expertise would create long-term, sustainable jobs for workers who have so far been employed in industries, which have been heavily relying on fossil fuels.

With regard to point nine of the UK Green Industrial Nature – forests and green spaces, a massive opportunity has been missed: one single hardwood tree (such as oak, beech and elm) [20, 21] sequesters up to 1 t/CO_2 in a 40-year lifetime [25, 26].

Interestingly, the UK government's Forest Research group found in 2021 that currently only 13% of available land has tree cover in the United Kingdom, estimated to be 3.21 million hectares, as confirmed by the Woodland Trust [27]. This represents 10% in England, 15% in Wales, 19% in Scotland and 9% in Northern Ireland. Interestingly, most of the UK's direct continental neighbours have an average tree cover of 37% [28]. We can rebuild natural forests as carbon sinks very easily by paying farmers to do so during environmental stewardship schemes [25, 26]. The 'Climate Action Tracker' has deemed the UK policies for net zero as insufficient and climate finance as highly insufficient [29].

Multiple areas with room for improvement have been identified within the UK Green Industrial Revolution agenda, which will be demonstrated in the following.

8.2.2.13 Research gaps

Hydrogen (H) In the 1960s, hydrogen was developed by a range of companies, and General Motors was one of the first ones to build a hydrogen-powered car back in the 1960s [30]. Hydrogen produced based on fossil fuels generates 9.3 kg CO_2, which is being produced per kg of hydrogen [31]. Hydrogen has just the same destructive energy balance as combustion engine fuel, which produces 9.1 kg of CO_2 per 5 l (equivalent to 1 gallon). Due to an abundance of base material (water), hydrogen generation has received increased attention in recent years [32], particularly since the Paris Agreement [1] was signed. Support from signee governments, global industry [33] and the public focused attention on green and sustainable renewable energy sources, which positioned green hydrogen generation in the spotlight. Global players and industry leaders such as Tata Steel are investing substantial sums into green steel production, such as at their hydrogen by electrolysis and hydrogen direct reduction plant in Ijmuiden [33]. As our planet consists of 80% water (H_2O), there is an abundance of resources for the production of green hydrogen, and a great proportion of all research and development efforts should be focused on this area of energy generation.

Solar and wind energy However, solar and wind energy are even more abundantly available forms of energy than water, and here is the opportunity: hydrogen produced using solar energy could potentially be a more sustainable, future-proof solution [34, 35] for energy security and decarbonisation of heavy industry, which marks out the need for further research into this direction.

Nuclear Besides the negative CO_2 balance of the plant construction, the threat of nuclear catastrophe is real, as Chernobyl has taught us [36, 37]. Additionally, problematic by-products, such as weapon-worthy plutonium, could be obtained and used for illegal purposes. Solar, wind and hydropower energy are abundantly available in the United Kingdom, and an additional nuclear power plant is neither in keeping with efforts to reduce carbon emissions, renewable energy currently being achieved and potential for achieving 100% renewable energy within the next four years, nor in fulfilling any government's duty to keep their citizens safe.

Electric vehicles Currently, there seems to be the belief that a large-scale rollout of electric vehicles is the answer to the decarbonisation of transport. Looking into the details of the transport sector, it has become clear that this might be a total misconception, as volatile organic compounds (VOCs) from road transport and fossil fuels and by-products have yet to be included in the emissions factors of the transport sector [38]. Emissions measurements retrieved from mobile air quality monitoring stations revealed that alcohol-based compounds and other matters in road transport caused additional emissions in inventories at 39–58 mg per vehicle and kilometre. The use of ancillary solvent products, such as screen wash, additives and de-icer, and the emissions caused by their production and during their use have yet to be included in combustion engine cars and electric vehicles alike. These emissions are independent of vehicle engine systems and are therefore applicable to all road vehicle types – this includes fossil-fuel-based engine-driven vehicles as much as battery- and energy-storage-cell-based electric vehicles and power trains. With increased necessary social mobility, vehicle VOC emissions are likely to increase, given a predicted growth in total vehicle kilometres driven [38].

Public transport and jet zero Road transport is one of the fastest growing sources of GHG emissions and is driving global climate change at an accelerating pace [39]. A change in thinking is required: meetings can easily be conducted online, via MS Teams, ZOOM and other means. Even teaching is possible online, i.e. via Google Classroom; it is not necessary to commute unsociable distances on a daily basis, gathering with people we are forced to socialise with, which we would not choose to do so given half the chance. Aviation is one of the biggest polluters in existence, and drastic measures are required: driving and flying less means improving public transport infrastructure, reducing road traffic, reducing the number of airlines and training airline staff to work in green industries, as set out by the UK government.

Aviation accounts for 2.5% of global CO_2 emissions [40]. Unfortunately, besides a drive to further develop hydrogen-based public transport (with hydrogen production largely based on using fossil fuels) and more environmentally friendly aviation fuel studies and conceptualisations, the UK Green Industrial Revolution agenda falls short of proposing real change.

Many authors and scientists [5–9, 40–42] clearly identified the mechanisms with which the GHGs are interfering with our climate. Elaborate research has resulted in clearly identified main drivers within this complex interconnected scenario, that is aviation. What really needs to happen is a lifestyle change, a change in attitude: attend meetings and educational events online wherever possible, cut down on commuting for workers across all industries and eradicate the use of combustion engines within the next five years. This change could happen easily if fossil fuel subsidies were being diverted to incentivise drivers to switch to electric vehicles and planes with environmentally friendly jet engines, and radically subject public transport to a green overhaul – away from fossil fuels, using renewable energy technology only.

Missed opportunity 1: Agriculture GHG emissions from animal agriculture are 25% of global land use/land-use change and forestry emissions driven by beef production, including the conversion of forests in the Brazilian Amazon [22, 43]. Stipulating very clearly in governmental guidelines and laws, the compulsory requirement to install AD at any animal production site is a vital step to eliminate between 11 and 18% of global CO_2 emissions from agriculture. An additional benefit would be the production of biogas, which can be used in its raw form as fuels and used in its refined form in the National Grid. This would be an almost carbon-neutral energy generation concept, and it would support the drive towards energy independence and decarbonisation of the energy sector.

Missed opportunity 2: Land Use/Forestry One of the main causes of deforestation is agriculture (poorly planned infrastructure is emerging as a big threat too), and the main cause of forest degradation is illegal logging [44, 45]. In 2019, the tropics lost close to 30 football fields' worth of trees every single minute [46]. Deforestation is a particular concern in tropical rainforests because these forests are home to much of the world's biodiversity. For example, in the Amazon, around 17% of the forest has been lost in the last 50 years, mostly due to forest conversion for cattle ranching. Reforestation is one of the main components in the author's 'BiSC', which requires

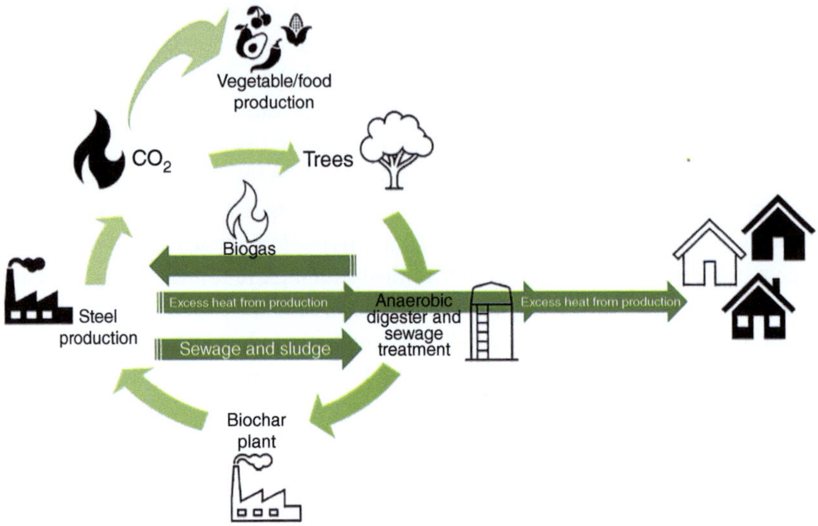

Figure 8.1 Simplified view of the Bio Steel Cycle.

the creation of a political and legal framework and possibly governmental intervention in the form of subsidies and incentives.

Missed opportunity 3: Industry There is a range of industries where the concepts developed by the author could be applied. One industry, which has been identified as one of the biggest emitters of GHG, is the steel industry. The author has developed the BiSC, which could easily be adapted to any industry – which provides scope for further research. Figure 8.1 will provide a simplified overview of the Bio Steel Cycle.

The basic principle is to apply circular economic processes, as the aim is to waste nothing and reuse or recycle most by-products. The underlying concept is that, already at the design stage, all products need to be produced with total reusability and recyclability in mind. If a product cannot be reused or recycled, it should not be made.

Recommendations Based on the findings of this work, a range of recommendations can be made as follows:

- Increasing offshore wind farms is highly commendable, but we should not stop short here: wind, hydro and solar energy are available in abundance and should all be utilised more and replace all fossil-fuel-based energy generation
- If hydrogen technology is to be utilised, it should be illegal to produce it using fossil fuels – which is mostly the case, at present. Instead, hydrogen produced by utilising renewable energy resources should be made compulsory
- Nuclear power plants are entirely unnecessary and bear unfathomable risks to mankind – any development in this direction should be stopped immediately, and the remaining resources should be invested into widening the renewable energy network

- The CO_2 emissions within transport and aviation can be made CO_2 neutral within a five-year window – which provides room for further research. Driving and flying less is key and also makes fossil fuel technologies redundant
- Buildings and industrial processes, existing infrastructure and future developments can easily achieve high amounts of CO_2 savings, facilitating the author's 'BiSC'
- Reforestation in the United Kingdom to a higher percentage than the current 13.2% of tree cover to the European geographical average of 39% can easily be achieved by working together with a range of industries and the agricultural industry, via subsidies, environmental stewardship schemes and reforestation incentives such as the Woodland Creation Grant and the Woodland Trust support. Each tree is able to absorb up to 1 t/CO_2 in a 40-year lifetime – with the highest absorption rate achieved once maturity has been reached. Bamboo species are able to absorb up to 17 t per hectare per year [47, 48]

All these mechanisms are already known, and we have the technology at our fingertips – all it needs now is the political willingness to do what is necessary to achieve a net-zero economy – for the survival of our planet and mankind.

8.2.3 CO_2 mitigation and sustainability adaptation – Brazil's evolving policy framework

Environmental Protection and Climate Change Policy Structure in Brazil – The National Policy on Climate Change (PNMC)

The PNMC, constituted through law No. 12.187/2009, is one of the main guidelines issued by the Brazilian government. It is instrumental for the implementation of environmental protection and reducing the effects of the looming climate catastrophe, with the objective of reducing the emissions of GHG at ca. 36.1–38.9%, based on 2020 emissions [49]. The PNMC is constructed of four main pillars:

i) Mitigation opportunities
ii) Impacts, vulnerabilities and adaptation
iii) Research and development
iv) Education, training and communication

The PNMC provides specific actions for each pillar and defines mitigation with a focus on arboreal businesses, agricultural businesses, energy generation, waste recycling and elimination, decarbonising construction, industry and transport, with CAT and CCUS technological changes. Additionally, the focus is on human and natural systems where everyone has a duty to act in the best interest of the natural environment, for the benefit of present and future generations, to reduce climate impacts. As Brazilian GHG emissions are mainly from deforestation, they therefore differ in their quantity of emissions in comparison with other countries; therefore, the efforts and guidelines of the PNMC focus strongly on controlling deforestation and encouraging reforestation.

Through the PNMC, Decree No. 9.578 contains action plans for the prevention of and controlling deforestation in specified habitats and sectors for mitigation of emissions and adaption to a changed climate:

I) Action Plan for the Prevention and Control of Deforestation in the Legal Amazon (PPCDAm)
II) Action Plan for the Prevention and Control of Deforestation and Burning in the Cerrado (PPCerrado)
III) Decennial Plan for the Expansion of Energy (PDE)
IV) Sectoral Plan for Mitigation and Adaptation to Climate Change for the Consolidation of a Low-Carbon Economy in Agriculture (ABC Plan)

It is especially tragic that the GHG emissions from the sector of change of use and use of land and forests grew between 1990 and 2004, caused by deforestation surrounding and in the Amazon and the Cerrado. The new Brazilian government under Lula is determined to halt deforestation and implement every measure that helps to support these plans [50].

8.2.4 CO_2 mitigation and sustainability adaptation – Russia's evolving policy framework

It could be argued that Russia is not prepared to or even interested in addressing the current and impending crises [51]. Russia's official scientific institutions are actively working on solutions to combat climate change, but their influence on implementing change is rather limited. Intellectual discourse and societies and organisations operating with critical knowledge and expertise regarding the climate crisis are facing an increasingly hostile environment, to the extent that people who have an opinion are being ignored – in the best case – or tend to fall from windows [52]. It seems that Russian elites currently lack political commitment to a carbon-neutral economy, partly due to their inability to establish international cooperations, due to sanctions imposed as a result of the war in Ukraine. The official line barely acknowledges the climate crisis, and no funding is being released to support key agencies or encourage intragovernmental and intergovernmental cooperations. The authorities do not even seem to acknowledge the thawing permafrost and rising sea levels, and the immeasurable damage this could cause to the natural environment [51]. While the Russian government's inaction can be seen at least as unfortunate, making the transition away from a hydrocarbon-based economy might cause short-term CAPEX costs for Russia; the long-term benefits for both Russia, as one of the biggest countries in terms of land mass, and the world are immense [50–52]. The situation within the economy has led to hydrocarbon exporters and lobbyists having considerable power over political decision-making and stifling political debate. It so seems that from the elitists' point of view, decarbonisation would result in demonopolisation of the energy sector, meaning tipping the scales of influence and political power against them. In the light of Russia's scientific and industrial advancement, this is the more tragic, as their research, development and achievement in nuclear power, hydrogen and other energy alternatives are immense [53]. These advancements could potentially allow Russia post-Ukraine war to rejoin international projects to fight climate change and re-establish itself on the international stage as a serious ally in fighting and averting the looming climate disaster.

8.2.5 CO_2 mitigation and sustainability adaptation – India's evolving policy framework

India is one of the emerging 'BRIC' (Brazil, Russia, India and China) economies and one of the fastest growing societies in the world – they are home to approximately one-sixth of humanity [54]. India's growth momentum is fundamental to meeting the set sustainable development goals. India has been one of the worst affected countries in the world, although their contribution to global warming is negligibly small. Nonetheless, India is committed to limiting their share of global cumulative emissions even further. They maintain that nurturing the principle of equity and climate justice implies that all countries adhere to the net-zero targets and that resources are being used in a circulatory way. Parties agreed, including India, under Article 4.19 of the Paris Agreement [1] 'to strive to formulate and communicate long-term low greenhouse gas emission development strategies (LT-LEDS), mindful of Article 2 taking into account their common but differentiated responsibilities and respective capabilities, in the light of different national circumstances'. Although India is one of the lesser emitters of GHGs, they are fully committed to 'The Summary for Policy Makers (SPM) of the Working Group III contribution to the Sixth Assessment Report (AR6) of the Intergovernmental Panel on Climate Change (IPCC)'[8] but have stated clearly that the entire southern Asian region is only responsible for ca. 4% of historical cumulative net anthropogenic CO_2 emissions between 1850 and 2019 – even though this geographical area is home to an estimated 24% of the global human population. In contrast: North America and Europe have contributed ~10 times more to global cumulative GHG emissions during this time, although they only have a share of ~13% of the global human population [54].

8.2.6 CO_2 mitigation and sustainability adaptation – China's evolving policy framework

China is undoubtedly the world's largest emitter of GHG: with 12.7 billion tonnes CO_2e (CO_2 equivalent, determination factor) (2019) and a 26% share of global GHG emissions. In comparison, the EU-27 together only had a 7% share of global harmful GHG emissions [55]. For the period 2005–2019, China's GHG emissions increased by a breathtaking (literally) 75%, catapulted into astronomical heights by strong economic growth and resulting increased energy demand. In contrast: the world's total global emissions grew by 24% in the same period, and simultaneously, the EU-27 reduced their GHG emissions by 20% [55]. On a positive note, it needs to be said that China's LULUCF sink (mainly forestry and changed land use) capacity almost doubled since 2005, mainly as a result of large-scale reforestation, habitat recreation and landscape restoration activities. In 2021, ahead of COP26, China published its long-term, low GHG emission development strategy, committing to achieve carbon neutrality before 2060, with the maximum of CO_2 emissions to peak before 2030. Based on the 2005 levels of GHG emissions, and the latest by 2030, China declared wanting to decrease their carbon intensity by over 65%. Additionally, they are working on achieving over 1200 GW of installed wind and solar power and a share of renewable energy fuels in primary energy consumption to reach 25% by 2030 at the latest. Furthermore, China is aiming to achieve the forest stock to be 6 billion m^3 above the 2005 level, by 2030. China's targets are published in the 14th Five-Year Plan (FYP), in which they declared that they will 'strictly control coal consumption' between 2021 and 2025 and encourage the phase-down of coal consumption over the same period.

In stark contrast to their published commitments, in 2021, China's annual coal production reached its highest ever level, and the consumption of coal also increased by 4.6%, pushed by an increase in energy demand. The energy-related CO_2 emissions in 2021 subsequently rose by 5.8% in comparison to 2020 levels. Based on these sobering results, the 'Climate Action Tracker' [56] rates China's climate policy as 'highly insufficient' and recommends that China should reduce its GHG emissions well before 2030, drastically reduce its coal and other fossil fuel consumption and set clear phase-out timelines. These are yet to materialise.

8.2.7 CO_2 mitigation and sustainability adaptation – Australia's evolving policy framework

The Australian Government has outlined its net-zero plan 2022 in the annual climate statement to Parliament, which is aligned with the stipulations of the Climate Change Authority (CCA) [57]. The plan has provision to reach net zero by 2050 at the latest. In the meantime, they aim to have reached by 2030 CO_2 emission levels of 43% below the 2005 recorded quantities.

Australia, like all signatories of the Paris Agreement [1], is committed to publishing its nationally determined contribution every five years. This means that new lower medium-term emissions targets for 2035 need to be set. To make the 2035 emissions reduction target, a structured approach is required to construct this action plan. Therefore, deep engagement with all sectors of society will be required, such as:

- First Nations
- Community
- Industry
- Academia
- Unions

The six sectoral decarbonisation plans consist of all major components of the economy as follows:

- Electricity and energy
- Industry
- Resources
- The built environment

- Agriculture and land
- Transport

Developing a circular economy is the underlying supporting principle in a cross-cutting effort across all societal sectors.

8.2.8 CO_2 mitigation and sustainability adaptation – Canada's evolving policy framework

Canada's enhanced 2030 emission reduction target plan engages with the provinces and territories to implement the Pan-Canadian Framework on Clean Growth and Climate Change [58]. This plan is set to achieve highly ambitious emissions reduction markers, provide impetus for growing the national economy and build coping mechanisms to cope with the effects of a changing climate. The actions to support climate change adaptation are as follows:

- Making information about climate change accessible to all
- Increasing investment in infrastructure and the natural environment, to enhance protection via the natural environment for the built environment
- Research and development into the creation of guidelines and legal structures to define building codes with the aim of improving the resiliency of the built environment and existing infrastructure
- Actively managing the negative effects of climate change on the physical and mental health of Canadians
- Supporting geographical regions particularly hard-hit by climate change, such as the north of the country and the natural coastline
- Investing time and effort to support the long-term health and resilience of the ecosystems and the natural environment [58]

The Canadian climate change action plan has been named the 'Pan-Canadian Framework on Clean Growth and Climate Change' and has been developed in cooperation with most of the Canadian provinces and territories, including the indigenous peoples [58]. The plan is to grow the economy, aggressively reducing CO_2 emissions and building resilience to the effects of climate change, as these will support the protection of the health, well-being and prosperity of all Canadians, and mitigate the risks to communities, businesses and the natural environment, via nature-based solutions. These nature-based solutions might consist of many forms, but it is evident [20, 21, 44, 50] that there is no far-reaching solution to tackle the effects of climate change without involving trees and more reforestation. Natural and managed forest systems generate economic, social, physical and mental health and environmental benefits for Canadians. These include the fact that trees or generally vegetation are GHG capture units [46, 47, 59] and erosion control in flood zones, which helps communities mitigate and adapt to climate change. That is why Canada will plant two billion trees over 10 years, as part of Canada's actions to advance the use of natural climate solutions [58]. These initiatives are supporting innovative forest management practices and are hoped to make certain particularly affected areas less vulnerable to floods, storms and wildfires. The Canadian government has acknowledged their awareness of large amounts of CO_2 being currently stored in trees and vegetation, exposed soils, natural wetlands, grasslands and steppes and the Earth's oceans. Therefore, effective land management (no-till) and effluent control practices (into bodies of water) are supporting CCUS mechanisms naturally performed by plants, and stored in the soil as well as bodies of water.

They are supporting green solutions and measures supporting the creation of resilient infrastructure projects to improve decarbonisation efforts and boost economic growth. A selection of efforts is as follows:

- $9.2 billion has been ring-fenced for providing the finance required for the bilateral project agreements with provinces and territories, particularly for those furthering adaptation and climate resilience
- $2 billion for a 'Disaster Mitigation and Adaptation Fund' [58] for built and natural, large-scale infrastructure projects

Building codes were adapted to accommodate the changed requirements of offices, and commercial and residential buildings to provide guides to those entities involved to ensure that new buildings and other infrastructure are able to fend off the impacts of climate change.

Other initiatives include:

- The Northern Transportation Adaptation Initiative
- The Transportation Assets Risk Assessment Program

The Canadian government's efforts are using the natural infrastructure as an effective adaptation option for communities, as it provides a range of benefits to all involved and supports healthy and diverse terrestrial and aquatic ecosystems, and federal, provincial and territorial governments are issuing guidance and best practice literature to encourage everyone to use natural infrastructure as an alternative to the built environment.

8.2.9 CO_2 mitigation and sustainability adaptation – EU's evolving policy framework

With the European Green Deal [60], the European Environment Agency (EEA) set the road map which should ensure achieving EU climate neutrality by 2050. This guidance paper addresses the threat of climate change whilst simultaneously growing economies across the regions and protecting people's well-being. Not only does this new European climate law anchor climate neutrality by 2050 firmly in the EU countries' law, as a commitment, but also as an action guidance the EU countries are legally bound to achieve. This includes the interim target of a physical, measurable 55% emissions reduction by 2030. Currently, there is work in progress for setting the 2040 emission reduction targets.

Within the European Green Deal, Europe's 2030 policy endeavours include:

- Decreasing net GHG emissions to at least 55% below the 1990 threshold levels by 2030
- Increasing the share of renewable energy technology to generate energy

- Improving energy efficiency
- Developing a more ambitious and economically viable pathway to achieving climate neutrality, at the latest by 2050
- Increasing momentum in the green sector for opportunity and job creation and continuing to build on the EU's track record of reducing GHG emissions whilst growing all economies
- Providing alternative power supply for cars, commercial vehicles, ships and planes and improving the refuelling network
- Continuing the encouragement of international partners to increase their country's ambition to ensure the rise in global temperature is kept at 1.5 °C to avert climate disaster
- Developing an effective waste hierarchy for waste management which includes prevention and prevention through design, reuse, recycling, recovery and disposal

To ensure the continuity of this positive trend across Europe, all member states need to honour their commitment and ensure that every effort is being made to make net zero a reality as soon as possible.

8.2.10 CO_2 mitigation and sustainability adaptation – Norway's evolving policy framework

In Norway, there is a growing public conscience of the fact that a rise in global GHG emissions in recent years is being accompanied by growing losses of biodiversity and ecosystems [61]. Although significant efforts are being made to reduce GHG emissions, these losses are threatening our capacity to feed an ever-increasing global human population. The Paris Agreement along with sustainable development goals, advances in research and development of technological solutions, sustainable business practice and emerging awareness that every person alive today must accept their share of responsibility – these are all indicators that change is not only possible but also already in progress. It is vitally important for the world to be shifting to a sustainability-aware and circular economic way of living that reduces GHG emissions, increases the CO_2 uptake of vegetation and soil and protects the natural environment and habitats, and changing to a circular use of resources across the whole of society. This would entail a constant reduction of emissions from energy generation and utilisation, the production of goods and provision of services, and the transport infrastructure. Norway is on course to becoming a low-emission society, but it is imperative that the whole of society is aware of its responsibility for Norway to achieve its climate targets. Norway has taken the political responsibility seriously in that they are aiming to ensure that new technologies, methods and processes are being implemented to help ease into the transition towards a sustainable economy and society.

8.2.11 CO_2 mitigation and sustainability adaptation – US' evolving policy framework

As highlighted in the Intergovernmental Panel on Climate Change's Special Report on Carbon Dioxide Capture and Storage and others, in order to accelerate CCS development, policies that increase demand and reduce costs will be needed [11, 62]. Several different types of policies have the potential to bring down the costs of CCS and encourage research, development and deployment, including carbon pricing policies, public investment and subsidies, and clean energy standards that credit companies generating electricity or other energy sources using CCS.

In the United States, multiple enacted policies aid and encourage the use of CCS technology. National tax credits for carbon sequestration are created through Section 45Q of the IRC. Adding to these national tax credits, several tax credits and other crediting mechanisms exist at the state level in California, Texas, Louisiana, Montana and North Dakota.

Furthermore, the 'Bipartisan Infrastructure Law' [63], which passed in 2021, allocated billions of dollars in funding for CCUS projects. The law expands existing programs, sets aside funding for pipeline construction and improves the permitting process for geological sequestration wells. It also establishes or expands major programs designed to support the development of this technology.

In mid-2021, congressional democrats introduced the 'Build Back Better Act' [64], which would increase the value of the tax credit associated with 45Q. The proposal to increase the subsidy rate from \$50/tonne of captured CO_2 to roughly \$85/tonne remains contentious, and as of this writing, the Build Back Better Act remains in limbo. The proposal would also create a minimum capture requirement for plants; any plant that captures less than 75% of its emissions would not be eligible for the credits. Bolstering the 45Q tax credits is especially attractive to decision makers who oppose the total phase-out of fossil fuels.

8.2.12 Sustainability, decarbonisation and carbon certifications

8.2.12.1 EU's CBAM

Regulation 2023/956 introduced the EU's Carbon Border Adjustment Mechanism (CBAM) [65]. It clearly sets out the objective of reducing CO_2 emissions, leveraging a price on the CO_2 emissions incurred during the production of carbon-intensive EU goods imports and creating momentum to promote cleaner industrial production throughout the EU via a set methodology for calculating the hidden emissions in accordance to the Paris Agreement and the EU Fit for 55 package (55% CO_2 emissions reduction target by 2030) [65].

CBAM will be implemented in three subsequent phases and is set to concur with the phase-out of the allocation of free allowances established in the EU Emissions Trading System (ETS).

This period of transition is supposed to serve the purpose of:

1) Serving as a pilot scheme and training opportunity for producers, importers, communities and authorities concerned
2) Gathering information on factual emissions to improve the methods for calculating hidden emissions

3) Aligning the price of carbon produced *in* the EU with the price of carbon-intensive goods imported *into* the EU

The obligations subject to the CBAM to be fulfilled by importers into the EU are stated as follows [65]:

- Submitting a comprehensive quarterly CBAM report:
  This entails the import of goods and declarations about the embedded (direct and indirect emissions) in their imports during a given quarter of a calendar year (no financial payments or adjustments are required to be made). The submission deadline is fixed to, at the latest, one month after the end of the quarter in question. The CBAM reporting obligation will be in effect for the whole of the transitional period from 01 October 2023 to 31 December 2025 [65].
- Registration in the CBAM Transitional Registry:
  The EU Commission developed the registry to assist traders in compiling and submitting the CBAM report, and this will also constitute a platform for communication between the Commission, national customs and other authorities, and traders.
  Access and registration to the registry are via the CBAM Portal (TAXUD) [65].

The EU's CBAM is set to be fully implemented by 2026, but in the start phase, it will only apply to a limited range of goods with an increased risk of carbon leakage such as:

- Iron/steel
- Cement
- Fertilisers
- Aluminium
- Hydrogen
- Electricity

The Regulation (EU) 2023/956 and Implementing Regulation (EU) 2023/1773 explicitly set out the rules for reporting obligations, particularly, on the quantity of imported goods and the hidden (direct and indirect) emissions, and any carbon price due to be paid for those emissions. The European Commission has prepared a number of written guidance documents to support importers in their endeavours to adhere to their obligations [65].

The carbon and sustainability certification market is an incredibly active sector, and new schemes and certifications are constantly emerging. An attempt was made to compile a list (Table 8.1, as follows) with the seemingly most decarbonisation-focused certifications and schemes, although – as mentioned – this is not an exhaustive list.

The decarbonisation and carbon certification sector is an innovative market segment and constantly evolving. It will presumably take some time for the most effective certifications to establish themselves and receive widespread acceptance.

Table 8.1 Some decarbonisation and sustainability certification and support schemes.

Country decarbonisation and sustainability certification provider	Certification scheme	References
EU CBAM	CBAM Transitional Registry	[65]
EU European Commission	EU ETS Union Registry	[65–68]
EU European Commission	Innovation Fund	[69]
EU European Commission	LIFE programme	[70]
EU European Commission	Horizon Europe programme	[71]
EU European Commission	EU certification framework for carbon removals	[72, 73]
EU European Commission	EU Environmental Technology Verification	[74]
EU European Commission	Eco-management and audit scheme (EMAS)	[75]
EU European Commission	Organisational environmental footprint	[76]
UK ETS	Established under the Greenhouse Gas Emissions Trading Scheme Order 2020	[66]
TÜV Austria	ClimatePartner	[77]
International	B Corporation	[78]
International	The Forest Stewardship Council	[78]
International	The Roundtable on Sustainable Palm Oil	[78]
International	Climate Neutral Certification Standards (CNCS)	[77]
International	ClimatePartner	[77]
International	Gold Standard	[77]
International Carbon Trust	Route to Net Zero Standard	[79]
International Carbon Trust	Carbon Trust	[79]
US sustainability and climate professional certifications	The International Society of Sustainability Professionals (ISSP)	[80]
US sustainability and climate professional certifications	The Association of Climate Change Officers (ACCO)	[80]
US sustainability and climate professional certifications	The Association of Energy Engineers (AEE)	[80]
US energy and decarbonisation certifications	Renewable Energy Institute (REI)	[80]
US energy and decarbonisation certifications	Association of Energy Engineers (AEE)	[80]
US energy and decarbonisation certifications	GHG Institute	[80]
US energy and decarbonisation certifications	Voiz Academy	[80]

Table 8.1 (Continued)

Country decarbonisation and sustainability certification provider	Certification scheme	References
US energy and decarbonisation certifications	Sustainability Academy	[80]
US sustainability and ESG reporting certifications	Global Reporting Initiative (GRI)	[80]
US sustainability and ESG reporting certifications	CDP	[80]
US supply chain and procurement	Sustainable Purchasing Leadership Council	[80]
US supply chain and procurement	Association for Supply Chain Management	[80]
US supply chain and procurement	International Supply Chain Education Alliance	[80]
US ESG and sustainable finance, governance and risk	Chartered Financial Analyst (CFA) Institute	[80]
US ESG and sustainable finance, governance and risk	Corporate Finance Institute (CFI)	[80]
US ESG and sustainable finance, governance and risk	The European Federation of Financial Analyst Societies (EFFAS)	[80]
US ESG and sustainable finance, governance and risk	Chartered Banker Institute	[80]
US ESG and sustainable finance, governance and risk	Global Association of Risk Professionals	[80]
US circular economy, product and packaging certifications	Competent Boards	[80]
US circular economy, product and packaging certifications	The Corporate Governance Institute	[80]
US circular economy, product and packaging certifications	Circular Economy Institute	[80]
US circular economy, product and packaging certifications	Circular Economy Alliance	[80]
US circular economy, product and packaging certifications	American Center for Life Cycle Assessment	[80]
US circular economy, product and packaging certifications	Product Stewardship Society	[80]
US circular economy, product and packaging certifications	U.S. Green Building Council Leadership in Energy and Environmental Design (LEED)	[80]
US facility, building and design standard certification	The Green Building Initiative	[80]
US facility, building and design standard certification	The International Living Future Institute (IFLI)	[80]

(Continued)

Table 8.1 (Continued)

Country decarbonisation and sustainability certification provider	Certification scheme	References
US facility, building and design standard certification	The International WELL Building Institute	[80]
US facility, building and design standard certification	The Center for Active Design	[80]
US facility, building and design standard certification	The Social Equity Assessment Method (SEAM)	[80]
US facility, building and design standard certification	Green Roofs for Healthy Cities Green Roof and Wall Program	[80]
US facility, building and design standard certification	The Total Resource Use and Efficiency (TRUE) Rating System	[80]
US water management	Association of Energy Engineers	[80]
US water management	Alliance for Water Stewardship	[80]
US water management	SDG Academy	[80]
US water management	Global Sustainable Tourism Council	[80]

8.3 Part II markets analysis

8.3.1 The United Kingdom

Taken at face value, the UK government's policies seem to promote decarbonisation across all industries and actively support carbon emissions reduction efforts. However, the analysis of the current policies resulted in the realisation that, even if all points were addressed, it would only lead to a less than 50% carbon emissions reduction overall, by 2050. The Climate Action Tracker [81] has investigated the UK decarbonisation efforts [29] and has come to the conclusion that overall, the UK rating is seen as 'Insufficient'. The climate finance section was even considered highly insufficient. This does not bode well for green steel or any efforts which aim to achieve a drastic reduction of CO_2 emissions, in line with the Paris Agreement [1]. Although this might sound negative, it is again private industry who is making every effort to remedy the environmental damage caused. Tata Steel in Port Talbot had announced at the beginning of 2024, that they are switching off their highly polluting blast furnaces and will be replacing these with 70% more environmentally friendly and efficient electric arc furnaces.

8.3.2 EU

Although Germany has recently switched off the country's last nuclear power reactor [82], the Climate Action Tracker has rated the overall CO_2 emissions reduction schemes and decarbonisation efforts as insufficient [83] – partly due to the inadequate funding provision for decarbonisation and green technology.

Across the EU, however, there is an immense effort underway to cut carbon emissions where possible, and the results are quite encouraging. Germany is on track as far as the target for 80% of its energy supply to be generated via renewables by 2030 – Germany was already at 23% in 2022 [84]. Heavy industry is being supported by incentives and green finance [70, 71, 73–76], which has resulted in a moderately speedy development of green production technologies and decarbonisation projects [4]. As heavy industry leaders, such as Thyssenkrupp and ArcelorMittal, are actively involved with immense investment, the future of green steel appears to be on a good footing for the European continent.

8.3.3 International

The BRIC economies, Brazil, Russia, India and China, can overall be seen as weak contenders for achieving net-zero targets and limiting the temperature rise to 1.5 °C – as well as their decarbonisation efforts, largely due to the fact that Russia is the largest country in the world with a total area of 17,098,242 Km^2 and its rating as 'critically insufficient' in most categories. China is a close second worst rated, with 'highly insufficient', although this does not seem to consider the immense effort China has made to decarbonise the energy sector and to increase the amount of energy generated with renewable energy technologies. In 2023 [85], China's brown economy was viewed as 'turning green', as the country has massively increased their energy generation via renewables and also most recently started cracking down on emissions reporting fraud.

8.4 PESTEL and SWOT analysis of green steel

8.4.1 Introduction

PESTEL analysis (Figure 8.2) of the United Kingdom has been carried out by attempting to add some information to answer some aspects of one of the research issues and to establish the current market situation for green steel, with a view to aligning production practice to circular tech economy processes. This aims to address the framework of the political, economic, social, technological, environmental and legal issues present in the United Kingdom today.

8.4.2 Political factors affecting the United Kingdom

The United Kingdom consists of England, Wales, Scotland and Northern Ireland and, until recently, was one of the most powerful countries in the world, with the capital London – which is still a globally powerful centre of business, finance and culture. Despite the latest controversy in political realms, it is still in theory a modern parliamentary democracy and a constitutional monarchy with the monarch being

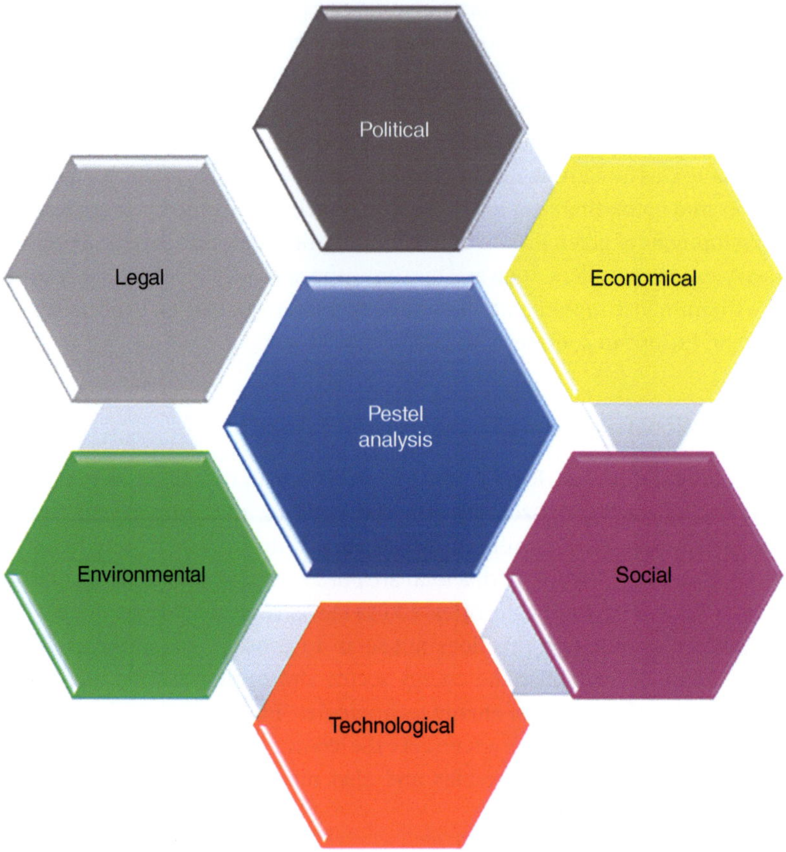

Figure 8.2 PESTEL analysis.

the head of state. The government is headed by the prime minister, who is the elected head of the government over a five-year term. The political landscape is mainly being dominated by four political parties: Labour, Conservative, Liberal Democrats and the Scottish Nationalist Party. Political stability used to be one of the great strengths of the country; however, politics of austerity, 'Brexit' (the exit of the United Kingdom from the EU framework) and the COVID-19 pandemic have created political uncertainties, cultural divide, rising unemployment, bringing the NHS (National Health Service) to the brink of collapse and caused unprecedented, partly questionable spending by the treasury – currently under investigation in the so-called COVID enquiry.

Some experts [86] and analysts believed that Brexit would result in unimaginable devastation across the economy, although some consider Brexit an immense opportunity for the country [87]. The polar opposite seems to unfold, though: the recent 25% rise in racially aggravated violence and crimes has caused serious concerns [88], and the 2022/2023 cost-of-living crisis demonstrates that Brexit has been very harmful to the United Kingdom, indeed. The United Kingdom continues to maintain good relations with the United States and many other countries across the world, although the relations with the EU and Commonwealth countries have come under some strain in 2022. As a permanent member of the United Nations (UN) with influence on global economic, cultural, military, scientific and political affairs, the country is aiming now to realign its global position in the wake of Brexit. The so-called Green Industrial Revolution [19] paper includes a Ten-Point-Plan, and all the efforts amount to a lamentable total of only 40.76% in carbon emission savings projected to have been achieved by 2030, if all measures described were to be implemented. Academic and scientific publications, governmental guidance papers and legislative frameworks influence the political landscape in any given country. In the following, Figure 8.3 shows a timeline of publications with significant influence on global politics with regard to the green agenda.

The key literature influencing the political landscape in the United Kingdom with regard to sustainability, circular economy and circular tech economy and carbon avoidance technology(CAT), CCS and CCUS has been thoroughly reviewed. The timeline (Figure 8.3) shows the key papers which constitute the foundation behind the thinking for creating the Bio Steel Cycle (BiSC) model and strategy.

The respiratory function of most living creatures is emitting CO_2, for a reason: most plants require CO_2 to grow. The natural carbon emissions cycle has been perfectly designed by nature, but anthropogenic production of CO_2 emissions beyond the requirements of living things for this – initially – naturally occurring gas has reached levels which are non-conducive to the life of oxygen-dependent organisms [1, 7, 8, 14, 41].

The IPCC published in quick succession two critical papers in February and April 2022, sounding the climate alarm bells: despite 192 country's commitment to the Paris Agreement [1] to do everything in their power to mitigate the impact of GHGs on our climate, industrial CO_2 emissions are at an all-time high [7, 9].

One of the results of this inaction we are currently experiencing is the climate crisis, as our planet is heating up to unprecedented temperature levels, and frequent fires and flooding across the globe are devastating large swatches of arable land, which is incompatible with human habitation.

In the following, a table with key guidance literature has been compiled to serve as an overview and orientation point (Table 8.2).

Going back more than 40 years, the beginning of a change in thinking towards a more sustainable way of using our resources and the environment has resulted in additional literature considered as direction-giving for developing more sustainable practices in the steel industry.

Common steelmaking practices and manufacturing protocols have been reviewed as matters of ongoing concern latest since the 1980s [95], utilising records dating back to 1890 and 1980, and economic reports in order to develop innovative practice, technological progress and economic performance-improving concepts for the British Steel industry. The findings pointed towards Britain being a net-steel importer, due to low labour productivity and no research and development investment towards more efficient and resource (environment)-saving practices.

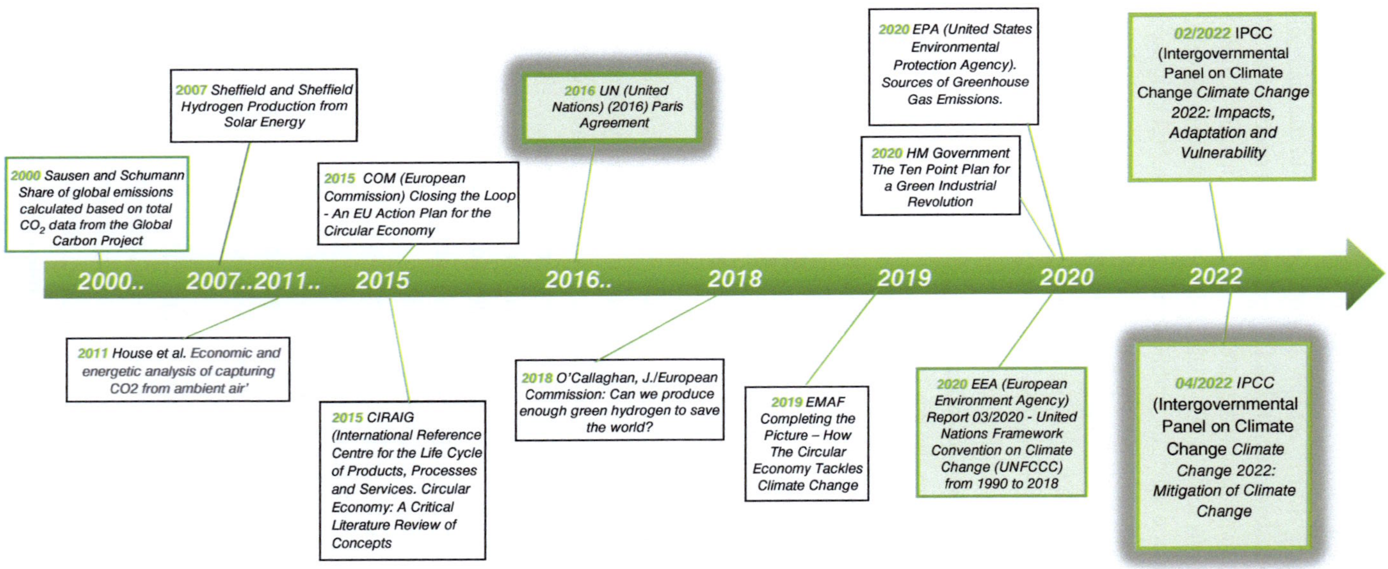

Figure 8.3 Timeline of key publications on sustainability.

Table 8.2 Circular tech economy and key guidance papers.

Year	Title	Description	Comments	References
1995	Carbon dioxide recovery from industrial Processes	One of the first comprehensive studies on the techno-economic performance of CO_2 capture in high-volume carbon-emitting industries for the iron and steel industry and the petrochemical sector.	First comprehensive study CCS/CCU; beware: in 1995, only the application of chemical absorption to capture CO_2 was considered the industry standard.	[89]
2000	Estimates of the climate response to aircraft CO_2 and NOx emissions scenarios	Investigated GHG and – specifically – aviation CO_2 emissions with data from the Global Carbon Project.	Aviation is one of the main contributors to global GHGs and on the list of future research.	[90]
2007	Assessment of hydrogen energy for sustainable development	Hydrogen production from solar energy.	First comprehensive analysis of green hydrogen.	[91]
2011	Economic and energetic analysis of capturing CO_2 from ambient air	Investigated the opportunities of the economic and energetic analysis of capturing CO_2 from ambient air.	DAC and hydropower are the most cost-efficient CCS at the lowest risk for humans and the environment.	[92]
2015	Closing the loop – an EU action plan for the circular economy	Identifying the sources of and potential ways to reduce GHG emissions and minimising waste, developing a resilient, sustainable, low-carbon, resource-efficient and competitive economic model.	Milestone guidance paper.	[93]
2016	Paris Agreement	An agreement binding signatory countries to secure a sustainable global economic recovery and net-zero emissions future and limiting global temperature rise to $<1.5°C$ to avert the worst ecological, financial and economic impacts of climate change.	193 Parties (192 countries plus the EU) out of 197 in total signed this commitment to more sustainable practice. Iran, Libya, Yemen and Eritrea have not yet ratified the agreement in June 2022.	[1]

(Continued)

Table 8.2 (Continued)

Year	Title	Description	Comments	References
2018	Can we produce enough green hydrogen to save the world?	Exploration of hydrogen produced exclusively using renewables as an alternative to fossil fuels.	Investigation of the technical readiness levels and implementation potential.	[94]
2020	Completing the picture – how the circular economy tackles climate change	EMAF carried out a thorough critical literature review of concepts concerning circular economic processes.	EMAF has the Butterfly model – the key concept to the circular economy.	[5]
2020	Report 03/2020 – United Nations Framework Convention on Climate Change (UNFCCC) from 1990 to 2018	Investigating the necessary steps and issuing guidance on how to achieve sustainability in Europe.	This constitutes the official inventory submission of the EU for 2020 under the UNFCCC, as well as the Kyoto Protocol (KP), covering the years between 1990 and 2018. Key guidance paper on knowledge for transition to a sustainable Europe.	[15]
2020	Greenhouse gas emissions – sources of greenhouse gas emissions	Identifying the sources of and potential ways to reduce GHG emissions.	US guidance paper.	[17]
2022	Climate change 2022: impacts, adaptation and vulnerability	Identifies effects on the environment and those populations most at risk and how to mitigate these effects.	Key report.	[7]
2022	Climate change 2022: mitigation of climate change	Advise on practical solution implementation.	Key report.	[9]

Farla and Hendriks [89] compiled in 1995 one of the first comprehensive studies on the techno-economic performance of CO_2 capture in high-volume carbon-emitting industries for the iron and steel industry and the petrochemical sector, although at the time only the application of chemical absorption to capture CO_2 was considered the industry standard. Then, since the late 1990s, a significant number of studies focused on a broad range of industrial sectors, with one of the main contributors being the European Commission [96], where a broad variety of carbon capture methods were investigated [93, 97], with the side effect of identifying technologies which were potentially more cost-effective than the chemical absorption/adsorption method, such as the European Cement Research Academy [98, 99]. They investigated CCS solutions for heavy industries and the cement industry.

Over the last 20 years, climate change and its effects on the global population, the natural world and our way of using up finite resources at an accelerating rate have been investigated and researched by a range of countries and numerous specialists and in the respective fields. The researcher's aim was to establish a viable, economically and ecologically sound circular tech economic concept, which could support the efforts of carbon avoidance, capture, utilisation and storage to avert the worst effects of climate change.

Sausen and Schumann had already investigated GHG in 2000 [90] and – specifically – aviation CO_2 emissions with data from the Global Carbon Project. Lee et al. [40] established how significant the impact of global aviation factually is in terms of CO_2 emissions produced, radiative forcing (RF) and effective radiative forcing (ERF) and the significant impact on climate change. The sums calculated refer to almost two decades and specifically to the years between 2000 and 2018.

Ture [34] researched the opportunities that lie within hydrogen production from solar energy, with hydrogen being a credible alternative to using fossil fuels in industrial applications. Effectively, CO_2 emissions were removed from hydrogen production, establishing that it is both economically and practically viable to produce hydrogen using solar energy – making hydrogen a renewable energy source.

In the context of the post-Kyoto era [10], where 192 nations signed an agreement and cooperation, the European Commission and 47 industrial partners in 15 European countries, in industry and academia, supported the ULCOS [97] project between 2004 and 2010 with the aim to develop ultra-low CO_2 steelmaking technology.

The approach was to develop:

1) New carbon-based smelting reduction concepts
2) Investigate the potential of new and less common reactors
3) Natural-gas-based prereduction reactors
4) Green hydrogen-based direct reduction using hydrogen
5) Direct steel production by electrolysis
6) The use of biomass
7) CCS

The ULCOS program investigated the techno-economic viability of a vast range of low-carbon ironmaking and steelmaking technologies, including and excluding CCS, in a variety of scenarios. Some of their findings suggest that carbon capture via chemical absorption/desorption is less economical than other commercially available CO_2 capture technologies.

Most of these aforementioned studies have been reviewed and compared in the UNIDO Industrial CCS Roadmap Project [100–102], which resulted in five sectoral assessment reports: (i) biomass, (ii) high-purity CO_2 sources, (iii) iron and steel, (iv) refineries and (v) cement, a technical synthesis report [103] and a policy brief for the 16th Conference of the Parties of the UNFCCC [104].

House et al. [92] investigated the opportunities of economic and energetic analysis of capturing CO_2 from ambient air. In conclusion, they established that the total levelised cost of $80.-/MWh natural gas advanced combined cycle is the lowest total levelised energy cost, whereas with regard to the $/($tCO_2$ avoided), hydropower is by far the most cost efficient and bears the lowest risk for humans and the environment, at $748/t of CO_2 captured.

Additional thorough critical literature review of concepts concerning circular economic processes was carried out by CIRAIG and EMAF [5, 6, 100], covering the following issues, presented in order of conceptual scale (from more to less encompassing):

- Sustainable development
- Ecological transition
- Green economy
- Functional economy
- Life cycle thinking
- Cradle-to-cradle thinking
- Shared value
- Industrial ecology
- Extended producer responsibility
- Eco design

In 2015, the European Commission issued a milestone guidance paper, an EU action plan for the circular economic process [93], which is quite similar to the five years later issued US Environmental Protection Agency (EPA) paper [18], identifying the sources of and potential ways to reduce rising GHG emissions.

One of the results of the EU countries' efforts towards averting climate disaster was the Paris Agreement, which was signed in 2016 by representatives of 192 member states of the UN. It binds all signing countries to secure a sustainable global economic recovery and net-zero emissions future (not only zero CO_2 emissions – but also GHG and other toxic emissions are ring-fenced within this agreement). Iran, Libya, Yemen and Eritrea have not yet ratified the agreement as of June 2022. One of the main goals of the Paris Agreement [1] is to limit global temperature rise to no more than 1.5 °C, as achieving this would avert the worst ecological, financial and economic impacts of climate change.

In 2018, O'Callaghan wrote for the European Commission about 'green' hydrogen – produced utilising exclusively renewable energy technologies – as an alternative to fossil fuels [94]. Possibly inspired by this, Sunny, Mac Dowell and Shah [35] investigated the issue of what might be needed to deliver carbon-neutral heat using hydrogen and (CCS, which led to the discovery that it is entirely possible to produce hydrogen based on using renewable energy technologies. O'Callaghan [94] found that, currently, around 20% of all GHG emissions are produced by industries such as steel and cement. It is therefore vital to develop promising technologies and concepts in these areas. Siemens was referred to as claiming that '...Hydrogen and fuels derived ... capable of reducing the CO_2 from fossil fuels ... down to zero'. O'Callaghan [94] and Sunny, Mac Dowell and Shah [35] referred to hydrogen technology as a concept based on generating energy exclusively utilising renewable energy technologies. The opportunity for transition to minimising waste and developing a resilient, sustainable, low-carbon, resource-efficient and competitive economic model must be taken seriously across Europe. As a side effect, energy will be saved and also irreversible damages avoided, which are currently caused by using up resources at an unsustainable rate, as it exceeds the Earth's capacity for renewal – as far as climate and biodiversity, air, soil and water pollution are concerned [7, 9, 14, 17, 20, 22, 41, 93]. One of the key drivers to achieve the set objectives is a major initiative called 'Horizon 2020', the purpose of which is to fund innovative projects and targeted action in areas such as plastics, food waste, construction, (critical) raw materials, industrial and mining waste, consumption and public procurement [105]. HM Government issued 'The Ten-Point-Plan for a Green Industrial Revolution' in 2020 [19], covering the issue that most emissions are being produced through the burning of fossil fuels for electricity, heat and transportation. The six sectors identified as being the most polluting are as follows:

- Electricity
- Transportation
- Industry
- Commercial/Residential
- Agriculture
- Land Use/Forestry

Unfortunately, the efforts and plans currently in place amount to only a total achievable CO_2 reduction of a mere 40.76%, with the potential to increase these efforts, based on the road map for further policies, to achieve a net-zero-carbon economy by 2050.

Omoregbe et al. [106] compiled a bibliometric analysis of the scientific discourse during 1998–2018, focusing on carbon capture and the three main types of technology, namely pre-combustion, post-combustion and oxy-fuel combustion carbon capture. The findings of the bibliometric analysis showed that post-combustion capture seems to be the most referenced carbon capture technology with ~80.9% of total publications retrieved and oxy-fuel combustion showed the least number of publications (3.4%). The EU was the single biggest contributor to publications relating to the development of carbon capture technology (CCT). Although the COREX process has been the dominating smelting reduction process commercially operating worldwide, Yi et al. [107] found that FINEX can be considered as having more potential for large-scale CO_2 removal, as the systematic development of the FINEX process is showing promise: energy recovery systems, binders for coal briquettes and a steel grid system in fluidised bed reactor (FBR) achieved reductions in economic terms and the overall environmental footprint. One of the biggest advantages of the FINEX over the COREX process is that the FINEX process enables the direct use of sinter feed fine ore (Yi et al. [107]). The scale of typical smelting reduction steelmaking plants is smaller than that of typical BF-based plants. The percentage of direct reduced iron has increased decade after decade from 2000.

The European Environment Agency Report 03/2020 constitutes the official inventory submission of the EU for 2020 under the UNFCCC, as well as the KP [10], covering the years between 1990 and 2018. The EU, party to the UNFCCC, issues these reports on GHG inventories annually, for emissions and removals within the area taking place within its territory. Due to the nature of this lag in reporting, the 2020 inventory report could not yet reflect the effects which the COVID-19 pandemic might possibly have had. Although the United Kingdom left the EU in 2020, EU law still applied until the end of the transition period. Moreover, key provisions of Regulation (EU) No 525/2013 ('Mechanism for Monitoring and Reporting GHG') and of Decision No. 406/2009/EC ('Effort Sharing') apply to the United Kingdom also in respect to GHGs emitted during 2019 and 2020. The EU and it have agreed to a quantified emission reduction commitment, which will limit their average annual GHG emissions to 80% of the sum of their base year emissions, reflected in the Doha Amendment [108]. Since then, GHG emissions decreased in the majority of sectors between 1990 and 2018, with the notable exception of transport, including international transport, refrigeration and air conditioning. Considering the aggregate level, the highest emission reductions were achieved in manufacturing industries, construction, electricity/heat production, iron/steel production and domestic combustion.

Allegedly, contrail cirrus (the condensed clouds one can observe as an air signature of a plane's flight path), consisting of linear contrails and the cirrus cloudiness arising from them, yields the most significant warming ERF, closely followed by CO_2 and NOx emissions. Interestingly, emissions of sulphate aerosol formation result in a negative (cooling) term. In conclusion, the mean contrail cirrus ERF/RF ratio of 0.42 indicated that contrail cirrus is less effective in surface and atmospheric warming than other terms. It was established that by 2018, the net aviation ERF is +100.9 milliwatts (mW) m^{-2}, meaning a 5–95% likelihood range (of 55, 145) with significant contributions from contrail cirrus (57.4 mW m^{-2}), CO_2 (34.3 mW m^{-2}) and also NOx (17.5 mW m^{-2}).

With the publications in February and April 2022, the IPCC sounded the climate alarm bells and had scientists and informed citizens reeling with despair: despite the Paris Agreement in 2016, where 192 out of 197 countries signed a firm commitment to doing everything to keep the global temperature rise to <1.5 °C, and scientists, industry and academics rallying to find solutions to mitigate GHG emissions and reduce the impact of anthropogenic CO_2 emissions, the situation only seems to get worse. The research and study of these key papers have led the author to develop the BiSC and connected models and strategies. It is worth considering that every component within the circular model of the BiSC represents a cyclical model in itself, which will be explored in future research.

Furthermore, the unfortunate issue of giving planning consent for the first coal mining plant creation [109] in 30 years and nuclear power plant creation [110] will be *adding* an incalculable amount of GHG and CO_2 emissions, through construction, building and operation of required infrastructure – besides the known risks of combusting coal as one of the largest emitters of CO_2, and operating nuclear power plants, as demonstrated by the Chernobyl disaster and the potential for illegal weaponising of by-products, such as plutonium.

Besides having issued the planning permission for a new EDF (Energy de France) nuclear power plant at Hinckley Point [110], the UK government has just issued planning permission for a coal mine in Cumbria [109] in December 2022, which bodes even worse for the UK's commitment to decarbonising manufacturing. However, some political manoeuvres give hope: British Steel and Tata Steel have secured government funding of £300 m, each, for decarbonising steel production, under the premise that they will secure a total of 7500 British jobs. British Steel employs ca. 4000 people across its operations, including at the plant in Scunthorpe, and India's Tata employs 3500 people in Port Talbot in South Wales [111].

This bodes well for the political climate towards green steel, bringing decarbonisation of the steel industry in the United Kingdom that little bit closer to reality.

8.4.3 Economic factors affecting the United Kingdom

The economy of the United Kingdom is the fifth largest national economy in the world (by GDP), the ninth largest by purchasing power parity (PPP) and the 22nd highest by GDP per capita, constituting 3.3% of nominal world GDP [112]. The United Kingdom has recently been overtaken by India as being the fifth largest economy, due to 14 years of successive UK conservative government policy and dismantling of the public sector [112, 113]. Although the recession in 2008 hit the UK economy hard and COVID-19 is adding to these woes, the government aimed to lessen the impact of the global recession. The attempt was made by implementing adequate measures, such as quantitative easing, the Bank of England lowering bank interest rates for UK banks borrowing money, improving redundancy protection, issuing renters eviction protection during the COVID-19 crisis and setting national minimum wages, to safeguard the economy and the UK population against any negative economic impact Brexit and COVID-19 might cause. Unfortunately, failed policies and government intervention have led to an inflation rate rise of 10.1% in November 2022 [114].

Foreign direct investment (FDI) by private investors and organisations from around the world has injected immense financial impetus into a variety of industries in the United Kingdom – such as sports, real estate, technology, the car industry and the grocery sector.

Although the cost-of-living crisis and inflation were at an all-time high of 10.1% in March 2023, the economy grew by 0.3% in January 2023. The country is expected to just about escape a recession and return to the net growth in 2024 [115].

As the current corporation tax rate for company profits was 19% in 2019, the government announced a reduction to the corporation tax from 1 April 2020 and to a flat rate of 19% as of April 2021 [116].

Considering the United Kingdom sometimes faces challenging weather conditions, millions of tourists from around the world come to visit the United Kingdom every year. Before the COVID-19 crisis, tourism was booming in the United Kingdom and contributed to ~£127 billion annually to the economy (by 2022) [117].

8.4.4 Social factors affecting the United Kingdom

The Office for National Statistics (ONS) released data in March 2021 stating that the strong consumer market in the United Kingdom held a population of 59.5 million [118], down from 66.4 million in 2018 [119], partly due to EU citizens leaving the UK's politically hostile environment. Allegedly, births continued to outnumber deaths, resulting in the population expected to reach 74 million by 2039. However, the COVID-19 and Brexit fallout had not yet been taken into consideration and future data will cast a better-informed light on the actual situation. In terms of demographics, it needs to be noted that the UK population is getting older, in comparison with our European neighbours, due to advancements in medicine, care provision and increased living standards. With young people staying longer in education, according to the ONS, in 2016, 18% of people were aged 65 and over, while 2.4% were aged 85 and over, with implications on the labour market, the health and care system and the wider economy. Interestingly, this poses a lot of opportunities for organisations, enabling them to cater for the needs of an older population. Industry sectors such as real estate, holiday, insurance, healthcare and many others have already benefitted from this shift within the population.

Historically, the United Kingdom has been strongly influenced by the concept of social class – where people now have to adjust their way of thinking, due to an increasingly multicultural population. Owing to this increased demographic variety, companies have developed a variety of strategies to accommodate these variants, providing new products and markets to cater for the needs of different races, cultures and religions.

Ernst and Young, 2023 [120] have issued a bleak report for the United Kingdom, stating that the cost-of-living pressures are set to intensify the UK's regional economic divide.

8.4.5 Technological factors affecting the United Kingdom

Overall, the United Kingdom is seen to be one of the most technologically advanced countries in the world; with the capital London being a focal point for technological, financial and real estate-based institutions, it offers entrepreneurs from all business pathways immense opportunities to develop their business in the United Kingdom.

The technology sector in business terms, in education and the manufacturing sector, is a major contributor to the UK GDP. FDI (Foreign Direct Investment), talented individuals and interested non-governmental organisations are moving into the sector, which attracted £28 billion in technology investment since 2011, compared to £11 billion in France and £9.3 billion in Germany. Despite all efforts, the Brexit referendum, the COVID-19 crisis and the fallout of the United Kingdom leaving the EU have the United Kingdom falling behind the United States and Asia–Pacific states in technological development. According to the ONS (2023):

- The value of the UK's *inward* FDI stock (position) (£1,929.2 billion) was higher than that of the outward position (£1660.9 billion) in 2020.
- Inner London West was the International Territorial Level 2 (ITL2) subregion with the highest inward FDI stock value in 2020 (£529.2 billion); the subregion with the lowest inward value was Cornwall and Isles of Scilly (£2.1 billion).
- UK manufacturing companies with foreign parent companies (inward FDI) accounted for the highest stock values in 18 of the 41 subregions in 2020; these were mainly located in Wales and the English Midlands.
- Financial and insurance companies in the United Kingdom that have companies abroad (outward FDI) accounted for the highest stock values in 19 of the 41 subregions in 2020; these were mainly located in the east of the United Kingdom, including eastern Scotland, east Yorkshire and North Lincolnshire, East Anglia and London.

Despite the FDI increase, the United Kingdom is in danger of falling behind its European and global counterparts in technological advancement, as the manufacturing sector is increasingly lacking the technical expertise of its workforce, due to skilled workers taking early retirement or being otherwise economically inactive as one of the effects of the COVID-19 pandemic, and EU citizens have left in droves after Brexit and less young people joining the manufacturing workforce. Investment in people, 'green' skills, and innovative and renewable energy technologies can be a once-in-a-generation opportunity for the United Kingdom to innovate itself out of the economic and technological downturn. As the examples of British Steel, Tata Steel and the newest hydropower project by SSE are demonstrating: the willingness to decarbonise production and transform into a green industry is abundant; it now requires the political powers-at-be to provide the legal framework, grants and incentives to the technical and manufacturing sector to transform this opportunity into a reality.

8.4.6 Environmental factors affecting the United Kingdom

Economic activities have a direct impact on ecological systems, like the natural environment. Although the United Kingdom has made efforts to reduce the negative impact of economic activity, vouching for 'Zero Net Carbon' by 2050 [19], local councils, newspapers and some charities have taken a variety of initiatives to create environmental awareness and reduce the negative impact of economic growth on the environment. Unfortunately, these efforts seem not to have yielded the desired effects, as of autumn 2022, 85% of UK rivers are heavily polluted by a 'chemical cocktail' of sewage, agriculture and road pollution. This has been identified to be the result of repeated breaches of environmental regulations on an industrial scale by water companies and farms alike [121].

Interestingly, at first glance, it so seems as if the three most polluting sectors in the United Kingdom are not sufficiently focused on within the UK Green Industrial Revolution agenda, with most CO_2-saving opportunities in transport/aviation, residential dwellings almost and (animal) agriculture completely being left out. It was found that animal agriculture, such as beef production, accounts for 25% of global land use, land-use change and forestry emissions – and these issues do not appear to have been factored into the UK 10-point agenda [14, 20–22, 122].

There appears to be a political willingness to implement the principles of circular economic processes, which promote circular economic production cycles – as opposed to linear production cycles, which have been in place since long even before the Industrial Revolution. Linear production processes and manufacturing have led to the destruction of almost 40% of flora and fauna, globally, but the efforts in the United Kingdom to create the framework for adequate environmental protection, rewilding and reforestation action to follow are inadequate.

8.4.7 Legal factors affecting the United Kingdom

The UK legal landscape follows common law principles, which have been adopted, besides the United Kingdom, in the United States and Commonwealth countries. England and Wales operate a common law system which combines the passing of legislation with the creation of precedents via case law. The laws are established by the passing of legislation by Parliament. This consists of the monarch (king or queen), the House of Commons and the House of Lords (the University of Oxford, no date). A detailed discussion of the UK legal environment would be beyond the boundaries of this analysis. Overall, following thoughts in the section discussing political factors, the United Kingdom follows democratic principles and has a framework of laws and policies in place, which are currently still aligned with the Human Rights Act 1998.

Nonetheless, at this point, it needs to suffice to mention the Employment Act 1996, which protects the rights of employees. There are laws and regulations in place regarding maternity and paternity leave, minimum wage, holiday pay, sick pay and other legally protected provisions. Likewise, the Equality Act 2010 protects people from any discrimination.

8.4.8 Emphasising strategic opportunity

The PESTEL landscape of the United Kingdom suggests that it lends itself to host future generations of workers with 'green' skills. Steel manufacturing has been one of the early industries to establish itself in the United Kingdom and has been an industry mainstay since the days of the Industrial Revolution in 1740. The evolved legal and political framework in the United Kingdom, geared towards industry, could make the United Kingdom the hub of decarbonised steel production and the accelerator for producing green steel and decarbonised production in heavy industries. The recent 2023 IPCC [8] report issued a clear warning and called for immediate action, so as not to lose any more time to decarbonise production and limit the overall – and not just the industrial – CO_2 emissions drastically, immediately.

8.5 SWOT analysis of the United Kingdom

8.5.1 Introduction

This SWOT (Strengths/Weaknesses/Opportunities/Threats) analysis of the United Kingdom explores some of the key aspects in the respective areas, which – given the gravity of their impact – were explored in more detail throughout the PhD project, as appropriate.

8.5.2 Strengths of the United Kingdom

Despite the latest controversy in political realms, it is still in theory a modern parliamentary democracy and a constitutional monarchy with the monarch being the head of state – hence, it can be argued that the United Kingdom is a country of political stability.

The government is headed by the prime minister, who is the elected head of the government over a five-year term. The political landscape is mainly being dominated by four political parties: Labour, Conservative, Liberal Democrats and the Scottish Nationalist Party.

The United Kingdom is a founding member and one of the five permanent members of the UN. Not only is the United Kingdom a key ally of the United States, but there are also good connections with the Commonwealth of Nations – therefore playing a significant role in world politics.

The United Kingdom is the fifth largest economy in the world, and the UK's GDP in 2019 was 2.827 trillion US dollars [112]. Although the recession in 2008 hit the UK economy hard and COVID-19 is adding to these woes, the government aimed to lessen the impact of the global recession by implementing adequate measures, such as quantitative easing, the Bank of England lowering bank interest rates (currently at 0.1%) for UK banks borrowing money, improving redundancy protection, issuing renters eviction protection during the COVID-19 crisis and setting national minimum wages. The minimum wage as of April 2019 for people aged 25 or over is £8.21 per hour which is likely to increase in the near future.

The United Kingdom is home to some of the greatest and global companies in the world: Shell, GlaxoSmithKline (GSK), BP, Vodafone, BAE Systems, ASOS, Lloyd's, Aviva, AstraZeneca, Unilever and Tesco, some of the brand names which are contributing vastly to the annual GDP and are forerunners for research and development leading the way in everything from tech and manufacturing, infrastructure to food, pharmaceuticals and fashion.

Healthcare in the United Kingdom is provided by the National Health Service (NHS), which is the publicly funded healthcare system of the United Kingdom and is universally free at the point of delivery.

Geographically, the United Kingdom is a European country with a temperate climate – although, with its insular location, the coastal areas are exposed to the elements – which can be experienced in local weather phenomena. The Uk as an extraordinary island with several world heritage sites, sites of special scientific interest (SSSI), national parks, iconic landmarks, quaint villages, vibrant coastlines, scenic countryside, castles and palaces, events and festivals, filming locations and many more. Although the United Kingdom sometimes faces challenging weather conditions, millions of tourists from around the world come to visit the United Kingdom every year. Before the COVID-19 crisis, tourism was booming in the United Kingdom and contributed ~£127 billion annually to the economy [117].

The ONS [123] stated that the strong consumer market in the United Kingdom held a population of 59 million in 2022. Allegedly, births continued to outnumber deaths, resulting in the population expected to reach 74 million by 2039.

The United Kingdom is also home to some of the best higher education institutions, such as St Andrews, Cambridge and Oxford University. Historically, many scientists, historians, inventors, laureates and intellectuals with international acclaim are British – such as Mary Currie, Elisabeth Townsend, Alfred Turing and Alexander Fleming.

8.5.3 Weaknesses of the United Kingdom

The United Kingdom seems on the surface to be one of the top destinations for foreign direct investment (FDI), confirmed by the value of the UK's inward FDI stock (position) (£1,929.2 billion) being higher than that of the outward position (£1,660.9 billion) in 2020 [123]. However, some experts argue that this current situation is one of the negative aspects of Brexit, as even some of the British household brand names, such as Dyson, have moved their production abroad. The number of UK businesses exporting goods to the EU fell by 33% to 18,357 (2021), from 27,321 (2020), according to 2023 data provided by the ONS [123].

Additionally, in the international context, the United Kingdom is one of the most expensive countries to live in. In order to provide a realistic and relatable example, the average monthly net rent (without services) for a one-bedroom flat even in poorer areas of the capital London may cost around £1000. Due to FDI, the housing market is extremely buoyant, and it forces native Londoners and young people out of the city, as it is very difficult for a first-time buyer to buy a property. Furthermore, generally, the cost of living, i.e. to buy food, clothing and consumables, is very expensive in the country, particularly in London.

Nevertheless, the United Kingdom is one of the richest countries in the world, with a classical north/south divide, as more employment opportunities have been created in southern cities than elsewhere across the United Kingdom. This has led to regional wealth inequality; the richest 10% of UK households owned 45% of the UK's national wealth in 2020, and the UK wealth of the richest 1% of households was on average more than £3.6 million, compared with £15,400 or less for the least wealthy 10% of the population [123].

Another area of concern is the lack of investment in research and development. Only recently, as an effect of Brexit, the United Kingdom has lost access to the international space research program 'Galileo', which has not yet been replaced with an equivalent in the United Kingdom. Additionally, access to European academics has been cut off, with the Erasmus programme not having been negotiated to continue and high-achieving university graduates seeking employment elsewhere instead of in the United Kingdom. As a growing concern, the United Kingdom has been falling behind many advanced nations as far as spending on research and development as a proportion of GDP is concerned [124].

8.5.4 Opportunities for the United Kingdom

Every literate person in the United Kingdom will by now have understood that Brexit has created great uncertainties and the sheer expanse of negative impact on the UK GDP is not yet fully known. However, the United Kingdom had been a leading industrial power and global player before joining the EU in 1974 and there is therefore every chance that – although it may take some time, given that COVID-19 has added to the UK's woes – the United Kingdom will be seen as a leader and economically strong once more. Freed from the binding powers of the EU membership, new doors to new possibilities have been opened. There is every chance that the United Kingdom can now build bilateral and free-trade relations with nations worldwide, which will in their own right create opportunities for British companies to expand. It is too early to say if access to the EU single market can be maintained. However, as 41% of the UK's export flows directly into Europe, countries and companies are all interested in continuing this positive relationship and access to the EU single market might yet be secured. Being steeped in history has made the United Kingdom a very desirable location which draws talents from all over the world. It is now key to attracting highly skilled talents, increasing productivity, investing in R and D, meeting domestic demands and increasing the influence in global affairs.

8.5.5 Threats to the United Kingdom

With the old world order changing and the BRIC (Brazil, Russia, India and China) economies advancing, many competitors are challenging the United Kingdom in a variety of sectors and adding significant competitive pressures on the United Kingdom. Additionally, owning and running businesses in the United Kingdom is a very expensive undertaking, depending on the sector, prohibitively expensive to the extent that often especially small businesses find them unbearable. But even household names, big businesses such as Dyson, have moved abroad, as the running cost in the United Kingdom was just in no positive relation to the potential profit and resulted in yet another big player taking their money elsewhere. The UK government is under great pressure to restart the economy and change these unfavourable conditions, as otherwise, many businesses may go bust, resulting in a wave of unemployment and subsequently decreased GDP.

8.5.6 Conclusions – emphasising strategic opportunity for the UK

On the surface, the commercial environment and the current Conservative government's drive to cut 'red tape' seem to be geared towards economic growth and a greener economy. As a result of a decade of underinvestment in infrastructure and public service, the 2020–2023 COVID-19 pandemic and the 2022/2023 ongoing cost-of-living crisis, along with a high inflation rate of 14.6% as of 20 June 2023, have left businesses and the population reeling with the aftermath of the aforementioned situations. One of the consequences is an ailing economy, and currently, 40% of businesses are said to be failing as a result of BREXIT and the increase in energy prices. The country is facing severe challenges at the time of writing, but this also provides considerable opportunity for improvement. The negative and positive impact factors are known – all that is required now is political willingness to make the change for a technical green industrial revolution with the effect of decarbonisation of production and environmental rehabilitation happens with an adequate legal and political framework.

8.6 CO_2 emissions for the entire steel production process

To date, there is no comprehensive overview of all CO_2 emissions associated with steelmaking published in the public domain. It has been common practice to merely account for the emissions produced during the ironmaking (BF) and steelmaking processes (BOF) or EAF operations. However, it makes logical sense that all processes involved, for all resources associated with the production of steel, need to be included in the determination of total emissions. Although there are currently six main GHGs resulting from anthropological activity, this work focused on CO_2 emissions only – for the time being. Figure 8.4 gives an overview of most CO_2 emissions levels at the different stages and quantities required for the respective resources. The investigations, calculations and process simulations, based on 330,000 t capacity BF/BOF, have resulted in the findings that 4,270,655 t of CO_2 are being emitted per steel production cycle, thus resulting in 12.94 t of CO_2 per tonne of steel produced – with all production stages from mining to marketing considered.

8.7 BiSC – steel production decarbonisation model and strategy

By designing a viable model and strategy, the main aims and objectives have been achieved. The BiSC model contains components which are classed as being at TRL 6–9 and can be readily implemented. Following the 7-step strategy to achieving net-zero carbon emissions steel production, zero-carbon emissions steel production can be achieved, in the short term. There are 59 projects underway, worldwide

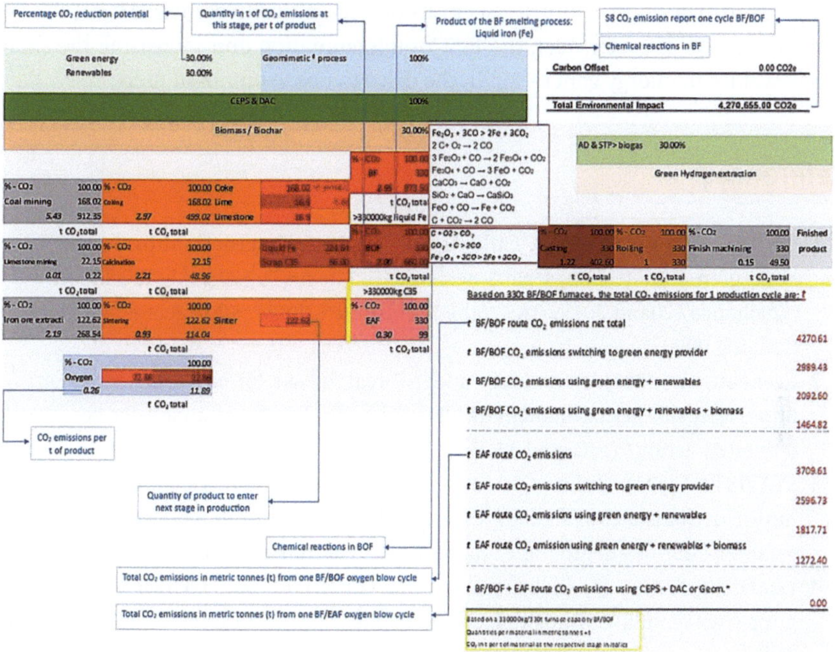

Figure 8.4 Overview of CO_2 emissions along the steelmaking process in a 330 t BF/BOF operation with report excerpt from S8.

(Appendix 6), which are individually investigating and operating the technology mentioned above, so the viability of the model has indirectly been vindicated by the operation of the selected technologies.

8.8 Higher degree of energy independence

Given that Tata Steel has set up a cooperation in Ijmuiden directly next to the North Sea shores gives hope to the scenario that hydrogen-based technologies will be implemented in the near future, in heavy industry. Implementing renewable energy technologies will not only support the decarbonisation of the steel industry, but the mechanisms are transferrable to most heavy industries. The currently available UK grants, subsidies and tax incentives make this move to greener production and a higher degree of energy independence a viable solution.

This work certainly has a UK bias, but other countries have been considered in the research, too. The EU, Australia, Norway, the United States, Canada and BRIC economies [122, 125] as well as other main steel-producing countries have been researched for information on their GHG emissions, decarbonisation efforts and plans to work towards keeping the global temperature rise below 1.5 °C of pre-industrial levels, according to the Paris Agreement [1].

Besides Russia, most countries seem to provide a positive environment for the implementation of green steel, the BiSC implementation and increasing the share of renewables in energy generation in the drive to phase out fossil fuels.

References

1 UN United Nations (2016). Paris Agreement 2016 [Internet]. https://www.un.org/en/climatechange/paris-agreement#:~:text=To%20tackle%20climate%20change%20and,2015%3A%20the%20historic%20Paris%20Agreement (accessed 2 March 2020).

2 Evennett, H. (2022). Democracy under threat: a case for co-ordinated action? https://lordslibrary.parliament.uk/democracy-under-threat-a-case-for-co-ordinated-action/ (accessed 15 February 2024).

3 STATISTA (2024). Größte Gefahren für die Demokratie in Deutschland 2024. https://de.statista.com/statistik/daten/studie/1337730/umfrage/groesste-gefahren-fuer-die-demokratie/ (accessed 15 February 2024).

4 Leadership Group for Industry Transition (LEADIT) (2023). Green steel tracker. https://www.industrytransition.org/green-steel-tracker/ (accessed 18 September 2023).

5 Ellen MacArthur Foundation (EMAF) (2019). Completing the picture – how the circular economy tackles climate change. [Internet]. https://www.ellenmacarthurfoundation.org/assets/downloads/Completing_The_Picture_How_The_Circular_Economy-_Tackles_Climate_Change_V3_26_September.pdf (accessed 9 October 2022).

6 Ellen MacArthur Foundation (EMAF) (2023). Summit 23 — redesigning the future: a circular economy showcase [Internet]. https://www.ellenmacarthurfoundation.org/summit/2023 (accessed 18 October 2023).

7 Intergovernmental Panel for Climate Change (IPCC) (2022). Climate change 2022: impacts, adaptation and vulnerability. [Internet]. https://www.ipcc.ch/report/ar6/wg2/ (accessed 3 January 2023).

8 Intergovernmental Panel for Climate Change (IPCC) (2023). AR6 synthesis report climate change 2023 [Internet]. Report No.: AR6. https://www.ipcc.ch/report/ar6/syr/downloads/report/IPCC_AR6_SYR_SPM.pdf (accessed 5 January 2023).

9 Intergovernmental Panel for Climate Change (IPCC) (2022a). Climate change 2022: mitigation of climate change [Internet]. https://www.un.org/en/climatechange/reports?gclid=EAIaIQobChMIxsyF3ZSl-AIVoo1oCR19Dw3IEAAYAiAAEgItyfD_BwE (accessed 30 April 2022).

10 UNFCCC (2005). What is the Kyoto protocol? | UNFCCC [Internet]. https://unfccc.int/kyoto_protocol (accessed 11 July 2023).

11 Gonzales, V., Krupnick, A., Dunlap, L., and Resources for the Future (2024). Carbon capture and storage 101. https://www.rff.org/publications/explainers/carbon-capture-and-storage-101/ (accessed 22 February 2024).

12 The White House (2021). The build back better framework. https://www.whitehouse.gov/build-back-better/ (accessed 29 February 2024).

13 World Steel Association (WSA) (2023). Fact-sheet-raw-materials-2023.pdf [Internet]. https://worldsteel.org/wp-content/uploads/Fact-sheet-raw-materials-2023.pdf (accessed 29 February 2024).

14 European Environment Agency (EEA) (2020). The European environment — state and outlook 2020: knowledge for transition to a sustainable Europe — European Environment Agency [Internet]. https://www.eea.europa.eu/soer/2020 (accessed 11 July 2023).

15 European Energy Agency (EEA) (2020). EEA Report No 3/2020 trends and drivers of EU greenhouse gas emissions — European Environment Agency [Internet]. https://www.eea.europa.eu/publications/trends-and-drivers-of-eu-ghg (accessed 6 July 2023).

16 United States Environmental Protection Agency (USEPA) (2023). Greenhouse gas (GHG) emissions and removals [Internet]. https://www.epa.gov/ghgemissions (accessed 2 January 2023).

17 United Nations Environmental Protection Agency (USEPA) (2015). Sources of greenhouse gas emissions [Internet]. https://www.epa.gov/ghgemissions/sources-greenhouse-gas-emissions (accessed 6 July 2023).

18 Desai, M. and Camobreco, V. (2020). Inventory of US greenhouse gas emissions and sinks 1990–2018 [Internet]. p. 733. https://www.epa.gov/sites/default/files/2020-04/documents/us-ghg-inventory-2020-main-text.pdf (accessed 7 June 2020).

19 HM Government UK (2020). The Ten Point Plan for a Green Industrial Revolution – building back better, supporting green jobs, and accelerating our path to net zero [Internet]. https://www.legislation.gov.uk/ukpga/2008/27/contents (accessed 7 July 2023).

20 Nix, J. (2020). *John Nix Pocketbook – For Farm Management*, 50e. Leicestershire: The Anderson Centre.

21 Nix, J. (2023). *John Nix Pocketbook – For Farm Management*, 53e. Leicestershire: The Anderson Centre.

22 World World Wildlife Fund (WWF) (2025). What is the sixth mass extinction and what can we do about it? | Stories | WWF. https://www.worldwildlife.org/stories/what-is-the-sixth-mass-extinction-and-what-can-we-do-about-it (accessed 7 July 2023).

23 Passivhaus Institute (PHI) (2015). Passivhaus Institute [Internet]. https://passiv.de/en/index.php (accessed 29 February 2024).

24 Union of Concerned Scientists (2022). What is sustainable agriculture? | Union of Concerned Scientists [Internet]. https://www.ucsusa.org/resources/what-sustainable-agriculture (accessed 29 February 2024).

25 Forestry Commission (2003). *Forests, Carbon and Climate Change: The UK Contribution 2003*, 1–12. Edinburgh: Forestry Commission.

26 Forestry Commission (2023). *Forests, Carbon and Climate Change: The UK Contribution [Internet]*, 1–60. Edinburgh: Forestry Commission https://assets.publishing.service.gov.uk/government/uploads/system/uploads/attachment_ data/file/1162830/Forestry-Commission-Key-Performance-Indicators-Report-2022-23.pdf (accessed 7 June 2023).

27 Woodland Trust (2021). State of the UK's woods and trees. https://www.woodlandtrust.org.uk/state-of-uk-woods-and-trees/ (accessed 29 February 2024).

28 EUROSTAT (2021). 39% of the EU is covered with forests [Internet]. https://ec.europa.eu/eurostat/web/products-eurostat-news/-/edn-20210321-1 (accessed 29 February 2024).

29 Climate Action Tracker (2023). United Kingdom [Internet]. https://climateactiontracker.org/countries/uk/ (accessed 5 March 2024).

30 HDI Global SE (2021). Hydrogen – all hype or fuel of the future? [Internet]. https://www.hdi.global/infocenter/insights/2021/hydrogen--all-hype-or-fuel-of-the-future/ (accessed 29 February 2024).

31 Rapier, R. (2012). The race to replace oil. In: *Power Plays: Energy Options in the Age of Peak Oil [Internet]* (ed. R. Rapier), 161–179. Berkeley, CA: Apress https://doi.org/10.1007/978-1-4302-4087-7_11 (accessed 29 February 2024).

32 Zainal, B.S., Ker, P.J., Mohamed, H. et al. (2024). Recent advancement and assessment of green hydrogen production technologies. *Renewable and Sustainable Energy Reviews* 189: 113941.

33 Hatch(2022). Tata Steel hydrogen-based steel manufacturing [Internet]. https://www.hatch.com/Projects/Metals-And-Minerals/Tata-Steel-hydrogen-based-steel-manufacturing (accessed 29 February 2024).

34 Ture, E. (2007). Hydrogen production from solar energy. In: *Assessment of Hydrogen Energy for Sustainable Development* (ed. J.W. Sheffield and Ç. Sheffield), 135–146. Dordrecht: Springer Netherlands (NATO Science for Peace and Security Series C: Environmental Security).

35 Sunny, N., Mac Dowell, N., and Shah, N. (2020). What is needed to deliver carbon-neutral heat using hydrogen and CCS? *Energy & Environmental Science* 13 (11): 4204–4224.

36 World Nuclear Association Chernobyl | chernobyl accident | chernobyl disaster – World Nuclear Association [Internet]. https://world-nuclear.org/information-library/safety-and-security/safety-of-plants/chernobyl-accident.aspx (accessed 29 February 2024).

37 Rapier, R. (2012). Nuclear power. In: *Power Plays: Energy Options in the Age of Peak Oil [Internet]* (ed. R. Rapier), 101–116. Berkeley, CA: Apress https://doi.org/10.1007/978-1-4302-4087-7_7 (accessed 29 February 2024).

38 Cliff, S.J., Lewis, A.C., Shaw, M.D. et al. (2023). Unreported VOC emissions from road transport including from electric vehicles. *Environmental Science & Technology* 57 (21): 8026–8034.

39 Charabi, Y., Al Nasiri, N., Al Awadhi, T. et al. (2020). GHG emissions from the transport sector in Oman: trends and potential decarbonization pathways. *Energy Strategy Reviews* 32: 100548.

40 Lee, D.S., Fahey, D.W., Skowron, A. et al. (2021). The contribution of global aviation to anthropogenic climate forcing for 2000 to 2018. *Atmospheric Environment (1994)* 244: 117834.

41 Intergovernmental Panel for Climate Change (IPCC) (2021). *Climate Change 2021 – The Physical Science Basis Summary for Policymakers [Internet]*. Cambridge: Cambridge University Press https://www.ipcc.ch/report/ar6/wg1/downloads/report/IPCC_AR6_WGI_SPM.pdf (accessed 7 July 2023).

42 Armstrong McKay, D., Lade, S., and Rockström, J. (2023). New report: tipping point threats and opportunities accelerate [Internet]. https://www.stockholmresilience.org/research/research-news/2023-12-06-new-report-tipping-point-threats-and-opportunities-accelerate.html (accessed 28 February 2024).

43 Vijn, S. and World World Wildlife Fund (WWF) (2014). The biogas solution: turning manure into profit | Blog Posts | WWF. https://www.worldwildlife.org/blogs/sustainability-works/posts/the-biogas-solution-turning-manure-into-profit (accessed 14 July 2023).

44 EUROSTAT (2023). Forests, forestry and logging [Internet]. https://ec.europa.eu/eurostat/statistics-explained/index.php?title=Forests,_forestry_and_logging (accessed 12 July 2023).

45 Hararuk, O., Kurz, W.A., and Didion, M. (2020). Dynamics of dead wood decay in Swiss forests. *Forest Ecosystems* 7 (1): 36.

46 Foundation AD and Lambe, D. (2022). *Now Is the Time for Trees: Make an Impact by Planting the Earth's Most Valuable Resource*, 391. Hachette UK.

47 Seethalakshmi, K.K., Jijeesh, C.M., and Balagopalan, M. (2016). *Bamboo Plantations: An Approach to Carbon Sequestration [Internet]*. Kerala/Trichur: Kerala Forest Research Institute https://www.researchgate.net/publication/215475397_Bamboo_plantations_An_approach_to_Carbon_sequestration.

48 Seethalakshmi, K.K., Balagopalan, M., and Jijeesh, C.M. (2009). Bamboo plantations: an approach to carbon sequestration. https://forest.kerala.gov.in/images/gwarm/17bmbopacs.pdf (accessed 5 March 2020).

49 de Souza Mendonça, A.K., Silva, S.A.d., Pereira, L.Z. et al. (2020). An overview of environmental policies for mitigation and adaptation to climate change and application of multilevel regression analysis to investigate the CO_2 emissions over the years of 1970 to 2018 in all Brazilian states. *Sustainability* 12 (21): 9175.

50 Paraguassu, L. (2023). Brazil's Lula unveils plan to stop deforestation in Amazon by 2030. Reuters [Internet]. https://www.reuters.com/world/americas/brazils-lula-launches-plan-stop-deforestation-amazon-by-2030-2023-06-05/ (accessed 5 March 2024).

51 Javeline, D., Orttung, R., Robertson, G. et al. (2023). Russia in a changing climate. *WIREs Climate Change* 15 (3): e872.

52 Rahman, K. and Newsweek (2022). Russians keep mysteriously falling from windows to their deaths. https://www.newsweek.com/russians-keep-mysteriously-falling-windows-deaths-1738954 (accessed 5 March 2022).

53 Hafner, M. and Tagliapietra, S. (ed.) (2020). *The Geopolitics of the Global Energy Transition [Internet]*. Springer Nature https://library.oapen.org/handle/20.500.12657/39553 (accessed 5 March 2024).

54 Ministry of Environment, Forest and Climate Change, Government of India (2022). India's long-term low-carbon development strategy [Internet]. https://

unfccc.int/sites/default/files/resource/India_LTLEDS.pdf) accessed 5 March 2023).

55 European Parliament (2022). China's climate change policies. https://www.europarl.europa.eu/RegData/etudes/BRIE/2022/738186/EPRS_BRI(2022)738186_EN.pdf (accessed 5 March 2024).

56 Climate Action Tracker (2023). China. https://climateactiontracker.org/countries/china/ (accessed 5 March 2024).

57 Australian Government DCCEEW (2022). Australia's climate change strategies – DCCEEW [Internet]. https://www.dcceew.gov.au/climate-change/strategies (accessed 5 March 2024).

58 Government of Canada E and CC (2020). Climate change adaptation plans and actions [Internet]. https://www.canada.ca/en/environment-climate-change/services/climate-change/adapting/plans.html (accessed 6 March 2024).

59 Sheil, D., Bargués-Tobella, A., Ilstedt, U. et al. (2019). Forest restoration: transformative trees. *Science* 366 (6463): 316–317.

60 European Environment Agency (2024). Climate change mitigation: reducing emissions [Internet]. https://www.eea.europa.eu/en/topics/in-depth/climate-change-mitigation-reducing-emissions (accessed 6 March 2024).

61 Environment M of C and. Government.no. regjeringen.no (2017). Better growth, lower emissions – the Norwegian Government's strategy for green competitiveness. https://www.regjeringen.no/en/dokumenter/bedre-vekst-lavere-utslipp--regjeringens-strategi-for-gronn-konkurransekraft-engelsk/id2575420/ (accessed 7 March 2024).

62 Kempken, T., Hauck, T., DeSantis, M., et al. (2021) Collection of possible decarbonisation barriers (deliverable D1.5). [Internet]. Centre for European Policy Studies (CEPS), Brussels (Belgium): Green Steel for Europe Consortium. https://www.estep.eu/assets/Uploads/D1.5-Collection-of-possible-decarbonisation-barriers.pdf (accessed 2 March 2024).

63 Federal Transit Administration (2021). Bipartisan Infrastructure Law | FTA [Internet]. https://www.transit.dot.gov/BIL (accessed 11 March 2024).

64 Schakowsky, J. (2021). Build Back Better Act | Congresswoman Jan Schakowsky [Internet]. http://schakowsky.house.gov/build-back-better-act (accessed 11 March 2024).

65 European Commission, Access2Markets (2023). Carbon border adjustment mechanism (CBAM) | Access2Markets. https://trade.ec.europa.eu/access-to-markets/en/news/carbon-border-adjustment-mechanism-cbam (accessed 20 February 2024).

66 HM Government UK. GOV.UK (2023). Participating in the UK emissions trading scheme (UK ETS). https://www.gov.uk/government/publications/participating-in-the-uk-ets (accessed 2 August 2023).

67 HM Government UK. GOV.UK (2016). EU ETS: carbon markets. https://www.gov.uk/guidance/eu-ets-carbon-markets (accessed 7 March 2024).

68 Fraunhofer Institute (2009). Methodology for the free allocation of emission allowances in the EU ETS post 2012 Sector report for the iron ore industry By order of the European Commission Study Contract:

07.0307/2008/515770/ETU/C2 Ecofys project Number: PECSNL082164 [Internet]. p. 1–15. https://climate.ec.europa.eu/system/files/2016-11/bm_study-iron_ore_en.pdf (accessed 7 June 2023).

69 European Commission Innovation fund – European Commission [Internet]. https://climate.ec.europa.eu/eu-action/eu-funding-climate-action/innovation-fund_en (accessed 7 March 2024).

70 European Commission (2024). Environment – life: about life [Internet]. https://wayback.archive-it.org/12090/20220915214907/https://ec.europa.eu/environment/archives/life/about/index.htm (accessed 7 March 2024).

71 COM (European Commission) (2023). Horizon Europe [Internet]. https://research-and-innovation.ec.europa.eu/funding/funding-opportunities/funding-programmes-and-open-calls/horizon-europe_en (accessed 6 July 2023).

72 European Commission (2022). Carbon removal certification – European Commission [Internet]. https://climate.ec.europa.eu/document/download/fad4a049-ff98-476f-b626-b46c6afdded3_en?filename=Proposal_for_a_Regulation_establishing_a_Union_certification_framework_for_carbon_removals.pdf (accessed 7 March 2024).

73 European Commission (2024). Carbon removal certification – European Commission [Internet]. https://climate.ec.europa.eu/eu-action/sustainable-carbon-cycles/carbon-removal-certification_en (accessed 8 March 2024).

74 European Commission (2024). Environmental technology verification (ETV) [Internet]. https://ec.europa.eu/newsroom/env/newsletter-archives/6030 (accessed 8 March 2024).

75 European Commission (2024). About EMAS [Internet]. https://green-business.ec.europa.eu/eco-management-and-audit-scheme-emas/about-emas_en (accessed 8 March 2024).

76 European Commission (2024). Environmental footprint methods – European Commission [Internet]. https://green-business.ec.europa.eu/environmental-footprint-methods_en (accessed 8 March 2024).

77 AKEPA (2024). A guide to sustainability certifications [Internet]. https://thesustainableagency.com/blog/sustainability-certifications-and-ecolabels-guide/#climateneutral (accessed 8 March 2024).

78 Provenance (2023). 10 key sustainability certifications for businesses | Provenance [Internet]. https://www.provenance.org/news-insights/10-key-sustainability-certifications-for-businesses (accessed 7 March 2024).

79 Carbon Trust (2021). Route to net zero standard. https://www.carbontrust.com/what-we-do/net-zero-transition-planning-and-delivery/route-to-net-zero-standard (accessed 8 March 2024).

80 Kenlon, T. (2024). Which of these 46 sustainability certifications is right for you? | GreenBiz [Internet]. https://www.greenbiz.com/article/which-these-46-sustainability-certifications-right-you (accessed 7 March 2024).

81 Climate Action Tracker (2024). Home [Internet]. https://climateactiontracker.org/ (accessed 8 March 2024).

82 Bundesamt für die Sicherung der nuklearen Versorgung (2023). BASE. Der Atomausstieg in Deutschland. https://www.base.bund.de/DE/themen/kt/ausstieg-atomkraft/ausstieg_node.html (accessed 8 March 2024).

83 Climate Action Tracker (2024). Germany [Internet]. https://climateactiontracker.org/countries/germany/ (accessed 8 March 2024).

84 EUROSTAT (2023). Renewable energy statistics [Internet]. https://ec.europa.eu/eurostat/statistics-explained/index.php?title=Renewable_energy_statistics (accessed 8 March 2024).

85 Hanson, I and Energy Monitor (2024). How China is quietly becoming a renewables powerhouse [Internet]. https://www.energymonitor.ai/features/explainer-how-china-is-quietly-becoming-a-renewables-powerhouse/ (accessed 8 March 2024).

86 Giles, C. (2022). Financial Times: Brexit and the economy: the hit has been 'substantially negative' [Internet]. https://www.ft.com/content/e39d0315-fd5b-47c8-8560-04bb786f2c13 (accessed 14 July 2023).

87 Tetlow, G. and Stojanovic, A. (2018). Understanding the economic impact of Brexit [Internet]. https://www.instituteforgovernment.org.uk/sites/default/files/publications/2018%20IfG%20%20Brexit%20impact%20[final%20for%20web].pdf (accessed 18 July 2020).

88 Nandi, A. and Luthra Reichl, R. (2023). How did levels of UK hate crime change during and after Covid-19? [Internet]. Economics Observatory. https://www.economicsobservatory.com/how-did-levels-of-uk-hate-crime-change-during-and-after-covid-19 (accessed 14 July 2023).

89 Farla, J.C.M. and Hendriks, C.A. (1995). Carbon dioxide recovery from industrial processes. *Climatic Change* 29: 439–461.

90 Sausen, R. and Schumann, U. (2000). Estimates of the climate response to aircraft CO_2 and NO_x-emission scenarios. *Climatic Change* 44: 27–58.

91 Sheffield, J.W. and Sheffield, Ç. (ed.) (2007). *Assessment of Hydrogen Energy for Sustainable Development [Internet].* Dordrecht: Springer Netherlands (NATO Science for Peace and Security Series C: Environmental Security). http://link.springer.com/10.1007/978-1-4020-6442-5 (accessed 5 February 2024).

92 House, K.Z., Baclig, A.C., Ranjan, M. et al. (2011). Economic and energetic analysis of capturing CO_2 from ambient air. *Proceedings of the National Academy of Sciences USA.* 108 (51): 20428–20433.

93 COM (European Commission) (2015). Communication from the Commission to the European Parliament, The Council, The European Economic and Social Committee and the Committee of the Regions Closing the loop – an EU action plan for the Circular Economy [Internet]. https://eur-lex.europa.eu/legal-content/EN/TXT/?uri=CELEX:52015DC0614

94 O'Callaghan, J. (2018). Can we produce enough green hydrogen to save the world? | Research and Innovation [Internet]. https://ec.europa.eu/research-and-innovation/en/horizon-magazine/can-we-produce-enough-green-hydrogen-save-world (accessed 11 July 2023).

95 Aylen, J. (1980). Innovation in the British Steel industry. In: *Technical Innovation and British Economic Performance [Internet]* (ed. K. Pavitt), 200–234.

London: Palgrave Macmillan UK http://link.springer.com/10.1007/978-1-349-04561-7_12 (accessed 2 November 2022).

96 Canadell, J.G., Dickinson, R., Hibbard, K. et al. (2003). Global Carbon Project (GCP) [Internet]. https://www.globalcarbonproject.org/science/sfi.htm (accessed 6 July 2023).

97 COM (European Commission) | CORDIS (2010). Ultra-Low CO_2 steelmaking | ULCOS Project | Fact Sheet | FP6. https://cordis.europa.eu/project/id/515960 (accessed 10 July 2023).

98 European Cement Research Academy (ECRA) (2009). *ECRA CCS roject and Report About Phase II [Internet].* Düsseldorf: ECRA https://ecra-online.org/fileadmin/redaktion/files/pdf/ECRA__Technical_Report_CCS_Phase_II.pdf (accessed 4 March 2020).

99 European Cement Research Academy (ECRA) Development of state of the art-techniques in cement manufacturing: trying to look ahead [Internet]. https://ecra-online.org/fileadmin/redaktion/files/pdf/CSI_ECRA_Technology_Papers_2009.pdf (accessed 2 January 2021).

100 United Nations Industrial Development Organization (UNIDO) (2010). *Carbon Capture and Storage in Industrial Applications: Technology Synthesis Report Working Paper – Working Paper – [Internet].* Vienna: https://www.unido.org/sites/default/files/2010-12/synthesis_final_0.pdf (accessed 7 July 2023).

101 United Nations Industrial Development Organization (UNIDO) (2011). Carbon capture and storage – Industrial Sector Roadmap | UNIDO [Internet]. https://www.unido.org/our-focus/safeguarding-environment/clean-energy-access-productive-use/industrial-energy-efficiency/selected-projects/carbon-capture-and-storage-industrial-sector-roadmap (accessed 14 July 2023).

102 Zakkour, P. and Cook, G. (2010). *UNIDO CCS Roadmap for Industry: High-Purity CO_2 Sources. Sectoral Assessment and Final Draft Report.* Vienna, Austria: United Nations Industrial Development Organization.

103 DeConinck, H. and Mikunda, T. (2014). Carbon Dioxide Capture and Storage: Issues and Prospects. *Annual Review of Environment and Resources* 39 (1): 243–270.

104 UNFCC (2010). COP 16 | UNFCCC [Internet]. https://unfccc.int/event/cop-16 (accessed 14 July 2023).

105 COM (European Commission) (2018). How Horizon Europe was developed [Internet]. https://research-and-innovation.ec.europa.eu/funding/funding-opportunities/funding-programmes-and-open-calls/horizon-europe/how-horizon-europe-was-developed_en (accessed 6 July 2023).

106 Omoregbe, O., Mustapha, A.N., Steinberger-Wilckens, R. et al. (2020). Carbon capture technologies for climate change mitigation: a bibliometric analysis of the scientific discourse during 1998–2018. *Energy Reports* 6: 1200–1212.

107 Yi, S.H., Choi, M.E., Kim, D.H. et al. (2019). FINEX® as an environmentally sustainable ironmaking process. *Ironmaking & Steelmaking* 46 (7): 625–631.

108 UNFCCC (2012). The Doha Amendment | UNFCCC [Internet]. https://unfccc.int/process/the-kyoto-protocol/the-doha-amendment?

gclid=EAIaIQobChMI8sDOgKiGgAMVtYZoCR3uuwx4EAAYASAAEgIhzPD_BwE (accessed 11 July 2023).
109 British Broadcasting (BBC) (2022). Cumbria coal mine: decision delayed until November. https://www.bbc.co.uk/news/uk-england-cumbria-62499981 (accessed 30 August 2022).
110 Watt, H. (2017). Hinkley point: the 'dreadful deal' behind the world's most expensive power plant. The guardian [Internet]. https://www.theguardian.com/news/2017/dec/21/hinkley-point-c-dreadful-deal-behind-worlds-most-expensive-power-plant (accessed 14 July 2023).
111 BBC News (2023). Government to offer £600m for green steel switch [Internet]. https://www.bbc.com/news/uk-64366998 (accessed 14 July 2023).
112 World Economic Forum (WEF) (2019). The $86 trillion world economy – in one chart. https://www.weforum.org/agenda/2019/09/fifteen-countries-represent-three-quarters-total-gdp/ (accessed 14 July 2023).
113 World Economic Forum (WEF) (2023). 5 ways to remove carbon and tackle the climate crisis. https://www.weforum.org/agenda/2023/09/carbon-removal-climate-crisis/ (accessed 15 November 2023).
114 Francis-Devine, B., Barton, C., Harari, D. et al. (2023). *Rising Cost of Living in the UK [Internet]*. London: House of Commons Library https://commonslibrary.parliament.uk/research-briefings/cbp-9428/ (accessed 14 July 2023).
115 British Chamber of Commerce (BCC) (2023). BCC economic forecast: upgrade to GDP but UK economy flatlining [Internet]. https://www.britishchambers.org.uk/news/2023/06/bcc-economic-forecast-upgrade-to-gdp-but-uk-economy-flatlining (accessed 14 July 2023).
116 HM Revenue and Customs. GOV.UK (2022). Corporation tax rates and allowances. https://www.gov.uk/government/publications/rates-and-allowances-corporation-tax/rates-and-allowances-corporation-tax (accessed 14 July 2023).
117 VisitBritain (2023). 2023 tourism forecast. https://www.visitbritain.org/2023-tourism-forecast (accessed 14 July 2023).
118 Office for National Statistics (2021). Population estimates – Office for National Statistics [Internet]. https://www.ons.gov.uk/peoplepopulationandcommunity/populationandmigration/populationestimates (accessed 14 July 2023).
119 Office for National Statistics (2021). Population estimates for the UK, England and Wales, Scotland and Northern Ireland – Office for National Statistics [Internet]. https://www.ons.gov.uk/peoplepopulationandcommunity/populationandmigration/populationestimates/bulletins/annualmidyearpopulationestimates/mid2018 (accessed 14 July 2023).
120 Ernst & Young (2023). Cost of living pressures set to intensify the UK's regional economic divide [Internet]. https://www.ey.com/en_uk/news/2023/02/cost-of-living-set-to-intensify-the-uks-regional-economic-divide (accessed 14 July 2023).
121 UK Parliament (2022). Water quality in rivers – Environmental Audit Committee [Internet]. https://publications.parliament.uk/pa/cm5802/cmselect/cmenvaud/74/report.html (accessed 4 July 2023).
122 IEA (2022). Overview – global methane tracker 2022 – analysis. https://www.iea.org/reports/global-methane-tracker-2022/overview (accessed 5 March 2024).

123 Office for National Statistics (ONS) (2023). Foreign direct investment, experimental UK subnational estimates: 2021 – Office for National Statistics [Internet]. https://www.ons.gov.uk/releases/foreigndirectinvestmentexperimentaluk-subnationalestimatesmarch2023 (accessed 14 July 2023).

124 Rhodes, C., Hutton, G., and Ward, M. (2023). Research and development spending. https://commonslibrary.parliament.uk/research-briefings/sn04223/ (accessed 14 July 2023).

125 CIRAIG (International Reference Centre for the Life Cycle of Products, Processes and Services) (2015). *Circular Economy: A Critical Literature Review of Concepts*. Canada: Polytechnique Montréal.

9

Skills set required within the different components and sectors

9.1 Introduction

The effective implementation of the Bio Steel Cycle model and connected technologies requires in parts specialised workers and staff to achieve the best efficiency. This chapter offers insight into the opportunities for retraining workers from fossil-fuel-based sectors, such as mining, energy generation and oil refinery to being environmentally benign sector employees. Renewable energy technologies and biomass applications including green hydrogen production are the key sectors which require skilled workers in the future. In order to ascertain the different opportunities for retraining workers, the different steps and components required to build and install the different renewable energy systems need to be explored within a low-carbon steel economy.

9.2 Solar

9.2.1 Stages of producing solar panels

It so seems that when any conversation concerns solar panels, most people appear to only recognise solar photovoltaic (PV) panels for generating electricity on their own roofs. But there are also solar thermal panels generating hot water to consider.

The manufacturing of solar PV and thermal panels is distinctly different, and some details will be provided in the following text.

9.2.1.1 Solar photovoltaic panels

Most solar manufacturing in the United States as per 2023 is focused on producing PV panel systems. These energy-generating installations consist of PV modules, frames and wiring, power electronics and system monitoring devices as well as battery storage components [1].

Commercially available PV modules in the United States rely on crystalline silicon as the absorber material [1], although manufacturers in the EU, China and the United States are increasingly using borosilicate plates due to their many specific

Circular Steel Production: Pathways to Net-Zero Carbon Emissions, First Edition.
Sandra Kiessling, Hamidreza Gohari Darabkhani, and Abdel-Hamid Soliman.

benefits, such as the resistance to thermal shocks, meaning it does not break during drastic temperature changes [2]. The PV modules are manufactured in several subsequent steps as follows:

1) Polysilicon production
2) Ingot and wafer production
3) Cell fabrication
4) Module assembly

9.2.1.2 Thin-film PV

The term thin-film PV refers to a range of different absorber materials. The most commonly used material is called cadmium telluride (CdTe), although a variety of materials has recently been researched for utilisation in thin-film PV modules, such as printed solar foils [3, 4]. Printed and manufactured thin-film solar foils can be retrofitted to almost any surface with exposure to sunlight or daylight. Besides being ultrathin and lightweight, they are also very efficient (>35%). Typically, solar PV panels and foils are manufactured in one facility, with all the steps mentioned taking place [1]. This reduces GHG emissions from transport and makes the production of solar panels a streamlined and very efficient manufacturing process, in line with net-zero policies. The first step in manufacturing usually starts with the float glass (top cover) being coated with a transparent and highly conductive layer, and the PV absorber material is then added in close-spaced sublimation. To pattern the cell strips, a network of interconnected pathways between the collection of arranged cells is created using laser scribing. Copper ribbons are applied; two sealed sheets of glass are placed on top, and the entire component is laminated to ensure that it is weatherproof. The finished modules are fitted with a junction box, and passing through this, electrical cables are connected to the copper ribbons through holes in the back of the cover-glass sheet [2].

9.2.1.3 Racking systems

Racking systems are the support structures which are custom-built, load-bearing PV modules on roofs or mounted on the ground, the manufacture of which varies significantly, depending on the site location parameters. For ground-mounted PV panels, racking is made from coated and galvanised steel frames which are attached to concrete foundations. Some large ground-mounted systems also feature one-axis tracking mechanisms, to capture the sunrays as the sun moves from east to west. These tracking devices require electrical and mechanical parts, such as sensors, timers, cables, motors and bearings. The constellation of roof-mounted racking systems depends on the pitch and type of roof at the installation site. Installation of flat roofs is relatively easy, and solar PV panels are usually mounted onto fixed-tilt steel racking. The installation of solar PV systems onto pitched residential roofs is individual, as the racking is designed to attach securely to the roof rafters and is supposed to keep the panels a few inches above the roof. This type of installation is vitally important, as this allows airflow around the panels, cooling the rear of the modules, which improves their performance [2].

9.2.1.4 Power electronics

The necessary power electronics for PV systems, including power optimisers and inverters, are preassembled on electronic circuit boards. Another vital component in any solar PV installation is battery storage, as the harvested solar energy needs to be stored for peak-time usage and when there is no sunshine or daylight (night-time). The connected inverter converts direct current (DC) electricity, generated by the solar panels, to alternating current (AC) electricity, which is used by the owner and surplus added to the national electrical grid [2].

9.2.1.5 Solar thermal panels

There are three main types of concentrating solar thermal power systems [5]:

- Linear concentrating systems, including parabolic troughs and linear Fresnel reflectors
- Solar power towers
- Solar dish/engine systems

Solar thermal panel systems are so-called wet closed systems, as they have water travelling through the pipework. They can produce between 40% and 60% of the hot water consumption of the average residential house. The average house produces ~4.54 t/CO_2/year via the heating system, with 30% of this apportioned for hot water generation. Therefore, for hot water, this equates to CO_2 emissions savings of ~0.9 t [5, 6].

The basic working principle of solar thermal panel installations is to utilise the sun's energy and simply convert it into heat. This heat is then transferred into your building as hot water. From there, it can either be used as hot water or can be transferred into space heating and underfloor heating system loops. The majority of systems consist of roof-mounted solar panels, which are connected via pipes with a central heating boiler, water storage device, collector or immersion heater. The solar thermal collector uses the sunrays to heat a transfer fluid which runs through the solar thermal panels mounted on the roof. From there, this fluid (a mixture of water and glycol [antifreeze]) is pumped to a heat exchanger, which is usually located inside the water storage tank in your dwelling. The absorbed heat from the heat exchanger will then be used to heat up the water inside the tank. Using gravitational force and circulation pumps, and after the liquid has released its heat to the heat exchanger, the water has cooled down and flows back to the collectors for reheating. And now, the cycle starts all over. A control panel and system internal sensors ensure that the fluid will circulate to the collector when there is sufficient heat available and/or with higher demand [6].

Now, for all these components, skilled and specialised workers are required, from soldering the control panels, overconstructing the sensors and connectors, manufacturing the plate heat exchangers and the pipes running through, welding the panel frames, and engineers carrying out the installation on site. Additionally, service engineers to ensure the maintenance of the installed solar thermal and PV panels once they have been installed.

There are further roles involved for the production, marketing, installation and maintenance. Further, skilled staff is needed to compile the amortisation plans, create the planning applications and see the applications through to the building stage, land agents to identify suitable sites, finance specialists and insurance companies. There are plenty of roles which would suit workers and staff from former fossil-fuel-based operations, which could be retrained or upskilled to work in these new environments.

9.3 Wind turbines

Materials used to make wind turbines are, according to a report from the National Renewable Energy Laboratory, depending on the make and model:

- Steel (66–79% of total turbine mass)
- Fibreglass, resin or plastic (11–16%)
- Iron or cast iron (5–17%)
- Copper (1%)
- Aluminium (0–2%) [7–9]

Wind turbines are usually built from the ground up or from the seabed up for offshore wind turbines. An average large-sized 70 KW vertical wind turbine contains c. 8000 parts, including the foundation, pillar (tower), nacelle (which contains the motor for the blades and much more) and rotor blades. In the United States, more than 500 facilities are manufacturing wind turbine components [8].

In order to erect a wind turbine, a concrete foundation needs to be constructed. The first step is to level the ground and lay down a concrete foundation, which could take anything between 600 and 1000 tonnes of concrete and 165 tonnes of steel.

For the installation of an offshore wind turbine, building the foundation follows a different pattern. Commercial construction ships sail to the sites which can be up to 70 m deep. The ship extends steel legs, rooting the ship to the seabed and elevating the ship out of the water. Then, an on-board crane lowers a steel tower monopile or a different anchoring structure, i.e. a jacket, tripod or gravity base, into the water, where mechanical drills fix these into the sea-floor. A transition piece is then lowered above the monopile or anchored foundation, onto which the tower will be fixed with stainless steel cables. For deeper than 70 m anchored turbines or floating offshore wind turbines, wind turbines are placed atop buoyant substructures at industrial port facilities and towed with tug-boats to their location. Mooring lines are then connected to the structure and connected to anchors which have been rooted into the seabed [8].

For onshore wind turbine installations, once the foundation has been installed, heavy goods vehicles (HGV) bring the preassembled tower horizontally transported to its final destination. Here, the steel pillar is finish assembled on site with the tower lying horizontally on the ground. When assembly has been completed, a crane pulls the tower upright and it is then fixed to the foundation [7–9].

For offshore wind, the preassembled turbine steel tower is transported by ship to the foundation. The tower is then guided into position from a crane, to arrive on top of the transition piece. It is then fixed either to the foundation on the seabed or to the floating substructure and secured and anchored to the seabed with stainless steel ropes [7–9].

The electrical box (nacelle) sitting in the middle of the blades houses the gearbox, the generator and the electronic control and transfer systems for each wind turbine. For both onshore and offshore wind turbines, the nacelle is lifted with a crane onto the top of the tower, which has been fixed to the foundation beforehand. The nacelle houses either a permanent magnet generator (PMG) or an induction generator. Although they are slightly more costly, 36% of onshore and 76% of offshore wind turbines use PMGs. They are simply much more efficient for larger turbines, and, as they do not need a gearbox, PMGs are much quieter and need less service or maintenance. The downside of PMGs is that they are made using a combination of rare earth elements [7–9]: neodymium, praseodymium, dysprosium and terbium.

The blades (rotors) had an average rotor diameter (the size of the full blade circle) of c. 140 m, per 2021, and this means they were at this point 164% bigger than compared with blades in 1999. In 2021, in the United States, the average wind turbine capacity was 319% in comparison with 1999. The blades are basically built using the sandwich principle: large blade-shaped moulds are filled with dry glass fibres, which are then injected with a resin, which will fill in all the air pockets between the fibres. This mass is then heated and becomes a carbon fibre structure, to the back of which a carbon spine 'spar' is added for stability and strength. Suitably shaped balsa wood, industrial expanding foam and sometimes recycled plastic are then added as the next layer, and they add strength without adding too much weight. Onto this structure, more dry fibres are added and again injected with resin. Strong supports (webs) are installed vertically within the mould, which will act as a supportive brace within the hollow centre of the blade, not unlike honeycomb. Two identical blade halves are produced this way and, as a final step, folded over and secured on top of each other. On average, it takes up to two days to manufacture wind turbine blades and up to 100 employees. Standard wind turbines require three blades, and once they have been delivered on site, computer-assisted operators will guide the three blades into the nacelle [7–9].

The eight stages of building a wind turbine [10–12] are, as follows:

1) Planning and method statement
2) Site risk assessment
3) Loading the wind turbine for transport onto an HGV. Pre-transport arrangements for vehicles with exceptional width are required and possibly an escort vehicle (police)
4) Transportation of the wind turbine to the site
5) Delivering the wind turbine on site
6) Crane-lifting the wind turbine components into position
7) Finish assembly of wind turbine components and installation on site
8) Mechanical and electrical fit out

In order to build a wind turbine, anywhere, there are certain administrative stages to go through, which are, as follows:

- Site and register screening and consultations
- Application surveys, preparation and submission
- Application management
- Post-consent and appeals [11, 12]

Wind turbine technicians should be familiar with all systems concerning the operation of the wind turbine, including structural, mechanical, electrical, controls, communications, electronic components, hydraulic systems and environmental safety and health systems. As a wind turbine technician, the following skills and abilities are required [13]:

- The ability to use, repair and maintain specialist tools and equipment
- Good mathematical skills and knowledge of engineering science and technology
- Problem-solving skills
- Being meticulous and conscientious
- Knowledge of computer operating systems, hardware and software

Skills needed for wind energy systems manufacture, installation, operation and maintenance are, as follows:

Wind turbine technicians (building and maintaining); civil engineers to plan, erect and maintain the turbines; surveyors to investigate, verify and achieve commencement of proceedings; builders to lay the concrete foundations; companies for the transport of the heavy and large wind turbine components (motor, motor housing, pillar segments and blades); agencies to apply for planning permissions, apply for grants, subsidies, incentives and loans, apply for SEGs and correspond with the National Grid for SEGs and other payments; insurers to secure the building and running of the site; and investors' roles in producing, marketing, installing and maintaining the wind turbine.

9.4 Hydro

Renewable energy technology generating electricity using either biomass or hydro turbines (water-based) and anaerobic digestion (AD) systems [14–17] are also included in some of the schemes currently supported by the 2023/2024 UK government's incentives to promote sustainable energy generation, as displayed in Chapter 5: Greater Energy Independence with Sustainable Steel Production [18]. Currently, there are closed-loop and open hydropower systems available. Given the volatility of UK river water levels and contamination with faeces released by water companies, closed-loop hydropower systems might be the technology of choice. The image in Chapter 4, Figure 4.5, demonstrates the layout of an open-loop system, and in Figure 9.1 below.

Additionally, the installed renewable energy technologies (solar, wind, geothermal and hydropower) and the back-feeding of biogas from AD all provide the industrial

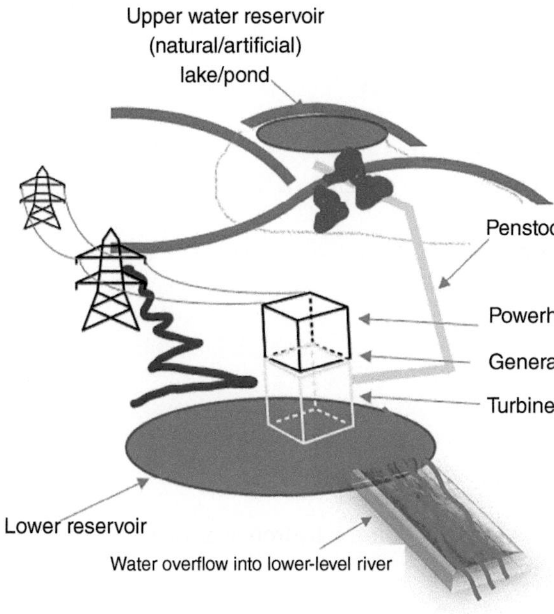

Figure 9.1 Open-loop hydropower system configuration.

processes with renewable energy, from their own infrastructure. Thus, the energy produced on site will make most industries inevitably more energy independent.

The principle of building a hydroelectric power plant has to undergo similar preliminary stages as the wind turbine building process and also the same administrative stages. Additionally, there are marine biologists and biologists required for the site assessment, to make sure that the planned hydropower plant is not having a detrimental impact on local flora and fauna and to receive recommendations for location, installation schedule and capacity to build.

The considerations to be undertaken before building a hydropower station give a clear indication of the skill sets required by workers and staff in these innovative industrial sectors [19]:

- Assessment of the hydrology
- Stream flow determination, hydrographs and flow duration curves
- Mass curve assessment and determination for storage capacity planning
- Site investigation
- Assessment of whether there are dams required for the power project
- Head and power
- Types of dam, if required
- Arrangement and location of hydroelectric station
- Determination of location, size and capacity of spillways, intake, fore-bay, penstock, water hammer and surge tank
- Power capacity to be developed
- Size of plant and choice of units

- Types of turbine and characteristics
- Runaway speed
- Design of the main dimensions of the turbines
- Main dimensions of reaction turbines
- Main dimensions of impulse wheel turbines
- Draft tubes: types, setting and preliminary dimensions
- Scroll-case
- Preliminary design of penstock
- Characteristics of generators
- Exciters for hydro generators [19]
- Power station control
- Main dimensions of generators
- Power station structure and layout
- Costs of hydroelectric stations
- Assessment of potential issues [19]

Taking the example of an open-loop system, the hydropower plant requires the main components of the following:

- A main water source at the base (pond)
- A pump station submerged in the pond operating the housed turbines and pumps
- Pipes running up to the elevated position of the second pond
- Converters and generators
- Control panels and sensors
- Connection cables to transfer the energy generated to the power grid
- Safety appliances
- Water release and shut-off barriers

There is a lot more to a hydropower station or plant, but for the purpose of illustrating the immense opportunities posed by renewable energy technologies, such as hydropower, this will have to suffice.

9.5 Geothermal

Geothermal energy is a complex science, but accessing it can be relatively simple. Although similarly complex preliminary procedures as with wind turbines and hydropower station building projects need to be followed, for which workers and staff are desperately needed, the technology is well established, tried and tested (Figure 9.2) [20].

The general principle can be explained easily:

1) Hot water under high pressure needs to be pumped up from deep underground
2) The water reaches the surface, the pressure drops, and the water turns into steam
3) This steam then spins a turbine, which is connected to a generator, producing electricity
4) The cooling tower condenses steam back to water

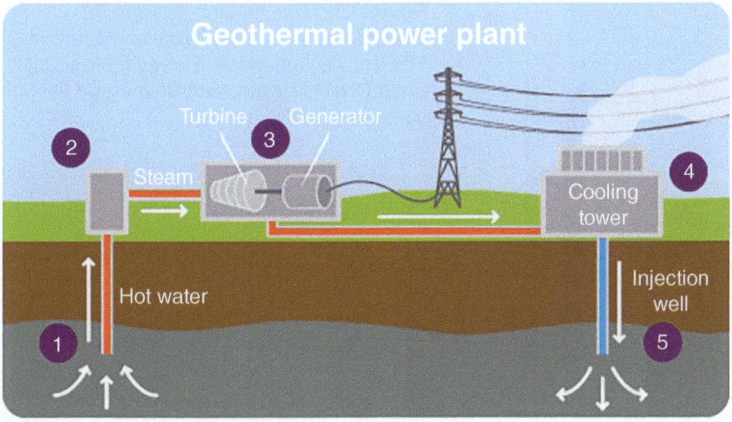

Figure 9.2 Geothermal power plant configuration. Source: US EPA O [20] / United States Environmental Protection Agency / Public domain.

5) The cooler water is injected back to where it came from to begin a full cycle again [20]

As can be observed in the image above, the components of a geothermal power plant are quite numerous and complex. Technicians currently involved in fossil-fuel-based energy generation would possibly find it quite easy to transition into this renewable energy technology.

9.6 Green hydrogen

Green hydrogen means hydrogen produced using exclusively renewable energy sources of electricity [21]. Electrolysis is the process which uses electricity to split water molecules (H_2O) into their components hydrogen (H_2) and oxygen (O_2), and the unit in which this reaction takes place is called an electrolyser. These can range from small, home appliance-size equipment which is well suited for small-scale distributed hydrogen production up to large-scale, central power stations and hydrogen production facilities which can be connected directly to renewables or other non-greenhouse-gas-emitting forms of energy generation (Figure 9.3).

Very similar to traditional fuel cells, electrolysers contain an anode and a cathode, separated by an electrolyte agent. Although a variety of electrolysers has been developed, the basic working principle is always the anode/cathode/electrolyte system.

At present, one of the biggest steel producers, Tata Steel, is working on a project in the Netherlands (Figure 9.4) to generate green hydrogen directly from seawater. The added complication of a high salt content is taken into consideration, and the results, so far, have been impressive.

It is entirely possible that there are even more employment opportunities at renewable energy power plants, such as hydrogen production plants and hydropower plants, than the equivalent in the fossil fuel sector. With most heavy

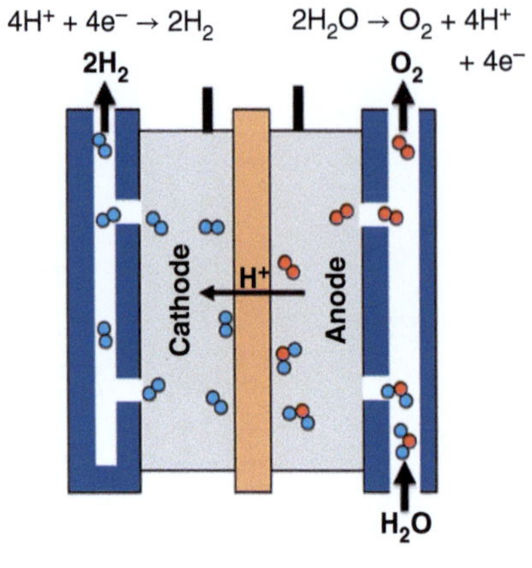

Figure 9.3 The working principle of electrolysis. Source: US Department of Energy Efficiency & Renewable Energy (EERE) [21] / United States Department of Energy / Public domain.

Figure 9.4 Tata Steel Ijmuiden (NL) location of planned project hydrogen production facility [24]. Source: Alf van Beem / Wikimedia / CC0 1.0.

industries making huge strides towards decarbonisation [22], an increasing number of opportunities are opening up [23], and it is hoped that the transition from a fossil-fuel-based society towards a net-zero society will provide the people depending on this work at least a living wage if not more.

9.7 Biomass

Replacing coal and coke in the steelmaking process with biomass [25] will quite possibly reduce the CO_2 emissions by at least 30% [26, 27]. Innovative technologies such as HIsarna [28] and GrynHy [29, 30] and hydrogen direct reduction (HDR) have CO_2-saving potential in their own right, as already explained in more detail in Chapter 4: The key components of the Bio Steel Cycle (BiSC). Removing coal as a primary energy source or using (green) HDR (essentially biomass), an immediate 30% CO_2 emissions reduction is possible [26, 31–37], and therefore, replacing coal with

Figure 9.5 Miscanthus grass. Source: Unsplash [39] Kin Chan / Unsplash.com.

green hydrogen or other biomass would reduce the CO_2 emissions from steelmaking potentially by the same percentage. According to Siemens [38], even a drastic 50% carbon emissions reduction is immediately possible via utilisation of green HDR. Biomass can be derived from energy crops, such as miscanthus grass, as in the image below (Figure 9.5).

Growing energy crops could be an opportunity for farmers seeking farm diversification opportunities and for owners of brownfield sites unsuitable for human habitation to generate income streams, and, thus, places of work. Ideally, these crops would be grown on brownfield sites which already are accommodating onshore wind turbines or parks, as these crops are quite hardy once established and are not affected by the whooshing of the nearby wind turbine blades.

9.8 Biogas

As already mentioned in Chapter 4: 4.8 Anaerobic digestion – biogas is a product as a result of processes taking place during AD. Biogas is the end-product of recycled waste or organic matter and a vital component of green steel, produced using the BiSC model and strategy. Biogas is being fed back into the steelmaking process, with the raw materials for its production resulting from debris from reforested areas utilising DAC, treating slags and waste water and organic matter from external sources. Particularly, the organic matter sourced from external third parties could potentially combine three or more industries and solve multiple issues in one sweeping step: the steel industry, agriculture and agri-waste and waste water treatment industries. By motivating landowners and farmers alike to dispose of their wastes at the nearby anaerobic digester facility, less nitrogen is being deposited on arable land and in bodies of water. More needs to be done to interconnect the various industries, such as agriculture and steel, which would naturally not have any overlapping interests. However, as demonstrated in this book, the technology is ready to be implemented.

9.9 Conclusion

It seems obvious that here are a lot of different skill sets required within the various renewable energy technology sectors which provide exciting opportunities for workers and staff who might find themselves in transition from fossil-fuel-based positions, such as steel-workers at steel plants, to replacing BF with the much more energy-efficient and less polluting EAF.

The transition to renewable energy sector employment in the steel industry has enormous potential for growth and to build up energy independence for businesses willing to transition towards decarbonisation of the steel industry.

It is now up to the individual countries' governments, industry leaders, employers and operators who are willing to transform their business to meet net-zero and decarbonisation targets to make these retraining and upskilling opportunities available and take along their loyal workforce on their journey to a net-zero economy. This chapter briefly explained the skills required and the low-carbon and renewable technologies for implementing the BiSC.

References

1 US EERE Energy Efficiency and Renewable Energy (2024). Solar photovoltaic manufacturing basics. Energy.gov. https://www.energy.gov/eere/solar/solar-photovoltaic-manufacturing-basics (accessed 11 March 2024).
2 SCIENCE Chemistry (2022). Borosilicate glass: global market analysis | Borates Today [Internet]. https://borates.today/borosilicate-glass/ (accessed 11 March 2024).
3 CSIRO (2024). Printed solar film [Internet]. CSIRO. https://www.csiro.au/en/work-with-us/ip-commercialisation/Marketplace/Printed-solar-film (accessed 11 March 2024).
4 Peacock, B. (2022). Printed solar panels developed in Newcastle to power Tesla's journey around Australia. pv magazine Australia. https://www.pv-magazine-australia.com/2022/04/26/printed-solar-panels-developed-in-newcastle-to-power-teslas-journey-around-australia/ (accessed 7 February 2024).
5 Energy Information Administration (EIA) (2022). Solar thermal power plants – U.S. Energy Information Administration (EIA) [Internet]. https://www.eia.gov/energyexplained/solar/solar-thermal-power-plants.php (accessed 3 April 2024).
6 The Renewable Energy Hub UK (2023). How do Solar Thermal Panels work [Internet]. https://www.renewableenergyhub.co.uk/main/solar-thermal-information/how-do-solar-thermal-panels-work (accessed 3 April 2024).
7 Letcher, T. (2023). *Wind Energy Engineering: A Handbook for Onshore and Offshore Wind Turbines*, 588p. Elsevier.
8 Union of Concerned Scientists (2022). The Equation. How are wind turbines made? https://blog.ucsusa.org/charlie-hoffs/how-are-wind-turbines-made/ (accessed 3 April 2024).

9 USGS (ND). What materials are used to make wind turbines? | U.S. Geological Survey [Internet]. https://www.usgs.gov/faqs/what-materials-are-used-make-wind-turbines (accessed 3 April 2024).

10 Terry (2023). The eight stages of wind turbine installation [Internet]. AIS Wind Energy. https://aiswindenergy.co.uk/the-eight-stages-of-wind-turbine-installation/ (accessed 3 April 2024).

11 Renewables First (2023). How much does a wind turbine cost ? – Renewables First. Renewables First – The Renewable Energy Company. https://www.renewablesfirst.co.uk/home/renewable-energy-technologies/windpower/windpower-learning-centre/how-much-does-a-wind-turbine-cost/ (accessed 7 July 2023).

12 Renewables First (2024). Wind turbine planning permission – Renewables First [Internet]. Renewables First – The Renewable Energy Company. https://renewablesfirst.co.uk/renewable-energy-technologies/windpower/wind-turbine-consenting/wind-turbine-planning-permission/ (accessed 3 April 2024).

13 National Careers Service (2024). Wind turbine technician | Explore careers | National Careers Service [Internet]. https://nationalcareers.service.gov.uk/job-profiles/wind-turbine-technician (accessed 3 April 2024).

14 Green City Times (2015). What is anaerobic digestion? | Green City Times [Internet]. https://www.greencitytimes.com/anaerobic-digestion-a-proven-solution-to-our-waste-problem/ (accessed 19 July 2023).

15 Al-Alawi, M. (2007). Biohydrogen production by anaerobic biological fermentation of agriculture waste. In: *Assessment of Hydrogen Energy for Sustainable Development* (ed. J.W. Sheffield and Ç. Sheffield), 177–185. Dordrecht: Springer Netherlands (NATO Science for Peace and Security Series C: Environmental Security).

16 HM Government UK (2023). Section 3: Anaerobic digestion. GOV.UK. https://www.gov.uk/government/statistics/area-of-crops-grown-for-bioenergy-in-england-and-the-uk-2008-2020/section-3-anaerobic-digestion (accessed 7 July 2024).

17 Nag, R., Auer, A., Markey, B.K. et al. (2019). Anaerobic digestion of agricultural manure and biomass – critical indicators of risk and knowledge gaps. *Science of the Total Environment* 690: 460–479.

18 Kiessling, S., Gohari Darabkhani, H., and Soliman, A.H. (2024). Greater energy independence with sustainable steel production. *Sustainability* 16 (3): 1174.

19 Deshpande, M.V. (2009). *Elements of Electrical Power Station Design*, 433p. PHI Learning Pvt. Ltd.

20 US EPA O (2017). Geothermal energy [Internet]. https://archive.epa.gov/climatechange/kids/solutions/technologies/geothermal.html (accessed 3 April 2024).

21 US Department of Energy Efficiency & Renewable Energy (EERE) (2024). Hydrogen production: electrolysis. Energy.gov. https://www.energy.gov/eere/fuelcells/hydrogen-production-electrolysis (accessed 4 April 2024).

22 Leadit (2023). Leadership Group for Industry Transition. Green Steel Tracker. https://www.industrytransition.org/green-steel-tracker/ (accessed 18 September 2023).

23 Tata Steel Europe (2023). Sustainability performance at our sites IJmuiden. Tata Steel in Europe. https://www.tatasteeleurope.com/construction/sustainability/performance-at-our-sites/ijmuiden (accessed 7 July 2023).

24 Tata Steel Nederland (2024). Green steel plant. Tata Steel. https://www.tatasteel.com/media/media-kit/gallery/ (accessed 3 April 2024).

25 Teagasc (2017). Miscanthus Energy Crop – Teagasc | Agriculture and Food Development Authority [Internet]. https://www.teagasc.ie/rural-economy/rural-development/diversification/miscanthus-energy-crop/ (accessed 21 February 2024).

26 Toktarova, A., Karlsson, I., Rootzén, J. et al. (2020). Pathways for low-carbon transition of the steel industry – a Swedish case study. *Energies* 13 (15): 3840.

27 Toktarova, A. (2021). The low-carbon steel industry-interactions between the hydrogen direct reduction of steel and the electricity system – ProQuest [Internet]. Gothenburg, Sweden: Department of Space, Earth and Environment Chalmers University of Technology. https://www.proquest.com/openview/c2a4d77d71403fdbffd33f250b13251d/1?pq-origsite=gscholar&cbl=2026366&diss=y (accessed 7 July 2023).

28 Jones, B. (2011). Tata Steel and Rio Tinto sign agreement on HIsarna™ [Internet]. Ijmuiden: COM, Tata Steel and Rio Tinto. https://www.tatasteeleurope.com/corporate/news/tata-steel-and-rio-tinto-hisarna-agreement (accessed 10 July 2023).

29 Blume, L. (2021). GrInHy2.0 – wasserstoff-niedersachsen.de [Internet]. hydrogen-lower-saxony.com. https://www.wasserstoff-niedersachsen.de/en/grinhy2-0/ (accessed 7 July 2023).

30 COM (European Commission) (2022). Green industrial hydrogen via reversible high-temperature electrolysis | GrInHy project | fact sheet | H2020. CORDIS | European Commission. https://cordis.europa.eu/project/id/700300 (accessed 6 July 2023).

31 Benetto E, Popovici EC, Rousseaux P, Blondin J. Life cycle assessment of fossil CO_2 emissions reduction scenarios in coal-biomass based electricity production. *Energy Conversion and Management*. 2004;45(18):3053–3074.

32 Worrell E, Blinde P, Neelis M, Blomen E, Masanet E. Energy efficiency improvement and cost saving opportunities for the U.S. Iron and Steel Industry an energy star(R) guide for energy and plant managers [Internet]. Report No.: LBNL-4779E, 1026806. http://www.osti.gov/servlets/purl/1026806/ (accessed 2 Novemeber 2022).

33 Kittipongvises, S. (2017). Assessment of environmental impacts of limestone quarrying operations in Thailand. *Environmental and Climate Technologies* 20 (1): 67–83.

34 Chen, Q., Gu, Y., Tang, Z. et al. (2018). Assessment of low-carbon iron and steel production with CO_2 recycling and utilization technologies: a case study in China. *Applied Energy* 220: 192–207.

35 Suopajärvi, H., Umeki, K., Mousa, E. et al. (2018). Use of biomass in integrated steelmaking – status quo, future needs and comparison to other low-CO_2 steel production technologies. *Applied Energy* 213 (C): 384–407.

36 Wilkerson, J. (2018). Harvard University Blog biomass over coal: burning different carbon to mitigate climate change [Internet]. Science in the News. https://sitn.hms.harvard.edu/flash/2018/biomass-over-coal-burning-different-carbon-to-mitigate-climate-change/ (accessed 7 July 2023).

37 Liu, Z., Ciais, P., Deng, Z. et al. (2020). Near-real-time monitoring of global CO_2 emissions reveals the effects of the COVID-19 pandemic. *Nature Communications* 11 (1): 5172.

38 SIEMENS (2022). Hydrogen solutions. siemens-energy.com. Global Website. https://www.siemens-energy.com/global/en/offerings/renewable-energy/hydrogen-solutions.html (accessed 7 July 2023).

39 Unsplash (2019). Photo by Kin Chan on Unsplash [Internet]. https://unsplash.com/photos/shallow-focus-photography-of-green-leafed-plant-meOMVubr1ck (accessed 3 April 2024).

10

The future of green steel

10.1 Introduction

The future of green steel looks promising, indeed. Based on the analysis in Chapter 8, the opportunities for green steel (by country) will be discussed in some detail, as well as the nationally determined contributions (NDC), the effects of implemented policy and decarbonisation efforts. Increasing government initiatives and partnerships with commercial entities are set to boost the production of sustainable products, and lower CO_2 emissions are expected to propel the growth of the global green steel market through 2030: the global green steel market is set to rise at a monumental compound annual growth rate (CAGR) of 122.9% during the period of assessment 2023–2030. The green steel market is estimated to reach an approximate market value of around US$47.2 billion by 2030 [1–3]. Being an essential resource for the engineering and construction industry, steel is one of the most popular industrial materials in the world, with almost 2 billion tonnes produced annually [4].

The steel production process has undergone significant changes throughout the course of using the BF–BOF route in steelmaking. Since the blast furnace (BF) operation has been identified as being one of the biggest CO_2 emission points within the production stages, there is now a movement in the steel industry towards replacing BF operation with electric arc furnaces (EAFs). Tata Steel has already announced plans to shut down the existing BF at Port Talbot and to replace these highly polluting BFs with EAFs [5]. Producing steel using EAFs as a market segment is seen as having the highest expected growth rates in the global green steel market. Using EAF to produce steel has two main advantages: the steel production with EAF generates only ca. 0.3 t of CO_2 emissions per tonne of steel produced, in comparison with the BF of up to 3.2 t of CO_2 emissions. The EAF route of producing steel can also use – currently – up to 45% of scrap steel. This adheres to the EMAF circular economic guiding principles: reduce, reuse and recycle, with conserving resources and minimising the requirement to use virgin raw materials [6, 7].

Besides replacing BFs with EAFs, green hydrogen utilisation (generated using renewable energy technology exclusively) for energy generation and hydrogen direct reduction (HDRI) is thought to be one of the dominating forces in the global marketplace. Green hydrogen is generated using electrolysis, the energy

Circular Steel Production: Pathways to Net-Zero Carbon Emissions, First Edition.
Sandra Kiessling, Hamidreza Gohari Darabkhani, and Abdel-Hamid Soliman.

for which is being generated exclusively utilising renewable energy technology, which generates H_2 and O_2. In EAF for steel production in conjunction with hydrogen utilisation for generating energy and also in HDRI, green steel has immense scope to reduce carbon emissions from steelmaking significantly – between 65 and 85% [8] – and support achieving net-zero carbon emissions production in the short term. Additionally, hydrogen has all the features to replace fossil fuels in steel production, as hydrogen combines with the liquid iron and generates water vapour (H_2O), instead of CO_2 from coal or coke.

10.2 Evaluation of the United Kingdom and its NDCs

With respect to images 10.1 and 10.2 and further images of similar nature, it needs to be emphasised that the rating was established as per the date shown (22/09/2023) in Figure 10.1 and projections based on the status quo covering the period between 2020 and 2030. The countries are being continually monitored and re-evaluated, and the rating might change at any time if the country's performance in the analysed sections changes.

It was observed that the United Kingdom updated its 2030 NDC to reduce GHG emissions by at least 68% below the benchmark 1990 levels including land use, land-use change and forestry (LULUCF) in 2020. These did not include international aviation and transport, such as shipping. Later, a 2035 target was also established with the commitment to reduce GHG emissions by 78% below 1990 levels, this time including both LULUCF and international aviation and shipping; although an adjustment to the UK's targets was made and currently, the aim to reduce GHG emissions stands at reduction by 68% (by 2030) and 82% below 1990 levels by 2035 [10–109].

The UK's net-zero emissions targets and action plans were evaluated and deemed overall 'acceptable' [9]. The Climate Change Act 2008 was revised in 2019, which anchored the net-zero targets to be achieved by 2050 in UK law. The United Kingdom was the first major economy to pass a legally binding

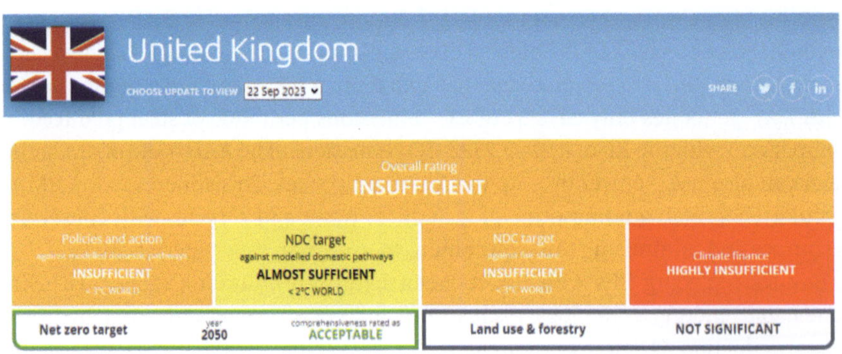

Figure 10.1 Overall rating of the United Kingdom. Source: Ref. [9] / with permission of Climate Action tracker.

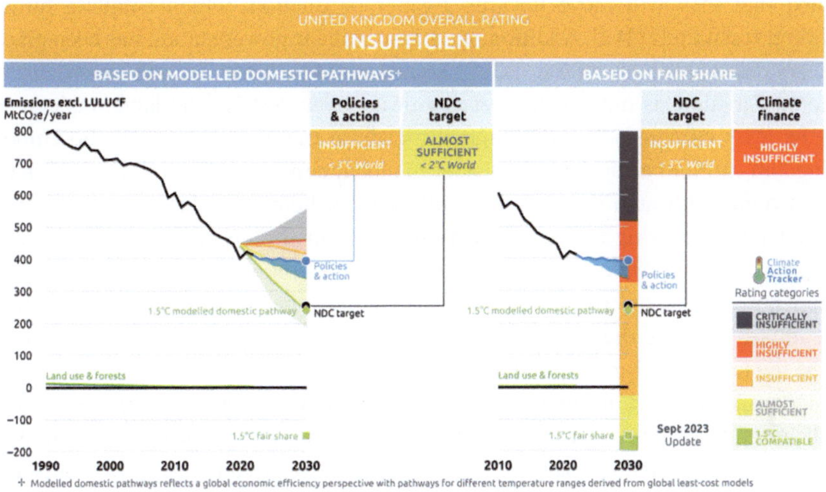

Figure 10.2 CAT rating explained. Source: Ref. [9] / with permission of Climate Action tracker.

framework for achieving a net-zero economy. This amendment was established in October 2021 and was submitted to the United Nations Framework Convention on Climate Change (UNFCCC), committing to a vision for the future much more geared towards decarbonising production and reducing greenhouse gas emissions as a nation. The revised Climate Change Act 2008 (2050 Target Amendment) from 2019 is a much more comprehensive action plan than the previously submitted '10 Point Plan for a Green Industrial Revolution' [10–109].

The revised Climate Change Act 2008 (2050 Target Amendment) from 2019 contains most key benchmarks which they are being evaluated against [9] and constitute key performance indicators for enhanced transparency, target structure and scope of the sectoral requirements set, although the practical application seems lacking behind the ambitious targets [10–109].

Within the revised Climate Change Act 2008 from 2019, the United Kingdom covers all sectors and GHG and even includes emissions from international aviation and shipping as a result of export activity. It can be said that the UK's net-zero targets' main strengths are the periodic carbon budgets and the associated intermediate benchmarks: the five-year statutory carbon budget sets out a clear action plan which might quite possibly facilitate the tracking of progress towards net zero. Partly due to the changeability of the UK policy development, the UK's current policies leading up to 2030 are seen as insufficient [110] to reach the targets agreed to within the Paris Agreement [111]. The absolute absence of development of new policies, guidelines or measures to mitigate GHG emissions implies that the United Kingdom is on a course to failing its own set climate targets. Even worse, new coal mines are being approved [112], although even the mine operator had to concede since being given permission to mine until 2049, that their business case is dead. The projected customers Tata and British Steel have clearly stated that they will move towards greener production, which entails a future without fossil fuels [112]. Additionally, a new nuclear power station has been given the go-ahead [113]. Using the IPCC AR6 report [114], it has become clear that the United Kingdom is not set to meet any of its given NDCs. The lamentable state of the UK's climate policy development leads to the conclusion that the United Kingdom cannot be considered a leader in global climate action, to avert a climate catastrophe. Although there are currently some schemes in place that one would consider incentivising industry leaders to invest in the decarbonisation of heavy industry (Chapter 6), the investment in people and green economy job skills is not happening under the current (2025) government in the United Kingdom. Chances are missed to make the United Kingdom the green energy and green skills hub or to be the leader in climate change action. The outlook for green steel in the United Kingdom looks more positive than what the current rating might suggest, as the policies and guidelines created, and climate finance are supporting the rollout of renewable energy technologies. Industry leaders have already taken the lead, such as Tata Steel, when they received more than £500m in support to shut down their highly polluting BFs and have started proceedings to install EAFs in their place. [10–109]. These only generate on average 0.3 t of CO_2/t of crude steel produced, in comparison to BFs, which can generate up to 3.2 t of CO_2/t of crude steel produced. Therefore, one could argue that the future of green steel in the United Kingdom has been set up for success.

10.3 The United States of America

In 2023, the United States were rated as 'insufficient' in their endeavours to reach targets set as a result of signing the Paris Agreement [111], as they had only achieved one-third of their 2030 GHG emissions reduction target. CAT current policy projections, which include the IRA, show a more pronounced reduction in emissions until 2030. Evaluation of a wide range of data and information [80, 82, 116–218] suggests that the United States will potentially achieve an additional GHG emissions reduction between 30 and 44% by 2030, but the currently active US target of 50–52% GHG emissions reduction below the 2005 levels by 2030 is not compatible with the aim to limit the global temperature rise to 1.5 °C (Figures 10.3 and 10.4).

Although the Biden–Harris administration has made their level best effort to widen the network of renewable energy technology installations for energy generation, further industry decarbonisation and phase-out of fossil fuel reliance, they are constrained by a legacy of existing legal frameworks and industry interests – which provide places of employment. Additionally, there are different sectors and their impact on the economy to consider, such as increasing employment, raising the living standard of all Americans, reducing poverty-related negative impacts on the economy and adherence to NDCs [80, 82, 116–218].

The United States currently seems to be on a trajectory to even increase instead of phase down/out their oil and gas production and exports, and they are currently planning to increase the export capacities for natural gas by more than 40% by 2026.

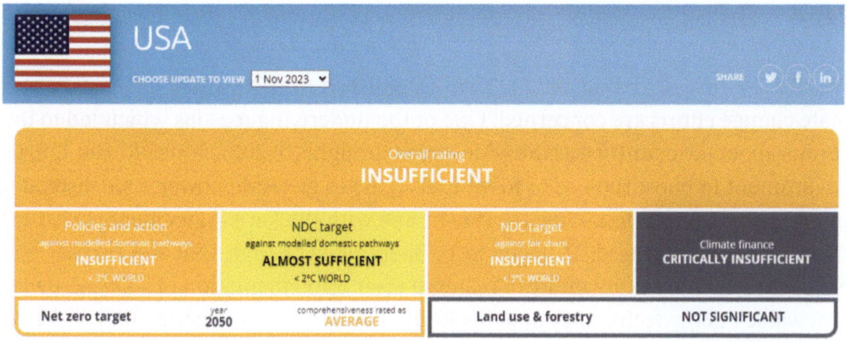

Figure 10.3 Overall rating of the United States of America. Source: Ref. [115] / with permission of Climate Action tracker.

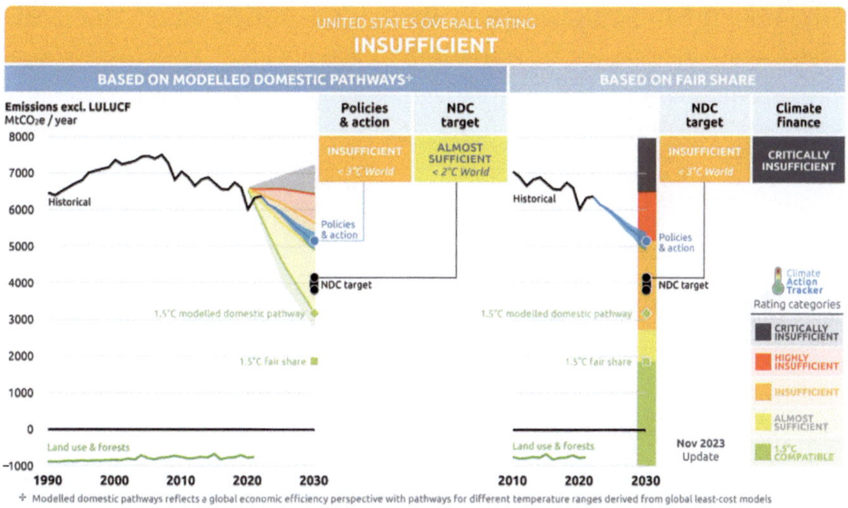

Figure 10.4 CAT rating explained. Source: Ref. [115] / with permission of Climate Action tracker.

Additionally, newly approved major oil drilling projects are taking place [172]. It was established that the United States is clearly not achieving its own NDCs, with its support for fossil fuels, simultaneously compromising its own integrity [80, 82, 116–218]. Although climate finance in the United States is considered critically insufficient, the future of green steel in the United States looks more positive than what this rating might suggest. Steel production and manufacturing is still a strong industrial sector, providing employment and income for many American families. Therefore, there is hope that the measures instated by the Biden–Harris administration and their legacy will bear fruit in years to come and enable the production of green steel. It remains to be hoped that their efforts are not going to be dismantled by future United States governments.

10.4 Brazil

Brazil's CAT overall rating per 2023 was established to be 'insufficient', as far as climate change efforts are concerned. One of the underlying reasons which led to this verdict about the country was the widespread droughts in 2021, which forced Brazil's government to boost the use of fossil fuels (gas) to generate power – so, instead of phasing down or out fossil fuels to stay in line with the Paris Agreement [111, 121], the usage was increased. It needs to be mentioned though that the current government is making every effort to increase their implementation of solar and wind renewable energy technology systems in the whole country (Figures 10.5 and 10.6) [184–244].

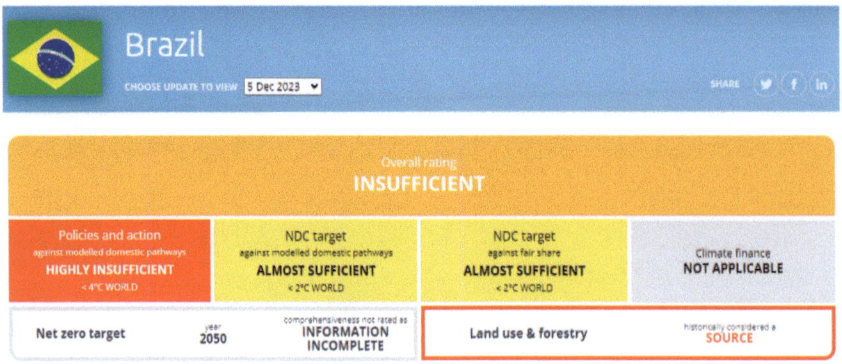

Figure 10.5 Overall rating of Brazil. Source: Ref. [183] / with permission of Climate Action tracker.

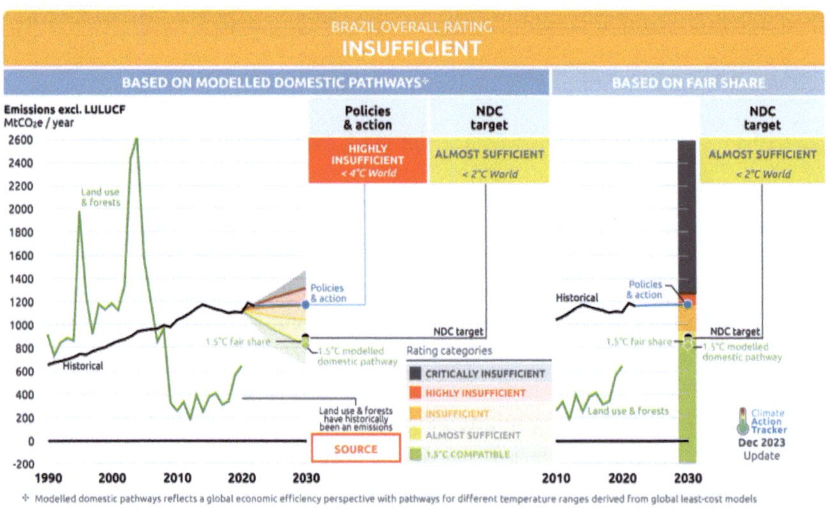

Figure 10.6 CAT rating explained. Source: Ref. [183] / with permission of Climate Action tracker.

Having conducted a thorough investigation of policies, governmental guidance and a variety of official records and reports, it was established that Brazil's policies and long-term energy outlook are entirely inadequate to stay in line with the Paris Agreement [111]. It was decided that the production of oil and gas and using fossil fuels for energy generation will even have to increase for the foreseeable future, as the rollout of renewable energy technologies is seen as not feasible to install at a reasonable capacity to meet current demand, despite Brazil's considerable potential for renewable power generation [184–244].

Distressingly, deforestation in Brazil has still not been halted, despite an immense international outcry and the newly elected president's considerable efforts to halt the destruction of one of the planet's green lungs. Deforestation even increased in 2020 and 2021 (before the election of Lula in 2022), largely as a result of weak law enforcement, illegal mining activity and animal husbandry. To achieve a balanced view of all aspects, however, it needs to be emphasised that also some positive aspects could be identified, which gave rise to hope as Brazil's Supreme Court declared in July 2022 that the climate fund must be reactivated. Additionally, the Paris Agreement was given the status of a human rights treaty, which established Brazil as the first country in the world to do so [184–244].

The CAT [183] rating of Brazil's climate targets and policies was 'insufficient' in 2023. It was hoped that this evaluation would send a clear message that Brazil's climate policies and commitments are in dire need of a systemic step change to achieve being anywhere near achieving the Paris Agreement 1.5 °C temperature increase limitation to before industrialisation levels. As it stands at the time of writing, Brazil's GHG emissions are set to stagnate at least at today's levels, and it is unlikely that the country will meet its 2030 NDC targets. Brazil's policies, internal action plans and domestic pathways have been rated as 'highly insufficient' against the IPCC (114) AR6 data set and additionally resulted in the finding and recommendation that Brazil needs to drastically cut its GHG emissions to meet its 2030 NDC target and for 1.5°C compliancy. However, a more positive outlook was established with regard to the NDC targets set against modelled domestic pathways, as Brazil's efforts were assessed and found to be 'almost sufficient' [183–244].

The aforementioned ratings give an indication that Brazil's NDC target for 2030 is currently not in line with limiting the temperature rise to 1.5 °C, but moderate improvements in national target setting and action plans could lead to significant improvements. In order to be 1.5 °C compatible and exclude the LULUCF sector, Brazil needs to reduce GHG emissions by at least 13% below 2005 levels. Interestingly though, Brazil is one of only a handful of countries to also set and work towards a 2025 target, although the current efforts are pointing towards Brazil achieving neither its 2025 targets nor its 2030 targets [184–244]. Considering Brazil's current framework, the development of green steel in the country seems rather difficult, given the numerous hurdles that are currently hampering progress towards reaching NDCs and climate-protective goals.

10.5 Russian Federation

Russia's overall ratings by CAT have been established overall as 'critically insufficient', as Russia seems to focus all efforts on integrating Western countries into the Russian Federation instead of tackling the effects of climate change with adequate measures. The country is still heavily reliant on fossil fuels, and the efforts towards decarbonisation and green energy technology remain very low. The scarcely developed policies and guidelines are of low value as they have not set out any clear climate targets or action plans to achieve any reasonable reduction in GHG emissions and there does not seem to be any real commitment to reducing GHG or CO_2 emissions [61, 63, 123, 205, 244, 259–289]. With the view of Russia's NDC targets from November 2020, the country could easily modify its policies and submit a stronger 2030 target with little effort. Seen as a positive development, the Russian government offered further details on the newly developed net-zero GHG emissions strategy and targets for 2060, in September 2020 (Figures 10.7 and 10.8) [61, 63, 123, 205, 244, 245, 259–289].

In a very optimistic description of the situation, the Russian government assumes that by 2050, Russian forests would be able to take up twice as much CO_2 than today and that would imply that, therefore, all other emissions also do not have to reach net zero but have to be reduced by only 50% to arrive at the net-zero target. The Russian government did not supply any detailed information as to the action plan which would induce achieving such enormous carbon uptake by the referred to national forests, and there was also no indication of how the increased CO_2 emissions and released methane quantities from the thawing Siberian permafrost would be tackled any time soon. There is also a distinct lack of measures addressing the impact of numerous substantial wildfires throughout the Siberian forests, which the world witnessed in the past five years since 2019. Although Russia adopted its law in 2021, aimed at limiting greenhouse gas emissions, all measures effecting any substantive emissions reductions were removed from the final text version. Additionally, the Energy Strategy for 2035 specifically refers to the promotion of fossil fuel extraction, increasing consumption and exports worldwide.

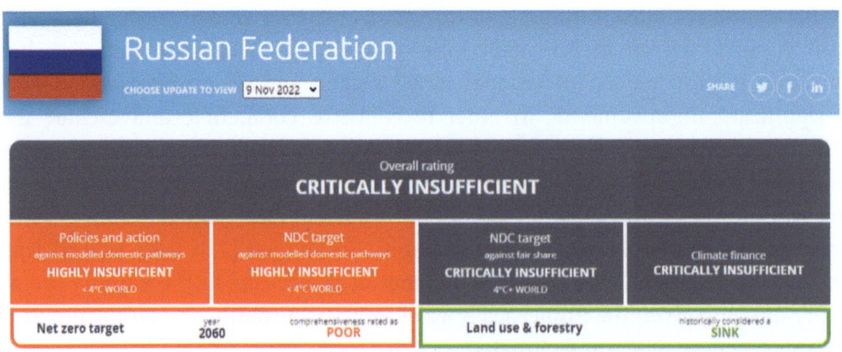

Figure 10.7 Overall rating of Russia. Source: Ref. [245] / with permission of Climate Action tracker.

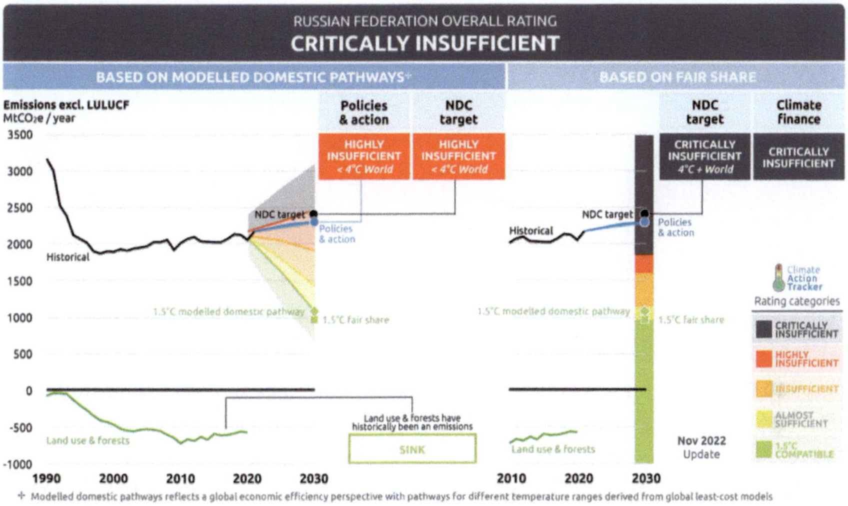

Figure 10.8 CAT rating explained. Source: Ref. [245] / with permission of Climate Action tracker.

The focus is clearly set on profit and not the planet and bears immense economic risk for the country, along with a future incompatible with limiting the temperature rise to 1.5 °C according to the Paris Agreement [111]. Russia failed to submit any considerable contribution, agenda or action plan towards achieving a net-zero scenario or achieving international climate finance goals, whilst setting insufficient domestic targets or adequate climate guidance and policies has therefore resulted in Russia being rated overall as 'critically insufficient'. This rating reflects the fact that Russia's submitted policies and plans reflect critically inadequate to no action towards climate change goals and are not in line with any component of the Paris Agreement [111] or the agreed efforts to limit global warming to 1.5 °C of before industrial activity [61, 63, 123, 205, 244, 245, 259–289].

Although an NDC update was provided by Russia in September 2020, they failed to submit an improved document as far as climate goals are concerned. The modelled domestic pathways, when compared to other countries' targets and commitments, look inadequate and highly insufficient. What Russia needs to do now is to set more adequate targets for GHG emissions reduction and follow through with implementing these, along with issuing GHG reporting standards and providing climate finance for empowering industry leaders in the country to improve their emission reduction efforts [61, 63, 123, 205, 244, 245, 259–289].

10.6 India

India is a country with a quite young overall population and, although their reliance on fossil fuels and their dependency on fossil fuels is still rising, they are not considered a big emitter of GHGs in a global context. However, their overall

Figure 10.9 Overall rating of India. Source: Ref. [277] / with permission of Climate Action tracker.

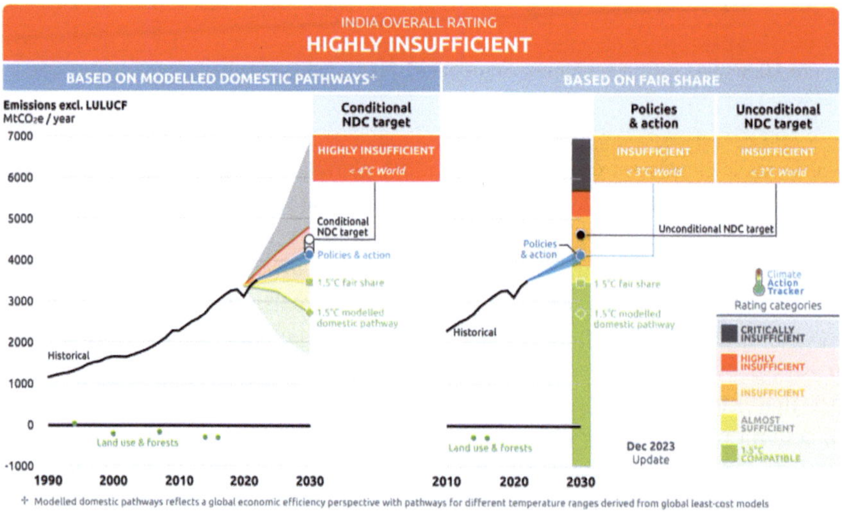

Figure 10.10 CAT rating explained. Source: Ref. [277] / with permission of Climate Action tracker.

rating was established to be 'highly insufficient', as their efforts to increase the number of hydropower stations and other renewable energy technologies across the country have been hampered by droughts, earthquakes, a rise in seasonal electricity demand and summer temperatures breaking every record (Figures 10.9 and 10.10) [2, 63, 121, 123–371].

This newly developed situation requires India to develop long-term plans to enable them to deal with the unpredictable energy demand and to respond adequately to changing weather patterns and climate change events. A definitive phase-down and phase-out of fossil fuels is imperative for all countries to achieve the NDCs and climate goals; so, the global temperature rise can be limited to within 1.5 °C of pre-industrial levels [2, 63, 121, 123–371].

The latest investigated data suggest a slightly more positive outlook, as a lower emissions projection of 2030 GHG emissions could be compiled, with the current policies in mind, which are partly due to India's efforts to increase the share of renewable energy technologies in the energy generation sector. Therefore, lower than estimated and projected power sector emissions were recorded. With international support, India would be able to facilitate a faster rate of emissions reduction and reach their climate targets much quicker than anticipated [2, 63, 121, 123–371].

Although the implementation of several policy measures has been prioritised, which aim to promote the uptake of renewable energy technology and increase energy generation capacity, the legacy of antiquated administrative processes with all their cultural nuances is hampering the nationwide rollout of solar energy installations and the increase of green hydrogen production. Additionally, the historic reliance on fossil fuels, particularly coal, is a damper for achieving the high ambitions. Although there were initially no plans to build further coal capacity beyond projects which are currently under development and to conclude by 2026–2027, the newest (December 2023) submitted electricity projections feature the addition of substantial new coal power capacity for the next five years [2, 63, 121, 123–371]. This is a lamentable development (on paper), particularly, as the falling cost of renewables is making generating energy from renewable sources much cheaper than generating energy from fossil fuels – and without the detrimental effects on the natural environment and the climate. However, several planned coal projects in recent years have been stopped before they even started, and India has also increased their liquefied natural gas (LNG) imports and is increasing the utilisation of gas power plants. Interestingly, the national electricity plan (NEP2023, from May 2023) eliminates again the planned new gas power plants which had been set to be built between 2022 and 2027. Additionally, the reliance on nuclear power has been markedly reduced for the next 10 years. The factual total increased solar capacity was 31 GW in comparison with the policy draft version, but the paper also contains a reduction in wind power (12 GW less) capacity. In contrast, 21 GW in non-hydro renewable energy capacity was also included to be papered. All these positive aspects as a result of renewed efforts to tackle the effects of climate change on the country by the new government under Modi give rise to hope that with newly submitted policies and action plans, India will overachieve its own NDC targets. Based on this, it could be argued India should set stronger targets and thus improve their overall ratings [2, 63, 121, 123–371].

Considering all the negative and positive aspects of India's efforts to tackle the effects of climate change and set NDCs to stay in line with the Paris Agreement [111], the future outlook for green steel looks promising, as India has a younger population than most countries, and they are encouraging the rollout of green hydrogen production, which is a key factor of making green steel according to the Bio Steel Cycle model and strategy [2, 63, 121, 123–371].

10.7 China

China's current policies contain projections for greenhouse gas emissions to peak by 2025. This would be ahead of the set 2030 target, but there are no plans or frameworks in sight to achieve meaningful emission reductions during the period before 2030. The energy and electricity demand projections forecast continue growth, in line with global development. It remains to be seen if China's record renewables deployment will be sufficient to provide an adequate amount of energy for the increased demand and whether this will be enough to accelerate its post-coal transition [293–311]. The Chinese government seems to provide continued support for their fossil fuel companies; although immense efforts for transitioning away from fossil fuels in its energy sector are well underway, the increased energy demand has caused an increase in fossil fuel production to meet current demand (Figures 10.11 and 10.12) [293–311].

China and the United States published a joint statement on climate cooperation on 15 November 2023 ahead of the COP28. Yet again, although countries seemingly supported the G20 Leaders' Declaration to triple the global renewable energy capacity by 2030, no substantive emissions reduction targets were agreed upon by the two world's largest current emitters. Although a pledge to accelerate renewable energy deployment was made, the efforts to reduce power sector emissions this decade are set within a non-binding timeline and reduction levels have yet to be defined [293–311]. In line with these non-committing agreements, China also published its Methane Action Plan, which set some directions to control methane emissions across sectors but failed to set concrete, specified reduction targets [293–311]. Therefore, the CAT rating for China's policies and targets remains 'Highly insufficient'. During the COVID-19 pandemic years, most countries' emissions were drastically reduced, but China's 2022 emissions only declined 0.2% to 14.4 Gt CO_2e. With this background and known emissions levels to peak in 2025, it is expected that emissions will at least plateau at high levels for the rest of the decade. There is, however, the potential for the acceleration of renewable

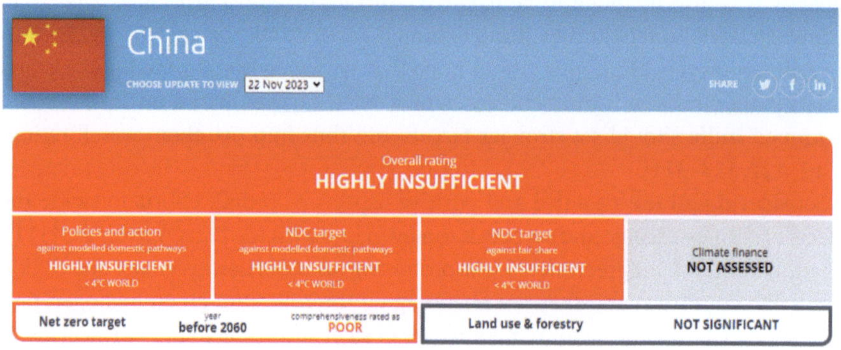

Figure 10.11 Overall rating of China. Source: Ref. [292] / with permission of Climate Action tracker.

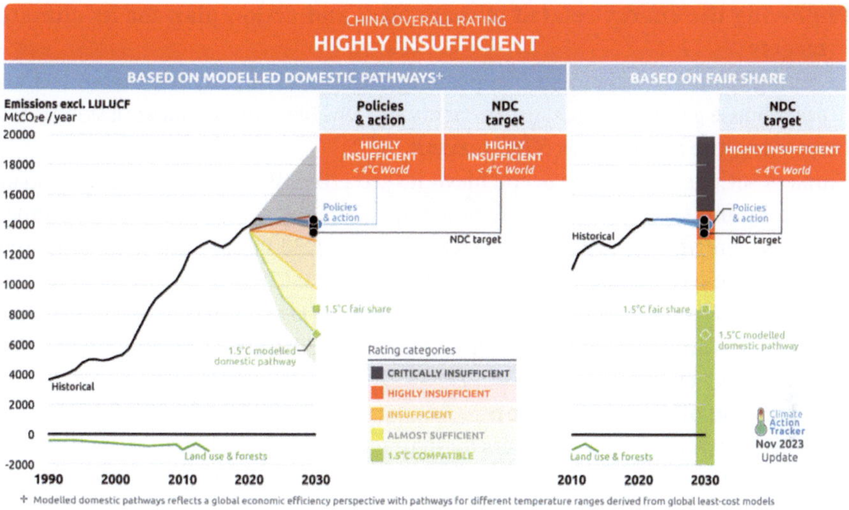

Figure 10.12 CAT rating explained. Source: Ref. [292] / with permission of Climate Action tracker.

energy deployment to cut annual emissions by 600 $MtCO_2$ or 4% of current emission levels [293–311]. China's climate, energy and renewable energy deployment policies are currently not seen as effective enough to reduce emissions in a meaningful way in the foreseeable future. In order to support worldwide efforts to limit global warming to 1.5 °C, China as one of the two world's largest emitters has the duty to decarbonise their society in the shortest time frame possible. Improved policies and action plans are required to be set and executed, to support the efforts of averting a climate catastrophe (2026–2030) [293–311]. Considering China's NDCs, they could easily overachieve the non-fossil energy share and renewable capacity targets, which suggests that China could achieve the set level of 1200 GW production of wind and solar capacity by 2030 at least five years ahead of their target date [293–311]. A circumstance hampering China's climate-protecting progress is the slow economy, which might render China unable to cut its fossil fuel dependency before 2030. This might cause China to miss its own carbon intensity NDC targets [293–311].

Although the 2060 target year is too far out to have any impact on policy and decarbonisation implementation now, China could improve on its climate policy and stay conform to its 2060 carbon neutrality targets, with easy-to-implement steps:

Reducing its fossil fuel dependency

President Xi Jinping's speech during the 20th CPC National Congress made it very clear that emphasis is on ensuring China's energy security and it will always be its first-order concern, and fossil fuels are an equal solution as is renewable energy technology implementation. Production and mining of coal in 2022 were at record levels and still the largest, measured by country [292].

Improving the energy-related climate NDC and action plan for its climate targets
by setting economy-wide fossil fuel peaking targets and directly covering non-CO_2 greenhouse gases (2.5 $GtCO_2e$/p.a. or almost 18% of China's total emissions).

Improving the scope of its NDC targets
China is set to significantly overachieve its set 1200 GW target of wind and solar capacity and 25% non-fossil share target by 2030; as by 2023, its year-on-year milestones were already achieved well ahead of the 2030 target year, with wind and solar capacity, reaching 921 GW in September 2023. China is expected to reach at least 1900 GW by 2030. The non-fossil fuel share is projected to reach ~30% by 2030, and China has all the potential to achieve 3200 GW capacity of wind and solar energy by 2030.

Acceleration of decarbonisation
in all high-emitting industry sectors by setting the carbon peak times to be reached much earlier, as they were found to be feasible for the critical industrial sectors of cement, chemicals, iron and steel and other metals.

Although it seems that China's dependency on fossil fuels and high emission levels, which are yet to peak, are here to stay in the foreseeable future, China's efforts in transitioning their economy away from fossil fuels and towards energy generated increasingly from renewable sources are gaining pace. The momentum is gathering for a post-coal era, which is emphasised by the increasing number of renewable energy installations, structurally effective reforms in the energy and power sector, and climate targets and policies improving decarbonisation efforts in high-emitting industrial sectors [290–389]. With China's President Xi Jinping's declaration of planning to reach carbon neutrality well before 2060 at the UN General Assembly in September 2020, China has followed through since with the submission of a long-term strategy to the UNFCCC in October 2021. As the long-term strategy is not in line with the Paris Agreement [111] or the UNFCCC guidelines, China's net-zero target evaluation was still established as 'Poor'.

However, with China's real-time efforts in the wind and solar energy installations breaking records almost on a monthly basis, the emissions reduction in the Chinese steel industry seems practical and a realistic prospect and is in line with recommendations for a decarbonised steel industry [8, 312]. It is therefore a realistic evaluation to consider the situation for Chinese green steel as a genuine prospect and certainly a growth market, as the demand for grey steel, worldwide, is set to decline [313].

10.8 Australia

The Australian government's declaration of intent to 'take the country forward on climate action' has yet to materialise. An over-reliance on fossil fuels and new policy measures for high-emitting industries are either insufficiently effective or not implemented in any meaningful way. Hence, the rating by the Carbon Action Tracker has established Australia's climate targets and action remaining as 'Insufficient' (Figures 10.13 and 10.14) [121, 314–378].

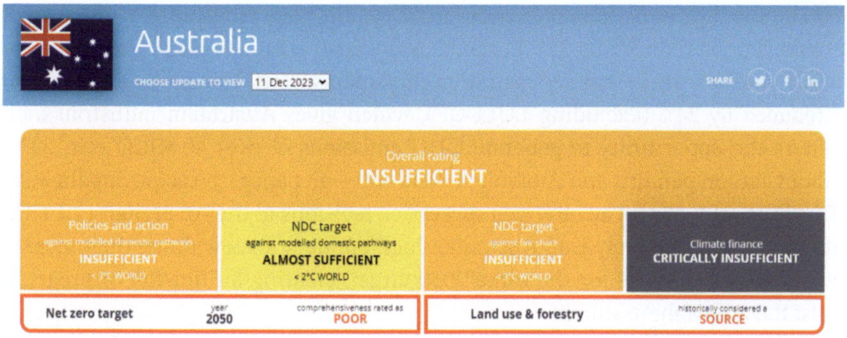

Figure 10.13 Overall rating of Australia. Source: Ref. [314] / with permission of Climate Action tracker.

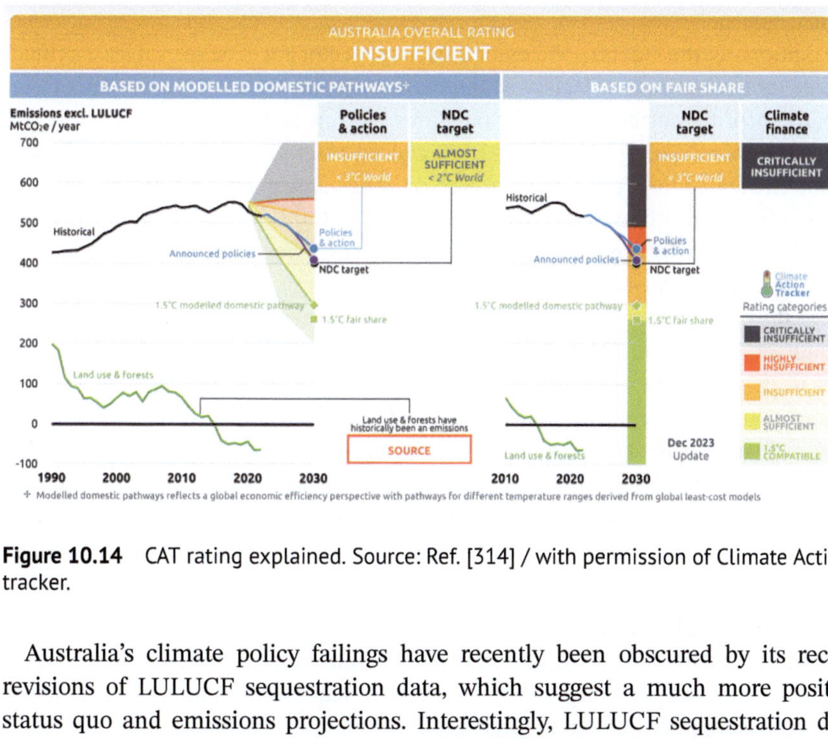

Figure 10.14 CAT rating explained. Source: Ref. [314] / with permission of Climate Action tracker.

Australia's climate policy failings have recently been obscured by its recent revisions of LULUCF sequestration data, which suggest a much more positive status quo and emissions projections. Interestingly, LULUCF sequestration data for the years up to 2020 have been revised upwards, and sequestration projections have seen a 200% increase between 2020 and 2021 and almost doubled again between 2022 and 2023. These adjustments had a significant impact on achieving Australia's 2030 and 2050 climate targets. The Australian 2030 net-zero emissions target of a 43% reduction below 2005 levels allows emissions determination within this target to be summarised using data on industrial greenhouse gas emissions and land-use sequestration, although it needs to be pointed out that an increase in projected land sequestration values by the target year 2030 actually decreases the reductions required under the guidelines from emissions *excluding* LULUCF.

In 2020, the 43% emissions reduction target would have required greenhouse gas emissions to be reduced by 32% at the latest by 2030 [121, 314–378]. With the new revised land-use emissions sequestration projections, GHG emissions only have to be reduced by 24% (excluding LULUCF), which gives Australian industrial high emitters the opportunity to generate GHG emissions of over 40 $MtCO_2e$ in 2030, without carbon penalty, and Australia would still – on paper – be achieving their set NDCs [121, 314–378]. The observed successive revisions to the Australian LULUCF data inventory effectively lead Australian high emitters to slow down their level of decarbonisation efforts for the same 2030 target, creating an illusion of progress, whilst the atmosphere suffers.

Contrary to the stipulations of the Paris Agreement [111], the Australian government continues to endorse the expansion of the fossil fuel industry and is actively pursuing the creation of regulations to enable high emitters to sequester CO_2 emissions beyond country borders – which is the polar opposite of its NDCs and climate goals.

Additionally, the federal policies for the energy sector are to be seen at least as inadequate, as the absence of legislated targets or other action plans requiring a faster expansion of renewable energy installations, and in addition administration and challenges of an ageing national grid, has caused a significant slowdown in deployment. The stated target of relying at 82% on renewable energy generation by 2030 is way off track. Although the 2023 revision of the Capacity Investment Scheme is demonstrating governmental efforts to address these shortcomings, targeted action plans and time frames for phasing out fossil fuel energy generation remain absent, hence the continued rating of the Australian government as far as net-zero policies and targets are concerned as 'Insufficient' [121, 314–378]. Lamentably, in comparison to its fair share of global action, Australia's overall rating was also deemed 'Insufficient', as the contributions to climate finance are currently inadequate.

In order to get anywhere near the NDCs in comparison with other countries and in line with the Paris Agreement [111], Australia will be required to drastically improve their climate policies and take concrete action towards decarbonisation to achieve a 1.5 °C Paris Agreement compatibility. Time is off the essence for the run-up to 2030, and to improve its climate action, Australia should:

Decrease the reliance on carbon emissions offsetting
by creating and implementing radical policies in line with the Paris Agreement [111] and to reduce dependency on domestic carbon sequestration international offsets to reach its carbon reduction targets.

Immediately stop supporting the use of fossil fuels for energy generation
starting with ceasing its support for any fossil fuel (coal and gas) projects and establishing a realistic timeline to phase out fossil fuel exports.

Accelerate the phase-out of fossil fuels for power generation
by shifting the support for fossil fuels to incentivising renewable energy installation, increasing storage capacity for energy from renewables and presenting a clear framework of how to accelerate the phasing out of coal and gas-fired power plants, including facilities currently not connected and off-grid facilities, in a feasible and timely manner.

Transition the Australian economy to creating new exports
by utilising the immense potential for energy generated from renewable energy technology (wind, solar and hydro), without utilising 'red-herring' solutions such as offsets and carbon capture and storage, which only delay dealing with the fallout of current industrial activity.

Foster the decarbonisation of the transport sector
by creating policies for accelerating the deployment of electric vehicles (EVs), improving fuel standards, improving public transport planning towards decarbonisation and creating incentives, legal framework and transition pathways for heavy mobility.

Creating policies and action plans to manage emissions from agriculture and waste and stationary energy sectors
to ensure a feasible and timely transition to drastically reduce CO_2 emissions.

Australia has seen some interesting developments in recent years, and positive developments should be highlighted as they occurred:

- In 2022, energy generated from renewable sources is a 31% share of the total electricity production, an increase of 4% of 2021 and twice the capacity of 2017.
- Several Australian states have taken the lead in the climate transition and are implementing effective measures to address the failings of federal-level policies.

Although there is some evidence of efforts towards net zero in Australia, the overall CAT rating, as far as climate targets, policies and climate finance are concerned, was established as 'Insufficient' [121, 314–378].

Australia's CAT rating of climate finance contributions has been established to be 'Critically insufficient', as a fair interpretation of their efforts in evaluation against UNFCCC standards, the Paris Agreement and their NDCs. At the recent COP28 (December 2023), Australia clearly stated they were going to immediately stop supporting and financing fossil fuel projects abroad which has led to the climate financing rating improving from previously 'Highly Insufficient' to now 'Insufficient'. To receive a better overall rating, Australia needs to increase the level of its international climate finance to promote the rollout of renewable energy technologies for energy generation. As Australia's climate finance is not sufficient to improve the rating of its NDC target against fair share, CAT rated Australia's overall fair share contribution also as 'Insufficient'.

Additionally, Australia is globally the only developed country which has been classified as a deforestation hot spot – and quite contradictory – the government intends to meet its NDC and net-zero targets by heavily relying on LULUCF as a carbon sink. Logic dictates that this is entirely impractical if the vegetation one relies on as carbon sinks is being decimated at a rate of knots [121, 314–378]. Hence, the CAT rating for Australia to achieve its net-zero target was established as 'Poor'.

As far as the evaluation of green steel potential is concerned, for production as well as utilisation and as a possible market, Australia has great potential to make this one of its desired new export articles. The country has immense potential to generate energy from renewable sources and has the capacity to decarbonise its steel production far better than most countries that have been evaluated, in direct comparison.

10.9 Canada

Positive developments were noted to happen in Canada, as its emissions were observed to decline as the government continues to implement its climate policy agenda, although there remains a deficit between current policies, its NDC target and 1.5 °C compatibility. The current published policies are supporting the reduction of these discrepancies, but concrete action is required to happen immediately (Figures 10.15 and 10.16) [345, 380–425].

Canada's Environment Commissioner published a scathing report in November 2021 in which he outlined three decades of government failures to meet its own targets and decrease GHG emissions. One of the issues the commissioner called the Canadian government out on was the need for strong action plans and timelines and highlighted policy incoherence, such as the Trans Mountain purchase [345, 380–425].

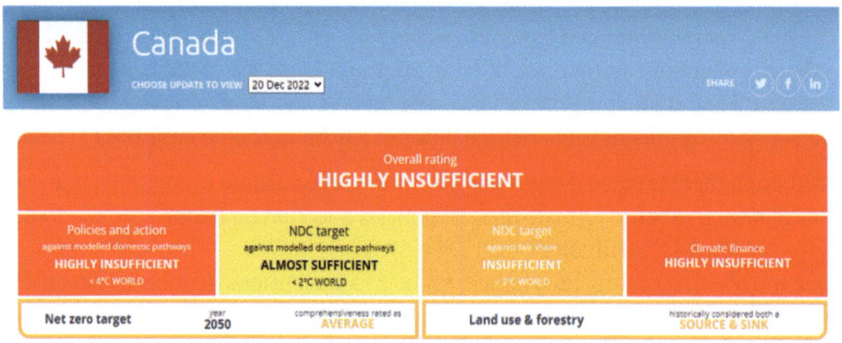

Figure 10.15 Overall rating of Canada. Source: Ref. [379] / with permission of Climate Action tracker.

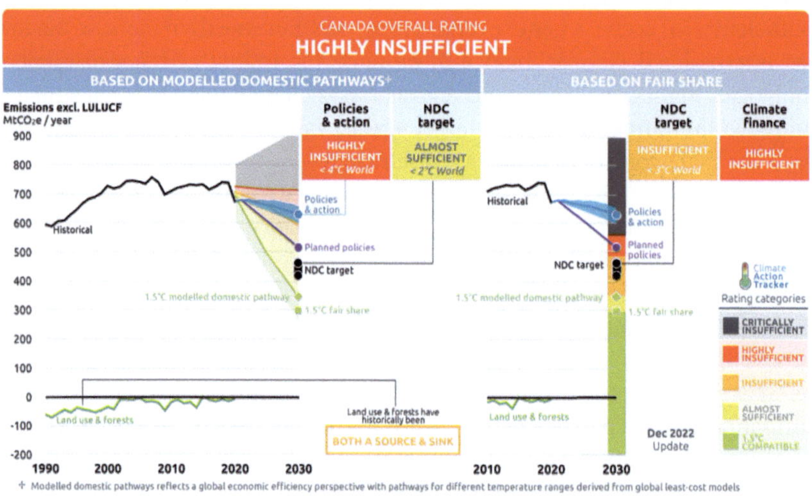

Figure 10.16 CAT rating explained. Source: Ref. [379] / with permission of Climate Action tracker.

The Canadian government continues to support the Trans Mountain pipeline, even though the internal budget watchdog has pointed out that the pipeline will no longer be profitable, as the installation costs have run out of control. Additionally, the Canadian government seems to have its mind set on exporting LNG from its east coast to Europe – even though developments of this scale and type are endangering the targets and efforts to limit the global temperature rise to 1.5 °C from pre-industrial levels [345, 380–425].

All scenarios, policies, NDC and domestic pathways taken into consideration, Canada will most likely still be producing and exporting oil and gas in 2050. Hence, Canada's overall CAT rating remained at 'Highly insufficient' [345, 380–425].

It was observed that the speed and scale of Canada's climate policy implementation is incongruent with the urgent need to avert more detrimental effects of the climate crisis and Canada needs to ramp up their efforts in this regard. 2030 GHG emissions are projected to be between 602 and 656 $MtCO_2e$ (without LULUCF emissions), meaning: between 11 and 19% below established 2005 levels. This is a positive development, as it is much lower than the last CAT evaluation but is almost double the acceptable target amount of 345 $MtCO_2e$ required to be 1.5 °C compatible [345, 380–425].

Since the last CAT evaluation in September 2021, Canada has made remarkable progress in some areas, which should be highlighted accordingly:

Canada has

- Continued to make progress on implementing its carbon pricing increases between now and 2030
- Adopted some of its outdated regulations and codes, such as the clean fuel standard, the GHG offset programmes, building standards and code revisions
- Published a methane strategy
- Guidance and time frame on phasing out of public finance for fossil fuels

There are still some areas which require attention, such as regulations and strategies for the following issues:

- EV sales mandate and strategic papers
- Proposals for a 2035 net-zero grid
- Actual enforcement of the thermal coal export ban
- Publishing standards, policies and guidance on constructing green buildings
- Establishing a fertiliser emissions reduction target
- Creating oil and gas methane targets and landfill methane targets

Canada's only commercial coal-fired power plant with integrated carbon capture and storage is continually plagued by technical and other issues and highlights the perils of over-reliance on CCS to achieve set climate targets [345, 380–425]. Canada's NDC target against modelled domestic pathways was considered by CAT as 'Almost Sufficient'. Canada submitted a stronger NDC target in July 2021 but has not done so since, despite the agreement to submit annually updated NDC targets under the Glasgow Climate Pact [129]. The Canadian government has committed to cut GHG emissions by at least 40-45% below 2005 levels by 2030, which means a 15%

rise in the best-case scenario of the emissions reduction target level, in comparison with the previous target of 30%. However, in order to be 1.5 °C compatible on a global least-cost basis, the target needs to be at least 54% below the 2005 levels. Due to this positive development and comparing Canada's NDC target against modelled domestic pathways, the CAT rating developed to 'Almost sufficient'. However, Canada is not on track to achieve these set targets with the currently implemented policies, pathways or action plans, nor do the planned policies entirely eliminate the gap between policy, targets and action required to limit the temperature rise. Hence, the CAT's assessment of Canada's total fair share contribution – considering its emissions reduction target and its climate finance – established a CAT rating of NDC target against fair share of 'Insufficient'. What Canada is required to do now is: (i) improving on its NDC target and their implementation and (ii) increasing support for GHG emission reduction efforts in developing countries to improve on this classification. Upon evaluation by CAT, Canada's international public climate finance contribution was established as 'Highly insufficient'. The Canadian government recently announced a 100% increase in its climate finance over the period of the next five years. As Canada's climate finance contributions to date have been low compared to its fair global share, and its continued support for fossil fuel developments domestically and abroad, the rating was set as 'Highly insufficient' [345, 380–425]. To improve its rating, Canada needs to stop funding domestic and international fossil fuel projects (such as the Trans Mountain pipeline and LNG exports to Europe) and increase efforts and commitments to improve climate finance. Currently, Canada's climate finance is not deemed adequate to improve its fair share target CAT rating, and therefore, Canada's overall fair share contribution was classed as 'Insufficient' [345, 380–425]. LULUCF in Canada is simultaneously a source and substantial GHG emissions sink. The Canadian government claims that net emissions for land use and forests in Canada are almost at net zero (National Inventory), based on harvested wood products being balanced by carbon sequestration in managed forests. A realistic evaluation of the LULUCF GHG emission sources and sinks is individually higher than 20% of emissions *excluding* LULUCF, and we therefore highlight the sector for Canada. Therefore, if either the source or sink components change – which means: wood products or managed forests – the net emissions will inevitably change, too, and there is an equal potential risk for either a stronger contributor to overall emissions or a stronger carbon removal component. In 2021, Canada passed the *Canadian Net-Zero Emissions Accountability Act*, which anchored its 2050 net-zero targets in country law [345, 380–425]. The Canadian long-term strategy was updated at the end of 2022 and lays out a range of different scenarios, each capable of achieving net-zero emissions in 2050. Unfortunately, it fails to demonstrate a particular domestic pathway for the country to follow or an action plan. The outline policies and measures needed to achieve its net-zero target are insufficiently clear, and, additionally, there is a distinct over-reliance on LULUCF and CDR, almost accounting for 45% of Canada's emissions in 2020 (or 301 Mt CO_2e). Considering all possible scenarios, Canada is still projected to be a producer–exporter of oil and gas in 2050. With the policies, domestic pathways, NDCs and climate targets still some way off from reaching compliance with the Paris Agreement [111] and limiting the global temperature rise to 1.5 °C from pre-industrial levels, the potential for green steel in Canada as a producer and market seems bleak. However, Canada has great potential for renewable energy technology implementation (solar, wind, hydro and geothermal) [345, 380–425]. Therefore, positive future development for decarbonised steel production and a healthy market for green steel consumption in Canada seems entirely possible.

10.10 Norway

Norway improved its 2030 emissions target to a reduction of at least 55% below 1990 levels in November 2022, which was compared against projected domestic pathways, and the minimum level of mitigation. It was established that Norway will need significantly improved policies to meet its climate targets (Figures 10.17 and 10.18) [40, 406, 427–474].

Norway has seen significant positive development towards reaching the net-zero targets by committing more climate finance for developing nations, but Norway's climate finance is still seen by CAT as 'Insufficient'. The reason for this is the fact that there is much scope for improving domestic pathways to ensure Norway's net-zero and climate targets are in line with the Paris Agreement's 1.5 °C temperature goal. Whilst current policies and action plans have been considered to be 'Almost sufficient' by CAT, Norway's NDC targets set against fair share and net-zero targets were still deemed 'Insufficient', and it needs to be emphasised at this point that the exported emissions from Norwegian oil and gas are not considered in the CAT's overall 'Almost sufficient' rating [40, 406, 427–474].

It is therefore lamentable that Norway as such a progressive country still continues to support the expansion of fossil fuel for energy generation, namely oil and gas exploration, in its already substantial fossil fuel sector. Norway's fossil fuel exports are an immense source of global GHG emissions, based on their combustion parameters [40, 406, 427–474]. The Norwegian government is currently

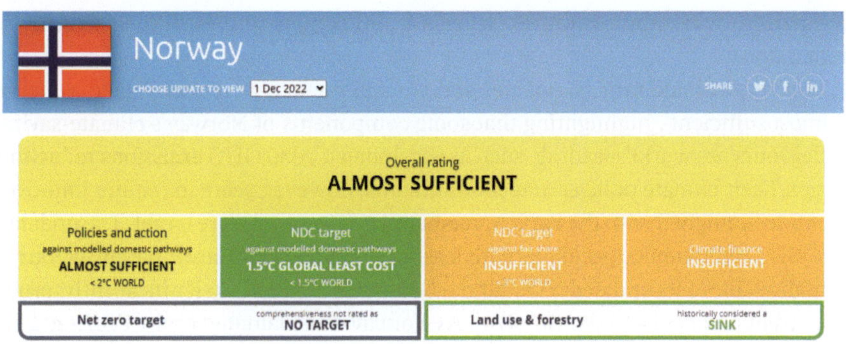

Figure 10.17 Overall rating of Norway. Source: Ref. [426] / with permission of Climate Action tracker.

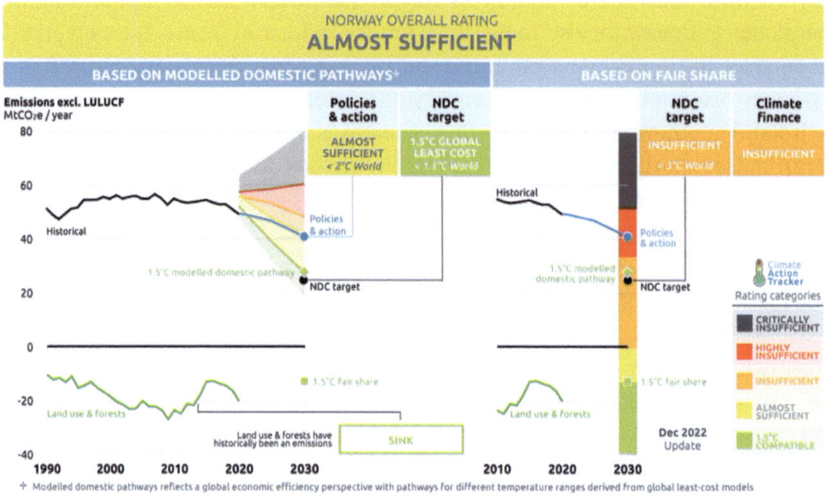

Figure 10.18 CAT rating explained. Source: Ref. [426] / with permission of Climate Action tracker.

fighting before the European Court of Human Rights (ECHR), as charges were filed which state that approval of new permits for exploration of new oil and gas would constitute a breach of fundamental human rights.

It can be seen as a substantial positive factor in terms of evaluation of rating Norway's transition to net zero as the country is world's leading on electric vehicle adoption, not in the least as the result of a long track record of government support, incentives and adequate regulation. Norway is well on track to achieve its 2025 phase-out of fossil fuel vehicle sales ahead of schedule, and the market share of sales in zero-emission vehicles and plug-in hybrid vehicles combined reached 86% in February 2022. Norwegian transport sector emissions reduction was seen as one of the top priorities, in the light that its power sector is already largely decarbonised [40, 406, 427–474]. Congruently, this also means that the electrification of transport in Norway led to a disproportionate reduction in emissions, in comparison with other countries which are still combusting fossil fuels for energy generation.

Norway's climate targets, policies and climate finance were established as being 'Almost sufficient', highlighting that some components of Norway's climate-saving endeavours are world's leading, such as the updated 2030 GHG emissions reduction target. Their climate policies and commitments, however, seem to require improvements to be aligned with the Paris Agreement 1.5 °C temperature target. The updated 55% NDC reduction target aligns with least-cost modelled domestic pathways, but to actually achieve its updated NDC target, Norway is required to significantly improve its current policies [40, 406, 427–474]. As Norway was evaluated against current 2 °C warming targets, it was rated as 'Almost sufficient'.

However, when Norway's NDC target was compared with its fair share contribution to climate action, it was rated as 'Insufficient' by CAT. To meet the fair share contribution target, Norway needs to provide additional, reliable climate finance to developing countries to support international decarbonisation. The CAT rating against Norway's NDC targets does not include emissions from exported fossil fuels and its policies and action were CAT rated as 'Almost sufficient', which suggests that its climate policies and action towards 2030 are not yet aligned with limiting global warming to 1.5 °C, but moderate efforts could achieve the desired outcome. Norway has almost provided a blueprint for decarbonising their economy, and if other countries were to follow the Norwegian example, global warming could at least be halted at 2 °C – not quite at 1.5 °C as required, but almost. The 2030 projections, based on current policies, domestic pathways and NDCs, are projected to lead to emissions of 41 Mt CO_2e, set at −21% against emission levels in 1990 [40, 406, 427–474].

A further positive aspect observed was that Norway's electricity generation is almost entirely based on renewable energy sources: In 2020, 98% of its electricity came from renewable energy sources, with 92% of electricity generated by hydropower plants, whilst 2.6% was generated by wind farms. Only 2% of electricity generation was the result of operating thermal power plants and mainly as a result of off-heat utilisation in industrial combustion processes. Being the country to be seen as the world leader in hydropower supply, Norway can provide the blueprint for other countries to utilise their large coastlines for energy generation and phase out fossil fuels [40, 406, 427–474]. Additionally, Norway's energy generation is so reliable that it is able to cooperate with the country's neighbours in balancing their electricity grids, providing favourable weather conditions.

Controversially, the biggest hydrocarbon reserves in Europe are located in Norway, which makes the country the world's fifth largest exporter of crude oil. Norway's offshore drilling projects have been subjected to having to pay considerable carbon tax since 1991, and in 2005, Norway joined the EU Emissions Trading System (ETS) [475].

Due to Norway's oil and gas sector being a high emitter, large amounts of carbon tax are paid each year. The sector was responsible for more than 25% of Norway's total annual GHG emission in 2020, although the vast majority had to be paid due to emissions from the combustion of fossil fuels abroad. Interestingly, this considerable amount of emissions is not currently included in Norway's emissions inventory, although it is much higher than the total of domestic post-combustion emissions and soils Norway's otherwise impressive climate record. The current Norwegian government has rebutted requests for reconsideration, so far, and stated that a phase-out of this sector is not currently even in the planning stage, indirectly confirming the continued support for fossil fuel exploration and exploitation [40, 406, 427–474]. Areas in the Barents Sea are of particular concern, although the decision to issue drilling and exploitation licences is currently being challenged in the ECHR.

Norway's National Transport Plan 2022–2033 includes the decarbonisation of the nation's transport system, and almost half of the NOK 1200 billion (USD 132bn) was allocated to measures still flowing into the construction and improvement of roads and highways.

Battery-based electric vehicle (BEV) sales reached a 65% market share in 2021, counting a total of ~110,000 vehicles. BEVs and plug-in hybrid electric vehicles (PHEVs) reached an 86% share of the total new car sales, and they were on track to reach 100% by the end of 2022.

Norway formally updated the country's 2030 NDC in November 2022 to at least a 55% GHG emissions reduction to below 1990 levels. This development was seen as compatible with modelled domestic pathways with the aim of limiting the global temperature increase to 1.5 °C on a globally, cost-effective basis.

Norway is truly in a unique position, as they are one of the few countries to generate its electricity almost entirely, at 92.6%, from renewable energy sources. Based on this fact and the considerable efforts the country has made over the years, it demonstrates how serious the country is taking the climate emergency and, as they are making every effort to decarbonise their economy, there is a positive outlook for Norwegian green steel and utilisation of green steel in Norway, as a future market for green steel [40, 406, 427–474].

10.11 European Union (EU) – Germany

The Climate Action Tracker only provided an evaluation of Germany, but not the EU or other EU countries. Germany as one of the strongest economies in the EU, besides France, will therefore have to suffice as a stand-in, as the only country representing the EU in the context of this evaluation.

The EU can be seen as one of the main drivers to decarbonise production and working towards a net-zero society, overall, with the climate goal to reach climate neutrality at the latest by 2050 [1–3]. The momentum behind the development and setting of these goals has caused significant investment in green steel technology, steel production process improvement [521] and a strong push towards decarbonising heavy industries [476–520, 522]. With Germany being one of the main drivers towards sustainability and net zero, they have demonstrated that it is possible to have a functioning society without polluting energy companies generating energy – as just demonstrated by switching off their last nuclear power station in April 2023 (Figures 10.19 and 10.20) [523].

The country's recent climate policies show decreased emission projections, but the divisions within Germany's government are making agreements on a comprehensive action plan across all industries very difficult and are making it difficult to reach the country's agreed-upon climate targets.

Although the solar energy sector lacks support and policies and incentives to increase a nationwide rollout, the industry has recently shown a significant increase in performance. The industry is up against lacking adequate short-term incentivising and implementation policies, and building laws are not showing what is commonly known as green building standards. The building and transport sector are very much trawling behind expectations and there are currently no comprehensive policies in place for the transport sector, at the time of publication. A distinctive step change is needed for Germany to be able to meet its GHG emissions reduction targets, which in themselves are not reaching far enough to

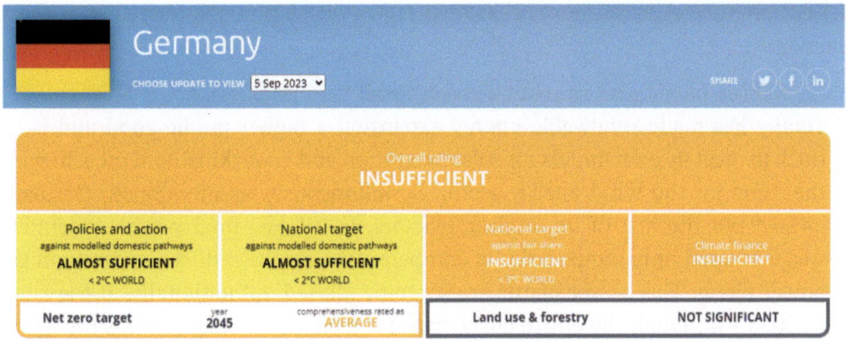

Figure 10.19 Overall rating of Germany. Source: Ref. [476] / with permission of Climate Action tracker.

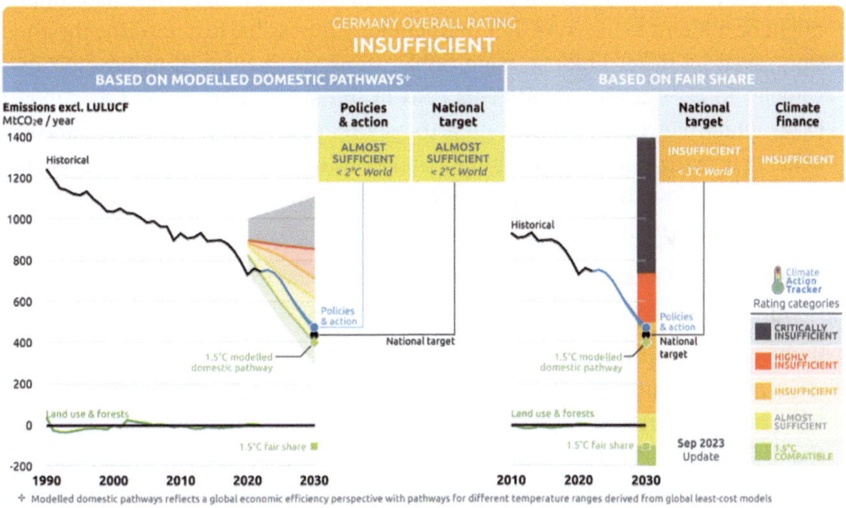

Figure 10.20 CAT rating explained. Source: Ref. [476] / with permission of Climate Action tracker.

be considered adequate to safeguard the 1.5 °C global temperature warming limit. Therefore, Germany's climate action was considered 'Insufficient' [476–520, 522]. This could be argued to be a bit harsh, considering that Germany is the first country in the world to completely walk away from harmful nuclear power generation and is one of the driving forces behind a Europe-wide decarbonisation effort.

Considering that an economy-wide and just transformation to a 1.5 °C compatible society is required, it is not acceptable that Germany is not adhering to its own set reporting requirements and is not implementing measures such as a general speed limit on freeways. However, as the driving on the Autobahn without a speed limit is one of the country's last remaining freedoms, and generally on motorways a 130 km/h (~80 mph) advisory speed limit for all vehicles under 3.5 tonnes is in place, politicians did not see a need to create policies for something that was already in place [476–520, 522].

Germany has seen some very positive policy developments, indeed, and these include:

Renewables-based and market-driven fossil fuel phase-out
Germany has made renewable energy expansion a higher public good and is on track to meet its own targets for solar expansion and is working hard on achieving the same for the wind energy sector. Simultaneously, as mentioned, Germany phased out the last of its nuclear reactors. Fossil fuel-fired power generation was increased only temporarily in compensation for the reduced capacity in gas power plants. Gas power plants are set to be phased out nationally, at the latest by 2038 – although the entire government is focused on reaching this milestone by 2030. The EU ETS is set to eliminate coal out of the German market by 2030.

Germany-wide subsidised public transport
The German government has created a ticket for the country-wide use of public transport at €49.- per month since May 2023. Although the immediate impact might be difficult to measure, it is mathematically possible to determine the emission saved by ticket holders using public transport instead of using their own vehicles to get out and about. This goes in the same direction as Luxembourg [524], which has recently made public transport free for everyone to use. This €49.- ticket may mark a shift to make public transportation in Germany more accessible to everyone. Now the ailing public infrastructure needs to be rejuvenated to be in line with its own set targets and initiatives.

In most areas, Germany requires drastic improvements to meet its climate and energy targets and go beyond them, towards a Paris-compatible trajectory:

Adhere to and implement its own governance system:
The German government proposed a world-first change in the climate law, making the whole government responsible for meeting the climate targets, as opposed to the individual ministries as before. However, it needs to be emphasised that it remained the responsibility of the individual ministries to achieve the sectoral targets, making them accountable for climate policy in their respective sector. This course effectively caused a dilution of responsibility and made it difficult for Germany to meet its targets. To date, the ministries of buildings and transport failed to submit an emergency programme as required by the new climate law.

Development and implementation of a transport sector strategy:
Currently, there does not seem to be much apparent effort in the transport ministry to implement policies to initiate the sector's transition towards net-zero GHG emissions.

Implementation of a building-sector-wide energy law:
The German government coalition contract has a very clear provision that as of 2024, all new heating systems have to use 65% of renewable energy technology. To date, there does not seem consensus over whether this rule was supposed to be implemented or even monitored. Measures like introducing a ban on new fossil fuel heating systems were introduced but were met with little enthusiasm or formal acceptance. An innovative draft law was since considerably slimmed down and still requires approval by Parliament, which is all but uncertain to be approved.

Taking a clear stand against fossil gas in Germany and abroad:

Due to the pressures of the energy crisis, the German government is still in some way supportive of new fossil fuel infrastructure in and outside of Germany, due to the loss of power generation capacity from switching off the nuclear power plants and the reduced supply of Russian gas. However, the LNG import capacity is larger than the quantity the country requires to function, and financing gas abroad is still being discussed, such as a cooperation with the Canadian government [476–520, 522].

10.12 Discussion and conclusions

In order to ensure the acceleration of decarbonising the steel industry, all steel-producing countries are required to be provided with the legal framework to protect and support their country's steel producers and provide them with the financial backing (climate finance) to implement the required changes.

Having evaluated some major steel-producing economies, it has become clear that in most countries, the opportunities for producing green steel and accessing existing markets with green steel are viable future prospects.

The hurdles to the implementation of decarbonisation processes and projects and producing green steel are:

- legacies of legal frameworks
- legacies of administrative challenges as a result of decision-making in the political realms hindering planning applications (renewables, heat recovery, fossil fuel phase-out and infrastructure)
- Shareholder interests (ROI) are still the first consideration for any decision-making: profit before climate
- High CAPEX expenditure and asset lifespan (20+ years)
- Socio-economic resistance to change towards green steel production: Workers need to be upskilled to work in the newly established industries for a smooth transition from fossil-fuel-based steel production processes to decarbonised and renewables-based steel manufacturing

Key areas where policy development can support green steel production:

- Creating steel-sector-specific strategies and guidance papers to drive system-wide decarbonisation:

  Although organisations in the steel industry, such as the World Steel Association, have stood up for their members and repeatedly approached political leaders and asked for the publication of such guidance, at present there are no such plans available at the time of print.
- Green industrial electrification:

  Country-specific business models and climate finance are needed for the rollout of renewables and integrating renewable energy technologies into steel production to support it.

- Behavioural and societal change:
  Although there are mechanisms in place, such as the EU ETS and Carbon Border Adjustment Mechanism (CBAM), it will take time before the CBAM will have established carbon worldwide as a new cost factor in international trade. Consumers and producers alike require a change of mindset and need to demand green steel over grey steel.

Although the actual implementation of decarbonising the steel industry is a challenging task, for any given country, there seems to be a lot of goodwill towards and markets for green steel, which heralds an exciting future of zero CO_2 emissions steel production.

References

1 Fairfield (2024). Green Steel Market Size $47.2Bn by 2030 [Internet]. https://www.fairfieldmarketresearch.com/report/green-steel-market (accessed 14 March 2024).

2 Fortune Business Insights (2024). Green Steel Market Size, Share | Growth Analysis Report [2030] [Internet]. https://www.fortunebusinessinsights.com/green-steel-market-108711 (accessed 14 March 2024).

3 Vantage Market Research (2024). 2024 Green Steel Market Size USD 120004.8 Million by 2030. Vantage Market Research. https://www.vantagemarketresearch.com (accessed 14 March 2024).

4 Worldsteel (2022). World steel in figures 2022 [Internet]. worldsteel.org. https://worldsteel.org/steel-topics/statistics/world-steel-in-figures-2022/ (accessed 15 March 2024).

5 Davies, R. (2024). Tata Steel to shut down Port Talbot blast furnaces, putting 3,000 jobs at risk. The Guardian [Internet]. https://www.theguardian.com/business/2024/jan/18/tata-steel-to-shut-down-port-talbot-blast-furnaces-3000-jobs-at-risk (accessed 15 March 2024).

6 Ellen MacArthur Foundation (EMAF) (2019). Completing the picture – how the circular economy tackles climate change. [Internet]. https://www.ellenmacarthurfoundation.org/assets/downloads/Completing_The_Picture_How_The_Circular_Economy-_Tackles_Climate_Change_V3_26_September.pdf (accessed 9 October 2022).

7 Ellen MacArthur Foundation (EMAF) (2023). Summit 23 – Redesigning the future: a circular economy showcase [Internet]. https://www.ellenmacarthurfoundation.org/summit/2023 (accessed 18 October 2023).

8 Kiessling, S., Gohari Darabkhani, H., and Soliman, A.H. (2024). Greater energy independence with sustainable steel production. *Sustainability.* 16 (3): 1174.

9 Climate Action Tracker (2023). United Kingdom [Internet]. https://climateactiontracker.org/countries/uk/ (accessed 5 March 2024).

10 Department for Energy Security and Net Zero (DESNZ) (2023). Green Homes Grant and Home Upgrade Grant statistics [Internet]. https://www.gov.uk/government/collections/green-home-grant-statistics (accessed 30 May 2025).
11 Climate Change Committee (2023). Delivering a reliable decarbonised power system [Internet]. https://www.theccc.org.uk/publication/delivering-a-reliable-decarbonised-power-system/#:~:text=Key%20messages,dependence%20on%20imported%20fossil%20fuels (accessed 30 April 2025).
12 Abdulla, A., Hanna, R., Schell, K.R. et al. (2021). Explaining successful and failed investments in U.S. carbon capture and storage using empirical and expert assessments. *Environmental Research Letters* 16: https://iopscience.iop.org/article/10.1088/1748-9326/abd19e.
13 Abnett, K. (2023). EU plans subsidies for hydrogen made using renewable energy -draft document | Reuters Online [Internet]. https://www.reuters.com/business/energy/eu-plans-subsidies-hydrogen-made-using-renewable-energy-draft-document-2023-03-16/ (accessed 30 April 2025).
14 Aviation Environment Federation (AEF) (2022). UK airport expansions – Aviation Environment Federation Online [Internet]. https://www.aef.org.uk/uk-airport-expansions/ (accessed 30 April 2025).
15 Bah, I. (2022). Germany to start gas phase-out from 2024 Online [Internet]. https://www.montelnews.com/news/1308596/germany-to-start-gas-phase-out-from-2024. (accessed 30 April 2025).
16 British Energy Security Strategy (BEIS) (2022a). British energy security strategy Dep. Business, Energy Ind [Internet]. https://www.gov.uk/government/publications/british-energy-security-strategy/british-energy-security-strategy#immediate-support-on-energy-bills (accessed 30 April 2025).
17 British Energy Security Strategy (BEIS) (2022b). Digest of UK Energy Statistics (DUKES): electricity Online [Internet]. https://www.gov.uk/government/statistics/electricity-chapter-5-digest-of-united-kingdom-energy-statistics-dukes (accessed 30 April 2025).
18 British Energy Security Strategy (BEIS) (2020). Contracts for difference for low carbon electricity generation consultation on proposed amendments to the scheme [Internet]. https://assets.publishing.service.gov.uk/government/uploads/system/uploads/attachment_data/file/885248/cfd-ar4-proposed-amendments-consultation.pdf (accessed 30 April 2025).
19 British Energy Security Strategy (BEIS) and Department for Energy Security & Net Zero (DESNZ) (2023). Final UK greenhouse gas emissions national statistics [Internet]. https://www.gov.uk/government/statistics/final-uk-greenhouse-gas-emissions-national-statistics-1990-to-2021 (accessed 30 April 2025).
20 British Energy Security Strategy (BEIS) and Department for Energy Security and Net Zero (DESNZ) (2023) Provisional UK greenhouse gas emissions national statistics 2022 Online [Internet]. https://www.gov.uk/government/statistics/provisional-uk-greenhouse-gas-emissions-national-statistics-2022 (accessed 30 April 2025).

21 Bloomberg (2023). EVs take record share of ailing UK auto market in 2022 Online [Internet]. https://europe.autonews.com/sales-market/evs-take-record-share-ailing-uk-auto-market-2022 (accessed 30 April 2025).

22 Calverley, D. and Anderson, K. (2022). Phaseout pathways for fossil fuel production within paris-compliant carbon budgets [Internet]. https://pure.manchester.ac.uk/ws/portalfiles/portal/213256008/Tyndall_Production_Phaseout_Report_final_text_3_.pdf (accessed 30 April 2025).

23 Climate Change Committee (2023). 2023 Progress report to parliament [Internet]. https://www.theccc.org.uk/publication/2023-progress-report-to-parliament/ (accessed 30 April 2025).

24 National Infrastructure Commisson (NIC) and Climate Change Committee (CCC) (2022). Act now to secure energy and climate security, NIC and CCC advise new Prime Minister – NIC [Internet]. https://nic.org.uk/news/nic-ccc-letter-to-prime-minister/ (accessed 30 April 2025).

25 Climate Change Committee (CCC) (2022). Progress in reducing emissions: 2022 Report to Parliament [Internet]. https://www.theccc.org.uk/publication/2022-progress-report-to-parliament/ (accessed 30 April 2025).

26 Climate Change Committee (CCC) (2020). The sixth carbon budget: The UK's path to net zero [Internet]. https://www.theccc.org.uk/publication/sixth-carbon-budget/ (accessed 30 April 2025).

27 Chuter, A. (2022). Three British nuclear programs are $1.67 billion over budget [Internet]. https://www.defensenews.com/global/europe/2020/05/12/three-british-nuclear-programs-are-167-billion-over-budget/ (accessed 30 April 2025).

28 Climate Analytics (2023). 2030 targets aligned to 1.5°C: evidence from the latest global pathways [Internet]. https://climateanalytics.org/publications/2023/2030-targets-aligned-to-15c-evidence-from-the-latest-global-pathways/ (accessed 30 April 2025).

29 HM Government UK (2023). £2 bus fare cap Online [Internet]. Department for Transport. https://www.gov.uk/guidance/2-bus-fare-cap (accessed 30 April 2025).

30 HM Government UK (2023). A zero emission vehicle (ZEV) mandate and CO_2 emissions regulation for new cars and vans in the UK Online [Internet]. Department for Transport. https://www.gov.uk/government/consultations/a-zero-emission-vehicle-zev-mandate-and-co2-emissions-regulation-for-new-cars-and-vans-in-the-uk (accessed 30 April 2025).

31 HM Government UK (2021). Bus back better national bus strategy for England [Internet]. Department for Transport. https://assets.publishing.service.gov.uk/government/uploads/system/uploads/attachment_data/file/980227/DfT-Bus-Back-Better-national-bus-strategy-for-England.pdf (accessed 30 April 2025).

32 HM Government UK (2022). Jet Zero strategy: delivering net zero aviation by 2050 [Internet]. London: UK Department of Transport; Department for Transport. https://www.gov.uk/government/publications/jet-zero-strategy-delivering-net-zero-aviation-by-2050 (accessed 30 April 2025).

33 HM Government UK (2021). Department for Transport [Internet]. https://www.gov.uk/government/publications/transport-decarbonisation-plan (accessed 30 April 2025).

34 Department of Transport (2020). Gear change a bold vision for cycling and walking [Internet]. https://assets.publishing.service.gov.uk/government/uploads/system/uploads/attachment_data/file/904146/gear-change-a-bold-vision-for-cycling-and-walking.pdf (accessed 30 April 2025).

35 Department for Energy Security and Net Zero (DESNZ) (2023). Carbon budget delivery plan [Internet]. https://assets.publishing.service.gov.uk/government/uploads/system/uploads/attachment_data/file/1147369/carbon-budget-delivery-plan.pdf (accessed 30 April 2025).

36 Department for Energy Security and Net Zero (DESNZ) (2023). Household energy efficiency statistics [Internet]. https://www.gov.uk/government/statistics/household-energy-efficiency-statistics-headline-release-june-2023 (accessed 30 April 2025).

37 Department for Energy Security and Net Zero (DESNZ) (2023). Net Zero Hydrogen Fund strands 1 and 2: Round 2 open to applications – GOV.UK [Internet]. https://www.gov.uk/government/publications/net-zero-hydrogen-fund-strand-1-and-strand-2 (accessed 30 April 2025).

38 Department for Energy Security and Net Zero (DESNZ) (2023). Planning for new energy infrastructure: revisions to National Policy Statements [Internet]. https://www.gov.uk/government/consultations/planning-for-new-energy-infrastructure-revisions-to-national-policy-statements (accessed 30 April 2025).

39 Department for Energy Security and Net Zero (DESNZ) (2023). Powering Up Britain [Internet]. https://www.gov.uk/government/publications/powering-up-britain/powering-up-britain (accessed 30 April 2025).

40 E3G (2023). Carbon capture and storage ladder [Internet]. https://www.e3g.org/publications/carbon-capture-and-storage-ladder/ (accessed 30 April 2025).

41 E3G (2023). UK Government continues to underspend on home retrofits [Internet]. https://www.e3g.org/news/uk-government-continues-to-underspend-on-home-retrofits/ (accessed 30 April 2025).

42 Energy & Climate Intelligence Unit (ECIU) (2023). Billpayers could miss out on £1bn a year in savings due to wind auction error. Energy & Climate Intelligence Unit Online [Internet]. https://eciu.net/media/press-releases/2023/billpayers-could-miss-out-on-1bn-a-year-in-savings-due-to-wind-auction-error (accessed 30 April 2025).

43 Energy & Climate Intelligence Unit (ECIU) (2023). PM's net zero changes could cost households up to £8bn. Energy & Climate Intelligence Unit Online [Internet]. https://eciu.net/media/press-releases/2023/pms-net-zero-changes-could-cost-households-up-to-8bn (accessed 30 April 2025).

44 Energy & Climate Intelligence Unit (ECIU) (2022). Poorly insulated homes to pay £1,000 more on gas bills this winter [Internet]. https://eciu.net/media/press-releases/2022/poorly-insulated-homes-to-pay-1-000-more-on-gas-bills-this-winter (accessed 30 April 2025).

45 Ember (2020). The Burning Question: Should the UK end tax breaks on burning wood for power? [Internet]. https://ember-climate.org/insights/research/the-burning-question/ (accessed 30 April 2025).

46 Energy Transitions Commission (2021). Energy Transitions Commission 2021. Bioresources within a Net-Zero Emissions Economy Online [Internet]. https://www.energy-transitions.org/publications/bioresources-within-a-net-zero-emissions-economy/ (accessed 30 April 2025).

47 Evans, C., Artz, R., Moxley, J. et al. (2017). Implementation of an Emissions Inventory for UK Peatlands A report to the Department for Business, Energy & Industrial Strategy Title Implementation of an Emissions Inventory for UK Peatlands Online [Internet]. https://uk-air.defra.gov.uk/assets/documents/reports/cat07/1904111135_UK_peatland_GHG_emissions.pdf (accessed 30 April 2025).

48 Evans, S. (2020). Analysis: fuel-duty freeze has increased UK CO_2 emissions by up to 5% Carbon Br [Internet]. https://www.carbonbrief.org/analysis-fuel-duty-freeze-has-increased-uk-co2-emissions-by-up-to-5-per-cent (accessed 30 April 2025).

49 Evans, S. (2022). Analysis: Why UK energy bills are soaring to record highs – and how to cut them – carbon brief [Internet]. https://www.carbonbrief.org/analysis-why-uk-energy-bills-are-soaring-to-record-highs-and-how-to-cut-them/ (accessed 30 April 2025).

50 Forest Research (2021). 1: woodland area & planting [Internet]. https://www.forestresearch.gov.uk/tools-and-resources/statistics/forestry-statistics/forestry-statistics-2021/1-woodland-area-planting/ (accessed 30 April 2025).

51 Forest Research (2022). Forestry statistics [Internet]. https://www.forestresearch.gov.uk/tools-and-resources/statistics/forestry-statistics/forestry-statistics-2022/ (accessed 30 April 2025).

52 Forest Research (2022). Woodland statistics – Forest Research Online [Internet]. https://www.forestresearch.gov.uk/tools-and-resources/statistics/statistics-by-topic/woodland-statistics/ (accessed 30 April 2025).

53 George, S. (2023). UK Government rejects MPs' calls for bolder action to phase down fossil fuels [Internet]. https://www.edie.net/ (accessed 30 April 2025).

54 Grant, N., Hawkes, A., Mittal, S., and Gambhir, A. (2021). The policy implications of an uncertain carbon dioxide removal potential. *Joule* 5: 2593–2605. https://doi.org/10.1016/j.joule.2021.09.004.

55 Grant, N., Hawkes, A., Napp, T., and Gambhir, A. (2021). Cost reductions in renewables can substantially erode the value of carbon capture and storage in mitigation pathways. *One Earth* 4: 1588–1601. https://doi.org/10.1016/j.oneear.2021.10.024.

56 Greene, T. (2022). UK has approved several fossil fuel projects since Cop26, analysis finds Online [Internet]. https://www.theguardian.com/environment/2022/may/19/uk-has-approved-several-fossil-fuel-projects-since-cop26-analysis-finds (accessed 30 April 2025).

57 Harvey, F. (2022). A guide to the UK energy security strategy. The Guardian Online [Internet]. https://www.theguardian.com/environment/2022/mar/19/

a-guide-to-the-uk-energy-security-strategy-cost-of-living-crisis (accessed 30 April 2025).

58 HH and PG (2023). Revealed: UK plans to drop flagship £11.6bn climate pledge | Climate crisis | The Guardian Online [Internet]. https://www.theguardian.com/environment/2023/jul/04/revealed-uk-plans-to-drop-flagship-climate-pledge-rishi-sunak (accessed 30 April 2025).

59 HM Government UK (2021). House of Commons Public Accounts Committee 2021. Achieving Net Zero [Internet]. https://committees.parliament.uk/publications/4921/documents/49419/default/ (accessed 30 April 2025).

60 The International Council on Clean Transportation (ICCT) (2022). Real-world usage of plug-in hybrid vehicles in Europe: A 2022 update on fuel consumption, electric driving, and CO_2 emissions [Internet]. https://theicct.org/publication/real-world-phev-use-jun22/ (accessed 30 April 2025).

61 The International Energy Agency (IEA) (2021). Net zero by 2050: a roadmap for the global energy sector [Internet]. https://www.iea.org/reports/net-zero-by-2050 (accessed 30 April 2025).

62 The International Energy Agency (IEA) (2022) The future of heat pumps – analysis online [Internet]. https://www.iea.org/reports/the-future-of-heat-pumps (accessed 30 April 2025).

63 The International Energy Agency (IEA) (2022). World Energy Outlook 2022 [Internet]. https://www.iea.org/topics/world-energy-outlook (accessed 30 April 2025).

64 The International Energy Agency (IEA) (2022). World Energy Outlook 2022 [Internet]. https://www.iea.org/reports/world-energy-outlook-2022 (accessed 30 April 2025).

65 The International Renewable Energy Agency (IRENA) (2023). Renewable capacity statistics 2023 Online [Internet]. https://www.irena.org/Publications/2023/Mar/Renewable-capacity-statistics-2023 (accessed 30 April 2025).

66 Holgate, J. (2022). Summary of Judgement: The Queen (on the application of (1) Friends of the Earth Limited (2) ClientEarth (3) Good Law Project and Joanna Wheatley v Secretary of State for Business, Energy and Industrial Strategy [Internet]. 2022. https://www.judiciary.uk/wp-content/uploads/2022/07/FoE-v-BEIS-judgment-180722.pdf (accessed 30 April 2025).

67 Lee, D.S., Fahey, D.W., Skowron, A. et al. (2021). The contribution of global aviation to anthropogenic climate forcing for 2000 to 2018. *Atmospheric Environment.* 244: 117834.

68 Merrick, R. (2021). Boris Johnson's £11.6bn climate fund to be swiped from aid budget Indep [Internet]. https://www.independent.co.uk/climate-change/news/climate-fund-boris-johnson-aid-cop26-b1877116.html (accessed 30 April 2025).

69 Oil and Gas Authority (2023). Publication of decision of the Secretary of State and the Oil and Gas Authority.

70 Pachauri, S., Pelz, S., Bertram, C. et al. (2022). Fairness considerations in global mitigation investments. *Science.* 378: 1057–1059.

71 Pickard, J., Mooney, A., Millard, R., and Campbell, P. (2023). Car industry reels from Sunak's retreat on net zero plans. *Financial Times.*

72 Reid, A. (2023). New North Sea oil and gas licences threaten UK energy security Energy Monitor Online [Internet]. https://www.energymonitor.ai/sectors/industry/opinion-new-north-sea-oil-and-gas-exploration-licences-pose-a-threat-to-uks-future-energy-security/ (accessed 30 April 2025).

73 Rosenow, J. and Gibb, D. (2023). Guest post: how the energy crisis is boosting heat pumps in Europe – Carbon Brief Online [Internet]. https://www.carbonbrief.org/guest-post-how-the-energy-crisis-is-boosting-heat-pumps-in-europe/ (accessed 30 April 2025).

74 Rowlatt, J. (2023). Memo reveals pressure on UK climate finance pledge. BBC News Online [Internet]. https://www.bbc.co.uk/news/uk-66105901 (accessed 30 April 2025).

75 Rutherford, D. (2022). Sustainable aviation fuels: What does real leadership look like? [Internet]. https://theicct.org/us-eu-saf-leadership-aug22/ (accessed 30 April 2025).

76 Savage, M. (2022). Steel boss dismisses claim that sector needs new Cumbrian coalmine Online [Internet]. https://www.theguardian.com/business/2022/may/01/steel-boss-dismisses-claim-that-sector-needs-new-cumbrian-coalmine (accessed 30 April 2025).

77 Savage, M. (2023). UK's net zero ambitions at risk after 'disastrous' offshore wind auction The Observer Online [Internet]. https://www.theguardian.com/environment/2023/sep/10/uks-net-zero-ambitions-at-risk-after-disastrous-offshore-wind-auction (accessed 30 April 2025).

78 Forestry Scotland (2021). Scotland showing leadership on climate forests Online [Internet]. Scottish Forestry. https://forestry.gov.scot/news-releases/scotland-showing-leadership-on-climate-forests (accessed 30 April 2025).

79 Shrestha, P. (2022). Netherlands to ban fossil fuel boilers from 2026. Energy Live News Online [Internet]. https://www.energylivenews.com/2022/05/19/netherlands-to-ban-fossil-fuel-boilers-from-2026/ (accessed 30 April 2025).

80 Skidmore, C. (2023). Mission zero: net zero review online [Internet]. https://www.gov.uk/government/news/net-zero-review-uk-could-do-more-to-reap-economic-benefits-of-green-growth (accessed 30 April 2025).

81 The Society of Motor Manufacturers and Traders (SMMT) (2022). Covid stalls 2021 UK new car market but record EV sales show future direction Online [Internet]. https://www.smmt.co.uk/2022/01/covid-stalls-2021-uk-new-car-market-but-record-ev-sales-show-future-direction/ (accessed 30 April 2025).

82 Stewart, C., Piernas, C., Cook, B., and Jebb, S.A. (2021). Trends in UK meat consumption: analysis of data from years 1–11 (2008–09 to 2018–19) of the National Diet and Nutrition Survey rolling programme. *Lancet Planet Health* 5: 699–708.

83 The Wildlife Trust (2023). The Wildlife Trusts 2023 UK Government confirms ban on all peat-based gardening products will not be implemented until 2030 | The Wildlife Trusts Online [Internet]. https://www.wildlifetrusts.org/news/uk-government-confirms-ban-all-peat-based-gardening-products-will-not-be-implemented-until (accessed 30 April 2025).

84 Thomas, N. (2022). UK government approves Jackdaw gas project in North Sea |. Financial Times Online [Internet]. https://www.ft.com/content/a650f12e-65d7-4cf1-9d95-568d63229597 (accessed 30 April 2025).

85 Thurston, A. (2023). Europe's leakiest homes: new study shames Britain, fails to plug gaps Online [Internet]. https://theenergyst.com/europes-leakiest-homes-new-study-fingers-britain-fails-to-plug-gaps/ (accessed 30 April 2025).

86 Tidman, Z. (2022). UK government approves new oil and gas field weeks after COP26 | The Independent Online [Internet]. https://www.independent.co.uk/climate-change/news/oil-gas-field-cop26-climate-b2005123.html (accessed 30 April 2025).

87 UK Government (2022). £200 million boost to rollout of hundreds more zero-emission HGVs Online [Internet]. https://www.gov.uk/government/news/200-million-boost-to-rollout-of-hundreds-more-zero-emission-hgvs (accessed 30 April 2025).

88 UK Government (2022). Biggest renewables auction accelerates move away from fossil fuels Online [Internet]. https://www.gov.uk/government/news/biggest-renewables-auction-accelerates-move-away-from-fossil-fuels (accessed 30 April 2025).

89 UK Government (2022). Climate change agreements extended until [Internet] 2025. https://www.gov.uk/guidance/climate-change-agreements--2 (accessed 30 April 2025).

90 UK Government (2022). Digest of UK Energy Statistics (DUKES): Energy Online [Internet]. https://www.gov.uk/government/statistics/energy-chapter-1-digest-of-united-kingdom-energy-statistics-dukes (accessed 30 April 2025).

91 UK Government (2022). Government food strategy Online [Internet]. https://www.gov.uk/government/publications/government-food-strategy/government-food-strategy (accessed 30 April 2025).

92 UK Government (2022). Government hits accelerator on low-cost renewable power Online [Internet]. https://www.gov.uk/government/news/government-hits-accelerator-on-low-cost-renewable-power (accessed 30 April 2025).

93 UK Government (2021). Industrial decarbonisation strategy.

94 UK Government (2019). Net zero emissions target: house of commons hansard online [Internet]. https://hansard.parliament.uk/Commons/2019-06-12/debates/A348AE4C-8957-42C8-8180-0F59E597E3EA/NetZeroEmissionsTarget (accessed 30 April 2025).

95 UK Government (2021). Net zero strategy: build back greener.

96 UK Government (2019). PM Theresa May: we will end UK contribution to climate change by 2050 Gov.uk Online [Internet]. https://www.gov.uk/government/news/pm-theresa-may-we-will-end-uk-contribution-to-climate-change-by-2050 (accessed 30 April 2025).

97 UK Government (2023). The environmental targets (woodland and trees outside woodland) (England) regulations 2022. King's Printer of Acts of Parliament.

98 UK Government (2021). The Future Homes Standard: changes to Part L and Part F of the Building Regulations for new dwellings Online [Internet]. https://www.gov.uk/government/consultations/the-future-homes-standard-changes-to-part-l-and-part-f-of-the-building-regulations-for-new-dwellings (accessed 30 April 2025).

99 UK Government (2019). UK aid to double efforts to tackle climate change Online [Internet]. https://www.gov.uk/government/news/uk-aid-to-double-efforts-to-tackle-climate-change (accessed 30 April 2025).
100 UK Government (2019). UK becomes first major economy to pass net zero emissions law Online [Internet]. https://www.gov.uk/government/news/uk-becomes-first-major-economy-to-pass-net-zero-emissions-law (accessed 30 April 2025).
101 UK Government (2021). UK enshrines new target in law to slash emissions by 78% by 2035.
102 UK Government (2022). UK's Nationally Determined Contribution – updated September 2022 Online [Internet]. https://assets.publishing.service.gov.uk/government/uploads/system/uploads/attachment_data/file/1106145/NDC_ICTU_MASTER_VERSION_20PT_tables.pdf (accessed 30 April 2025).
103 United Nations Framework Convention on Climate Change (UNFCCC) (2011). Report of the Conference of the Parties on its sixteenth session, held in Cancun from 29 November to 10 December 2010 (accessed 30 April 2025).
104 Walker, P. (2023) New North Sea oil and gas licences will send 'wrecking ball' through climate commitments. The Guardian.
105 Welsh Government (2021). Deputy Minister issues 'national call to arms' [Internet]. https://gov.wales/deputy-minister-issues-national-call-arms-plant-more-trees (accessed 30 April 2025).
106 Winkler, L., Pearce, D., Nelson, J., and Babacan, O. (2023). The effect of sustainable mobility transition policies on cumulative urban transport emissions and energy demand. *Nature Communications.* 14: 1–14.
107 Wintour, P. (2021). Britain's aid cuts: what's been announced so far. The Guardian Online [Internet]. https://www.theguardian.com/global-development/2021/apr/30/britains-aid-cuts-whats-been-announced-so-far (accessed 30 April 2025).
108 Woodland Trust (2021). State of the UK's Woods and Trees. https://www.woodlandtrust.org.uk/state-of-uk-woods-and-trees/ (accessed 29 February 2024).
109 The World Wide Fund (WWF) (2020). COP26: The UK's 2030 climate target to cut emissions Online [Internet]. https://www.wwf.org.uk/updates/cop26-climate-target-cut-emissions (accessed 30 April 2025).
110 Carbon Action Tracker (2023). Policies & action [Internet]. https://climateactiontracker.org/countries/uk/policies-action/ (accessed 15 March 2024).
111 United Nations (UN) (2016). Paris Agreement 2016 [Internet]. United Nations. https://www.un.org/en/climatechange/paris-agreement#:~:text=To%20tackle%20climate%20change%20and,2015%3A%20the%20historic%20Paris%20Agreement (accessed 2 March 2020).
112 Farron, T. (2023). West Cumbria coal mine's business case "is dead", MP claims. BBC News. https://www.bbc.com/news/articles/cx71wzp9qego (accessed 15 March 2024).

113 Somerset Council (2024). Hinkley Point C planning [Internet]. https://www.somerset.gov.uk/business-economy-and-licences/major-developments/hinkley-point-c-planning/ (accessed 15 March 2024).

114 Intergovernmental Panel for Climate Change (IPCC) (2023). AR6 Synthesis Report Climate Change 2023 [Internet]. Report No.: AR6. https://www.ipcc.ch/report/ar6/syr/downloads/report/IPCC_AR6_SYR_SPM.pdf (accessed 5 January 2023).

115 Climate Action Tracker (2023). USA [Internet]. https://climateactiontracker.org/countries/usa/ (accessed 19 March 2024).

116 United States Department of Agriculture (USDA) (2023). Mountain Valley Pipeline and the Jefferson National Forest [Internet]. https://www.fs.usda.gov/detail/gwj/landmanagement/projects/?cid=stelprd3827827 (accessed 30 April 2025).

117 The Beyond Oil & Gas Alliance (2021). At COP26, 11 National and Subnational Governments Launch The Beyond Oil & Gas Alliance [Internet]. https://beyondoilandgasalliance.org/wp-content/uploads/2021/11/11-10-21-BOGA-Press-Release.pdf (accessed 30 April 2025).

118 The Beyond Oil & Gas Alliance (2022). BOGA announces support fund for fossil phase out [Internet]. https://beyondoilandgasalliance.org/wp-content/uploads/2022/12/BOGA_COP27_Press_Release_updated.pdf (accessed 30 April 2025).

119 California Air Resources Board (2021). Governor Newsom's Zero-Emission by 2035 Executive Order (N-79-20) [Internet]. https://ww2.arb.ca.gov/resources/fact-sheets/governor-newsoms-zero-emission-2035-executive-order-n-79-20 (accessed 30 April 2025).

120 California Air Resources Board (2023). States that have Adopted California's Vehicle Regulations [Internet]. https://ww2.arb.ca.gov/our-work/programs/advanced-clean-cars-program/states-have-adopted-californias-vehicle-regulations (accessed 30 April 2025).

121 Climate Action Tracker (2023). Clean electricity within a generation: Paris-aligned benchmarks for the power sector.

122 Climate Action Tracker (2016). The ten most important short-term steps to limit warming to 1.5°C [Internet]. Climate Action Tracker. https://climateactiontracker.org/publications/the-ten-most-important-short-term-steps-to-limit-warming-to-15c/ (accessed 30 April 2025).

123 Climate Action Tracker (2020). Paris Agreement Compatible Sectoral Benchmarks: Elaborating the decarbonisation roadmap [Internet]. https://climateanalytics.org/media/cat_2020-07-10_parisagreementbenchmarks_fullreport.pdf (accessed 30 April 2025).

124 Climate Action Tracker (2021). To show climate leadership, US 2030 target should be at least 57–63% [Internet]. Climate Action Tracker. https://climateactiontracker.org/documents/846/2021_03_CAT_1.5C-consistent_US_NDC.pdf (accessed 30 April 2025).

125 Climate Power (2023). One year of our clean energy boom (accessed 30 April 2025).

126 Climate Trace (2023). Climate Trace Inventory [Internet]. https://climatetrace.org/inventory (accessed 30 April 2025).

127 US Energy Information Administration (EIA) (2021). U.S. Energy Information Administration – EIA – Independent Statistics and Analysis – International Energy Outlook 2021 [Internet]. https://www.eia.gov/outlooks/aeo/data/browser/#/?id=1-IEO2021&sourcekey=0 (accessed 11 July 2023).

128 Hasanbeigi, A.S.C. (2019). How clean is the U.S. Steel Industry? An international benchmarking of energy and CO_2 intensities. *Global Efficiency Intelligence* 30 November 2019.

129 UK Government and UNFCCC (2021). COP26 Presidency Compilation of 2021–2025 Climate Finance Commitments [Internet]. https://ukcop26.org/wp-content/uploads/2021/11/Table-of-climate-finance-commitments-November-2021.pdf (accessed 30 April 2025).

130 Copley, M. (2023). Biden pledged to stop funding fossil fuels overseas. It's not stopping one agency [Internet]. NPR. https://www.npr.org/2023/05/12/1175798493/biden-pledged-to-stop-funding-fossil-fuels-overseas-its-not-stopping-one-agency (accessed 30 April 2025).

131 DeLauro, R. (2022). DeLauro Introduces Government Funding Legislation [Internet]. https://appropriations.house.gov/news/press-releases/delauro-introduces-government-funding-legislation (accessed 30 April 2025).

132 Dellasala, D.A., Gorelik, S.R., and Walker, W.S. (2022). The Tongass National Forest, Southeast Alaska, USA: a natural climate solution of global significance. *Land.* 11 (5): https://doi.org/10.3390/land11050717.

133 Donor Tracker (2021). United States – Climate [Internet]. https://donortracker.org/united-states/climate (accessed 30 April 2025).

134 Donor Tracker (2023). Donor Tracker SEEK Development Donors Sector Climate. https://donortracker.org/topics/climate (accessed 30 April 2025).

135 Douglas, L. (2023). Biden administration pledges fight to keep climate-friendly farming funds. Reuters [Internet]. https://www.reuters.com/business/sustainable-business/biden-administration-pledges-fight-keep-climate-friendly-farming-funds-2023-05-02/ (accessed 30 April 2025).

136 E2 (2023). Clean Economy Works: IRA One-Year Review [Internet]. https://e2.org/wp-content/uploads/2023/08/E2-Clean-Economy-Works-IRA-One-Year-Review_FINAL.pdf (accessed 30 April 2025).

137 EndCoal (2020). Global Coal Public Finance Tracker. Last updated [Internet]. https://endcoal.org/global-coal-finance-tracker/ (accessed 30 April 2025).

138 Europe Commission (2020). nzeb – Energy [Internet]. https://ec.europa.eu/energy/content/nzeb-24_en (accessed 30 April 2025).

139 The Financial Times (2023). 'Transformational change': Biden's industrial policy begins to bear fruit [Internet]. https://www.ft.com/content/b6cd46de-52d6-4641-860b-5f2c1b0c5622 (accessed 30 April 2025).

140 Forest Declaration Platform (2022). Transforming commitments into action [Internet]. https://forestdeclaration.org/ (accessed 30 April 2025).

141 Friedman, L. (2023). Biden Bans Roads and Logging in Alaska's Tongass National Forest. The New York Times [Internet]. https://www.nytimes.com/

2023/01/25/climate/alaska-tongass-national-forest.html (accessed 30 April 2025).

142 Sachs, G. (2023). The US is Poised for an Energy Revolution [Internet]. https://www.goldmansachs.com/intelligence/pages/the-us-is-poised-for-an-energy-revolution.html (accessed 30 April 2025).

143 Guillén, A. (2023). The nerd's guide to Biden's new climate rule.

144 Energy Information Administration (EIA) (2023). 2023 Net zero roadmap: a global pathway to keep the 1.5°C goal in reach – analysis. IEA. https://www.iea.org/reports/net-zero-roadmap-a-global-pathway-to-keep-the-15-0c-goal-in-reach (accessed 18 October 2023).

145 EVs I (2023). US: All-Electric Car sales surged In January 2023 - 7% market share. https://insideevs.com/news/657660/us-electric-car-sales-january2023/ (accessed 30 April 2025).

146 Intergovernmental Panel for Climate Change (IPCC) (2022). Climate change 2022: mitigation of climate change [Internet]. https://www.un.org/en/climatechange/reports?gclid=EAIaIQobChMIxsyF3ZSl-AIVoo1oCR19Dw3IEAAYAiAAEgItyfD_BwE (accessed 30 April 2022).

147 Joselow, M. and Montalbano, V. (2023). 3 trends to watch in US climate policy in 2023. The Washington Post [Internet]. https://www.washingtonpost.com/politics/2023/01/03/3-trends-watch-us-climate-policy-2023/ (accessed 30 April 2025).

148 Kühne, K., Bartsch, N., Driskell, R., and Higson, J. (2022) Carbon bombs [Internet]. https://doi.org/10.1016/j.enpol.2022.112950.

149 Lu, X., Jacob, D.J., Zhang, Y. et al. (2023). Observation-derived 2010–2019 trends in methane emissions and intensities from US oil and gas fields tied to activity metrics [Internet]. p. 1–9. https://doi.org/10.1073/pnas.2217900120.

150 Mass.gov (2022). Stretch energy code development 2022 [Internet]. https://www.mass.gov/info-details/stretch-energy-code-development-2022#new!-release-of-draft-code-language-for-stretch-code-update-and-new-specialized-stretch-code (accessed 30 April 2025).

151 Monke, J., Bracmort, K., Stubbs, M., and Hoover, K. (2022). Inflation reduction act: agricultural conservation and credit, renewable energy, and forestry [Internet]. https://crsreports.congress.gov/product/pdf/IN/IN11978 (accessed 30 April 2025).

152 National Highway Traffic Safety Administration (2022). Corporate average fuel economy standards for model years 2024–2026 passenger cars and light trucks [Internet]. https://www.nhtsa.gov/sites/nhtsa.gov/files/2022-04/Final-Rule-Preamble_CAFE-MY-2024-2026.pdf (accessed 30 April 2025).

153 Organisation for Economic Co-operation and Development (OECD) (2022). Climate change: OECD DAC external development finance statistics [Internet]. https://www.oecd.org/dac/financing-sustainable-development/development-finance-topics/climate-change.htm (accessed 30 April 2025).

154 Powering Past Coal Alliance (2019). Powering past coal alliance: our mission [Internet]. https://poweringpastcoal.org/about/Powering-Past-Coal-Alliance-Mission (accessed 30 April 2025).

155 Quinn, M. (2023). EPA agrees to reevaluate MSW landfill emissions measurements in effort to settle lawsuit. Waste Dive [Internet]. https://www.wastedive.com/news/epa-consent-decree-msw-landfill-emissions/642606/ (accessed 30 April 2025).
156 Raceto Zero (2021). The Breakthrough Agenda – Climate Champions [Internet]. https://racetozero.unfccc.int/system/breakthrough-agenda/ (accessed 30 April 2025).
157 Rhodium Group (2022). A turning point for US climate progress: assessing the climate and clean energy provisions in the inflation reduction act [Internet]. https://rhg.com/research/climate-clean-energy-inflation-reduction-act/ (accessed 30 April 2025).
158 Steinberg, D.C., Brown, M., Wiser, R. et al. (2023). Evaluating impacts of the inflation reduction act and bipartisan infrastructure law on the US power system [Internet]. https://www.nrel.gov/docs/fy23osti/85242.pdf.
159 West Virginia v (2022). EPA [Internet]. https://www.supremecourt.gov/opinions/21pdf/20-1530_n758.pdf (accessed 30 April 2025).
160 The White House (2021). Fact sheet: President Biden announces steps to drive American leadership forward on clean cars and trucks [Internet]. FACT SHEET: President Biden Announces Steps to Drive American Leadership Forward on Clean Cars and Trucks | The White House (accessed 30 April 2025).
161 The White House (2021). U.S. methane emissions reduction action plan [Internet]. https://www.whitehouse.gov/wp-content/uploads/2021/11/US-Methane-Emissions-Reduction-Action-Plan-1.pdf (accessed 30 April 2025).
162 The White House (2021). Executive order on catalyzing clean energy industries and jobs through federal sustainability [Internet]. The White House Briefing Room. https://www.whitehouse.gov/briefing-room/presidential-actions/2021/12/08/executive-order-on-catalyzing-clean-energy-industries-and-jobs-through-federal-sustainability/ (accessed 30 April 2025).
163 The White House (2022). Delivering on the U.S. methane emissions reduction action plan [Internet]. https://www.whitehouse.gov/wp-content/uploads/2022/11/US-Methane-Emissions-Reduction-Action-Plan-Update.pdf (accessed 30 April 2025).
164 The White House (2022). Fact sheet: Biden-Harris administration launches initiative to modernize building codes, improve climate resilience, and reduce energy costs [Internet]. The White House Briefing Room. https://www.whitehouse.gov/briefing-room/statements-releases/2022/06/01/fact-sheet-biden-harris-administration-launches-initiative-to-modernize-building-codes-improve-climate-resilience-and-reduce-energy-costs/ (accessed 30 April 2025).
165 The White House (2022). Fact sheet: Biden-Harris administration takes more than 100 actions in 2022 to strengthen energy efficiency standards and save families money [Internet]. The White House Briefing Room. https://www.whitehouse.gov/briefing-room/statements-releases/2022/12/19/fact-sheet-biden-harris-administration-takes-more-than-100-actions-in-2022-to-strengthen-energy-efficiency-standards-and-save-families-money/ (accessed 30 April 2025).

166 The White House (2023). Budget of the U.S. Government Fiscal Year [Internet]. https://www.whitehouse.gov/wp-content/uploads/2023/03/budget_fy2024.pdf (accessed 30 April 2025).

167 The White House (2023). Building a clean energy economy: a guidebook to the inflation reduction act's investments in clean energy and climate action [Internet]. https://www.whitehouse.gov/wp-content/uploads/2022/12/Inflation-Reduction-Act-Guidebook.pdf (accessed 30 April 2025).

168 The White House (2023). Fact sheet: one year in, President Biden's inflation reduction act is driving historic climate action and investing in America to create good paying jobs and reduce costs [Internet]. https://www.whitehouse.gov/briefing-room/statements-releases/2023/08/16/fact-sheet-one-year-in-president-bidens-inflation-reduction-act-is-driving-historic-climate-action-and-investing-in-america-to-create-good-paying-jobs-and-reduce-costs/ (accessed 30 April 2025).

169 The White House Council on Environmental Quality (2022). The White House Council on Environmental Quality [Internet]. https://www.sustainability.gov/pdfs/EO_14057_Implementing_Instructions.pdf (accessed 30 April 2025).

170 Economic Analysis USB (2023). Gross domestic product, first quarter 2023 [Internet]. https://www.bea.gov/news/2023/gross-domestic-product-first-quarter-2023-advance-estimate (accessed 30 April 2025).

171 Land Management USB (2023). Willow master development plan: supplemental environmental impact statement. p. 1–5.

172 Interior USD (2023). Interior department substantially reduces scope of willow project [Internet]. Interior Press. https://www.doi.gov/pressreleases/interior-department-substantially-reduces-scope-willow-project (accessed 30 April 2025).

173 Ocean Energy Management USB (2022). 2023–2028 National Outer Continental Shelf Oil and Gas Leasing Proposed Program [Internet]. https://www.boem.gov/sites/default/files/documents/oil-gas-energy/national-program/2023-2028_Proposed (accessed 30 April 2025).

174 Ocean Energy Management USB (2023). Gulf of Mexico OCS Oil and Gas Lease Sales 259 and 261: Final Supplemental Environmental Impact Statement [Internet]. https://www.boem.gov/sites/default/files/documents/renewable-energy/state-activities/GOM_LS259-261_SEIS_FINAL.pdf (accessed 30 April 2025).

175 US Department of Energy (n.d.). Hydrogen shot. Office of Energy Efficiency & Renewable Energy [Internet]. https://www.energy.gov/eere/fuelcells/hydrogen-shot (accessed 30 April 2025).

176 US Department of Energy (2022). The inflation reduction act drives significant emissions reductions and positions America to reach our climate goals [Internet]. The Inflation Reduction Act Drives Significant Emissions Reductions and Positions America to Reach Our Climate Goals | Department of Energy (accessed 30 April 2025).

177 US Department of Energy (2022). Biden-Harris administration announces availability of $350 million in grants to states to cut methane emissions from oil and gas sector [Internet]. https://www.energy.gov/articles/biden-harris-administration-announces-availability-350-million-grants-states-cut-methane (accessed 30 April 2025).

178 US Department of Energy (2022). Biden administration announces investments to make homes more energy efficient and lower costs for American families [Internet]. https://www.energy.gov/articles/biden-administration-announces-investments-make-homes-more-energy-efficient-and-lower (accessed 30 April 2025).
179 US Department of Energy (2023). DOE finalizes efficiency rules for room air conditioners and portable air cleaners [Internet]. https://www.energy.gov/articles/doe-finalizes-efficiency-rules-room-air-conditioners-and-portable-air-cleaners (accessed 30 April 2025).
180 US Department of Energy (2023). DOE proposes new standards for residential appliances to save Americans billions in annual energy and water bill. Department of Energy News [Internet]. https://www.energy.gov/articles/doe-proposes-new-standards-residential-appliances-save-americans-billions-annual-energy (accessed 30 April 2025).
181 US Environmental Protection Agency (2022). U.S. State-level non-CO_2 greenhouse gas mitigation potential. p. 2025–2050.
182 Lab ZERO (2023). Preview: Final REPEAT project findings on the emissions impacts of the inflation reduction act and infrastructure investment and jobs act [Internet]. https://repeatproject.org/docs/REPEAT_2023_Preview.pdf (accessed 30 April 2025).
183 Carbon Action Tracker (2023). Brazil [Internet]. https://climateactiontracker.org/countries/brazil/ (accessed 20 March 2024).
184 Angelo, C. and Werneck, F. (2021). O teste de fidelidade de Joaquim Leite. Observatório Do Clima [Internet]. https://www.oc.eco.br/o-teste-de-fidelidade-de-joaquim-leite/ (accessed 30 April 2025).
185 ANP (2020). Dados estatísticos: Vendas de derivados de petróleo e etanol [Internet]. http://www.anp.gov.br/dados-estatisticos (accessed 30 April 2025).
186 Araújo, G. (2021). Brazil raises mandatory biodiesel blend to 12%, below expectations. Reuters News. July 2021.
187 Argus Media (2021). Brazil auction brings LNG-to-power with high costs [Internet]. Lawless land in no man's land_ The undesignated public forests in the Brazilian Amazon (accessed 30 April 2025).
188 Azevedo-Ramos, C., Moutinho, P., Laísa, V. et al. (2020). Lawless land in no man's land: the undesignated public forests in the Brazilian Amazon. *Land Use Policy.* 99: 104863.
189 Brito, B. (2021). Brazil's Amazon is under threat from proposed land-use laws. Americas Quarterly [Internet]. https://www.americasquarterly.org/article/brazils-amazon-is-under-threat-from-proposed-land-use-laws/ (accessed 30 April 2025).
190 Business Wire (2021). Brazilian natural gas market growth opportunities 2021: biomethane takes off as an alternative gas supply – ResearchAndMarkets.com. Business Wire [Internet]. https://www.businesswire.com/news/home/20210917005154/en/Brazilian-Natural-Gas-Market-Growth-Opportunities-2021-Biomethane-Takes-Off-as-an-Alternative-Gas-Supply---ResearchAndMarkets.com (accessed 30 April 2025).

191 Climate Policy Initiative (2019). Amendments to a provisional measure threaten the implementation of Brazil's new Forest Code [Internet]. https://climatepolicyinitiative.org/publication/amendments-of-a-provisional-measure-threaten-the-implementation-of-brazils-new-forest-code/ (accessed 30 April 2025).

192 Climate Policy Initiative (2019). Decreto no 9.760/2019 altera regras de conversão de multa ambiental em prestação de serviços ambientais – CPI [Internet]. https://climatepolicyinitiative.org/publication/decreto-no-9-760-2019-altera-regras-de-conversao-de-multa-ambiental-em-prestacao-de-servicos-ambientais/ (accessed 30 April 2025).

193 Presidente da República (2010). Decreto no 7.390 de 9 de dezembro de [Internet]. p. 16–20. https://doi.org/10.1590/S0101-73301999000300018.

194 Climate Transparency (2021). Brazil: climate transparency report. https://www.climate-transparency.org/wp-content/uploads/2021/10/CT2021Brazil.pdf (accessed 30 April 2025).

195 Ministry of Energy of Brazil (2018). Conselho nacional de política energética.

196 Cuartas, L.A. (2022). Recent hydrological droughts in brazil and their impact on hydropower generation. *Water (Basel).* 14: 601.

197 Fellet, J. (2022). 36 indígenas e aliados foram ameaçados de morte em 2021; saiba quem são. BBC NEWS.

198 Ferraz, A. (2021). Brazil holds emergency auction to mitigate impact of water crisis. The Brazilian Report.

199 Fragoso, R. (2021). Analise das emissoes brasileiras de Gases de Efeito Estufa e suas implicacaoes para as metas climaticas do Brasil 1990–2020 [Internet]. https://seeg-br.s3.amazonaws.com/Documentos%20Analiticos/SEEG_9/OC_03_relatorio_2021_FINAL.pdf (accessed 30 April 2025).

200 Global Carbon Project (2020). Global Carbon Project [Internet]. https://www.globalcarbonproject.org (accessed 30 April 2025).

201 Paraguassu, L. (2023). Brazil's Lula unveils plan to stop deforestation in Amazon by 2030. Reuters [Internet]. https://www.reuters.com/world/americas/brazils-lula-launches-plan-stop-deforestation-amazon-by-2030-2023-06-05/ (accessed 5 March 2024).

202 Government of Brazil (2020). Brazil's Updated Nationally Determined Contribution (NDC) [Internet]. https://www4.unfccc.int/sites/ndcstaging/PublishedDocuments/Brazil (accessed 30 April 2025).

203 Government of Brazil. Brazil presents its new climate target aligned with Mission 1.5°C. Brazil presents its new climate target aligned with Mission 1.5°C — Planalto.

204 Government of Brazil (2024). Brazil NDC updated [Internet]. https://www4.unfccc.int/sites/ndcstaging/PublishedDocuments/Brazil%20First/Updated%20-%20First%20NDC%20-%20%20FINAL%20-%20PDF.pdf (accessed 30 April 2025).

205 Gütschow, J. and Pflüger, M. (2022) The PRIMAP-hist national historical emissions time series [Internet]. https://doi.org/10.5281/zenodo.7179775.

206 Imazon (2021). Deforestation in the Brazilian Amazon reached 2,095 km^2 in July, and the last 12 months cumulative is the highest in 10 years. Imazon [Internet]. https://imazon.org.br/en/imprensa/deforestation-in-the-brazilian-amazon-reached-2-095-km2-in-july-and-the-last-12-months-cumulative-is-the-highest-in-10-years/ (accessed 30 April 2025).

207 IMF (2020). *World Economic Outlook Update*. Washington, DC: IMF.

208 IMF (2022). Brazil country data [Internet]. https://www.imf.org/en/Countries/BRA (accessed 30 April 2025).

209 Nacional I, DE DEABRIL (2020). INSTRUÇÃO NORMATIVA No 9, DE 16 [Internet]. https://www.in.gov.br/en/web/dou/-/instrucao-normativa-n-9-de-16-de-abril-de-2020-253343033 (accessed 30 April 2025).

210 INPE-EM (2020). INPE Emission Module: Land cover changes greenhouse gases [Internet]. http://inpe-em.ccst.inpe.br/en/home/ (accessed 30 April 2025).

211 INPE (2020). TerraBrasilis PRODES (deforestation [Internet]. http://terrabrasilis.dpi.inpe.br/app/dashboard/deforestation/biomes/legal_amazon/rates (accessed 30 April 2025).

212 Ionova, A. (2021). As soy frenzy grips Brazil, deforestation closes in on Indigenous lands. Mongabay [Internet]. https://news.mongabay.com/2021/07/as-soy-frenzy-grips-brazil-deforestation-closes-in-on-indigenous-lands/ (accessed 30 April 2025).

213 Kimbrough, L. (2021). Soy and cattle team up to drive deforestation in South America: Study [Internet]. Mongabay. https://news.mongabay.com/2021/07/study-shows-how-soy-cattle-team-up-to-drive-deforestation-in-south-america/ (accessed 30 April 2025).

214 Minister of Agriculture, Livestock and Food Supply (2021). ABC+ plan for adaptation and low carbon emission in agriculture – strategic vision for a new cycle [Internet]. https://www.gov.br/agricultura/pt-br/assuntos/sustentabilidade/plano-abc/arquivo-publicacoes-plano-abc/abc-english.pdf (accessed 30 April 2025).

215 Brasil, M. (2021). CNPE reduz de 13% para 10% a adição de biodiesel ao óleo diesel. MME Noticias.

216 Mongabay (2019). In surprise move, Brazil has removed restrictions on Amazon sugarcane production. Mongabay [Internet]. https://news.mongabay.com/2019/11/in-a-surprise-move-even-to-the-sugarcane-industry-president-bolsonaro-has-removed-restrictions-on-sugarcane-production-in-the-brazilian-amazon-experts-expect-land-speculators-to-benefit/ (accessed 30 April 2025).

217 Mongabay (2020). Brazil's past finance ministers defend environment against Bolsonaro [Internet]. https://news.mongabay.com/2020/07/brazils-past-finance-ministers-defend-environment-against-bolsonaro/ (accessed 30 April 2025).

218 Mongabay (2020). Brazil drastically reduces controls over suspicious Amazon timber exports [Internet]. https://news.mongabay.com/2020/03/brazil-drastically-reduces-controls-over-suspicious-amazon-timber-exports/ (accessed 30 April 2025).

219 Mongabay (2020). International investors urge Brazil to take real action to stop deforestation [Internet]. https://news.mongabay.com/2020/07/international-investors-urge-brazil-to-take-real-action-to-stop-deforestation/ (accessed 30 April 2025).

220 Morais, L. (2021). Brazil sets price caps for 25 Oct reserve energy auction. Renewables Now [Internet]. https://renewablesnow.com/news/brazil-sets-price-caps-for-oct-25-reserve-energy-auction-756402/ (accessed 30 April 2025).

221 Neves, R. (2022). Sob Bolsonaro, Amazônia viu subir desmatamento, garimpo e violência. UOL noticias.

222 Institute NC (2020). Changes in the biodiesel blending mandate 2020–2023. Brazil: Climate Policy Database.

223 Observatório do Clima (2019). Forest Facts: myths about land use in Brazil, debunked. Youtube [Internet]. https://www.youtube.com/watch?v=pX9s_eznvpY (accessed 30 April 2025).

224 Observatório do Clima (2019). Is Brazil on its way to meet its climate targets? [Internet]. http://www.observatoriodoclima.eco.br/wp-content/uploads/2019/09/Is-Brazil-on-the-way-to-meet-its-climate-targets_-1.pdf (accessed 30 April 2025).

225 Observatório do Clima (2019). Bolsonaro inaugura era da multa zero [Internet]. http://www.observatoriodoclima.eco.br/bolsonaro-multa-zero/ (accessed 30 April 2025).

226 Observatório do Clima (2020). How threatened is the environment under the bolsonaro administration, so far? A brief context about environmental policy in brazil, sixteen months into the bolsonaro administration [Internet]. http://www.observatoriodoclima.eco.br/wp-content/uploads/2020/06/Doc-Response-OC2.pdf (accessed 30 April 2025).

227 Observatório do Clima (2020). Um dia depois de reunião com investidores, alertas de desmatamento têm novo pico [Internet]. http://www.observatoriodoclima.eco.br/um-dia-depois-de-reuniao-com-investidores-alertas-de-desmatamento-tem-novo-pico/ (accessed 30 April 2025).

228 Observatório do Clima (2021). Civil society organizations reject the Ministry of Economy's anti-environment package.

229 PetroNoticias (2020). In: 10 year energy plan provides resources up to $ 454 billion and 39 new platforms by 2030 in Brazil. Brazil Energy Insight [Internet]. https://brazilenergyinsight.com/2020/12/15/10-year-energy-plan-provides-resources-up-to-454-billion-and-39-new-platforms-by-2030-in-brazil/ (accessed 30 April 2025).

230 Rajão, B.R., Soares-Filho, B., Nunes, F. et al. (2020). The rotten apples of Brazil's agribusiness. *Science.* 369: 246.

231 Pierry, F. (2021). Brazil pushing gas-fired power as hydro sources dry up [Internet]. Argus Media. https://www.argusmedia.com/en/news/2203894-brazil-pushing-gasfired-power-as-hydro-sources-dry-up (accessed 30 April 2025).

232 Rochedo, P.R.R., Soares-Filho, B., Schaeffer, R. et al. (2018). The threat of political bargaining to climate mitigation in Brazil. Nature. *Climate Change.* 1–4. https://doi.org/10.1038/s41558-018-0213-y.

233 Reuters (2020). Fires in Brazil's Amazon rainforest surge in July, worst in recent days [Internet]. https://www.reuters.com/article/us-brazil-environment-idUSKBN24X3SW (accessed 30 April 2025).

234 Sistema de Estimativas de Emissoes e Remocoes de gases de efeito Estufa SEEG (2022). Mudanca de Uso da Terra e Florestas [Internet]. https://plataforma.seeg.eco.br/sectors/mudanca-de-uso-da-terra-e-floresta# (accessed 30 April 2025).

235 Federal, S. (2021). PL 1539 Law project.

236 Silva Junior, C.H.L., Pessôa, A.C.M., Carvalho, N.S. et al. (2021). The Brazilian Amazon deforestation rate in 2020 is the greatest of the decade. *Nature Ecology and Evolution.* 5 (2): 144–145.

237 Soterroni, A.C., Mosnier, A., Carvalho, A.X.Y. et al. (2018). Future environmental and agricultural impacts of Brazil's Forest Code. *Environmental Research Letters.* 13 (7): 074021.

238 Spatuzza, A. (2021). Renewables sweep board as low demand leaves Brazil power auction shadow of former glory [Internet]. https://www.rechargenews.com/markets/renewables-sweep-board-as-low-demand-leaves-brazil-power-auction-shadow-of-former-glory/2-1-1076108 (accessed 30 April 2025).

239 Spring, J. and Paraguassu, L. (2021). Brazil's Bolsonaro, under U.S. pressure, vows climate neutrality by 2050 [Internet]. https://www.reuters.com/business/environment/bolsonaro-says-brazil-will-reach-climate-neutrality-by-2050-2021-04-22/ (accessed 30 April 2025).

240 Spring, J., Samora, R., and Nogueira, M. (2021). Brazil minister warns of deeper energy crisis amid worsening drought. Reuters [Internet]. https://www.reuters.com/world/americas/brazil-minister-warns-deeper-energy-crisis-amid-worsening-drought-2021-08-31/ (accessed 30 April 2025).

241 Stabile, M.C.C., Guimarães, A.L., Silva, D.S. et al. (2020). Solving Brazil's land use puzzle: Increasing production and slowing Amazon deforestation. *Land Use Policy.* 91: 104362.

242 The New York Times (2019). Jair Bolsonaro, on Day 1, Undermines Indigenous Brazilians' Rights [Internet]. https://www.nytimes.com/2019/01/02/world/americas/brazil-bolsonaro-president-indigenous-lands.html (accessed 30 April 2025).

243 The World Bank (2022). World Development Indicators [Internet]. The World Bank Group. http://databank.worldbank.org/data/views/reports/tableview.aspx (accessed 30 April 2025).

244 U.S. Environmental Protection Agency (2019). Global non-CO_2 greenhouse gas emission projections & mitigation potential: 2015–2050 [Internet]. https://www.epa.gov/global-mitigation-non-co2-greenhouse-gases/global-non-co2-greenhouse-gas-emission-projections (accessed 30 April 2025).

245 Carbon Action Tracker (2024). Russian federation [Internet]. https://climateactiontracker.org/countries/russian-federation/ (accessed 1 April 2024).

246 The International Energy Agency (IEA) (2023). World Energy Outlook 2023 [Internet]. Paris, France. https://www.iea.org/reports/world-energy-outlook-2023 (accessed 30 April 2025).

247 The International Energy Agency (IEA) (2022). Energy Strategy of the Russian Federation for the period up to 2035 [Internet]. https://www.iea.org/policies/14855-energy-strategy-of-the-russian-federation-for-the-period-up-to-2035 (accessed 30 April 2025).

248 The International Energy Agency (IEA) (2022). Energy fact sheet: why does Russian oil and gas matter? [Internet]. https://www.iea.org/articles/energy-fact-sheet-why-does-russian-oil-and-gas-matter (accessed 30 April 2025).

249 Interfax (2022). RSPP asked to postpone the adoption of a low-carbon development plan. Interfax [Internet]. https://www.interfax.ru/russia/853905 (accessed 26 July 2022).

250 Intergovernmental Panel for Climate Change (IPCC) (2019). 2019 refinement to the 2006 IPCC guidelines for national greenhouse gas inventories [Internet]. https://www.ipcc.ch/report/2019-refinement-to-the-2006-ipcc-guidelines-for-national-greenhouse-gas-inventories/ (accessed 5 January 2020).

251 Intergovernmental Panel for Climate Change (IPCC), Masson-Delmotte, P.Z., Pörtner, H.O. et al. (2018) Summary for policymakers. In In the context of strengthening the global response to the threat of climate change (ed. Waterfield, V.). IPCC [Internet]. https://www.ipcc.ch/sr15/chapter/spm/ (accessed October 2018).

252 International Renewable Energy Agency (IRENA) (2017). REmap 2030 renewable energy prospects for the Russian Federation [Internet]. https://www.irena.org/publications/2017/Apr/Renewable-Energy-Prospects-for-the-Russian-Federation-REmap-working-paper (accessed 30 April 2025).

253 Kalinin, M., Ishmukhametov, R., and Pankova, O. (2021). Russia: Jump-start of a national electric vehicles industry. Global Compliance News [Internet]. https://www.globalcompliancenews.com/2021/09/16/russia-jump-start-of-a-national-electric-vehicles-industry-09082021/ (accessed 16 September 2021).

254 Korppoo, A. (2018). Russian associated petroleum gas flaring limits: Interplay of formal and informal institutions. *Energy Policy.* 116: 232–241.

255 Light, F. (2021). Russia says its forests neutralize billions of tons of greenhouse gases. Scientists have their doubts. The Moscow Times [Internet]. https://www.themoscowtimes.com/2021/07/05/russia-says-its-forests-neutralize-billions-of-tons-of-greenhouse-gases-scientists-have-their-doubts-a74428 (accessed 7 September 2021).

256 Myllyvirta, L., Thieriot, H., Lietava, J. et al. (2022). Financing Putin's war: Fossil fuel exports from Russia in the first six months of the invasion of Ukraine [Internet]. https://energyandcleanair.org/publication/financing-putins-war-fossil-fuel-exports-from-russia-in-the-first-six-months-of-the-invasion-of-ukraine/ (accessed 30 April 2025).

257 Persily, L. (2021). Russia plows ahead on oil, natural gas. Alaska Journal of Commerce [Internet]. https://www.alaskajournal.com/2021-09-22/russia-plows-ahead-oil-natural-gas (accessed 22 September 2021).

258 Proskuryakova, L. (2022). Russia's renewable energy sector: policy recommendations [Internet]. https://www.centrumbalticum.org/files/5176/BSR_Policy_Briefing_3_2022.pdf (accessed 30 April 2025).

259 Reuters (2021). Russia plans to subsidise electric cars to spur demand. Reuters [Internet]. https://www.reuters.com/business/autos-transportation/russia-plans-subsidise-electric-cars-spur-demand-2021-08-04/ (accessed 4 August 2021).

260 Staff, R. (2021). Russian oil producers struggle to contain gas flaring. Reuters [Internet]. https://www.reuters.com/article/russia-flaring-idUSL8N2M23D0 (accessed 4 September 2021).

261 Russia Federation (2009). Decree No. 7 regarding measures for stimulation of reduction of air pollution by products of burning of associated petroleum gas at flare units [Internet]. https://leap.unep.org/countries/ru/national-legislation/ministerial-decree-no-7-regarding-measures-stimulation-reduction (accessed 30 April 2025).

262 Russia Federation (2019). Russian federation fourth biennial report [Internet]. https://unfccc.int/sites/default/files/resource/124785_Russian (accessed 30 April 2025).

263 Russia Federation (2020) Decree No. 1523-r: Energy Strategy of the Russian Federation until 2035 [Internet]. http://static.government.ru/media/files/w4sigFOiDjGVDYT4IgsApssm6mZRb7wx.pdf (accessed 30 April 2025).

264 Russia Federation (2020). Mikhail Mishustin approves Energy Strategy to 2035 [Internet]. Russian Federation Government Website. http://government.ru/en/docs/39847/ (accessed 30 April 2025).

265 Russia Federation (2021). Strategy for the socio-economic development of russia with low greenhouse gas emissions until 2050 [Internet]. http://static.government.ru/media/files/ADKkCzp3fWO32e2yA0BhtIpyzWfHaiUa.pdf (accessed 30 April 2025).

266 Russia Federation (2021). Order of August 23, 2021 No 2290-r: concept for the development of the production and use of electric road transport in Russia until 2030 [Internet]. Russian Federation. http://static.government.ru/media/files/bW9wGZ2rDs3BkeZHf7ZsaxnlbJzQbJJt.pdf (accessed 30 April 2025).

267 Russia Federation (2021). Decision of November 27 [Internet]. http://static.government.ru/media/files/7enYF2uL5kFZlOOpQhLl0nUT91RjCbeR.pdf (accessed 30 April 2025).

268 Russia Federation (2022). Andrey Belousov held a meeting with the business community on the implementation plan for the low-carbon development strategy for Russia [Internet]. Russian Federation Government Website. http://government.ru/news/44557/ (accessed 30 April 2025).

269 Russia Federation (2022). Russian National Inventory Report: Common Reporting Format (CRF) Table [Internet]. https://unfccc.int/documents/461969 (accessed 30 April 2025).

270 TASS (2022). The Ministry of Economic Development will update the implementation plan for the low-carbon development strategy of Russia [Internet]. TASS. https://tass.ru/ekonomika/16121487 (accessed 30 April 2025).

271 The Coal Hub (2021). Putin approves increase in Russian coal exports eastwards. The Coal Hub [Internet]. https://thecoalhub.com/putin-approves-increase-in-russian-coal-exports-eastwards.html (accessed 30 April 2025).

272 The Moscow Times (2019). Russia rejects climate change plan after business uproar. The Moscow Times [Internet]. https://www.themoscowtimes.com/2019/10/17/russia-rejects-climate-change-plan-after-business-uproar-a67780 (accessed 17 October 2019).

273 The Moscow Times (2021). Russia passes first law to limit greenhouse gas emissions. The Moscow Times [Internet]. https://www.themoscowtimes.com/2021/06/03/russia-passes-first-law-to-limit-greenhouse-gas-emissions-a74095 (accessed 7 September 2021).

274 The Moscow Times (2022). Toxic wildfire smog blankets Russia's Far Northeastern City Yakutsk. The Moscow Times [Internet]. https://www.themoscowtimes.com/2022/07/19/toxic-wildfire-smog-blankets-russias-far-northeastern-city-yakutsk-a78345 (accessed 19 June 2022).

275 Zelenaya, O. (2022). Russia slashes environmental protections as war rages, economic crisis looms. The Moscow Times [Internet]. https://www.themoscowtimes.com/2022/06/25/russia-slashes-environmental-protections-as-war-rages-economic-crisis-looms-a77562 (accessed 1 July 2022).

276 Zhelenin, A. (2020). Our doctrine is the rejection of progress. Rosbalt [Internet]. https://www.rosbalt (accessed 30 April 2025).

277 Carbon Action Tracker (2024). India [Internet]. https://climateactiontracker.org/countries/india/ (accessed 1 April 2024).

278 Ministry of Environment, Forest and Climate Change, Government of India (2022). India's long-term low-carbon development strategy [Internet]. https://unfccc.int/sites/default/files/resource/India_LTLEDS.pdf (accessed 30 April 2025).

279 The Global Wind Energy Council (2022). India wind energy market outlook 2022–2026 [Internet]. https://gwec.net/wp-content/uploads/2022/08/India-Outlook-2026.pdf (accessed 30 April 2025).

280 Hindustani Times (2022). Govt may extend coal cess for green mission [Internet]. https://www.hindustantimes.com/india-news/govt-may-extend-coal-cess-for-green-mission-101671472333050.html (accessed 30 April 2025).

281 Hydrogen Insight (2023). India puts plans for industrial green hydrogen mandates back on the table as it eyes 70% exports [Internet]. https://www.hydrogeninsight.com/industrial/india-puts-plans-for-industrial-green-hydrogen-mandates-back-on-the-table-as-it-eyes-70-exports/2-1-1483337 (accessed 30 April 2025).

282 The International Energy Agency (IEA) (2020). Average age of existing coal power plants in selected regions in 2020 [Internet]. https://www.iea.org/data-and-statistics/charts/average-age-of-existing-coal-power-plants-in-selected-regions-in-2020 (accessed 30 April 2025).

283 The International Energy Agency (IEA) (2020). Industrial electricity prices in India and selected countries, 2005–2019 – Charts – Data & Statistics - IEA [Internet]. https://www.iea.org/data-and-statistics/charts/industrial-electricity-prices-in-india-and-selected-countries-2005-2019 (accessed 30 April 2025).

284 The Institute for Energy Economics and Financial Analysis (IEEFA) (2022). IEEFA India: Solar tariffs projected to increase by one-fifth over the next year | IEEFA [Internet]. https://ieefa.org/articles/ieefa-india-solar-tariffs-projected-increase-one-fifth-over-next-year (accessed 30 April 2025).

285 The Institute for Energy Economics and Financial Analysis (IEEFA) (2023). Analysis Solar Energy Security Energy Policy South Asia India India's rising prominence in solar photovoltaic manufacturing [Internet]. https://ieefa.org/resources/indias-rising-prominence-solar-photovoltaic-manufacturing (accessed 30 April 2025).

286 The Institute for Energy Economics and Financial Analysis (IEEFA) (2023). Proper policy support will drive green hydrogen proliferation in India [Internet]. https://ieefa.org/resources/proper-policy-support-will-drive-green-hydrogen-proliferation-india (accessed 30 April 2025).

287 The Institute for Energy Economics and Financial Analysis (IEEFA) (2023). Bidding process begins for 310 megawatts of floating and ground-mounted solar in India [Internet]. https://ieefa.org/articles/bidding-process-begins-310-megawatts-floating-and-ground-mounted-solar-india (accessed 30 April 2025).

288 The Institute for Energy Economics and Financial Analysis (IEEFA) (2023). From intent to action: India well poised to become a global leader in green hydrogen [Internet]. https://ieefa.org/resources/intent-action-india-well-poised-become-global-leader-green-hydrogen (accessed 30 April 2025).

289 The Institute for Energy Economics and Financial Analysis (IEEFA) (2023). How India can effectively electrify public transportation [Internet]. https://ieefa.org/resources/how-india-can-effectively-electrify-public-transportation (accessed 30 April 2025).

290 The Institute for Energy Economics and Financial Analysis (IEEFA) (n.d.). How to mobilise green finance for Indian micro, small and medium enterprises [Internet]. https://ieefa.org/resources/how-mobilise-green-finance-indian-micro-small-and-medium-enterprises (accessed 30 April 2025).

291 The Institute for Energy Economics and Financial Analysis (IEEFA) (n.d.). India could become the world's second-largest solar photovoltaic manufacturer by 2026 [Internet]. https://ieefa.org/articles/india-could-become-worlds-second-largest-solar-photovoltaic-manufacturer-2026 (accessed 30 April 2025).

292 Carbon Action Tracker (2024). China [Internet]. https://climateactiontracker.org/countries/china/ (accessed 1 April 2024).

293 Aitken, G., Langenbrunner, B., and Zimmermann, S. (2022). Europe gas tracker report 2022 [Internet]. https://globalenergymonitor.org/report/europe-gas-tracker-2022/ (accessed 30 April 2025).

294 Asia Pacific Energy (n.d.). CHINA: energy supply and consumption revolution strategy (2016–2030) [Internet]. Asia Pacific Energy. https://policy.asiapacificenergy.org/node/3587 (accessed 30 April 2025).

295 Bloomberg (2020). China's 40 year, billion tree project is a lesson for the world. Bloomberg Green [Internet]. https://www.bloomberg.com/news/articles/2020-09-14/china-s-40-year-billion-tree-project-is-a-lesson-for-the-world (accessed 30 April 2025).

296 Bloomberg (2021). Peak aluminium looms for China as Xi pushes new green mission. The Business Times [Internet]. https://www.businesstimes.com.sg/energy-commodities/peak-aluminium-looms-for-china-as-xi-pushes-new-green-mission (accessed 30 April 2025).

297 Caixin (2022). China overtakes Japan as Top LNG Importer. Caixin Global [Internet]. https://www.caixinglobal.com/2022-01-07/china-overtakes-japan-as-top-lng-importer-101826804.html (accessed 30 April 2025).

298 Cao, Y., Xu, C., Kamaruzzaman, S.N., and Azuz, N.M. (2022). A systematic review of green building development in china: advantages, challenges and future directions. *Sustainability.* 14 (12293): https://doi.org/10.3390/su141912293.

299 China Electricity Council (CEC) (2023). Analysis and forecast of china power demand-supply situation in the first quarter of 2023. China Electricity Council [Internet]. https://english.cec.org.cn/#/newsdetails?id=1658663989908951042 (accessed 30 April 2025).

300 China Electricity Council (CEC) (2023). China electricity sector annual development report 2023 [Internet]. https://cec.org.cn/detail/index.html?3-322624 (accessed 30 April 2025).

301 China Electricity Council (CEC) (2023). National power supply and demand situation analysis and forecast report for Q3 [Internet]. http://www.chinapower.com.cn/xw/zyxw/20231026/221922.html (accessed 30 April 2025).

302 China Global Television Network (CGTN) (2023). China aims to increase EV share in public sector to 80% in pilot cities. CGTN [Internet]. https://news.cgtn.com/news/2023-02-04/China-aims-to-raise-EV-share-in-public-sector-to-80-in-pilot-cities--1h9BV3tIXOE/index.html (accessed 30 April 2025).

303 China Dialogue (2022). Carbon peaking goal for steel sector toned down. China Dialogue [Internet]. https://chinadialogue.net/en/digest/carbon-peaking-goal-for-steel-sector-toned-down/ (accessed 30 April 2025).

304 China Energy News (2022). National Energy Administration: In principle, no new coal power projects purely for power generation will be built. China Energy News [Internet]. https://mp.weixin.qq.com/s/eTw2Lvg0-wfJ0blP8-tbiw?utm_campaign=China (accessed 30 April 2025).

305 Chow, E. and Singh, S. (2021). China to use more natural gas in energy mix to 2035 – CNPC. Reuters [Internet]. https://www.reuters.com/business/sustainable-business/china-use-more-natural-gas-energy-mix-2035-cnpc-2021-06-24/ (accessed 30 April 2025).

306 Client Earth (2020). China introduces new law to safeguard forests and improve governance. Client Earth [Internet]. https://www.clientearth.org/china-introduces-new-law-to-safeguard-forests-and-improve-governance/ (accessed 30 April 2025).

307 Yu, Z., Ciais, P., and Piao, S. (2022). Forest expansion dominates China's land carbon sink since 1980. *Nature Communications.* 13 (1): 5374.

308 Deng, Y., Fuentes, U., and Hare, B. (2020). U.S. and China climate goals: scenarios for 2030 and mid-century [Internet]. https://climateanalytics.org/media/aspi-ca-us_chinaclimategoals-2020.pdf (accessed 30 April 2025).

309 Zhang, S.C., Yang, X.Y., Xu, W., and Fu, Y.J. (2021). Contribution of nearly-zero energy buildings standards enforcement to achieve carbon neutral in urban area by 2060. *Advances in Climate Change Research.* 12 (5): 734–743.

310 Zhang, S.C. and Fu, Y.J. (2021). Assessment of mid-to-long term energy saving impacts of nearly zero energy building incentive policies in cold region of China. *Energy and Buildings.* 241 (2): 110938.

311 Zhang, Y. (2021). Li Chen: China's cement industry expected to reach peak carbon ahead of schedule (Chinese) [Internet]. https://www.thepaper.cn/newsDetail_forward_10787817 (accessed 30 April 2025).

312 Kiessling, S., Darabkhani, H.G., and Soliman, A.H. (2022). Energies | Free Full-Text | The bio steel cycle: 7 steps to net-zero CO_2 emissions steel production [Internet]. https://www.mdpi.com/1996-1073/15/23/8880 (accessed 11 July 2023).

313 Dr Zoyrk, A. and Deloitte Deutschland (2022). EU Green Steel Transformation – why, what, and how? | Deloitte Germany. https://www2.deloitte.com/de/de/pages/sustainability1/articles/green-steel-transformation.html (accessed 1 April 2024).

314 Carbon Action Tracker (2024). Australia [Internet]. https://climateactiontracker.org/countries/australia/ (accessed 1 April 2024).

315 Australian Government (2021). Australia's long-term emissions reduction plan: a whole-of-economy plan to achieve net zero emissions by 2050 [Internet]. https://unfccc.int/sites/default/files/resource/Australias_LTS_WEB.pdf (accessed 30 April 2025).

316 ABC (2023). Australia to re-join UN's green climate fund after Morrison government withdrawal.

317 ABC (2023). Australia will fall well short of 82 per cent renewable energy by 2030 [Internet]. https://www.abc.net.au/news/2023-08-06/australia-likely-to-fall-short-of-82pc-renewable-energy-target/102689392 (accessed 30 April 2025).

318 ABC (2023). Federal Senate inquiry to examine NT government's proposed Middle Arm industrial precinct near Darwin [Internet]. https://www.abc.net.au/news/2023-09-05/middle-arm-precinct-darwin-nt-senate-inquiry-established/102817580 (accessed 30 April 2025).

319 ABC (2023). Fracking regulator disputes NT government's claim it met all Pepper Inquiry promises [Internet]. https://www.abc.net.au/news/2023-05-04/nt-beetaloo-analysts-question-pepper-inquiry-progress/102299456 (accessed 30 April 2025).

320 ABC (2023). WA premier says fracking policies won't change as environmental groups push for Kimberley ban [Internet]. https://www.abc.net.au/news/2023-08-18/wa-premier-backs-fracking-kimberley-policies/102743938 (accessed 30 April 2025).

321 Rainforest Foundation (2021). Open Letter to the COP26 Presidency and Parties to the UNFCCC and CBD on Nature-based Solutions.

322 Government ACT (2020). ACT zero emissions vehicles strategy 2022–30.

323 The Australian Energy Market Operator (AEMO) (2020). 2020 integrated system plan for the national electricity market.

324 The Australian Energy Market Operator (AEMO) (2022). Integrated system plan for the national electricity market.

325 The Australian Energy Market Operator (AEMO) (2022). Western Australia gas statement of opportunities.

326 The Australian Energy Market Operator (AEMO) (2022). AEMO suspends NEM wholesale market.

327 The Australian Energy Market Operator (AEMO) (2023). Inputs, assumptions and scenarios [Internet]. https://aemo.com.au/consultations/current-and-closed-consultations/2023-inputs-assumptions-and-scenarios-consultation (accessed 30 April 2025).

328 Australian Labor Party (ALP) (2022). Powering Australia. Australian Labor Party.

329 Australian National University (ANU) (2022). Australia's carbon market a "fraud on the environment" [Internet]. Australian National University. https://law.anu.edu.au/news-and-events/news/australia (accessed 30 April 2025).

330 Australian Institute (2023). A fair COP31 [Internet]. www.australiainstitute.org.au (accessed 30 April 2025).

331 Mackenzie, W. (2021). Putting coal mine emissions under the microscope [Internet]. https://www.woodmac.com/news/opinion/putting-coal-mine-emissions-under-the-microscope/ (accessed 7 July 2023).

332 Australian Energy Council (2023). 82 The per cent national renewable energy target – where did it come from and how can we get there?

333 Australian Government (2020). Estimating greenhouse gas emissions from bushfires in Australia's temperate forests: focus on 2019–20.

334 Australian Government (2020). Australia's Biennial Communication. UNFCCC [Internet]. https://www4.unfccc.int/sites/SubmissionsStaging/Documents/202012221040---Australia (accessed 30 April 2025).

335 Australian Government (2021). Australia's long-term emissions reduction plan – modelling and analysis.

336 Australian Government (2021). Australia's nationally determined contribution communication 2021. Department of Industry, Science, Energy and Resources.

337 Australian Government (2022). Australia's nationally determined contribution communication 2022.

338 Australian Institute (2023). Fossil fuel subsidies in Australia 2023: Federal and state government assistance to fossil fuel producers and major users in 2022–23 [Internet]. www.australiainstitute.org.au (accessed 30 April 2025).

339 Australian Taxation Office (2017). Fuel tax credits – business [Internet]. https://www.ato.gov.au/business/fuel-schemes/fuel-tax-credits---business/ (accessed 30 April 2025).

340 Bowen, C. (2023). Safeguard Mechanism one step closer to Parliamentary passage [Internet]. https://minister.dcceew.gov.au/bowen/media-releases/safeguard-mechanism-one-step-closer-parliamentary-passage?_gl=1*103nm0m*_ga*NjM5NTc1MjYyLjE2NzgzNjU0Mjg.*_ga_1M2TBC9WWS*MTY4MDQxNTE1Ny4zMS4xLjE2ODA0MTUyMjcuMC4wLjA.&_ga=2.46596462.858034206.1680415158-639575262.1678365428 (accessed 30 April 2025).

341 Bowman, D.M.J.S., Williamson, G.J., Price, O.F. et al. (2021). Australian forests, megafires and the risk of dwindling carbon stocks. *Plant, Cell & Environment.* 44 (2): 347–355.

342 Beyond Zero Emissions (2019). 10 The Gigwawatt Vision. How renewable energy can power jobs and investment in the Northern Territory.

343 Center of International Environmental Law (2023). Deep trouble – the risks of offshore carbon capture and storage.

344 Global Australia (2023). Australian Budget commits A$25bn to clean energy and renewables projects [Internet]. https://www.globalaustralia.gov.au/news-and-resources/news-items/australian-budget-commits-a25bn-clean-energy-and-renewables-projects (accessed 30 April 2025).

345 Clean Energy Regulator (2023). Emissions Reduction Fund project register [Internet]. https://www.cleanenergyregulator.gov.au/ERF/project-and-contracts-registers/project-register (accessed 30 April 2025).

346 Clean Energy Regulator (2023). Postcode data for small-scale installations [Internet]. https://www.cleanenergyregulator.gov.au/RET/Forms-and-resources/Postcode-data-for-small-scale-installations (accessed 30 April 2025).

347 Climate Action Tracker (2020). CAT scaling up climate action series. Australia.

348 Climate Action Tracker (2023). Countdown to COP28: time for world to focus on oil and gas phase-out, renewables target, not distractions like CCS. Climate Analytics, NewClimate Institute [Internet]. https://climateactiontracker.org/publications/countdown-to-COP28/ (accessed 30 April 2025).

349 Climate Analytics (2018). Australia Climate Factsheets – Industry [Internet]. https://climateanalytics.org/publications/2018/australia-climate-factsheets-industry/ (accessed 30 April 2025).

350 Climate Analytics (2019). Australia's proposed "Kyoto carryover" – nature, scale, implications, legal issues and environmental integrity of the Paris Agreement.

351 Climate Analytics (2021). Warming Western Australia, how Woodside's Scarborough and Pluto project undermines the Paris agreement [Internet]. https://climateanalytics.org/media/climateanalytics_scarboroughpluto_dec2021.pdf (accessed 30 April 2025).

352 Climate Analytics (2023). Emissions as usual: implications for the safeguard mechanism of LNG and coal mine projects [Internet]. https://climateanalytics.org/publications/2023/emissions-as-usual-implications-for-the-safeguard-mechanism-of-lng-and-coal-mine-projects/ (accessed 30 April 2025).

353 Climate Analytics (2023). Why offsets are not a viable alternative to cutting emissions [Internet]. www.climateanalytics.org (accessed 30 April 2025).

354 Climate Analytics (2023). Emissions impossible: unpacking CSIRO GIS-ERA Beetaloo Middle Arm fossil gas emissions estimates [Internet]. https://climateanalytics.org/publications/emissions-impossible# (accessed 30 April 2025).

355 Climate Choices (2023). What the ACT Government is doing [Internet]. https://www.climatechoices.act.gov.au/energy/what-the-act-government-is-doing (accessed 30 April 2025).

356 Climateworks Australia (2014). Pathways to deep decarbonization in 2050 – how Australia can prosper in a low carbon world.
357 Climateworks Australia (2020). Decarbonisation futures: solutions, actions and benchmarks for a net zero emissions Australia [Internet]. https://www.climateworksaustralia.org/resource/decarbonisation-futures-solutions-actions-and-benchmarks-for-a-net-zero-emissions-australia/ (accessed 30 April 2025).
358 Department of Climate Change (2023). Australia's National Hydrogen Strategy [Internet]. https://www.dcceew.gov.au/energy/publications/australias-national-hydrogen-strategy (accessed 30 April 2025).
359 COP28 (2023). Global renewable and energy efficiency pledge [Internet]. https://www.cop28.com/en/global-renewables-and-energy-efficiency-pledge (accessed 30 April 2025).
360 COP28 (2023). UAE declaration on sustainable agriculture, resilient food systems and climate action (accessed 30 April 2025).
361 GISERA CSIRO (2023). Mitigation and offsets of Australian life cycle greenhouse gas emissions of onshore shale gas in the northern territory.
362 Department of Climate Change (2021). Australia's emissions projections 2021 [Internet]. https://www.industry.gov.au/sites/default/files/October (accessed 30 April 2025).
363 Department of the Environment and Energy (DEE) (2019). Australia's emissions projections 2019.
364 Department of the Environment and Energy (DEE) (2019). Energy efficient communities program. Department of the Environment and Energy [Internet]. https://www.energy.gov.au/government-priorities/energy-programs/energy-efficient-communities-program (accessed 30 April 2025).
365 Department of Agriculture and Water Resources (2017). Carbon Farming Futures [Internet]. http://www.agriculture.gov.au/ag-farm-food/climatechange/carbonfarmingfutures (accessed 30 April 2025).
366 Department of Climate Change (2015). National energy productivity plan [Internet]. https://www.energy.gov.au/publications/national-energy-productivity-plan (accessed 30 April 2025).
367 Department of Climate Change (2022). Threatened species action plan 2022–2023 [Internet]. www.dcceew.gov.au/ (accessed 30 April 2025).
368 Department of Climate Change (2022). Capacity Investment Scheme to power Australian energy market transformation [Internet]. https://www.energy.gov.au/news-media/news/capacity-investment-scheme-power-australian-energy-market-transformation (accessed 30 April 2025).
369 Department of Climate Change (2022). Heating and cooling [Internet]. https://www.energy.gov.au/households/heating-and-cooling (accessed 30 April 2025).
370 Department of Climate Change (2022). National energy transformation partnership [Internet]. https://www.energy.gov.au/government-priorities/energy-and-climate-change-ministerial-council/priorities/national-energy-transformation-partnership (accessed 30 April 2025).

371 Department of Climate Change (2022). National waste policy – action plan annexure [Internet]. https://www.dcceew.gov.au/sites/default/files/documents/national-waste-policy-action-plan-annexure-2022.pdf (accessed 30 April 2025).

372 Department of Climate Change (2023). Capacity investment scheme consultation paper.

373 Department of Climate Change (2023). National inventory report 2021: the Australian government submission to the United Nations Framework Convention on Climate Change.

374 Department of Climate Change (2023). Consultation hub – powering the regions fund [Internet]. https://consult.dcceew.gov.au/powering-the-regions-fund (accessed 30 April 2025).

375 US Energy Information Administration (2023). Natural gas weekly update natural gas [Internet]. https://www.eia.gov/naturalgas/weekly/archivenew_ngwu/2023/03_16/ (accessed 30 April 2025).

376 Department of Climate Change (2023). Establishing offshore renewable energy infrastructure [Internet]. https://www.dcceew.gov.au/energy/renewable/establishing-offshore-infrastructure (accessed 30 April 2025).

377 Department of Climate Change (2023). Government launches environmental offsets crackdown [Internet]. https://minister.dcceew.gov.au/plibersek/media-releases/government-launches-environmental-offsets-crackdown# (accessed 30 April 2025).

378 Department of Climate Change (2023). Incorporating an emissions reduction objective into the national energy objectives [Internet]. https://www.energy.gov.au/government-priorities/energy-and-climate-change-ministerial-council/working-groups/governance-working-group/incorporating-emissions-reduction-objective-national-energy-objectives# (accessed 30 April 2025).

379 Carbon Action Tracker (2024). Canada [Internet]. https://climateactiontracker.org/countries/canada/ (accessed 1 April 2024).

380 Agriculture and Agri-Food Canada (2020). Helping the dairy sector mitigate the impact of COVID-19 [Internet]. https://www.canada.ca/en/agriculture-agri-food/news/2020/05/helping-the-dairy-sector-mitigate-the-impact-of-covid-19.html (accessed 30 April 2025).

381 Agriculture and Agri-Food Canada (2020). Surplus food rescue program [Internet]. https://www.canada.ca/en/agriculture-agri-food/news/2020/08/surplus-food-rescue-program.html (accessed 30 April 2025).

382 Billinger, N. (2022). Reducing food waste would help food banks [Internet]. CJWW Radio. https://www.cjwwradio.com/2022/10/27/reducing-food-waste-would-help-food-banks/ (accessed 30 April 2025).

383 Canada Energy Regulator (2022). Provincial and territorial energy profiles. https://www.cer-rec.gc.ca/en/data-analysis/energy-markets/provincial-territorial-energy-profiles/provincial-territorial-energy-profiles-explore.html. (accessed 30 April 2025).

384 Cascade Power (2021). Cascade power project [Internet]. https://cascadepower.ca/ (accessed 30 April 2025).

385 CBC News (2018). SaskPower abandons carbon capture at Boundary Dam 4 and 5 [Internet]. CBC News. https://www.cbc.ca/news/canada/saskatoon/saskpower-abandons-carbon-capture-at-boundary-dam-4-and-5-1.4739107 (accessed 30 April 2025).

386 CBC News (2022). Bay du Nord may not be last oil project off Newfoundland, energy minister says. CBC News [Internet] https://www.cbc.ca/news/canada/newfoundland-labrador/jonathan-wilkinson-offshore-oil-1.6511480 (accessed 30 April 2025).

387 Chan, E., Worthy, D.E.J., Chan, D. et al. (2020). Eight-year estimates of methane emissions from oil and gas operations in Western Canada are nearly twice those reported in inventories. *Environmental Science & Technology.* 54: 14899–14909.

388 Christensen, L.T. and Dusyk, N. (2022). Why Canadian liquefied natural gas is not the answer for the European Union's Short-Term Energy Needs [Internet]. https://www.iisd.org/articles/deep-dive/canadian-lng-not-eu-energy-crisis-solution (accessed 30 April 2025).

389 Climate Analytics (2022). 1.5 °C national pathway explorer: Canada – power sector [Internet]. https://1p5ndc-pathways.climateanalytics.org/countries/canada/sectors/power/ (accessed 30 April 2025).

390 UNFCCC (2024). About carbon pricing | UNFCCC [Internet]. https://unfccc.int/about-us/regional-collaboration-centres/the-ciaca/about-carbon-pricing (accessed 28 February 2024).

391 Government of Canada (2021). Federal greenhouse gas offset system [Internet]. https://www.canada.ca/en/environment-climate-change/services/climate-change/pricing-pollution-how-it-will-work/output-based-pricing-system/federal-greenhouse-gas-offset-system.html (accessed 30 April 2025).

392 Government of Canada (2016). Pan-Canadian framework on clean growth and climate change. Annex I: federal investments and measures to support the transition to a low-carbon economy [Internet]. https://www.canada.ca/en/services/environment/weather/climatechange/pan-canadian-framework/annex-federal-investments-measures.html (accessed 30 April 2025).

393 Government of Canada (2022). Carbon pollution pricing: considerations for facilitating indigenous participation in the federal greenhouse gas offset system [Internet]. https://www.canada.ca/en/environment-climate-change/services/climate-change/pricing-pollution-how-it-will-work/output-based-pricing-system/federal-greenhouse-gas-offset-system/considerations-indigenous-participation.html (accessed 30 April 2025).

394 Department of Finance Canada (2022). Government announces next steps on trans mountain expansion project. News Release [Internet]. https://www.canada.ca/en/department-finance/news/2022/02/government-announces-next-steps-on-trans-mountain-expansion-project.html (accessed 30 April 2025).

395 Department of Finance Canada (2022). Update on trans mountain expansion project [Internet]. https://www.canada.ca/en/department-finance/news/2022/05/update-on-trans-mountain-expansion-project.html (accessed 30 April 2025).

396 Dyer, E. (2021). Ottawa's promising a tax credit for carbon capture – but is the tech worth the money? CBC News [Internet]. https://www.cbc.ca/news/politics/carbon-capture-storage-fossil-fuel-climate-change-2021-budget-1.6003427 (accessed 30 April 2025).

397 Ecojustice (2022). Coalition of environmental groups calls for a federal ban on thermal coal exports on Earth Day [Internet]. https://ecojustice.ca/pressrelease/coalition-of-environmental-groups-calls-for-a-federal-ban-on-thermal-coal-exports-on-earth-day/ (accessed 30 April 2025).

398 Efficiency Canada and Carleton University (2022). The municipal guide to net zero energy ready building codes [Internet]. 2022-Municipal-Guide-to-Net-Zero.pdf (accessed 30 April 2025).

399 Environment and Climate Change Canada (2018). Technical backgrounder: federal methane regulations for the upstream oil and gas sector [Internet]. https://www.canada.ca/en/environment-climate-change/news/2018/04/federal-methane-regulations-for-the-upstream-oil-and-gas-sector.html (accessed 30 April 2025).

400 Environment and Climate Change Canada (2020). A healthy environment and a healthy economy: Canada's strengthened climate plan to create jobs and support people, communities and the planet [Internet]. https://www.canada.ca/content/dam/eccc/documents/pdf/climate-change/climate-plan/healthy_environment_healthy_economy_plan.pdf (accessed 30 April 2025).

401 Environment and Climate Change Canada (2020). Carbon pollution pricing: considerations for protocol development in the federal GHG offset system [Internet]. https://www.canada.ca/en/environment-climate-change/services/climate-change/pricing-pollution-how-it-will-work/output-based-pricing-system/carbon-pollution-pricing-considerations-protocol-development.html (accessed 30 April 2025).

402 Environment and Climate Change Canada (2021). Canada confirms its support for the Global Methane Pledge and announces ambitious domestic actions to slash methane emissions. News Release [Internet]. https://www.canada.ca/en/environment-climate-change/news/2021/10/canada-confirms-its-support-for-the-global-methane-pledge-and-announces-ambitious-domestic-actions-to-slash-methane-emissions.html (accessed 30 April 2025).

403 Environment and Climate Change Canada (2021). National inventory report 1990–2019: greenhouse gas sources and sinks in Canada [Internet]. https://unfccc.int/documents/271493 (accessed 30 April 2025).

404 Environment and Climate Change Canada (2022). Canada's greenhouse gas and air pollutant emissions projections 2021 [Internet]. https://publications.gc.ca/collections/collection_2022/eccc/En1-78-2021-eng.pdf (accessed 30 April 2025).

405 Environment and Climate Change Canada (2022). Engagement on federal regulations for reducing methane emissions from Canada's municipal solid waste landfills [Internet]. https://www.canada.ca/en/environment-climate-change/services/managing-reducing-waste/municipal-solid/waste-greenhouse-gases-canada-actions/engagement-landfill-emission.html (accessed 30 April 2025).

406 Environment and Climate Change Canada (2022). Faster and further: Canada's methane strategy [Internet]. https://publications.gc.ca/collections/collection_2022/eccc/En4-491-2022-eng.pdf (accessed 30 April 2025).

407 Environment and Climate Change Canada (2022). National inventory report 1990–2020: greenhouse gas sources and sinks in Canada [Internet]. https://unfccc.int/documents/461919 (accessed 30 April 2025).

408 Equinor (2022). The Equinor share [Internet]. https://www.equinor.com/investors/the-equinor-share (accessed 30 April 2025).

409 EduCanada (2021). Climate change policy [Internet]. https://www.edc.ca/content/dam/edc/en/non-premium/climate_change_policy_board_final_en.pdf (accessed 30 April 2025).

410 EduCanada (2022). EDC net zero 2050: 2022 update [Internet]. https://www.edc.ca/content/dam/edc/en/non-premium/edc-net-zero-emissions-2050-update.pdf (accessed 30 April 2025).

411 Ernst and Young (2022). Nearly half of Canadians say they plan to buy an electric vehicle as cost concerns fade. Cision [Internet]. https://www.newswire.ca/news-releases/nearly-half-of-canadians-say-they-plan-to-buy-an-electric-vehicle-as-cost-concerns-fade-807453417.html (accessed 22 June 2022).

412 United Kingdom (2021). Carbis Bay G7 summit communiqué: our shared agenda for global action to build back better [Internet]. https://www.consilium.europa.eu/media/50361/carbis-bay-g7-summit-communique.pdf (accessed 30 April 2025).

413 Global Affairs Canada (2022). Export controls [Internet]. https://www.international.gc.ca/controls-controles/about-a_propos/expor/before-avant.aspx?lang=eng (accessed 30 April 2025).

414 Government of Canada (2010). Renewable fuels regulations [Internet]. https://laws-lois.justice.gc.ca/eng/regulations/SOR-2010-189/index.html (accessed 30 April 2025).

415 Government of Canada (2016). Pan-Canadian framework on clean growth and climate change [Internet]. https://www.canada.ca/content/dam/themes/environment/documents/weather1/20170125-en.pdf (accessed 30 April 2025).

416 Government of Canada (2018). Greenhouse gas pollution pricing act [Internet]. https://laws-lois.justice.gc.ca/eng/acts/G-11.55/ (accessed 30 April 2025).

417 Supreme Court of Canada (2021). Case in brief: references re greenhouse gas pollution pricing act [Internet]. https://www.scc-csc.ca/case-dossier/cb/2021/38663-38781-39116-eng.aspx (accessed 30 April 2025).

418 Government of Canada (2018). Regulations respecting reduction in the release of methane and certain volatile organic compounds (upstream oil and gas sector) [Internet]. https://laws-lois.justice.gc.ca/PDF/SOR-2018-66.pdf (accessed 30 April 2025).

419 Transports Canada (2021). Building a green economy: Government of Canada to require 100% of car and passenger truck sales be zero-emission by 2035 in Canada [Internet]. https://www.canada.ca/en/transport-canada/news/2021/06/building-a-green-economy-government-of-canada-to-require-100-of-car-and-passenger-truck-sales-be-zero-emission-by-2035-in-canada.html (accessed 30 April 2025).

420 Transports Canada (2022). Light-duty zero-emission vehicles [Internet]. https://tc.canada.ca/en/road-transportation/innovative-technologies/zero-emission-vehicles/light-duty-zero-emission-vehicles (accessed 30 April 2025).

421 Transports Canada (2022). Medium and heavy-duty zero-emission vehicles [Internet]. https://tc.canada.ca/en/road-transportation/innovative-technologies/zero-emission-vehicles/medium-heavy-duty-zero-emission-vehicles (accessed 30 April 2025).

422 Tuttle, R. (2022). BP's oilsands exit may not be the last as big oil revises its image. Financial Post [Internet]. https://financialpost.com/commodities/energy/oil-gas/bps-oil-sands-exit-may-not-be-the-last-as-big-oil-revises-image (accessed 30 April 2025).

423 US Department of Energy (2021). Joint Statement on Establishing a Net-Zero Producers Forum between the Energy Ministries of Canada, Norway, Qatar, Saudi Arabia, and the United States [Internet]. https://www.energy.gov/articles/joint-statement-establishing-net-zero-producers-forum-between-energy-ministries-canada (accessed 30 April 2025).

424 UK Government (2021). Statement on International Public Support for the Clean Energy Transition – UN Climate Change Conference (COP26) at the SEC – Glasgow 2021 [Internet]. UK COP26. https://ukcop26.org/statement-on-international-public-support-for-the-clean-energy-transition/ (accessed 30 April 2025).

425 Willberg, D. (2022). Lots of questions remain about the future of power production in Estevan [Internet]. SaskToday. https://www.sasktoday.ca/south/local-news/lots-of-questions-remain-about-the-future-of-power-production-in (accessed 30 April 2025).

426 Carbon Action Tracker (2024). Norway [Internet]. https://climateactiontracker.org/countries/norway/ (accessed 1 April 2024).

427 Adomaitis, N. (2019). Norway's CO_2 seabed storage project gets boost from industry, EU [Internet]. Reuters. https://www.reuters.com/article/us-norway-ccs/norways-co2-seabed-storage-project-gets-boost-from-industry-eu-idUSKCN1VQ1SP (accessed 30 April 2025).

428 Buli, N. and Adomaitis, N. (2021). Norway's plans to raise carbon tax draw oil industry ire [Internet]. Reuters. https://www.reuters.com/article/us-climate-change-norway-idUSKBN29D1BD (accessed 30 April 2025).

429 Ambrose, J. (2019). World's biggest sovereign wealth fund to ditch fossil fuels. The Guardian [Internet]. https://www.theguardian.com/business/2019/jun/12/worlds-biggest-sovereign-wealth-fund-to-ditch-fossil-fuels (accessed 30 April 2025).

430 Avinor (2018). Avinor and Norwegian aviation 2018 [Internet]. https://avinor.no/globalassets/_konsern/om-oss/rapporter/en/avinor-and-norwegian-aviation-2018_4.pdf (accessed 30 April 2025).

431 Bellona (2021). Bellona.org. Hydrogen in steel production: what is happening in Europe – part two. https://bellona.org/news/eu/2021-05-hydrogen-in-steel-production-what-is-happening-in-europe-part-two (accessed 6 July 2023).

432 Bellona (2021). Norway grants €100 million for clean hydrogen [Internet]. https://bellona.org/news/ccs/2021-12-norway-grants-e100-million-for-clean-hydrogen (accessed 30 April 2025).

433 Bjørnestad, S. (2020). Sanner endrer oljeselskapenes skatteregler, får kritikk av SSB-direktør [Internet]. https://e24.no/olje-og-energi/i/b5xwee/sanner-endrer-oljeselskapenes-skatteregler-faar-kritikk-av-ssb-direktoer (accessed 30 April 2025).

434 Brekke, T., Isachsen, O., and Marton, I. (2020). Implementation of the EPBD Norway: status in 2020 [Internet]. https://epbd-ca.eu/wp-content/uploads/2021/07/Implementation-of-the-EPBD-in-Norway- (accessed 30 April 2025).

435 Brown, M. (2018). Norway just mandated zero-emission fjords to lead electric boat charge | Inverse [Internet]. https://www.inverse.com/article/44507-norway-zero-emission-fjords?refresh=8 (accessed 30 April 2025).

436 Buildup (2015). Implementation of the EPBD in Norway [Internet]. https://buildup.eu/en/practices/publications/epbd-implementation-norway-status-november-2015-0 (accessed 30 April 2025).

437 Climate Analytics (2022). 1.5 °C National Pathways Explorer – Norway [Internet]. https://1p5ndc-pathways.climateanalytics.org/countries/norway/sectors/buildings/ (accessed 30 April 2025).

438 Daily Scandinavian (2021). World's largest electric ferry now operational in Norway [Internet]. https://www.dailyscandinavian.com/worlds-largest-electric-ferry-now-operational-in-norway/ (accessed 30 April 2025).

439 Direktoratet for Byggkvalitet (2017). Sammenstilling av forskriftene TEK10 og TEK17 [Internet]. https://dibk.no/globalassets/34.-byggereglene/tek17-superside/sammenstilling-av-forskriftene-tek10-og-tek17-11.07.17.pdf (accessed 30 April 2025).

440 Dunn, G. (2020). Norway to extend airline loan guarantee scheme [Internet]. https://www.flightglobal.com/airlines/norway-to-extend-airline-loan-guarantee-scheme/138603.article (accessed 30 April 2025).

441 European Environment Agency (2021). CO_2 performance of new passenger cars in Europe. European Environment Agency [Internet]. https://www.eea.europa.eu/ims/co2-performance-of-new-passenger (accessed 30 April 2025).

442 European Environment Agency (2021). Greenhouse gas emission intensity of electricity generation in Europe [Internet]. https://www.eea.europa.eu/data-and-maps/indicators/overview-of-the-electricity-production-3/assessment#:~:text=In%20non-EU%20EEA%20countries,e%2FkWh%20in%20Norway (accessed 30 April 2025).

443 Faschevsky, S. (2016). Policy framework – energy efficiency in buildings in Norway [Internet]. http://abconsulting-bg.com/training/wp-content/uploads/sites/4/2016/04/1_Norwegian_policy_EE_buildings.pdf (accessed 30 April 2025).

444 Ferris, N. (2022). Weekly data: Why Norway leads the world for electric vehicles. Energy Monitor [Internet]. https://www.energymonitor.ai/sectors/transport/weekly-data-why-norway-leads-the-world-for-electric-vehicles#:~:text=The%20impact%20of%20EVs%E2%80%99%20increasing,%25%20BEV%20and%206%25%20hybrid (accessed 30 April 2025).

445 Gavenas, E., Rosendahl, K.E., and Skjerpen, T. (2015). CO_2-emissions from Norwegian oil and gas extraction [Internet]. https://www.ssb.no/en/forskning/discussion-papers/_attachment/225118?_ts=14de17b6918 (accessed 30 April 2025).

446 Norway Government (2019). Norway increases contribution to Green Climate Fund [Internet]. https://www.norway.no/en/ethiopia/norway-ethiopia/news-events/news2/press-release-norway-increases-contribution-to-green-climate-fund/ (accessed 30 April 2025).

447 Norway Government (2019). Politisk plattform [Internet]. https://www.regjeringen.no/contentassets/7b0b7f0fcf0f4d93bb6705838248749b/plattform.pdf (accessed 30 April 2025).

448 Norway Government (2020). Norway's long-term low-emission strategy for 2050. p. 1–49.

449 Norway Government (2020). One step closer to CO_2 storage in the North Sea [Internet] https://www.regjeringen.no/en/aktuelt/ett-skritt-narmere-et-co2-lager-i-nordsjoen/id2692396/ (accessed 30 April 2025).

450 Norway Government (2020). Update of Norway's nationally determined contribution [Internet]. https://unfccc.int/sites/default/files/NDC/2022-06/Norway_updatedNDC_2020%20%28Updated%20submission%29.pdf (accessed 30 April 2025).

451 Norway Government (2021). National Transport Plan 2022–2033 [Internet]. https://www.regjeringen.no/contentassets/117831ad96524b9b9eaadf72d88d3704/en-gb/pdfs/stm202020210020000engpdfs.pdf (accessed 30 April 2025).

452 Norway Government (2021). Norway to double climate finance to NOK 14 billion [Internet]. https://www.regjeringen.no/en/aktuelt/norway-to-double-climate-finance-to-nok-14-billion/id2881477/ (accessed 30 April 2025).

453 Norway Government (2021). Norway's comprehensive climate action plan [Internet]. https://www.regjeringen.no/en/historical-archive/solbergs-government/Ministries/kld/news/2021/heilskapeleg-plan-for-a-na-klimamalet/id2827600/ (accessed 30 April 2025).

454 Norway Government (2021). Oversikt over alle regjeringa vil-punkta i meldinga [Internet]. p. 1–7. https://www.regjeringen.no/contentassets/202fec60ac844d4ca7d53d65b6b9ac9c/alle-regjeringa-vil-punkt-i-meldinga.pdf (accessed 30 April 2025).

455 Norway Government (2022). Common Reporting Format (CRF) Table [Internet]. https://unfccc.int/documents/461704 (accessed 30 April 2025).

456 Holland, M. (2022). Norway above 86% plugin EV share in february, ioniq 5 leads [Internet]. https://cleantechnica.com/2022/03/03/norway-above-86-plugin-ev-share-in-february-ioniq-5-leads/ (accessed 30 April 2025).

457 Holter, M. (2019). Norway's oil industry must become emissions-free, labor says. Bloomberg [Internet]. https://www.bloomberg.com/news/articles/2019-04-04/norway-s-oil-industry-must-become-emissions-free-labor-says (accessed 30 April 2025).

458 Klesty, V. (2022). Electric cars hit 65% of Norway sales as Tesla grabs overall pole. Reuters [Internet]. https://www.reuters.com/business/autos-transportation/electric-cars-take-two-thirds-norway-car-market-led-by-tesla-2022-01-03/ (accessed 30 April 2025).

459 Losnegård, A., Befring, Å.M., and Lydersen, T. (2020). Økonomiprofessor kritisk til oljeforliket: Dette vil forlenge oljealderen [Internet]. NRK. https://www.nrk.no/norge/okonomiprofessor-kritisk-til-oljeforliket_-_-dette-vil-forlenge-oljealderen-1.15045774 (accessed 30 April 2025).

460 Climate Change Act (2017). Act relating to Norway's climate targets.

461 Masnadi, M.S., El-Houjeiri, H., Schunack, D. et al. (2018). Global carbon intensity of crude oil production. *Science.* 361: 851–853.

462 McKinsey&Company (2018). The market for unlisted renewable energy infrastructure [Internet]. https://www.regjeringen.no/contentassets/7fb88d969ba34ea6a0cd9225b28711a9/20181219-the-market-for-unlisted-renewable-energy-infrastructure_final.pdf (accessed 30 April 2025).

463 Milne R. (2020). Norway's oil fund sells out of Glencore, Anglo American and RWE. Financial Times [Internet]. https://www.ft.com/content/ca33f62e-242a-4304-9895-f7cea36a497e (accessed 30 April 2025).

464 Ministry of Finance (2019). Allowing for unlisted renewable energy infrastructure [Internet]. https://www.regjeringen.no/en/aktuelt/allowing-for-unlisted-renewable-energy-infrastructure/id2640386/ (accessed 30 April 2025).

465 Ministry of Petroleum and Energy (2018). The Norwegian Government continues with the planning of a demonstration project for CO_2 capture, transport and storage [Internet]. https://www.regjeringen.no/en/aktuelt/the-norwegian-government-continues-with-the-planning-of-a-demonstration-project-for-co2-capture-transport-and-storage/id2601399/ (accessed 30 April 2025).

466 NewScientist (2019). Norway is starting the world's biggest divestment in oil and gas [Internet]. https://www.newscientist.com/article/2196024-norway-is-starting-the-worlds-biggest-divestment-in-oil-and-gas/ (accessed 30 April 2025).

467 Nikel, D (2019). Norway wealth fund to dump fossil fuel stocks worth billions in environmental move. Bloomberg [Internet]. https://www.forbes.com/sites/davidnikel/2019/06/12/norway-wealth-fund-to-dump-fossil-fuel-stocks-worth-billions-in-environmental-move/#3d817c2848a3 (accessed 30 April 2025).

468 Norges Bank (2020). Exclusion and observation of coal companies [Internet]. https://www.nbim.no/en/the-fund/news-list/2020/exclusion-and-observation-of-coal-companies/ (accessed 30 April 2025).

469 Norges Bank (2021). Work on climate risk in the Government Pension Fund Global [Internet]. https://www.nbim.no/contentassets/c4353362c1904b88a0387e2485ae4681/work-on-climate-risk-in-the-government-pension-fund-global.pdf (accessed 30 April 2025).

470 Norsk elbilforening (2020). Utslippstallene i Norge synker og drivstoffsalget stuper: Elbiler har mye av æren for begge deler [Internet]. https://elbil.no/utslippstallene-i-norge-synker-og-drivstoffsalget-stuper-elbiler-har-mye-av-aeren-for-begge-deler/ (accessed 30 April 2025).

471 Norsk elbilforening (2021). Norwegian EV policy [Internet]. https://elbil.no/english/norwegian-ev-policy/ (accessed 30 April 2025).
472 Oil Change International (2022). The aggressive explorer: how Norway's Rapid ramp-up of oil and gas licensing is incompatible with climate leadership [Internet]. https://priceofoil.org/2022/02/08/norway-briefing-2022/ (accessed 30 April 2025).
473 Reuters (2017). Oil producer Norway bans use of heating oil in buildings [Internet]. https://www.reuters.com/article/us-climatechange-norway/oil-producer-norway-bans-use-of-heating-oil-in-buildings-idUSKBN1961VL (accessed 30 April 2025).
474 Reuters (2019). Norway awards Equinor license to build CO_2 storage under seabed [Internet]. https://www.reuters.com/article/equinor-carbon-idUSL8N1ZB3BN (accessed 30 April 2025).
475 European Commission (2024). EU Emissions Trading System (EU ETS) – European Commission [Internet]. https://climate.ec.europa.eu/eu-action/eu-emissions-trading-system-eu-ets_en (accessed 1 April 2024).
476 Carbon Action Tracker (2023). Germany [Internet]. https://climateactiontracker.org/countries/germany/ (accessed 1 April 2024).
477 ADAC (2023). Pkw-Neuzulassungen im Juli 2023: VW bei Elektroautos von [Internet]. https://www.adac.de/news/neuzulassungen-kba/ (accessed 30 April 2025).
478 Agora Energiewende (2019). The German Coal Commission – a roadmap for a just transition from coal to renewables.
479 ADAC (2023). Förderung für Elektroautos 2023: So kommen Sie an den Umweltbonus [Internet]. https://www.adac.de/rund-ums-fahrzeug/elektromobilitaet/kaufen/foerderung-elektroautos/ (accessed 30 April 2025).
480 BMWK (2022). Eröffnungsbilanz Klimaschutz [Internet]. https://www.bmwk.de/Redaktion/DE/Downloads/Energie/220111_eroeffnungsbilanz_klimaschutz.pdf?__blob=publicationFile&v=22 (accessed 30 April 2025).
481 BMWK (2023). Die Nationale Wasserstoffstrategie [Internet]. https://www.bmwk.de/Navigation/DE/Wasserstoff/wasserstoffstrategie.html (accessed 30 April 2025).
482 BMWK (2023). Förderprogramm Klimaschutzverträge [Internet]. https://www.bmwk.de/Redaktion/DE/Artikel/Klimaschutz/klimaschutzvertraege.html (accessed 30 April 2025).
483 BMWK (2023). Fortschreibung der Nationalen Wasserstoffstrategie [Internet]. https://www.bmbf.de/SharedDocs/Downloads/de/2023/230726-fortschreibung-nws.pdf?__blob=publicationFile&v=1 (accessed 30 April 2025).
484 BMWK (2023). Gesetz zur Entwicklung und Förderung der Windenergie auf See [Internet]. Federal Ministry for Economic Affairs and Climate Action. https://www.gesetze-im-internet.de/windseeg/WindSeeG.pdf (accessed 30 April 2025).
485 BMWK (2023). Klimapolitische Sektorleitlinien für Exportkreditgarantien – Entwurf [Internet]. https://www.exportkreditgarantien.de/de/nachhaltigkeit/nachhaltigkeit/klimastrategie.html (accessed 30 April 2025).

486 BMWK (2023). Plattform Klimaneutrales Stromsystem – im Dialog für ein neues Marktdesign [Internet]. https://www.bmwk.de/Redaktion/DE/Dossier/plattform-klimaneutrales-stromsystem.html (accessed 30 April 2025).

487 BMWK (2023). Renewable energy law. Federal Ministry for Economic Affairs and Climate Action [Internet]. https://www.gesetze-im-internet.de/eeg_2014/EEG_2023.pdf (accessed 30 April 2025).

488 BMWK (2023). Working paper on electricity pricing for the industrial sector (Wettbewerbsfähige Strompreise für die energieintensiven Unternehmen in Deutschland und Europa sicherstellen Arbeitspapier des BMWK zum Industriestrompreis für das Treffen Bündnis Zukunft der) [Internet]. https://www.bmwk.de/Redaktion/DE/Downloads/W/wettbewerbsfaehige-strompreise-fuer-die-energieintensiven-unternehmen-in-deutschland-und-europa-sicherstellen.pdf?__blob=publicationFile&v=6 (accessed 30 April 2025).

489 Bundesamt für Wirtschaft und Ausfuhrkontrolle (2023). Bundesförderung für effiziente Gebäude - Anlagen zur Wärmeerzeugung (Heizungstechnik [Internet]. https://www.bafa.de/DE/Energie/Effiziente_Gebaeude/Sanierung_Wohngebaeude/Anlagen_zur_Waermeerzeugung/anlagen_zur_waermeerzeugung_node.html (accessed 30 April 2025).

490 Bundesanstalt für Landwirtschaft und Ernährung (2023). Presseinformation - Fleischverzehr 2022 auf Tiefstand [Internet]. https://www.ble.de/SharedDocs/Downloads/DE/Pressemitteilungen/2023/230403_Fleisch.pdf;jsessionid=67F1AFAFA7F42C39BECFFD1B924785C8.internet982?__blob=publicationFile&v=2 (accessed 30 April 2025).

491 Wirtschaft Brandenberg (2019). Energie. Referentenentwurf des Bundesministeriums für Wirtschaft und Energie – Gesetz zur Reduzierung und zur Beendigung der Kohleverstromungund zur Änderung weiterer Gesetze (Kohleausstiegsgesetz) [Internet]. https://www.klimareporter.de/images/dokumente/2019/11/referentenentwurf-kohleausstiegsgesetz-11-11-2019.pdf (accessed 30 April 2025).

492 Bundesnetzagentur (2021). Ergebnisse der dritten Ausschreibung zum Kohleausstieg.

493 Bundesnetzagentur (2022). Kohleausstieg [Internet]. https://www.bundesnetzagentur.de/DE/Fachthemen/ElektrizitaetundGas/Kohleausstieg/start.html (accessed 30 April 2025).

494 Bundesnetzagentur (2023). Elektromobilität: Öffentliche Ladeinfrastruktur [Internet]. https://www.bundesnetzagentur.de/DE/Fachthemen/ElektrizitaetundGas/E-Mobilitaet/start.html (accessed 30 April 2025).

495 Climate Action Tracker (2021). Germany's proposed 2030 national target not yet 1.5 °C-compatible [Internet]. https://climateactiontracker.org/publications/germanys-proposed-2030-national-target-not-yet-15c-compatible/ (accessed 30 April 2025).

496 Climate Action Tracker (2018). What's on the Table? Mitigating agricultural emissions while achieving food security [Internet]. https://climateactiontracker.org/documents/68/CAT_2018-01-23_DecarbAgriculture_CATAnalysis.pdf (accessed 30 April 2025).

497 Deutsche Windguard GmbH (2023). Status des Windenergieausbaus an Land in Deutschland [Internet]. https://www.wind-energie.de/fileadmin/redaktion/dokumente/publikationen-oeffentlich/themen/06-zahlen-und-fakten/20230718_Status_des_Windenergieausbaus_an_Land_Halbjahr_2023.pdf. (accessed 30 April 2025)

498 Ember (2023). Carbon price tracker [Internet]. https://ember-climate.org/data/data-tools/carbon-price-viewer/ (accessed 30 April 2025).

499 European Commission (2021). State aid: commission approves competitive tender mechanism to compensate for early closure of hard coal-fired power plants in Germany [Internet]. https://ec.europa.eu/commission/presscorner/detail/en/ip_20_2208 (accessed 30 April 2025).

500 Expertenrat für Klimafragen (2023). Prüfbericht 2023 für die Sektoren Gebäude und Verkehr.

501 Expertenrat für Klimafragen (2023). Prüfbericht zur Berechnung der deutschen Treibhausgasemissionen für das Jahr 2022 [Internet]. https://expertenrat-klima.de/content/uploads/2023/04/ERK2023_Pruefbericht-Emissionsdaten-des-Jahres-2022.pdf (accessed 30 April 2025).

502 Federal Environment Agency (2017). Klimaverträgliche Abfallwirtschaft [Internet]. https://www.umweltbundesamt.de/daten/ressourcen-abfall/klimavertraegliche-abfallwirtschaft#textpart-1 (accessed 30 April 2025).

503 Federal Environment Agency (2023). Projections report 2023 for Germany (Projektionsbericht 2023 für Deutschland [Internet]. https://www.umweltbundesamt.de/sites/default/files/medien/11740/publikationen/2023_08_21_climate_change_39_2023_projektionsbericht_2023_0.pdf (accessed 30 April 2025).

504 Federal Ministry for Environment Nature Conservation (2016). Climate action plan 2050 – principles and goals of the German government's climate policy [Internet]. Berlin, Germany. https://www.bmu.de/fileadmin/Daten_BMU/Pools/Broschueren/klimaschutzplan_2050_en_bf.pdf (accessed 30 April 2025).

505 Federal Ministry of Economics and Climate Protection (2022). Überblickspapier Osterpaket. Berlin.

506 Federal Ministry of Economics and Climate Protection (2023). Entwurf eines Gesetzes zur Änderung des Entwurf eines Gesetzes zur Änderung des Gebäudeenergiegesetzes, zur Änderung der Heizkostenverordnung und zur Änderung der Kehr_und Überprüfungsordnung.

507 Fraunhofer ISE (2023). Monatlicher Anteil Erneuerbarer Energien an der öffentlichen Nettostromerzeugung in Deutschland 2023 [Internet]. https://energy-charts.info/charts/renewable_share/chart.htm?l=de&c=DE (accessed 30 April 2025).

508 Fraunhofer ISE (2023). Net installed electricity generation capacity in Germany in 2023 [Internet]. https://energy-charts.info/charts/installed_power/chart.htm?l=en&c=DE&chartColumnSorting=default&expansion=installed_power (accessed 30 April 2025).

509 Fraunhofer ISE (2023). Newly installed electricity generation capacity per month in Germany 2023 [Internet]. https://energy-charts.info/charts/installed_power/chart.htm?l=en&c=DE&chartColumnSorting=default&expansion=installation_decommission&interval=month&legendItems=000100 (accessed 30 April 2025).

510 German Government (2019). Klimaschutzprogramm 2030 zur Umsetzung des Klimaschutzplans 2050' [Internet]. Klimaschutzprogramm 2030 der Bundesregierung zur Umsetzung des Klimaschutzplans 2050.

511 German Government (2020). Grundlage für CO_2-Preis steht [Internet]. https://www.bundesregierung.de/breg-de/themen/klimaschutz/nationaler-emissionshandel-1684508 (accessed 30 April 2025).

512 German Government (2020). The National Hydrogen Strategy [Internet]. https://www.bmbf.de/files/bmwi_Nationale (accessed 30 April 2025).

513 Government G (2021). Bundes-Klimaschutzgesetz [Internet]. https://www.gesetze-im-internet.de/ksg/index.html (accessed 30 April 2025).

514 German Government (2021). Klimaschutzgesetz 2021 – Generationenvertrag für das Klima [Internet]. https://www.bundesregierung.de/breg-de/themen/klimaschutz/klimaschutzgesetz-2021-1913672 (accessed 30 April 2025).

515 German Government (2021). Lesefassung des Bundes-Klimaschutzgesetzes 2021 mit markierten Änderungen zur Fassung von 2019 [Internet]. https://www.bmu.de/fileadmin/Daten_BMU/Download_PDF/Klimaschutz/ksg_aendg_2021_3_bf.pdf (accessed 30 April 2025).

516 German Government (2022). Update to the long-term strategy for climate action of the Federal Republic of Germany [Internet]. https://unfccc.int/sites/default/files/resource/Anlage (accessed 30 April 2025).

517 German Government (2023). Ein Plan fürs Klima. 2023.

518 German Government (2023) Gesetzentwurf der Bundesregierung - Entwurf eines Gesetzes für die Wärmeplanung und zur Dekarbonisierung der Wärmenetze [Internet]. https://www.bmwsb.bund.de/SharedDocs/gesetzgebungsverfahren/Webs/BMWSB/DE/Downloads/kabinettsfassung/kommunale-waermeplanung.pdf;jsessionid=AB59717D777407BF0CF9000CD3DE187F.1_cid322?__blob=publicationFile&v=1 (accessed 30 April 2025).

519 Global Energy Monitor CREA and E3G (2022). Boom and bust coal 2022 [Internet]. https://globalenergymonitor.org/report/boom-and-bust-coal-2022/ (accessed 30 April 2025).

520 Harthan, R.O. and Repenning, J. (2022). Klimaschutzbeitrag verschiedener CO_2-Preispfade in den BEHG-Sektoren Verkehr, Gebäude und Industrie [Internet]. Germany: Dessau-Roßlau. https://www.umweltbundesamt.de/sites/default/files/medien/1410/publikationen/2022-05-04_climate-change_19-2022_co2-preissensitivitaet-behg_0.pdf (accessed 30 April 2025).

521 Leadership Group for Industry Transition (2023). Green Steel Tracker. https://www.industrytransition.org/green-steel-tracker/ (accessed 18 September 2023).

522 Höhne, N., Marquardt, M., Fekete, H. (2023). Pläne für deutsche Flüssigerdgas-Terminals nicht im Einklang mit dem deutschen Klimaschutzgesetz [Internet]. Deutschland Annex.pdf. https://newclimate.org/sites/default/files/2023-02/LNG (accessed 30 April 2025).

523 Federal Office for the Safety of Nuclear Waste Management (2024). BASE. 2024 The nuclear phase-out in Germany. https://www.base.bund.de/EN/ns/nuclear-phase-out/nuclear-phase-out_node.html (accessed 1 April 2024).

524 Luxembourg (2024). Public transport [Internet]. http://luxembourg.public.lu/en/living/mobility/public-transport.html (accessed 1 April 2024).

11

An idealised timeline of possibilities for steel decarbonisation

11.1 Introduction

In this chapter, the selection of the most suitable options for industrial decarbonisation is being discussed. By integrating technological, economic, and policy considerations, the findings from all previous chapters will be used to map out a pathway for decarbonisation of the steel industry worldwide.

11.2 Political decision-making and legislative foundations

To incentivise industry leaders across all sectors to implement the required climate actions and changes, the financial support to carry out the necessary projects and processes would potentially require a change in law across all countries: net-zero and decarbonisation measures need to be set before shareholders' interests, and companies need the legal assurance that they will be able to do so, without shareholders' interests getting in the way, climate security before shareholders' profits. The creation of a legally binding framework for all countries to adhere to is necessary, with a clear action plan to be implemented, without the need for country-specific plans to be submitted – as it currently seems the case, simply put: one plan, binding for all countries. The system currently in place, where countries have to submit their updated NDCs annually to the UNFCCC, does not seem to trigger decarbonisation efforts and climate action towards keeping global warming to 1.5 °C. Steel is the second most abundantly used building component after concrete, and every metric tonne of crude steel produced is emitting ~13 metric tonnes of CO_2 emissions, with all steps in the production process considered from mining to marketing. It is therefore crucial that all efforts are being made to avoid, reduce and capture CO_2 emissions at all stages of steel production. Creating a green steel passport or specific

Circular Steel Production: Pathways to Net-Zero Carbon Emissions, First Edition.
Sandra Kiessling, Hamidreza Gohari Darabkhani, and Abdel-Hamid Soliman.

carbon certificate of origin would be a helpful mechanism, where all steps in production need to be declared, and how much CO_2 has been emitted along the whole of the steelmaking process, not just the last step of converting liquid iron into steel. An additional instrument for the acceleration of the decarbonisation process in the steel industry is the global provision of tax rebates, incentives, grants and climate finance for measures reducing GHG emission in steelmaking. Already now, there is ample provision for incentives and grants available in the United Kingdom, as discussed in Chapter 6: Greater Energy Independence with Sustainable Steel Production. More countries should aim to follow this example to create momentum for the net-zero climate targets.

11.3 All-encompassing industrial response

An action plan, specific for the steel industry, needs to be created by the respective governments to drive change and support in achieving the net-zero targets and decarbonising industrial processes at the latest by 2030. Specific for the steel industry, there are easily implementable steps, as explained in Chapter 4: The Key Components of the Bio Steel Cycle (BiSC). Steel-industry-specific implementable carbon avoidance technologies should include:

- Switching to 100% green energy providers
- Using green HDR instead of sintering
- Replacing blast furnaces (BF) with electric arc furnaces (EAF)
- Increasing the EAF process utilisation percentage of steel scrap
- Compulsory off-heat harvesting and upgrading existing infrastructure to avoid energy loss in operation

The decarbonisation of the steelmaking process as such can be seen through by following the simple seven-step strategy [1, 2], as explained in detail in Chapter 4 and Chapter 6, and the following images will demonstrate in more detail (Figures 11.1 and 11.2).

A steel industry action plan is needed, ideally including the seven steps to net-zero steel [1, 2] to drive process innovation and decarbonisation. The following timeline items should be implemented latest from the date of publication as a starting point as follows (Figure 11.3).

Additionally, a legally binding industry-wide framework of compulsory actions needs to be created and implemented as soon as possible. In order to create a meaningful carbon cost model, similar to the EU ETS system, an all-sector-encompassing standard format for carbon emissions reporting is required to be created as a starting point for fair trade and promoting green before grey steel.

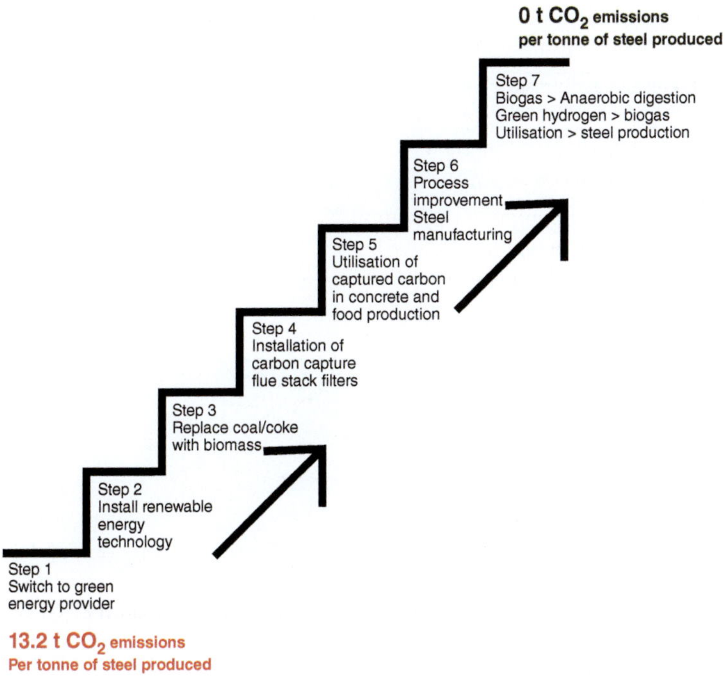

Figure 11.1 The Bio Steel Cycle (BiSC) – 7 steps to net zero carbon emissions steel production.

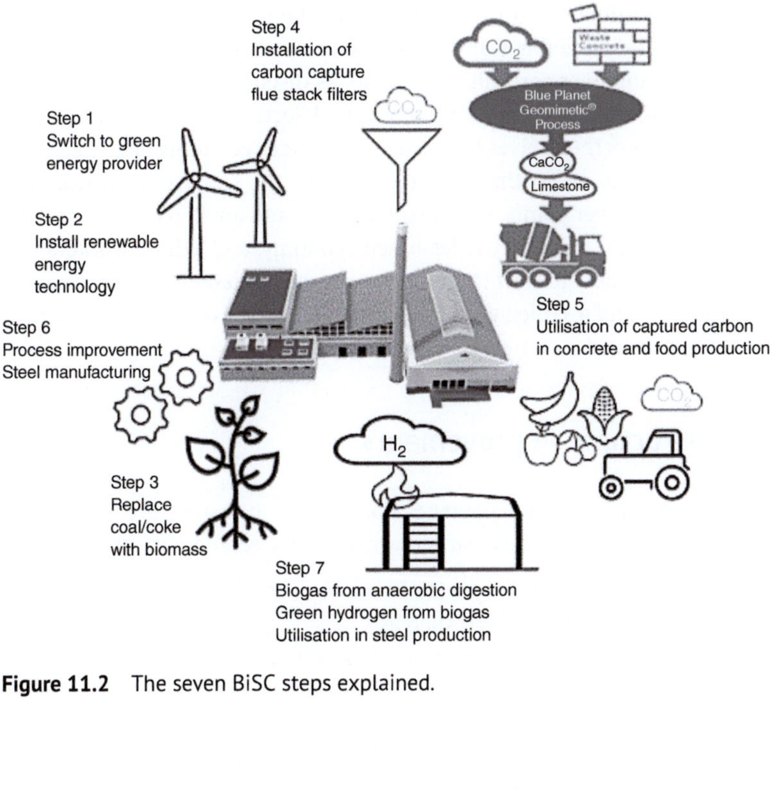

Figure 11.2 The seven BiSC steps explained.

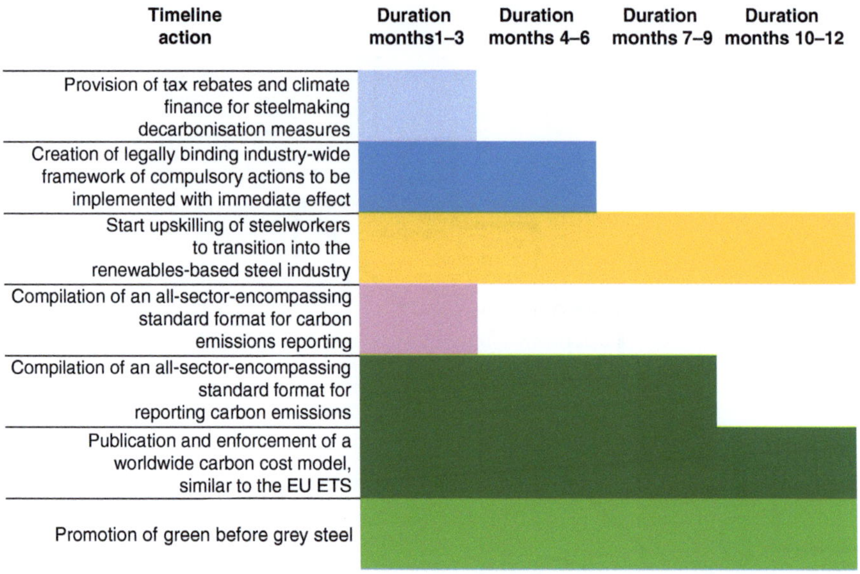

Figure 11.3 Idealised GANTT chart timeline decarbonisation of the steel industry.

11.4 Investment in people

In order to overcome the socio-economic hurdles of population resistance to change, as recently experienced in Port Talbot (Wales, UK), steel-workers currently employed in the fossil-fuel-based industries need to be upskilled and retrained to be economically active in the renewable energy sector and a renewables-based steel industry. There are ample opportunities to do so, as discussed in Chapter 9: Skills Sets Required Within the Different Components and Sectors. It is vitally important not only for the workers themselves but also for the wider economy and GDP to be training local workers to attain the skills required for the successful operation of the new plants. New power plants, steel plant sections and ancillary sectors will be affected by the transition to a renewables-based economy, and skilled workers will be needed. The time is now to start training people, so the challenges ahead can be met with enthusiasm and a skilled workforce proud to be part of positive change on the path to a decarbonised society.

11.5 Infrastructural improvements

As already discussed in detail in Chapter 6, the existing infrastructure needs to be upgraded to meet future challenges. Particularly for the steel industry, this means:

- creation of legally binding industry-wide framework of compulsory actions to be implemented with immediate effect, such as
  - Upgrade the existing infrastructure to eliminate energy loss during BF/BOF and EAF operation

- Integrate site-suitable renewable energy technologies where technically feasible
- Employ best practice innovative steel production processes to reduce steelmaking-related GHG emissions: replace BF with EAF
- Improve the steelmaking process in the existing EAF plants to increase the % share utilisation of scrap steel to 100%. Some are currently operating only on utilising 42% steel scrap of the total charge capacity
- Upgrade the existing network of pipes to be able to cope with the demands of hydrogen utilisation
- Use hydrogen as a replacement for coal and coke where feasible, i.e. in hydrogen direct reduction of iron ore instead of sintering
- Replace coal and coke with biomass and biochar for ironmaking
- Create a circular economic process of steelmaking based on the suggestions of the Bio Steel Cycle [1, 2]

The actions required as described and displayed in the idealised timeline are all perfectly simple as far as the implementation is concerned. All they require is the political willingness to prepare the industry for the required changes, support industry leaders in their endeavours to achieve climate targets and remember to take the workers along on this journey, because without them, all these perfectly feasible ideas will not come to pass.

We need skilled workers today more than ever before, and it is high time for all industries to recognise this and train, retrain or upskill their workforce and provide incentives for school leavers to be active in the emerging industries and sectors.

References

1 Kiessling, S, Darabkhani, H.G., and Soliman, A.H. (2022). Energies | Free Full-Text | The Bio Steel Cycle: 7 steps to net-zero CO_2 emissions steel production [Internet]. https://www.mdpi.com/1996-1073/15/23/8880 (accessed 11 July 2023).

2 Kiessling, S., Gohari Darabkhani, H., and Soliman, A.H. (2024). Greater energy independence with sustainable steel production. *Sustainability* 16 (3): 1174.

12

Concluding remarks and suggestions

12.1 Discussion of findings

The aim of this chapter was to present the findings from all previous chapters in an easily accessible way in order to support decarbonisation efforts, provide guidance on how to decarbonise the iron and steel sector and achieve net-zero carbon emissions in steel production. As a commercial commodity based on iron [1], steel contains carbon in amounts of up to about 1.7% as an essential alloying constituent [2] and is malleable when exposed to suitable conditions. It is distinguished from cast iron by its malleability and lower carbon content. Although evidence of early efforts in steelmaking was discovered in the thirteenth century CE [3] in the Arab world, it is argued that actually, the earliest known use of the crucible steel production technique occurred in India and central Asia in the early first millennium CE [4–9]. Documented early attempts to refine the process do not seem to have had a great impact on the craft, until the nineteenth century CE. In c. 1865 [4–6], Robert F. Mushet discovered that by adding *manganese* to Bessemer steel [10], the casting of ingots free of blowholes was now possible, and adding manganese also reduced the metal brittleness forced by the presence of *sulphur* [4]. Mushet [4–6, 9] produced the first commercial alloy steel in 1868 when he found that adding *tungsten* to steel [6], particularly after air cooling, significantly increased its hardness, which lead to this material further on forming the basis of the subsequent development of tool steels for the machining of metals [7].

The ironmaking and steelmaking process has undergone continuous development over the centuries, and improvements have been made to the steelmaking process, focusing on the quality of steel produced, reducing the quantity and differentiating the quality of coal and coke involved. The determination factor for greenhouse gases is the quantities and volumes of resources used in the blast furnace (BF) and basic oxygen furnace (BOF) and based on their capacities, each standing at 330 t or 330,000 kg. The quantities and values have been discussed in detail in Chapter 2: Steelmaking Processes, and in Figure 2.3, the quantities of iron ore, limestone, lime, coke and 1000 °C oxygen required to be sent to the BF as the first furnace-based process were displayed in detail. The product of the BF, the liquid iron, is sent to the BOF, where the hot-metal alongside lime, scrap steel and oxygen at 25 °C is being converted to liquid steel. The results displayed in Figure 2.3 summarise the quantities

Circular Steel Production: Pathways to Net-Zero Carbon Emissions, First Edition.
Sandra Kiessling, Hamidreza Gohari Darabkhani, and Abdel-Hamid Soliman.

of CO_2 emissions during the individual stages preceding the steelmaking process, including mining, calcification, coking, sintering, pressurising oxygen and through the BF/BOF route to casting, rolling and finally finish machining. In the absence of a global standard for reporting GHG and CO_2 emissions, the gathered information needs to be seen with caution. At present, to our best knowledge, the total CO_2 emissions per metric tonne of steel were found to be ~13 t per tonne of crude steel, although they could easily be twice and thrice this amount, as the different sources of information used different data extraction points, methods and timings and it is not entirely clear if a universally applicable standard measure, i.e. m^3 of gas volume investigated, has been applied when the various authors stated the applicable values of CO_2 emissions to the stages throughout the steel production process. As an example, we took the value of around ~5 t of CO_2 emissions for coal extraction given by BHP [8], but mathematical analysis has resulted in a much higher CO_2 emissions value, at 70,000,000 t of coal mined:5,680,000 t CO_2 emissions total coal mining = 12.323943662 or 12.32 t CO_2/t of coal mined. As good practice, we compared this with other sources of information regarding coal extraction, and the average of CO_2 emissions for coal extraction was around 5 metric tonnes. However, in all these observations, even by Bogunovic [9], we were not given the emissions caused by manufacturing the huge drilling rigs and bore heads used for coal extraction. To date, there does not seem to be any data available to that effect. These CO_2 emissions would have to be added to the total of emissions, per tonne of coal extracted. Steel production, due to its usage of coal and coke for the preceding steps of iron-making, sintering, coking and making lime from limestone, was known quite early to be detrimental to the environment. Evidence for this can be found in some newspaper articles, which are already pointing out the negative impact on the natural environment and people's health as early as 1912 [11].

The urgent requirement to drastically reduce CO_2 emissions has triggered the author's research into models and strategies for the decarbonisation of the steelmaking process. The publications of the Kyoto Protocol, the Paris Agreement and recent 2023 reports from the IPCC and EIA have kick-started the efforts to create a feasible, easily implementable process to follow and to do so as soon as possible. The aforementioned reports have clearly set out the impact which the highest ever recorded anthropogenic CO_2 emissions are having on our environment and climate. The iron and steel industry is thought to be responsible for between 7% and 11%, although voices are stating that the industry share is as high as 30% of global CO_2 emissions [12–16] – and China is responsible for 50% of these GHG emissions [15]. To provide an engineering solution to remedy this dire situation, the Bio Steel Cycle (BiSC) was designed, with a strategy to boot. However, it was found that there is a distinct absence of global CO_2 reporting standards and formats, and it was therefore quite difficult to determine the absolute share of CO_2 emissions, as the detection points and reference frameworks vary wildly. The factual level of CO_2 emissions for every tonne of steel produced currently stands at more than 4.6 t of CO_2 emissions [17] for the BF/BOF route alone and 13.2 t of CO_2/t of crude steel, from mining to marketing [18–40], although, as mentioned, this figure could be much higher than what has been documented, so far. The onus is on industry now to remedy the environmental damage caused in the past two centuries. The anthropogenic carbon emissions are at an all-time high with reported ~37.4 billion t or 37.4 Gt of CO_2 emissions in 2023 [41, 42]. To remedy future environmental damage caused by CO_2 emissions, the innovative concept, model and strategy 'The Bio Steel Cycle' is created, as shown in Figure 3.1.

The 64 steel-producing countries produced between January and December 2021 [43] 17 Gt CO_2 equivalent of CO_2 emissions as a result of the current linear steel manufacturing. The importance of significantly reducing GHGs and eliminating fossil fuel combustion and usage has never been greater, and fast, practical solutions implemented on a global scale are needed.

Already in 1912, it was recognised that coal consumption is an environmental hazard and incompatible with keeping global temperatures at a balanced level to sustain life. Industrial processes have for more than 200 years polluted the air we breathe, and already in 1912, this was recognised as displayed in an Australian newspaper clip [11]. It is worth pointing out that in this article, merely the in-furnace coal combustion process is mentioned in connection with carbon emissions. CO_2 emissions from energy consumption, mining, pelletising, coking, sintering, steel smelting, casting, rolling, annealing, rolling, finish machining and surface treatment and related processes have not been considered at that time, as far as was possible to ascertain. But these authors more than 100 years ago already established – possibly based on the carbon content of coal (78–95%; [44]) and the release of CO_2 into the atmosphere by combustion, the factor to be 3.5 (7,000,000,000/CO_2 emissions / 2,000,000,000 t of steel = 3.5), resulting in 3.5 t of CO_2 emissions in the BF operations alone, by 1912. Although the steelmaking process has undergone significant improvements and the BOF operation has allegedly reduced the CO_2 emissions to between 3.2 t and 1.6 t/CO_2/t of steel produced [12, 45–50], it begs the question of why this knowledge has not been used to establish the true CO_2 emissions over the past two centuries – including all necessary processes required to produce steel. The objectives of this multidisciplinary and multi-industry-overarching study are to identify the most efficient implementation opportunities of the chosen processes and technologies to reduce the current BF–BOF route 4.6 t CO_2 emissions per tonne of steel produced to factual '0', ordered in seven easy steps, from short-term to long-term solution implementation.

Within the BiSC model, everything is based on the concept that nothing is going to waste, and waste and by-products are seen as resources, which are either reused or recycled, an example of circular economic principles in practical application. The foundation for this concept is the basic circular economic principle integrated into an easy-to-follow model including processes such as carbon avoidance and carbon utilisation over to carbon sequestration processes, including the implementation of advanced renewable energy technologies and biomass utilisation. In order to prove the one hypothesis, the most likely implementable scenarios have been considered and simulated using standard mathematical principles, MS Excel calculations and process engineering software. The resulting reports are encouraging as they are providing strong evidence for the technical viability of the principles of the BiSC model and that it can possibly be applicable to most heavy industries: such as cement, chemical, glass, paper and transport. It includes using renewable energy technologies, avoiding CO_2 emissions by incorporating process improvement technologies, recycling waste and by-products and capturing post-combustion emissions where possible [51–58], and is displayed in more detail in Figure 4.1.

Analyses of the technical and economical long-term potential of novel steel production technologies CAT, CCS and CCUS in the United Kingdom [59], Germany [60] and beyond used techno-economic models to compile three research-stage [12, 47, 54, 61–65], ore-based steelmaking routes versus BF–BOF route [16, 66–68]. It was concluded that in comparison with the BF with CCS1 (BFCCS) [60, 62, 69–71], hydrogen direct reduction (H-DR) [72–76], and iron ore electrolysis (EW) [77–79], energy and raw material efficiency is significantly higher for H-DR and EW [60, 68, 72, 80–82], and the 80% reduction target by 2050 [83] was thought to be perfectly achievable in the scenario, as per Tata Steel Europe's vision map 2050 [81, 82, 84]. As shown in Table 4.1, it was found that there are a sufficient number of viable CAT, CCS and CCUS technologies, methods and strategies at TRL 6–9 available and already installed, which are suitable for immediate BiSC implementation and achieving short- to medium-term significant reduction of CO_2 emissions in steelmaking. The urgency for sufficient prioritisation throughout all industries and political willingness (subsidies) cannot be emphasised enough [83] – the need to create a viable commercial environment, due to the required high capital investment [45–47, 52, 85–90] and a significant dependency on electricity prices.

To implement renewable energy technologies, grants, incentives and climate finance are required to support industry leaders to implement change. In the United Kingdom, some climate finance in the form of incentives, tax credits, grants and loans is being provided and provides the foundation to build the industrial decarbonisation efforts upon. The solutions are not all-encompassing and need a substantial increase to have any real impact, but it is a positive step in the right direction and constitutes one that should be copied worldwide.

The move for renewables installation needs to be made more attractive by incentivising through monetary means and making the planning application process lean and easy. The cost for installation of energy-generating components varies from c. £189.000 for one 70 KW vertical axis wind turbine to almost £2.5 million for one 3.5 MW horizontal axis turbine. The cost for solar PV panel installations depends on the number and installation point of the panels, and the costs are calculated per project. Hydropower stations are equally very site dependent and can only be estimated on a case-by-case basis. With the UK climate finance available, most of the above-named installations are rendered cost neutral or very affordable for the new owner, as these even partially provide an additional source of income, such as solar PV panels. Additionally, creating woodlands for direct air capture (DAC) is a profitable business in itself, due to the grants provided and the earning potential in addition to the environmental benefits.

During steelmaking, the bulk of emissions are being produced during the coking process, energy generation from fossil fuel and the BF/BOF process route of steelmaking. However, there are various opportunities to decarbonise this process. The standard steel production (SSP) process in combination with the currently operational newly developed technologies can achieve a reduction of more than 50% [91, 92] with successive implementation and of almost 100% CO_2 emission reduction by full implementation of all steps outlined in the BiSC model.

The post-combustion capture of CO_2 (CCS) and other GHGs and the exploration of carbon scrubbing of flue gases have been explored. There are several possible technologies and processes to be considered for post-combustion carbon capture [12, 53, 54, 71, 93–95]:

- Mechanical capture
- Compression and dehydration
- Membrane installation
- Guiding off-gas through troughs of physical solvents/solid sorbents (such as Zeolite 13X) and chemical solvents
- As well as utilising metal–organic frameworks

Suitable literature was thoroughly investigated with regard to applied and innovative steelmaking procedures, CAT, CCS and CCUS processes and improved management systems such as I4.0 [96, 97]. The key components within the BiSC are based on a circular production process and are functioning in an interactive manner. The basis for this system is the BF/BOF route and involves the aforementioned CAT, CCS and CCUS and process improvements where possible. Table 4.1 provides an overview of the process improvement technologies which are in development at TRL 6–9 or are already in industrial application.

Incorporation of sustainable and renewable energy components into any production cycle is possible, in the short term, such as DAC to capture off-gas carbon and produce biomass for the production of biogas in AD, and carbon enrichment for plant stimulation (CEPS) (promoting growth in greenhouses for food production) and AD to produce biogas and hydrogen, and capturing and utilising excess heat can individually make significant contributions to a higher degree of energy independence for the individual commercial entity. Table 5.3 provides more clarity to that effect by displaying the relevant percentage changes.

Using MS Excel has been a gift for this research, due to its versatility. As an example, the MS Excel extrapolations established how the different components of the '7 steps to net-zero carbon emissions steelmaking' strategy can be implemented and installed. Installing renewable energy technologies can not only help to reduce CO_2 emissions but also supply the production site with renewable energy, where 30% of energy does not have to be imported from third parties and paid for, bringing the reduction in CO_2 emissions down to – 49%. At this point in production, where biomass has already replaced the use of coke, and renewable energy technologies (solar, solar PV, wind and hydro) have been installed, an additional 30% of energy can be saved and thus does not have to be bought in, using biogas, bringing the reduction in CO_2 emissions to −65.7% [98]. This will have been produced in the link-connected anaerobic digester, which produces biogas from connected agribusinesses.

The negative percentage reduction in emissions means, in reverse, that at these points in the (steel) production process, a greater degree of energy independence can be achieved at 30%, 49% and 65.7% [17, 98]. This implies that energy at the same percentage point levels is not required to be imported from external sources, as it is produced either on site or link connected.

Besides energy and heat saving, generating their own energy will inevitably lead to achieving a higher level of independence, at least by 30%, and ideally, at 65.7% and more, there are savings to be had by not being forced to import and pay for energy from conventional suppliers, energy derived from renewable sources or not. We have been made painfully aware that private and business users of energy and fuel are at the mercy of corporate stakeholder interests and thus vulnerable to high price velocity. Retrofitting the existing industrial building infrastructure with renewable energy technology, with components of the '7 step to net-zero carbon emissions steelmaking' strategy, can support achieving: (i) greater energy efficiency and independence, (ii) turbocharge the decarbonisation of energy production, (iii) decarbonise steel and industrial production and (iv) provide savings opportunities via excess heat recovery.

Observing economic principles, it can be assumed that, overall, there is a third of the cost involved, with two-thirds of savings over a 30-year investment period. Additional, significant positive milestones can be reached, such as investment into a workforce with 'green' skills, future-proofing the business against energy price hikes, besides the positive effects of greater energy independence, decarbonisation of production and rehabilitation of the natural world which has been disrupted beyond recognition by the Anthropocene. Figure 5.13 shows the current level of country policy guidance versus actual requirements [99].

As discussed in Chapter 11: An Idealised Timeline of Possibilities, the legal landscape and the country governmental guidance will have to change dramatically, in order to meet the targets, set by the Paris Agreement [83] and COP15 [100] and the dire warnings issued by the recent IPCC reports [41, 57, 58, 101, 102]. By improving their carbon capture and off-heat utilisation capabilities, and investing in renewable energy technology, businesses are as follows:

(a) Supporting achieving the decarbonisation of production
(b) Reaching a higher degree of energy independence
(c) Achieving a higher level of asset efficiency
(d) Training workforce in required 'green' skills
(e) Reducing their energy costs
(f) Creating a viable additional income stream for their present and future operations

Not only will a higher degree of energy independence achieve an economic advantage in monetary terms, but any installation of renewable energy systems will also support limiting the global temperature rise to below 2 °C and thus support avoiding climate disaster. The CO_2 emissions by industry sector were analysed [54, 64, 97, 103], and besides the iron and steel industry, there are other industries which are CO_2 emitters, which are also generating copious amounts of off-heat, co-products and therewith resources, which could be harvested and used to power energy-dependent devices: transport, energy supply, business (including Heavy industry), residential sector, agriculture and waste management. Achieving a higher degree of energy independence is within reach of all sectors of society, made possible by technological progress and incentives and grants provided by the respective governments and countries, besides the United Kingdom. The opportunities are manifold; they merely require political willingness and implementation across all industries.

The research preceding the compilation of the current paper has provided answers which are reaching over into multiple other heavy industries, and additionally resulting in a further opportunity for a range of research directions, which are listed as follows:

(1) The CO_2 captured in the iron and steel industry can be used in the building industry (concrete – Geomimetic® process), agriculture (CEPS), food and drinks industry (CO_2 for beverages) and pharmaceutical applications.
(2) Waste products from steelmaking, i.e. sludge and slag, can be used in infrastructural projects, such as road and bridge-building.
(3) Other GHGs captured from post-combustion processes can be used in the chemical industry and be reused or recycled, eliminating the need for waste management.

The upgrading and retrofitting of existing and new steel plants and other industrial production sites provide an immense opportunity for further research, as there is a vast range of implementation points at active and abandoned production plants, not limited to:

(1) Adding solar foil, panels, tiles and shingles to buildings and car parks.
(2) Adding a hydrogen transfer network for HDR.
(3) Adding wind turbines on suitable onshore brownfield sites.

The carbon avoidance impact of the BiSC key elements in percentage changes of the individual components, with details on how the implementation will either avoid or reduce carbon emissions, has been displayed in Figure 5.34 and Table 6.5.

The steelmaking process has been thoroughly investigated [104], starting with the BF/BOF route – as it dominates the production of steel worldwide. C. 70% is produced by integrated BF/BOF route plants and only 30% by the EAF route (71.5 and 28.0%, respectively, in 2017 [104]). With a 'BF', a vertical shaft furnace is usually referred to, which produces liquid metals by the reaction of hot air being added under pressure into the bottom of the furnace to a mixture of metallic ore, coke and flux fed into the top section. Continued immediate combustion is maintained by the continuous addition of air under pressure. Furthermore, the BF relies on coke and injected coal as a source of energy and reducing agent. One tonne of coke production requires 1.2–1.6 tonnes of coal [61, 105–107]. Preheating coal before coke making reduces coal consumption and increases coke oven productivity. The production of coke oven gas (COG) is about 300–360 m^3/t coke. COG has a low calorific value (6–8 GJ/t coke) and contains heavy hydrocarbons like tar which is subsequently removed in such a way as to minimise downstream disturbances [61, 105–107]. The BF is a highly efficient thermal reactor. It operates with two countercurrent streams, namely descending solids and ascending gases, and behaves as a heat exchanger. The BF produces its own reducing gas, CO. Reduction reactions of iron oxide by CO produce

$$CO_2 \cdot C_{(s)} + CO_{2(g)} = 2CO_{(g)}Fe_3O_{4(s)} + 4CO_{(g)} = 3Fe_{(s)} + 4CO_{2(g)0-/*870}$$

The BF generates about 70% of the total amount of CO_2 in the integrated mill, and most of the CO will be burned downstream into CO_2. The BF top gas contains ~25% of CO_2. A medium-sized BF with a production rate of 4500 tonnes/day produces about 1600 Nm^3/t of off-gas. To reduce the consumption of chemical energy in the BF, one should decrease the amount of fuel. There are several approaches to decreasing the consumption of coke, but there are two sustainable options: (i) recycling the off-gas and (ii) using biomass to replace coal injection. The consumption of coke in a modern BF is of the order of 300–350 kg/t of hot-metal, and including pulverised coal injection (PCI), the total is about 500 kg/t of hot-metal [61, 105–107].

The CO_2 data accumulated as a result of the literature review, mathematical analysis of aforementioned, MS Excel data compilation, evaluation, modelling and analysis, statistical validation and the process software simulations in Aspen, Simul8 and INOSIM have provided congruent data and results for verification and validation purposes. The BiSC components were applied, and the total CO_2 emissions reduction results were exported from the three process simulation software packages Simul8, Aspen V12.1 and INOSIM Expert Edition and sent to MS Excel. The values were then modelled and expressed in a graph, as displayed in Figure 5.34. It is safe to say that the different simulations provided proof of concept and rendered the one hypothesis as valid: it is possible to reduce CO_2 emissions in steel production to almost '0' when the BiSC model and strategy is being implemented, in sections or entirely.

Global energy market price volatility and an upward trajectory of prices per unit of electricity have sent all industrial sectors and many economies to the brink of recession. Alongside the urgent need for decarbonisation of all industries, achieving a globally higher level of energy independence across all sectors seems imperative. A multidisciplinary approach with a proposed system of CO_2 emissions reduction and capture technologies has the potential for short-term emissions reduction to near zero in the steel industry – although some of the mechanisms can be implemented across most heavy industries. The findings of this research show a CO_2 emissions reduction of ~30% from 977 t of CO_2 to 684 t in one single BF production cycle (based on 330 tonnes of liquid iron production capacity, with the mean of 2.1–3.2 tonnes CO_2/t of steel and chemical reactions emissions applied), by switching the electricity provider for operating the electric heaters to providers generating energy exclusively from renewable sources. Replacing coal with biomass and adding post-combustion capture units to the BF operation will add carbon neutrality into the process – resulting in CO_2 emissions reduction to near zero. Carbon capture from biomass utilisation (BECCS) will add the benefit of carbon-negative emissions to the cycle. Simultaneously, energy-saving and process improvement measures implementation (up to 60% efficiency increase), excess heat recovery <30% of energy savings and retrofitting renewable energy technology resulted in an energy independence of 88%. Engineering solutions, partly subsidised in the United Kingdom, are readily available for implementation in the iron and steel manufacturing industry. The volatility of the global energy market and recent price hikes by energy producers have caused never-before-seen levels of profit for energy companies and untold pressures for businesses and the population in the most developed economies. Numerous countries are on the brink of recession across geographical Europe at the time of writing, and energy price increases have made a strong case for the urgent need to achieve greater energy independence – the demonstration of the feasibility being the purpose of this paper. This could be considered one of the foremost important contemporary endeavours. The energy sector and the iron and steel industry, along with heavy industry and petroleum refineries [1], are by far the largest emitters of CO_2 emissions, due to their high fuel and energy demand. The steel industry accounts for between 7% and 11% [1–3, 10, 108–110] of global CO_2 emissions as a result of steelmaking, and China is responsible for 50% of these GHGs [10], due to their heavy reliance on coal. The increased use of coal in energy generation, due to imposed oil and gas shortages, was found to be the main factor [1, 4] driving up global energy-related anthropogenic CO_2 emissions by over two billion tonnes, their largest ever rise in absolute terms. This research focused on the technical solutions currently available for the decarbonisation of the steel industry and how decarbonisation could lead to a higher degree of energy independence, based on the currently known energy consumption trends. The consumption trend evaluation or influence of renewables implementation on consumption trends on a global scale was not the key focus of this project, as these are two completely independent, separate issues altogether, but pose an interesting challenge for future research.

Achieving a higher degree of energy independence is within reach of all sectors of society, made possible by technological progress and incentives and grants provided by the respective governments and countries, including the United Kingdom. The opportunities are manifold; they merely require political willingness and implementation across all industries. The process simulations compiled within the systems S8, Aspen V.12.1 and INOSIM aligned with the mathematical analysis and modelling in MS Excel and provided almost matching reports, stating a reduction of 30% from 977 t of CO_2 emissions to 684 t of CO_2 flow in one single BF production cycle (based on 330 tonnes liquid iron production capacity and a mean of 2.1–3.2 tonnes CO_2/t of steel and emissions from chemical reactions considered), by switching the electricity provider for operating the electric heaters to providers generating energy exclusively from renewable sources. Replacing coal with biomass added carbon neutrality into the process. Adding the post-combustion capture unit to the BF resulted in a CO_2 emissions reduction to near zero. BECCS added the benefit of carbon-negative emissions to the cycle. Simultaneously, energy-saving and process improvement measures implementation (up to 60% efficiency increase), excess heat recovery <30% of energy savings and retrofitting renewable energy technology resulted in an energy independence of 88%. Engineering solutions that are partly subsidised in the United Kingdom – as described in Table 4.1 – are readily available for implementation in the iron and steel manufacturing industry.

Furthermore, the application points of renewable energy technology within the steel production process were established where achieving a 30%, 49% and up to 88% higher energy independence is achievable, in the short term. The cost risk of installing renewable energy and decarbonisation technology was minimised to less than a third of the initial outlay in individual cases, as the UK government is providing grants, funding and loan schemes to encourage all industrial sectors to work towards net zero. The research preceding the compilation of the current paper has provided answers that are reaching over into multiple other heavy industries and additionally resulting in further opportunity for a range of research directions, which are listed as follows:

- Utilisation of CO_2 in the building industry (CCUS), agriculture (CEPS), chemical industry, food and drinks industry and pharmaceuticals
- Utilisation of waste products from steelmaking in infrastructural projects
- Other GHGs (CO and $H_{2)}$ captured in TGRB can also be reused in the chemical industry, thus eliminating the need for waste management

The upgrading and retrofitting of existing and new steel plants and other industrial sites provide an immense opportunity for further research, as there is a vast range of active and abandoned production plants available in the United Kingdom, not limited to:

- Adding solar foil, panels, tiles and shingles to buildings and car parks
- Adding a biogas, replacing fossil fuel and a hydrogen network from AD for hydrogen direct reduction of iron ore (DRI) in steel production, in addition to reformers and electrolysers
- Installing wind turbines at brownfield sites not suitable for human habitation
- Developing filters and capture units to be retrofitted to existing production plants
- Emitting GHGs

There are undoubtedly technological challenges to and opportunities for the BiSC concept implementation to consider, and the technological challenges to the integration of the individual improved processes, as well as barriers and hurdles, were discussed, with details on retrofitting existing steel plants and how to conceptualise newly conceived projects. The particular focus was on renewable energy technologies, biomass and specifically: green hydrogen and HDR.

Top gas recycling (TGR) in BF or BOF is a promising technology to significantly reduce the CO_2 emissions from BF/BOF operations [111]. It includes the injection of carbon content reducing top gas components CO_2 and H_2 into the shaft via tuyeres, lowering coke rates and using 99% of pure O_2 instead of hot blast air through the tuyeres. Additionally, the capture of CO_2 at 400 ppm [93] and CEPS from the off-gas stream is a useful technology for both the BF and BOF operation. The traditional BF is, for these purposes, retrofitted with a TGR unit, where the CO_2 is separated from the post-combustion off-gas and recycled – ideally in a CEPS component or as base material to produce recycled aggregate concrete (RAC). Furthermore, some remaining CO_2-stripped gas is diverted back into the BF, after being reheated using off-heat from the BF operation – as displayed in Figure 7.1. Reusing the CO_2-stripped off-gas in the BF operation reduces direct CO_2 emissions from the BF by 26%, and an additional 52% can be avoided (CAT) by combining the TGR with CCUS. The CO_2 reduction potential of the TGRBF is therefore set at 78% [112]. The TGRBF operation works in several stages, beginning with the capture of the post-combustion off-gas. The blast furnace gas (BFG) is cleaned and fed into a mixing unit, where water is added to the mixture. In the CEPS/CCUS unit, the BFG is stripped of its CO_2 contents and fed into an off-heat utilising heater element, from where the BFG (stripped of the CO_2 contents) is being fed back into the BF. The CO_2 capture and utilisation costs were estimated to be between £28 and £65/t CO_2 [1], depending on the end-use being either reusage at the plant or further recycling in RAC or agri-food production. The ULCOS TGRBF with CCS is included in the ultra-low CO_2 steel (ULCOS [112–115]) EU programme, and in an early phase of the ULCOS experimental BF in Luleå, the heat and mass balance as well as the tuyeres parameters were mathematically modelled; process simulations were compiled and resulting modifications influenced the subsequent two test campaigns. Interestingly, the campaign results obtained were similar to the simulated and calculated test configurations [47, 116–120]. Commercial deployment is ongoing [121]. As the cost has reduced significantly since the conception of the technology, the financial barriers to the implementation of TGR have been largely removed. The hurdles to overcome are now the willingness of industry to accept their responsibilities for the environmental devastation caused and to decarbonise all heavy industries to remedy the inflicted damages to the natural environment and all creatures within, including us humans.

The technological challenges of the BiSC component implementation seem to be manifold, but the research required to compile this book has shown the following positive outcomes:

- The technological challenges are mostly overcome through considerable efforts made in research and development
- Many technologies are now at TRL9 or already in industrial application
- In most industries, there seems to be a willingness to implement BiSC elements and any measures which aim to reduce GHG emissions, to achieve a net-zero socio-economic situation
- In many countries, a range of incentives, grants and subsidies is making it easier for entities willing to 'go green' and the ones who promote renewable energy technologies and carbon capturing or reducing GHG emissions are making it easier for all industries to overcome the individual technological challenges

It is therefore safe to say that there are no technological challenges which industry leaders cannot overcome, and with the right support, implementation of the BiSC and carbon-free iron and steel production sounds not so far-fetched as it did maybe 30 years ago.

The significance of implementation hurdles cannot be underestimated, such as the political will of the major steel-producing countries to decarbonise steel production and other industrial sectors. This work has a clear UK focus with attention directed towards the issues which require addressing to achieve successful implementation of the innovative BiSC concept. The political and legal framework, adherence to the Paris Agreement [83] sustainability goals and amortisation budgets as well as other industry-specific financial implications were being considered.

It was observed that politics in any country in recent years seem only to serve the purpose of either (i) keeping those currently in government in place or (ii) creating a legal and political situation where those on the far right of the political spectrum lobby and influence those in power to achieve their own objectives. Democracy is at risk in many developed countries [122, 123]. Therefore, the need for sustainability and GHG emissions reduction is insofar only recognised and acted upon as far as it is serving the political parties' agendas. However, the outlook is not generally bleak; the general populace and free industry have recognised the urgent need for immediate action, and there are a lot of efforts on the way to globally reduce GHG emissions [57, 58, 83, 101, 124–127] and ensure the world economy achieves remaining within the 1.5 °C temperature rise limit in comparison with pre-industrial temperatures. To that effect, sustainability certifications have been developed, which provide the framework for the industry to act within and responsibly respond to the challenges at hand.

This work certainly has a UK bias, but for data collection purposes, a global approach was taken. The EU, Australia, Norway, the United States, Canada and BRIC economies as well as other main steel-producing countries have been researched for information on their GHG emissions, decarbonisation efforts and plans to work towards keeping the global temperature rise at below 1.5 °C of pre-industrial levels, according to the Paris Agreement [83]. Besides Russia, most countries seem to provide a positive environment for the implementation of green steel production and are providing openings in international markets for the promotion of green steel, instead of grey steel (produced using fossil fuels). The BiSC implementation can significantly increase the share of renewables in energy generation in the drive to phase out fossil fuels. Immense opportunities were also identified for the renewable energy technology sector in itself, as retraining current workers, upskilling the existing staff and building training structures for school leavers to enter positions in this innovative sector could not only prevent job losses but also provide young people with the opportunities that are desperately needed after the collapse of many traditional industries, particularly in the United Kingdom.

The effective implementation of the BiSC model and connected technologies requires in parts specialised workers and staff to achieve the best efficiency. There are immense opportunities for retraining workers from fossil-fuel-based sectors, such as mining, energy generation and oil refinery to being environmentally benign sector employees.

Renewable energy technologies and biomass applications including green hydrogen production are the key sectors which require skilled workers in the future. In order to ascertain the different opportunities for retraining workers, the different steps and components required to build and install the different renewable energy systems need to be explored.

An action plan, specific for the steel industry, needs to be created by the respective governments to drive change and support in achieving the net-zero targets and decarbonising industrial processes at the latest by 2030. Specific for the steel industry, there are easily implementable steps, as explained in Chapter 4: The Key Components of the Bio Steel Cycle (BiSC). Steel-industry-specific implementable carbon avoidance technologies should include switching to 100% green energy providers, using green HDR instead of sintering, replacing BFs with EAFs, increasing the EAF process utilisation percentage of steel scrap, and compulsory off-heat harvesting and upgrading existing infrastructure to avoid energy loss in operation. As already discussed in detail in Chapter 6, the existing infrastructure needs to be upgraded to meet future challenges. Particularly for the steel industry, this entails the creation of a legally binding industry-wide framework of compulsory actions to be implemented with immediate effect. These measures include upgrading the existing infrastructure to eliminate energy loss during BF/BOF and EAF operation, integrating site-suitable renewable energy technologies where technically feasible, employing best practice innovative steel production processes to reduce steelmaking-related GHG emissions: replace BF with EAF, and improving the steelmaking process in the existing EAF plants to increase the % share utilisation of scrap steel to 100%, as some are currently operating only on utilising 42% steel scrap of the total charge capacity. Furthermore, upgrading the existing network of pipes is certainly necessary, so the infrastructure is able to cope with the demands of hydrogen utilisation. Generally, hydrogen should be used as a replacement for coal and coke where feasible, i.e. in hydrogen DRI instead of sintering. Replacing coal and coke with biomass and biochar for ironmaking is an immense opportunity, as it not only reduces carbon emissions by at least 30% but also provides incentives for hill farmers to produce energy crops instead of farming loss-making herds of sheep. Creating a circular economic process of steelmaking based on the suggestions of the BiSC [1, 2] is a technically relatively easy process to implement, and it is hoped that with the widening of climate finance offerings in the respective countries, the model and strategy will provide some real impetus to bring about change and – hopefully – cause the decarbonisation of the steel industry.

The actions required as described and displayed in the idealised timeline are all perfectly simple as far as the implementation is concerned. All they require is the political willingness to prepare the industry for the required changes, support industry leaders in their endeavours to achieve climate targets and remember to take the workers along on this journey, because without them, all these perfectly feasible ideas will not come to pass.

We need skilled workers today more than ever before, and it is high time for all industries to recognise this and train, retrain or upskill their workforce and provide incentives for school leavers to be active in the emerging industries and sectors. Equally crucial is sustained investment in fundamental research and development, particularly at the lab and pilot scale and demonstration to validate the Bio Steel Cycle technologies and bridge the gap between scientific innovation and full-scale industrial deployment.

References

1 Acharya, M. (2018). Iron facts, symbol, discovery, properties, uses [Internet]. Chemistry Learner. https://www.chemistrylearner.com/iron.html (accessed 19 July 2023).

2 Senkov, O.N., Miracle, D.B., Chaput, K.J., and Couzinie, J.P. (2018). Development and exploration of refractory high entropy alloys—a review. *Journal of Materials Research* 33 (19): 3092–3128.

3 Güder, Ü., Çeken, M., Yavaş, A. et al. (2022). First evidence of crucible steel production in Medieval Anatolia, Kubadabad: a trace for possible technology exchange between Anatolia and Southern Asia. *Journal of Archaeological Science* 137: 105529.

4 Royal Society of Chemistry (RSC) (2023). Sulfur – element information, properties and uses | Periodic Table [Internet]. https://www.rsc.org/periodic-table/element/16/sulfur (accessed 20 July 2023).

5 Warren, K. (1964). The Sheffield Rail Trade, 1861–1930: an episode in the locational history of the British Steel industry. *Transactions and Papers (Institute of British Geographers)* 34: 131.

6 Royal Society of Chemistry (RSC) (2023). Tungsten – element information, properties and uses | Periodic Table [Internet]. https://www.rsc.org/periodic-table/element/74/tungsten (accessed 20 July 2023).

7 Guthrie, R.I.L. and Isac, M.M. (2022). Continuous casting practices for steel: past, present and future. *Metals* 12 (5): 862.

8 BHP (2019). BHP annual report 2019 [Internet]. https://www.bhp.com/-/media/documents/investors/annual-reports/2019/bhpannualreport2019.pdf (accessed 3 March 2020).

9 Bogunovic, D., Kecojevic, V., Lund, V. et al. (2009). Analysis of energy consumption in surface coal mining. *Transactions of the Society for Mining, Metallurgy, and Exploration, Inc.* 326: 79–87.

10 Royal Society of Chemistry (RSC) (2023). Manganese – element information, properties and uses | Periodic Table [Internet]. https://www.rsc.org/periodic-table/element/25/manganese (accessed 20 July 2023).

11 The Rodney and Otamatea Times (1912). Coal consumption affecting climate. Rodney and Otamatea Times, Waitemata and Kaipara Gazette. 7. https://paperspast.natlib.govt.nz/newspapers/ROTWKG19120814.2.56.5 (accessed 7 December 2023).

12 Griffin, P. and Hammond, G. (2019). Industrial energy use and carbon emissions reduction in the iron and steel sector: a UK perspective. *Journal of Applied Energy* 249: 109–125.

13 Hasanbeigi, A. (2019). How clean is the U.S. steel industry? An international benchmarking of energy and CO_2 intensities - An International Benchmarking of Energy and CO2 Intensities. *Global Eciency Intelligence* 1–42. https://canadiansteel.ca/files/resources/HowCle.

14 Baena-Moreno, F., Cid-Castello, N., Arellano-García, H., and Reina, T. (2021). Towards emissions free steel manufacturing – exploring the advantages of a

CO_2 methanation unit to minimize CO_2 emissions. *Science of the Total Environment* 781 (7): 146776.

15 Ren, L., Zhou, S., Peng, T., and Ou, X. (2021). A review of CO_2 emissions reduction technologies and low-carbon development in the iron and steel industry focusing on China. *Renewable & Sustainable Energy Reviews* 143: 110846.

16 Wang, R. (2022). Energy saving technologies and optimisation of energy use for decarbonised iron and steel industry [Internet] [Doctoral]. Durham University. http://etheses.dur.ac.uk/14289/ (accessed 7 July 2023).

17 Kiessling, S., Darabkhani, H.G., and Soliman, A.H. (2022). Energies | Free Full-Text | The Bio Steel Cycle: 7 steps to net-zero CO_2 emissions steel production [Internet]. https://www.mdpi.com/1996-1073/15/23/8880 (accessed 11 July 2023).

18 Liu, Z., Ciais, P., Deng, Z. et al. (2020). Near-real-time monitoring of global CO_2 emissions reveals the effects of the COVID-19 pandemic. *Nature Communications* 11 (1): 5172.

19 McKinsey & Company (2021). Metals & mining practice: the future of the European steel industry. A road map toward economic and environmental sustainability 2021 [Internet]. https://www.mckinsey.com/~/media/mckinsey/industries/metals%20and%20mining/our%20insights/the%20future%20of%20the%20european%20steel%20industry/the-future-of-the-european-steel-industry_vf.pdf (accessed 7 July 2023).

20 Griffin, P. and Hammond, G. (2021). The prospects for 'green steel' making in a net-zero economy: a UK perspective. *Global Transitions* 3: 72–86.

21 Hegemann, K.R. and Guder, R. (2020). Linde-Fränkl-Verfahren der Sauerstoff-Gewinnung. In: *Stahlerzeugung: Integrierte Hüttenwerks- und Gasreinigungsanlagen [Internet]* (ed. K.R. Hegemann and R. Guder), 47–49. Wiesbaden: Springer Fachmedien https://doi.org/10.1007/978-3-658-29091-7_5 (accessed 8 July 2023).

22 Variny, M., Jediná, D., Rimár, M. et al. (2021). Cutting oxygen production-related greenhouse gas emissions by improved compression heat management in a cryogenic air separation unit. *IJERPH* 18 (19): 10370.

23 Intergovernmental Panel for Climate Change (IPCC) (1996). *Revised 1996 IPCC Guidelines for National Greenhouse Gas Inventories: Workbook [Internet].* Geneva: IPCC https://www.ipcc-nggip.iges.or.jp/public/gl/guidelin/ch2wb1.pdf (accessed 7 July 2023).

24 Lin, S., Kiga, T., Wang, Y., and Nakayama, K. (2011). Energy analysis of $CaCO_3$ calcination with CO_2 capture. *Energy Procedia* 4: 356–361.

25 Kittipongvises, S. (2017). Assessment of environmental impacts of limestone quarrying operations in Thailand. *Environmental and Climate Technologies* 20 (1): 67–83.

26 Shan, Y., Liu, Z., and Guan, D. (2016). CO_2 emissions from China's lime industry. *Applied Energy* 166: 245–252.

27 Riesbeck, J., Hooey, L., Kinnunen, K., et al. (2013). Global effects of closing down sinter plant at Ruukki Raahe integrated steel works [Internet].

http://ltu.diva-portal.org/smash/record.jsf?pid=diva2%3A1004013&dswid=-4624 (accessed 7 July 2023).

28 Fraunhofer Institute (2009). Methodology for the free allocation of emission allowances in the EU ETS post 2012 Sector report for the iron ore industry By order of the European Commission Study Contract: 07.0307/2008/515770/ETU/C2 Ecofys project Number: PECSNL082164 [Internet]. p. 1–15. https://climate.ec.europa.eu/system/files/2016-11/bm_study-iron_ore_en.pdf (accessed 7 June 2023).

29 Office of ENERGY EFFICIENCY & RENEWABLE ENERGY (EERE) (US) Energy.gov (2018). Manufacturing energy and carbon footprints (2018 MECS). https://www.energy.gov/eere/iedo/manufacturing-energy-and-carbon-footprints-2018-mecs (accessed 6 July 2023).

30 Mandova, H., Leduc, S., Wang, C. et al. (2018). Possibilities for CO_2 emission reduction using biomass in European integrated steel plants. *Biomass and Bioenergy* 115: 231–243.

31 Pandit, J.K., Qader, A., and Lim, S. (2021). *Cross-Technology Scheme Options to Reduce Greenhouse Gas Emissions in a Steel Industry [Internet].* Rochester, NY: https://papers.ssrn.com/abstract=3821438 (accessed 7 July 2023).

32 Prakash, R. and Muller, M. (2007). Industrial oxygen: its use and generation [Internet]. https://www.eceee.org/library/conference_proceedings/ACEEE_industry/2007/Panel_6/P6_13/ (accessed 7 July 2023).

33 US Department of Energy Efficiency & Renewable Energy (EERE). (2021). Hydrogen production: natural gas reforming [Internet]. https://www.energy.gov/eere/fuelcells/hydrogen-production-natural-gas-reforming (accessed 6 July 2023).

34 Wang, G., Li, R., Dan, J. et al. (2023). Preparation of biomass hydrochar and application analysis of blast furnace injection. *Energies* 16 (3): 1216.

35 Wang, C., Walsh, S., Haynes, M.W. et al. (2022). From Australian iron ore to green steel: the opportunity for technology-driven decarbonisation [Internet]. https://ecat.ga.gov.au/geonetwork/srv/api/records/5377741d-ece0-4e01-b87e-6bf49da11c03 (accessed 7 July 2023).

36 Garvey, A., Norman, J.B., and Barrett, J. (2022). Technology and material efficiency scenarios for net zero emissions in the UK steel sector. *Journal of Cleaner Production* 333: 130216.

37 Wänerholm, M. (2016). Rapport nr 2016–008 – climate impact of metal-casting [Internet]. https://www.diva-portal.org/smash/get/diva2:1140576/FULLTEXT01.pdf (accessed 7 August 2023).

38 Schmitz, N., Sankowski, L., Kaiser, F. et al. (2021). Towards CO_2-neutral process heat generation for continuous reheating furnaces in steel hot rolling mills – a case study. *Energy* 224 (15): 120155.

39 Sun, Y., Tian, S., Ciais, P. et al. (2022). Decarbonising the iron and steel sector for a 2 °C target using inherent waste streams. *Nature Communications* 13 (1): 297.

40 Svenskt Stål AB (SSAB) (2022). Sustainability at SSAB. https://www.ssab.com/en/company/sustainability/first-in-fossil-free-steel/environmental-benefits (accessed 7 July 2023).

41 Intergovernmental Panel for Climate Change (IPCC) (2023). Summary for policymakers. In: *Climate Change 2021 – The Physical Science Basis: Working Group I Contribution to the Sixth Assessment Report of the Intergovernmental Panel on Climate Change [Internet]*, 3–32. Cambridge: Cambridge University Press https://www.cambridge.org/core/books/climate-change-2021-the-physical-science-basis/summary-for-policymakers/8E7A4E3AE6C364220F3B76A189CC4D4C (accessed 12 July 2023).

42 International Energy Agency (IEA) (2023). Executive summary – CO_2 emissions in 2023 – analysis. https://www.iea.org/reports/co2-emissions-in-2023/executive-summary (accessed 3 April 2024).

43 World Steel Association (WSA) (2022). World steel in figures 2022 a healthy economy needs a healthy steel industry [Internet]. https://worldsteel.org/wp-content/uploads/World-Steel-in-Figures-2022-infographic.pdf?x65430 (accessed 7 July 2023).

44 International Energy Agency (IEA) (2023a). coal_documentation.pdf [Internet]. http://wds.iea.org/wds/pdf/coal_documentation.pdf (accessed 12 July 2023).

45 Worrell, E., Blinde, P., Neelis, M. et al. (2010). Energy efficiency improvement and cost saving opportunities for the U.S. Iron and Steel Industry An ENERGY STAR(R) Guide for Energy and Plant Managers [Internet]. p. LBNL-4779E, 1026806. Report No.: LBNL-4779E, 1026806. http://www.osti.gov/servlets/purl/1026806/ (accessed 2 November 2022).

46 Li, Y. and Zhu, L. (2014). Cost of energy saving and CO_2 emissions reduction in China's iron and steel sector. *Applied Energy* 130: 603–616.

47 Toktarova, A., Karlsson, I., Rootzén, J. et al. (2020). Pathways for low-carbon transition of the steel industry—a Swedish case study. *Energies* 13 (15): 3840.

48 Liu, X., Peng, R., Bai, C. et al. (2022). Technological roadmap towards optimal decarbonization development of China's iron and steel industry. *The Science of the Total Environment* 850: 157701.

49 Suer, J., Ahrenhold, F., and Traverso, M. (2022). Carbon footprint and energy transformation analysis of steel produced via a direct reduction plant with an integrated electric melting unit. *Journal of Sustainable Metallurgy* 8 (4): 1532–1545.

50 Tian, B., Zhu, R., Zhou, Y. et al. (2022). A static balance model and analysis of 430 stainless steel produced by basic oxygen furnace–argon oxygen decarburization furnace process. *Wiley Online Library. Steel Research International [Internet]* 93 (7): https://onlinelibrary.wiley.com/doi/abs/10.1002/srin.202100852 (accessed 7 July 2023).

51 International Renewable Energy Agency (IRENA) (2021). World energy transitions outlook: 1.5 °C pathway [Internet]. https://www.irena.org/publications/2021/Jun/World-Energy-Transitions-Outlook (accessed 11 August 2022).

52 International Renewable Energy Agency (IRENA) (2012). Renewable energy technologies: cost analysis series [Internet]. https://www.irena.org/-/media/

Files/IRENA/Agency/Publication/2012/RE_Technologies_Cost_Analysis-HYDROPOWER.pdf (accessed 7 July 2023).
53 Bataille, C., Åhman, M., Neuhoff, K. et al. (2018). A review of technology and policy deep decarbonization pathway options for making energy-intensive industry production consistent with the Paris Agreement. *Journal of Cleaner Production* 187: 960–973.
54 Mandova, H., Patrizio, P., Leduc, S. et al. (2019). Achieving carbon-neutral iron and steelmaking in Europe through the deployment of bioenergy with carbon capture and storage. *Journal of Cleaner Production* 218: 118–129.
55 National Renewable Energy Laboratory (NREL) (2019). 2019 year in review – NREL's top 20 stories [Internet]. https://www.nrel.gov/news/program/2019/2019-year-in-review-nrel-top-20-stories.html (accessed 11 July 2023).
56 Waisman, H., Bataille, C., Winkler, H. et al. (2019). A pathway design framework for national low greenhouse gas emission development strategies. *Nature Climate Change* 9: 261.
57 Intergovernmental Panel for Climate Change (IPCC) (2022). Climate change 2022: impacts, adaptation and vulnerability. [Internet]. https://www.ipcc.ch/report/ar6/wg2/ (accessed 3 January 2023).
58 Intergovernmental Panel for Climate Change (IPCC) (2022a). Climate change 2022: mitigation of climate change [Internet]. https://www.un.org/en/climatechange/reports?gclid=EAIaIQobChMIxsyF3ZSl-AIVoo1oCR19Dw3IEAAYAiAAEgItyfD_BwE (accessed 30 April 2022).
59 Office for National Statistics (ONS) (2023). Foreign direct investment, experimental UK subnational estimates: 2021 – Office for National Statistics [Internet]. https://www.ons.gov.uk/releases/foreigndirectinvestmentexperimentaluksubnationalestimatesmarch2023 (accessed 14 July 2023).
60 Vogl, V., Åhman, M., and Nilsson, L.J. (2018). Assessment of hydrogen direct reduction for fossil-free steelmaking. *Journal of Cleaner Production* 203: 736–745.
61 Benetto, E., Popovici, E.C., Rousseaux, P., and Blondin, J. (2004). Life cycle assessment of fossil CO_2 emissions reduction scenarios in coal-biomass based electricity production. *Energy Conversion and Management* 45 (18): 3053–3074.
62 Zakkour, P. and Cook, G. (2010). *UNIDO CCS roadmap for industry: high-purity CO2 sources. Sectoral assessment and Final Draft Report*. Vienna, Austria: United Nations Industrial Development Organization.
63 Morrow, W.R., Hasanbeigi, A., Sathaye, J., and Xu, T. (2014). Assessment of energy efficiency improvement and CO_2 emission reduction potentials in India's cement and iron & steel industries. *Journal of Cleaner Production* 65: 131–141.
64 Chen, Q., Gu, Y., Tang, Z. et al. (2018). Assessment of low-carbon iron and steel production with CO_2 recycling and utilization technologies: a case study in China. *Applied Energy* 220: 192–207.
65 Bhaskar, A., Abhishek, R., Assadi, M., and Somehesaraei, H.N. (2022). Decarbonizing primary steel production: techno-economic assessment of a hydrogen

based green steel production plant in Norway. *Journal of Cleaner Production* 350: 131339.

66 Ananthanarayanan, P.S., Balachandran, G., Balasubramanian, N. et al. (2014). Contributors to volume 3. In: *Treatise on Process Metallurgy [Internet]* (ed. S. Seetharaman), xxvii–xxx. Boston: Elsevier https://www.sciencedirect.com/science/article/pii/B9780080969886099879 (accessed 5 April 2023).

67 Yang, Y., Raipala, K., and Holappa, L. (2014). Chapter 1.1 – ironmaking. In: *Treatise on Process Metallurgy [Internet]* (ed. S. Seetharaman), 2–88. Boston: Elsevier https://www.sciencedirect.com/science/article/pii/B9780080969886000171 (accessed 7 July 2023).

68 Babich, A. and Senk, D. (2015). Recent developments in blast furnace iron-making technology. In: *Iron Ore [Internet]*, 505–547. Elsevier https://linkinghub.elsevier.com/retrieve/pii/B9781782421566000174 (accessed 2 November 2022).

69 Cumicheo, C., Mac Dowell, N., and Shah, N. (2019). Natural gas and BECCS: a comparative analysis of alternative configurations for negative emissions power generation. *International Journal of Greenhouse Gas Control* 90: 102798.

70 Tanzer, S.E., Blok, K., and Ramírez, A. (2021). Decarbonising Industry via BECCS: promising sectors, challenges, and techno-economic limits of negative emissions. *Current Sustainable/Renewable Energy Reports* 8 (4): 253–262.

71 Williams, R., Jack, C., Gamboa, D., and Shackley, S. (2021). Decarbonising steel production using CO_2 capture and storage (CCS): results of focus group discussions in a Welsh steel-making community. *International Journal of Greenhouse Gas Control* 104: 103218.

72 COM (European Commission) (2019). BIONICO: a pilot plant for turning biomass directly into hydrogen | BIONICO Project | Results in brief | H2020 | CORDIS | European Commission [Internet]. https://cordis.europa.eu/article/id/394984-bionico-a-pilot-plant-for-turning-biomass-directly-into-hydrogen (accessed 6 July 2023).

73 Colla, V. and Matino, I. (2021). Editorial for the special issue: overview, state of the art, recent developments and future trends regarding hydrogen route for a green steel making process. *Matériaux & Techniques* 109 (3–4): E301.

74 Garcia-Herrero, A., Tagliapietra, S., and Vorsatz, V. (2021). Hydrogen development strategies: a global perspective – document – Gale Academic OneFile [Internet]. https://go-gale-com.ezproxy.staffs.ac.uk/ps/i.do?id=GALE%7CA673932617&sid=googleScholar&v=2.1&it=r&linkaccess=abs&issn=&p=AONE&sw=w&userGroupName=anon%7E4246e29b&aty=open+web+entry (accessed 6 July 2023).

75 Grasa, G., Díaz, M., Fernandez, J.R., et al. (2022). Blast furnace gas decarbonisation through calcium assisted steel-mill off-gas hydrogen production. Experimental and modelling approach. *SSRN Journal [Internet]*. https://www.ssrn.com/abstract=4171691 (accessed 2 November 2022).

76 Öhman, A., Karakaya, E., and Urban, F. (2022). Enabling the transition to a fossil-free steel sector: the conditions for technology transfer for hydrogen-based steelmaking in Europe. *Energy Research and Social Science* 84: 102384.

77 COM (European Commission). CORDIS | European Commission (2022). Green industrial hydrogen via reversible high-temperature electrolysis | GrInHy Project | Fact Sheet | H2020. https://cordis.europa.eu/project/id/700300 (accessed 6 July 2023).

78 Galitskaya, E. and Zhdaneev, O. (2022). Development of electrolysis technologies for hydrogen production: a case study of green steel manufacturing in the Russian Federation. *Environmental Technology and Innovation* 27: 102517.

79 ArcelorMittal (2023). Siderwin: reducing iron ore via electrolysis [Internet]. https://corporate.arcelormittal.com/corporate-library/reporting-hub/siderwin-reducing-iron-ore-via-electrolysis (accessed 9 July 2023).

80 TATA Steel UK (2020). Tata steel in the UK fact sheet [Internet]. https://www.tatasteeleurope.com/sites/default/files/Tata%20Steel%20UK%20Factsheet%202020%20(1).pdf (accessed 7 July 2023).

81 Tata Steel Europe (2020). Tata steel in Europe sustainability report 2019/2020 [Internet]. https://www.tatasteeleurope.com/sites/default/files/TSE%20Sustainability%20report%202019-20%20(EN).pdf (accessed 7 July 2023).

82 Tata Steel Europe (2023). Sustainability home page. https://www.tatasteeleurope.com/sustainability (accessed 11 July 2023).

83 United Nations (UN) (2016). Paris Agreement 2016 [Internet]. https://www.un.org/en/climatechange/paris-agreement#:~:text=To%20tackle%20climate%20change%20and,2015%3A%20the%20historic%20Paris%20Agreement (accessed 2 March 2020).

84 Tata Steel (2024). Tata Steel in Europe. Zeremis. https://www.tatasteeleurope.com/sustainability/green-steel-solutions/zeremis (accessed 9 February 2024).

85 The Institute for Industrial Productivity (IETD) (2021). Basic oxygen furnace [Internet]. http://www.iipinetwork.org/wp-content/Ietd/content/basic-oxygen-furnace.html (accessed 11 August 2022).

86 The Institute for Industrial Productivity (IETD) (2021). Electric arc furnace [Internet]. http://www.iipinetwork.org/wp-content/Ietd/content/electric-arc-furnace.html (accessed 11 August 2022).

87 Sendi, M., Bui, M., Mac Dowell, N., and Fennell, P. (2022). Geospatial analysis of regional climate impacts to accelerate cost-efficient direct air capture deployment. *One Earth* 5 (10): 1153–1164.

88 International Energy Agency (IEA) (2000). IEA greenhouse gas R&D programme (IEA GHG) [Internet]. https://ieaghg.org/docs/General_Docs/Reports/PH3-30%20iron-steel.pdf (accessed 7 July 2000).

89 International Energy Agency (IEA) (2023). IEA data and statistics. https://www.iea.org/data-and-statistics (accessed 7 July 2023).

90 International Energy Agency (IEA) (2018). Global energy demand grew by 2.1% in 2017, and carbon emissions rose for the first time since 2014 – news. https://www.iea.org/news/global-energy-demand-grew-by-21-in-2017-and-carbon-emissions-rose-for-the-first-time-since-2014 (accessed 7 July 2023).

91 Voestalpine (2023). voestalpine, Siemens and VERBUND are building a pilot facility for green hydrogen at the Linz location. https://www.voestalpine.com/group/en/media/press-releases/2017-02-07-voestalpine-siemens-and-verbund-are-building-a-pilot-facility-for-green-hydrogen-at-the-linz-location/ (accessed 10 July 2023).

92 Kiessling, S., Gohari Darabkhani, H., and Soliman, A.H. (2024). Greater energy independence with sustainable steel production. *Sustainability* 16 (3): 1174.

93 Bao, J., Lu, W.H., Zhao, J., and Bi, X.T. (2018). Greenhouses for CO_2 sequestration from atmosphere. *Carbon Resources Conversion* 1 (2): 183–190.

94 Lu, X., Tian, W., Li, H. et al. (2022). Decarbonization options of the iron and steelmaking industry based on a three-dimensional analysis [Internet]. http://ijmmm.ustb.edu.cn/article/doi/10.1007/s12613-022-2475-7 (accessed 7 July 2023).

95 Singh, V., Buelens, L.C., Poelman, H. et al. (2022). Decarbonisation of steel mill gases in an energy-neutral chemical looping process. *Energy Conversion and Management* 254: 115248.

96 Franciosi, C., Voisin, A., Miranda, S., and Iung, B. (2020). Integration of I4.0 technologies with maintenance processes: what are the effects on sustainable manufacturing? *IFAC-PapersOnLine* 53 (3): 1–6.

97 Jahani, N., Sepehri, A., Vandchali, H.R., and Tirkolaee, E.B. (2021). Application of industry 4.0 in the procurement processes of supply chains: a systematic literature review. *Sustainability* 13 (14): 7520.

98 Devlin, A. and Yang, A. (2022). Regional supply chains for decarbonising steel: energy efficiency and green premium mitigation. *Energy Conversion and Management* 254: 115268.

99 United Nations Environment Program (UNEP). UN Environment (2020). The six-sector solution to the climate crisis. https://www.unep.org/interactive/six-sector-solution-climate-change/ (accessed 13 July 2023).

100 United Nations Environment Program (UNEP) (2022). UN biodiversity Conference (COP 15). http://www.unep.org/un-biodiversity-conference-cop-15 (accessed 13 July 2023).

101 Intergovernmental Panel for Climate Change (IPCC) (2023). AR6 synthesis report climate change 2023 [Internet]. Report No.: AR6. https://www.ipcc.ch/report/ar6/syr/downloads/report/IPCC_AR6_SYR_SPM.pdf (accessed 5 January 2023).

102 Intergovernmental Panel for Climate Change (IPCC) (2021). *Climate Change 2021 – The Physical Science Basis Summary for Policymakers [Internet].* Cambridge: Cambridge University Press https://www.ipcc.ch/report/ar6/wg1/downloads/report/IPCC_AR6_WGI_SPM.pdf (accessed 7 July 2023).

103 ONS Office for National Statistics. GOV.UK (2019). Provisional UK greenhouse gas emissions national statistics 2019. https://www.gov.uk/government/statistics/provisional-uk-greenhouse-gas-emissions-national-statistics-2019 (accessed 11 July 2023).

104 Conejo, A.N., Birat, J.P., and Dutta, A. (2020). A review of the current environmental challenges of the steel industry and its value chain. *Journal of Environmental Management* 259: 109782.

105 US Energy Information Administration (EIA) (1994). Carbon dioxide emission factors for coal [Internet]. https://www.eia.gov/coal/production/quarterly/co2_article/co2.html (accessed 6 July 2023).

106 Yang, W., Pudasainee, D., Gupta, R. et al. (2021). An overview of inorganic particulate matter emission from coal/biomass/MSW combustion: sampling and measurement, formation, distribution, inorganic composition and influencing factors. *Fuel Processing Technology* 2013: 106657.

107 Yi, Q., Zhao, Y., Huang, Y. et al. (2018). Life cycle energy-economic-CO_2 emissions evaluation of biomass/coal, with and without CO_2 capture and storage, in a pulverized fuel combustion power plant in the United Kingdom. *Applied Energy* 225: 258–272.

108 Russel, B. (2007). *The British Steel Industry*. London: Kluwer Law International.

109 Russel, B. (1968). *The British Steel Industry*. London: Kluwer Law International.

110 Speer, J.G. (2018). 2017 Bessemer lecture: progress in steel product research to advance the industry and society. *Ironmaking & Steelmaking* 45 (10): 858–866.

111 Elf, J.S., Espinosa, K.W., Martin, A., and Nemanova, V. (2017). Koldioxidåtervinning inom stålindustrin. https://www.diva-portal.org/smash/get/diva2:1201583/FULLTEXT01.pdf

112 Directorate-General for Research and Innovation (European Commission), Feiterna, A., Zagaria, A.M. et al. (2014). *ULCOS Top Gas Recycling BLAST Furnace Process (ULCOS TGRBF): Final Report [Internet]*. Luxembourg: Publications Office of the European Union https://data.europa.eu/doi/10.2777/59481 (accessed 12 February 2024).

113 Birat, J.P. *ULCOS Program: Status & Progress*. Brussels: European Commission; 2010 p. 1–30. https://www.eesc.europa.eu/sites/default/files/resources/docs/estep_ulcos_nov_2010.pdf (accessed 4 April 2020).

114 COM (European Commission). CORDIS | European Commission (2010). Ultra-low CO_2 steelmaking | ULCOS Project | Fact Sheet | FP6. https://cordis.europa.eu/project/id/515960 (accessed 10 July 2023).

115 Quader, A.M., Ahmed, S., Dawal, S.Z., and Nukman, Y. (2016). Present needs, recent progress and future trends of energy-efficient ultra-low carbon dioxide (CO_2) steelmaking (ULCOS) program. *Renewable and Sustainable Energy Reviews* 55: 537–549.

116 Toktarova, A. (2021). *The Low-Carbon Steel Industry-Interactions Between the Hydrogen Direct Reduction of Steel and the Electricity System – ProQuest [Internet]*. Gothenburg, Sweden: Department of Space, Earth and Environment CHALMERS UNIVERSITY OF TECHNOLOGY https://www.proquest.com/openview/c2a4d77d71403fdbffd33f250b13251d/1?pq-origsite=gscholar&cbl=2026366&diss=y (accessed 7 July 2023).

117 Johansson, M.T. and Söderström, M. (2011). Options for the Swedish steel industry – energy efficiency measures and fuel conversion. *Energy* 36 (1): 191–198.

118 Perpiñán, J., Peña, B., Bailera, M. et al. (2023). Integration of carbon capture technologies in blast furnace based steel making: a comprehensive and systematic review. *Fuel* 336: 127074.

119 Shahabuddin, M., Brooks, G., and Rhamdhani, M.A. (2023). Decarbonisation and hydrogen integration of steel industries: recent development, challenges and technoeconomic analysis. *Journal of Cleaner Production* 395: 136391.

120 Ren, L., Zhou, S., and Ou, X. (2023). The carbon reduction potential of hydrogen in the low carbon transition of the iron and steel industry: the case of China. *Renewable & Sustainable Energy Reviews* 171: 113026.

121 World Steel Association (WSA) (2023). *Carbon Capture and Storage (CCS) [Internet]*, 2. Brussels: WSA https://worldsteel.org/wp-content/uploads/Carbon-Capture-Storage_2023.pdf (accessed 15 February 2024).

122 Evennett, H. (2022). Democracy under threat: a case for co-ordinated action? https://lordslibrary.parliament.uk/democracy-under-threat-a-case-for-co-ordinated-action/ (accessed 15 February 2024).

123 STATISTA (2024). Größte Gefahren für die Demokratie in Deutschland 2024. https://de.statista.com/statistik/daten/studie/1337730/umfrage/groesste-gefahren-fuer-die-demokratie/ (accessed 15 February 2024).

124 Leadership Group for Industry Transition (LEADIT) (2023). Green steel tracker. https://www.industrytransition.org/green-steel-tracker/ (accessed 18 September 2023).

125 EMAF Ellen MacArthur Foundation (2019). Completing the picture – how the circular economy tackles climate change. [Internet]. https://www.ellenmacarthurfoundation.org/assets/downloads/Completing_The_Picture_How_The_Circular_Economy-_Tackles_Climate_Change_V3_26_September.pdf (accessed 9 October 2022).

126 Ellen MacArthur Foundation (EMAF) (2023). Summit 23 – redesigning the future: a circular economy showcase [Internet]. https://www.ellenmacarthurfoundation.org/summit/2023 (accessed 18 October 2023).

127 UNFCCC (2005). What is the Kyoto protocol? | UNFCCC [Internet]. https://unfccc.int/kyoto_protocol (accessed 11 July 2023).

Index

Circular Steel Production: Pathways to Net-Zero Carbon Emissions, First Edition.
Sandra Kiessling, Hamidreza Gohari Darabkhani, and Abdel-Hamid Soliman.

c

d